Controlled and Modified Atmospheres for Fresh and Fresh-Cut Produce

Controlled and Modified Atmospheres for Fresh and Fresh-Cut Produce

Edited by

Maria Isabel Gil
CEBAS-CSIC, Campus Universitario de Espinardo, Murcia, Spain

Randolph Beaudry
Department of Horticulture, Michigan State University, East Lansing, MI, United States

ELSEVIER

ACADEMIC PRESS
An imprint of Elsevier

Academic Press is an imprint of Elsevier
125 London Wall, London EC2Y 5AS, United Kingdom
525 B Street, Suite 1650, San Diego, CA 92101, United States
50 Hampshire Street, 5th Floor, Cambridge, MA 02139, United States
The Boulevard, Langford Lane, Kidlington, Oxford OX5 1GB, United Kingdom

Library of Congress Cataloging-in-Publication Data
A catalog record for this book is available from the Library of Congress

British Library Cataloguing-in-Publication Data
A catalogue record for this book is available from the British Library

ISBN 978-0-12-804599-2

For information on all Academic Press publications
visit our website at https://www.elsevier.com/books-and-journals

Publisher: Andre Gerhard Wolff
Acquisitions Editor: Nina Bandeira
Editorial Project Manager: Laura Okidi
Production Project Manager: Paul Prasad Chandramohan
Cover Designer: Mark Rogers
Cover Image Courtesy: Juan Antonio Tudela

Typeset by SPi Global, India

Dedication

This book is dedicated in memory of Adel A. Kader, the most relevant expert in Postharvest Technology of Horticultural Crops of all times. He spent his entire career at UC Davis starting as a postgraduate researcher in 1972, after as a Professor and Extension Specialist in Postharvest Physiology in the Department of Plant Sciences, and taking the emeritus status in 2007 until his death on December 10th, 2012.

Adel played a key role in the development of the UC Davis Postharvest Technology Center and the world's best source for postharvest information and education. He had the vision to develop collaborative research projects between scientists from multidisciplinary domains to cover plant physiology, biochemistry, and technology. Adel was convinced that improved postharvest practices would improve the quality, taste, and nutritional properties of fruits and vegetables. His global efforts aimed at reducing postharvest losses, improving the understanding of postharvest biology with great advances in postharvest technology of controlled and modified atmosphere storage of fruits and vegetables. Particularly relevant for him was 'flavor quality' as he always said 'the longer the time between harvest and eating, the greater the losses of characteristic flavor (taste and aroma).' He emphasized that it is essential to understand first the biological and environmental factors involved in deterioration and second, to use postharvest techniques that delay senescence and maintain the best possible quality between harvest and consumption.

He has been the leader, teacher, mentor, colleague, and friend of several generations in the postharvest community who always will maintain his spirit and memories in all conferences, courses, and events related to postharvest ever since. The many accomplishments of Adel's life and career are exemplified by the milestones of recognition and the words that his colleagues Michael S. Reid, James F. Thompson, and Mary E. Reed wrote in his 'In Memoriam.' "Adel was a constant and inspiring role, always organized, always prepared, and always ready to take time to listen to anyone or extend a helping hand. He was a model for every one of us who had the good fortune to work closely with him for the maintenance of the highest ethical standards in research, which challenged everyone with whom he worked to perform to their highest possible level."

Contents

18.2 Subtropical fruits: Cherimoyas

Reinaldo Campos-Vargas, and Patricio Olmedo

18.3 Subtropical fruits: Citrus

María B. Pérez-Gago, and Lluís Palou

18.4 Subtropical fruits: Dates

Nihad Alsmairat

19.1 Fresh-cut fruits: Apples and pears

Robert Soliva-Fortuny, and Olga Martín-Belloso

19.2 Fresh-cut fruits: Mangoes

Jeffrey K. Brecht

19.3 Fresh-cut fruits: Melons

Victor Rodov, and Ranjeet Shinde

Contributors

Numbers in parentheses indicate the pages on which the authors' contributions begin.

Gustavo Henrique de Almeida Teixeira (435), Universidade Estadual Paulista (UNESP), Faculdade de Ciências Agrárias e Veterinárias (FCAV), Departamento de Produção Vegetal, Jaboticabal, SP, Brazil

Eva Almenar (211), School of Packaging, Michigan State University, East Lansing, MI, United States

Nihad Alsmairat (421), Horticulture and Crop Science Department, University of Jordan, Amman, Jordan

Maria Luisa Amodio (131, 427, 557, 567), Department of Science of Agriculture, Food and Environment, University of Foggia, Foggia, Italy

Mary Lu Arpaia (389), Extension Subtropical Horticulturist, Department of Botany and Plant Sciences, University of California, Riverside, CA, United States

Jinhe Bai (109), USDA, ARS US Horticulture Research Laboratory, Fort Pierce, FL, United States

Elizabeth Baldwin (109), USDA, ARS US Horticulture Research Laboratory, Fort Pierce, FL, United States

Konstantinos Batziakas (147), Department of Horticulture and Natural Resources, Kansas State University, Olathe, KS, United States

Randolph Beaudry (265, 335), Department of Horticulture, Michigan State University, East Lansing, MI, United States

Ruth Ben-Arie (463), Emeritus, The Institute of Postharvest and Food Sciences, ARO, Volcani Center, Bet Dagan, Israel

David Bishop (265), Storage Control Systems, Inc., Sparta, MI, United States

John P. Bower (389), Consultant, Agassiz, BC, Canada

Jeffrey S. Brandenburg (185), The JSB Group, LLC, Greenfield, MA, United States

Jeffrey K. Brecht (147, 363, 495, 629), Horticultural Sciences Department, University of Florida, Gainesville, FL, United States

Jeremy Burdon (447), The New Zealand Institute for Plant & Food Research Limited, Auckland, New Zealand

Reinaldo Campos-Vargas (399), Centro de Biotecnologia Vegetal, Facultad Ciencias de la Vida, Universidad Andres Bello, Santiago, Chile

Nancy Jung Chen (381), Department of Tropical Plant and Soil Sciences, University of Hawaii at Manoa, Honolulu, HI, United States

Xi Chen (277), MirTech, Inc., Somerset, NJ, United States

Giancarlo Colelli (131, 427, 557, 567), Department of Science of Agriculture, Food and Environment, University of Foggia, Foggia, Italy

Carlos H. Crisosto (311), Department of Plant Sciences, University of California, Davis, Davis, CA, United States

Jennifer R. DeEll (293), Ontario Ministry of Agriculture, Food and Rural Affairs, Simcoe, ON, Canada

Bruno G. Defilippi (389), Depto. de Mejoramiento Genético y Biotecnología, Instituto de Investigaciones Agropecuarias—CRI La Platina, Unidad de Postcosecha, Santiago, Chile

Adem Dogan (477), Department of Horticulture, Akdeniz University, Antalya, Turkey

Merete Edelenbos (587, 593, 597, 605, 609, 615, 619), Department of Food Science, Aarhus University, Aarhus, Denmark

Mustafa Erkan (477), Department of Horticulture, Akdeniz University, Antalya, Turkey

Yolanda Garrido (167, 527, 551), Group on Quality and Safety, Department of Food Science and Technology, CEBAS-CSIC, Murcia, Spain

Maria Isabel Gil (131, 167, 521, 527, 537, 545, 551, 577, 583, 625), Group on Quality and Safety, Department of Food Science and Technology, CEBAS-CSIC, Murcia, Spain

José Angel Guerrero-Beltrán (469), Food and Chemical Engineering Department, Universidad de las Américas Puebla, Cholula, Puebla, Mexico

Hidemi Izumi (75), Faculty of Biology-Oriented Science and Technology, Kindai University, Kinokawa, Japan

Eduardo Kerbel (353), Postharvest Technology Business Manager, Carrier Transicold, Syracuse, NY, United States

Michele la Zazzera (557, 567), Department of Science of Agriculture, Food and Environment, University of Foggia, Foggia, Italy

Yong-Biao Liu (65), USDA-ARS, Salinas, CA, United States

Alexandru Luca (587, 593, 597, 605, 609, 615, 619), Department of Food Science, Aarhus University, Aarhus, Denmark

Luis Luchsinger (347), Postharvest Study Center (CEPOC), University of Chile, Santiago, Chile

Manuel Madrid (329, 335, 353, 455), Fruit Profits, Valencia, Spain

George A. Manganaris (311, 323), Department of Agricultural Sciences, Biotechnology and Food Science, Cyprus University of Technology, Lemesos, Cyprus

Olga Martín-Belloso (487, 511), Department of Food Technology, University of Lleida—Agrotecnio Center, Lleida, Spain

Nazir Mir (277), MirTech, Inc., Somerset, NJ, United States

Marta Montero-Calderón (511), Postharvest Technology Lab., Center for Agronomic Research, University of Costa Rica, San Jose, Costa Rica

Jurandi G. Oliveira (373), Universidade Estadual do Norte Fluminense, Rio de Janeiro, Brazil

Patricio Olmedo (399), Centro de Biotecnologia Vegetal, Facultad Ciencias de la Vida, Universidad Andres Bello, Santiago, Chile

Lluís Palou (411), Postharvest Technology Center (CTP), Valencian Institute of Agrarian Research (IVIA), Valencia, Spain

Robert E. Paull (373, 381), Department of Tropical Plant and Soil Sciences, University of Hawaii at Manoa, Honolulu, HI, United States

María B. Pérez-Gago (411), Postharvest Technology Center (CTP), Valencian Institute of Agrarian Research (IVIA), Valencia, Spain

Eleni D. Pliakoni (147), Department of Horticulture and Natural Resources, Kansas State University, Olathe, KS, United States

Anne Plotto (109), USDA, ARS US Horticulture Research Laboratory, Fort Pierce, FL, United States

Victor Rodov (501), Agricultural Research Organization, The Volcani Center, Rishon LeTsiyon, Israel

Mikal E. Saltveit (3), Department of Plant Sciences/Mann Lab, University of California, Davis, CA, United States

James Schaefer (265), Storage Control Systems, Inc., Sparta, MI, United States

Ranjeet Shinde (501), Department of Plant Agriculture, University of Guelph, Guelph, ON, Canada

Robert Soliva-Fortuny (487, 511), Department of Food Technology, University of Lleida—Agrotecnio Center, Lleida, Spain

Peter M.A. Toivonen (45, 323), Agriculture and Agri-Food Canada, Summerland Research and Development Centre, Summerland, BC, Canada

Juan A. Tudela (521, 537, 545, 577, 583, 625), Group on Quality and Safety, Department of Food Science and Technology, CEBAS-CSIC, Murcia, Spain

Yan Wang (299), Mid-Columbia Agricultural Research and Extension Center, Oregon State University, Hood River, OR, United States

Christopher B. Watkins (23), School of Integrative Plant Science, Horticulture Section, Cornell University, Ithaca, NY, United States

Justyna Wieczynska (587, 593, 597, 605, 609, 615, 619), Department of Food Science, Aarhus University, Aarhus, Denmark

Anne-Berit Wold (587, 593, 597, 605, 609, 615, 619), Department of Plant Sciences, Faculty of Biosciences, Norwegian University of Life Sciences, As, Norway

Allan B. Woolf (389), The New Zealand Institute for Plant & Food Research Limited, Auckland, New Zealand

Chenyi Xu (277), MirTech, Inc., Somerset, NJ, United States

Preface

The selection of optimal postharvest storage conditions and treatments to minimize postharvest losses and maintain the quality and safety of fresh produce is one of the main interests in postharvest. For achieving this purpose, there are different strategies, including the use of optimal postharvest technology procedures such as controlled and modified atmospheres (CA/MA) during storage and distribution chain. The results on CA/MA have enabled to translate fundamental knowledge on biological and physiological mechanisms into knowledge transfer of technological developments for the industry to extend the shelf-life and maintain product quality and safety.

This book is very useful as it contents relevant and applicable knowledge of CA/MA with color pictures illustrating the benefits and defects caused that are extremely useful and make this book unique. Fifty-seven chapters are included, of which 13 chapters cover aspects related to basic principles of CA/MA including the biological bases; 26 chapters cover specific recommendations and requirements for whole and fresh-cut fruits; and 18 chapters for fruit vegetables, leafy vegetables, floral vegetables, roots, and tubers. For each commodity, a summary with the requirements and recommendations is presented. This book is designed to benefit those who are currently working in the fresh produce industry including those people involved in the production, handling, packing, processing, storage, transport, wholesaling and retailing of fresh and fresh-cut produce. Efficient technologies to current industrial practices are combined for quality managers and technicians, machine engineers, packaging designers, researchers, and students. This book is very useful in courses such as those dealing with fresh and fresh-cut products as well as for bachelor degrees in Food Science and Plant Biology. This manual of guidelines and recommendations of CA/MA for the industry sector, researcher and students, is needed to understand the adequate conditions and the correct application for specific commodities. The technical personnel in the produce industry need concise but detailed information easily accessible to verify if the atmospheres could be potentially useful or damaging. This book can be a reference as their process requirements and packaging change.

The team of authors made their contributions for each selected commodity, sharing their knowledge and advances for the application in CA/MA. We wish to thank the authors who disinterestedly assisted us in making possible this book.

Part I

Basic principles of CA/MA and future trends

Chapter 1

Biological basis for CA and MA

Mikal E. Saltveit
Department of Plant Sciences/Mann Lab, University of California, Davis, CA, United States

Plants require aerobic respiration to produce the energy (i.e., ATP) and small organic molecules that are the energy and substrates used for subsequent anabolic reactions needed for continued growth and development (Saltveit, 2016b). Aerobic respiration is comprised of three sequential groups of enzymatically driven reactions: briefly they are glycolysis (conversion of hexoses like glucose to pyruvate), tricarboxylic acid cycle (the TCA cycle converts pyruvate to CO_2 with the capture of reducing power in NADH), and electron transport (transfer of electrons from NADP to O_2 with the production of ATP). Reduced O_2 availability (hypoxia) or a total lack of O_2 (anoxia) first leads to a reduced rate of electron transport and tricarboxylic acid cycle activity as NADH accumulates, and later to a metabolic rearrangement (i.e., anaerobic or fermentative respiration) that regenerates NAD^+ so that glycolysis can continue to produce an adequate supply of ATP to maintain key metabolic processes. Glycolysis is, in effect, an anaerobic pathway found in the cytoplasm, while the TCA cycle and electron transport are obligate aerobic pathways located in the mitochondria.

Oxygen is the terminal electron acceptor in mitochondrial oxidative phosphorylation (i.e., the production of ATP from NADH by chemiosmosis). In the final step of electron transport, the enzyme cytochrome *c* oxidase (COX) facilitates transfer of electrons to O_2 with the concomitant formation of water. The regeneration of NAD^+ from NADH during electron transport is needed for the continued functioning of the glycolytic pathway. In the absence of sufficient O_2 for aerobic respiration to continue, the NAD^+ regenerated during fermentative respiration allows glycolysis to continue and to extract a small portion of the energy contained in the glucose molecule; only 2 ATPs are produced during anaerobic respiration (i.e., fermentation) in comparison to the 38 ATPs produced during aerobic respiration. However, a 5- to 9-fold increase in the flux through the glycolytic pathway (termed the "Pasteur effect") compensates for this inefficiency so that ATP production remains at about one-third of that under aerobic conditions (Geigenberger, 2003).

The rate of respiration of many commodities begins to decline as the external oxygen concentration falls below around 10% (v/v) (Fig. 1). The consumption of

Controlled and Modified Atmospheres for Fresh and Fresh-Cut Produce
https://doi.org/10.1016/B978-0-12-804599-2.00002-8

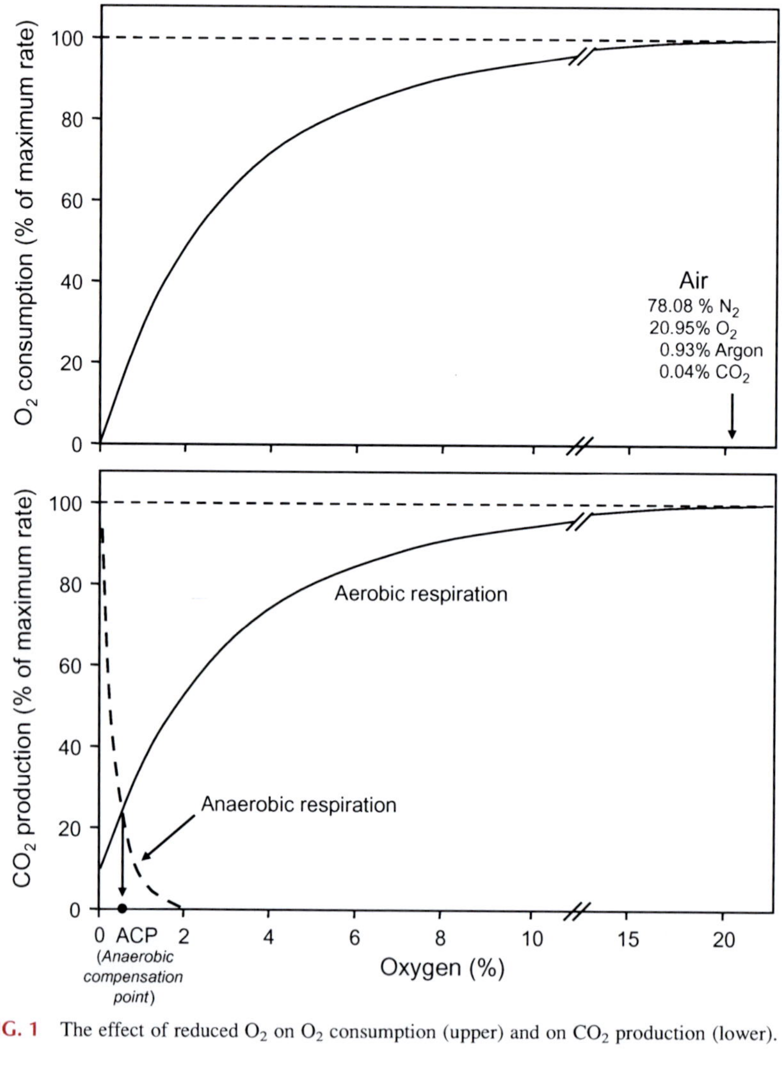

FIG. 1 The effect of reduced O_2 on O_2 consumption (upper) and on CO_2 production (lower).

oxygen and the production of carbon dioxide both decline in a quadratic fashion; slowly at first and then more rapidly as the external oxygen concentration continues to decrease linearly. At very low oxygen levels, the consumption of oxygen approaches zero as there is no oxygen available to consume. In contrast, at an external oxygen concentration below about 2% the production of carbon dioxide starts to increase as respiration in some parts of the commodity shifts from aerobic to anaerobic respiration. As the external oxygen concentration continues to decrease, more and more tissue within the commodity experiences anaerobiosis and the production of carbon dioxide rapidly increases. The cross-over point when the rate of carbon dioxide production from aerobic respiration equals that from anaerobic respiration is called the anaerobic compensation point (ACP). This point is easily identified as it occurs when the production of carbon dioxide abruptly changes from a downward to an upward slope as the oxygen concentration continues to decline. The ACP is useful because it approximates the optimal oxygen concentration that minimizes both respiration and possibility of low oxygen injury.

Inhibition of respiration at low O_2 concentrations (e.g., below 8% O_2 in the surrounding atmosphere) used in CA and MAP cannot be easily explained by substrate (O_2) limitation. The reported K_m value for the cytochrome oxidase (COX) enzyme is below 0.1 μM O_2 (an O_2 concentration in water in equilibrium with 0.011% O_2, or 0.04% air saturation), which is more than two orders of magnitude lower than the 270-μM concentration of O_2 (equal to 8.5 mg O_2/L) found in water saturated with the 20.9% O_2 found in air (Drew, 1997). The reduction in respiration at low O_2 concentrations maybe independent from the induction of fermentation, which is usually induced at O_2 levels close to zero (Geigenberger, 2003).

As the available O_2 declines, pyruvate, the end product of glycolysis, may start to accumulate as there is insufficient O_2 for the TCA cycle and oxidative phosphorylation to continue to oxidize all the pyruvate produced. To remove this accumulating pyruvate and to regenerate NAD^+, pyruvate can be converted to lactic acid by the enzyme lactate dehydrogenase (LDH) without CO_2 being released. However, this reaction acidifies the cytoplasm, which suppresses LDH activity, leading to the activation of pyruvate decarboxylase (PDC) and its catalysis of the conversion of pyruvate to acetaldehyde with the release of carbon dioxide. The enzyme alcohol dehydrogenase (ADH) rapidly converts the toxic acetaldehyde to ethanol. These two fermentative pathways give rise to yogurt and beer, respectively. Pyruvate can also be converted into carbohydrates via gluconeogenesis, to fatty acids via acetyl-CoA, or to the amino acid alanine, but NAD^+ is not regenerated in any of these reactions (Toro and Pinto, 2015).

Since most plant cells lack the buffering capacity to counter the acidification of the cytoplasm caused by the production of lactic acid, plants predominately produce ethanol and not lactic acid during fermentation (Davies et al., 1974). These fermentative pathways also produce other compounds that can impart an undesirable aroma and/or flavor to the commodity. Acidosis of the cytoplasm produced either by the accumulation of lactic acid or the products of carbonic acid (i.e., dissolving and dissociation of CO_2 in cellular water) during anaerobic respiration can cause loss of tonoplast integrity and cell death.

Plants may have a sensing mechanism that balances the metabolic demand for O_2 with its availability to prevent the induction of injurious fermentative pathways and limit the production of superoxide radicals. This sensing mechanism could involve changes in cellular pH, changes in cellular energy balance (i.e., ATP to ADP ratio), changes in redox potential (i.e., reduction/oxidation reactions), or some other mechanisms not yet elucidated (Geigenberger, 2003; Zabalza et al., 2009). For example, one plant O_2-sensing pathway involves aerobic degradation of constitutively expressed group VII of the ethylene response factors (EFR-VII). The ERF-VII gene product is not degraded under hypoxia and goes on to induce hypoxia-responsive genes which shift metabolism from aerobic to fermentative respiration (Banti et al., 2013; Kosmacz and Weits, 2014).

Besides having a significant effect on respiration, endogenous levels of O_2 and CO_2 also have a significant effect on the production and action of the plant growth regulator ethylene (C_2H_4). Low O_2 levels suppress the synthesis of ACC (1-amino-cyclopropane-1-carboxylic acid; the immediate precursor of C_2H_4) and its oxidation to C_2H_4, while high levels of CO_2 inhibit C_2H_4 action and the activities of the final two enzymes of C_2H_4 synthesis (ACC synthase and ACC oxidase) (Saltveit, 2016a).

Growing plants have adaptive responses to hypoxia and anoxia that are not available to harvested horticulture commodities. For example, modification of the diffusivity of the cuticle, the proliferation of lenticels, and the formation of adventitious roots increase O_2 diffusion into tissue, while formation of aerenchyma facilitates diffusion of O_2 to deficient tissues.

Other oxidative pathways may also be modulated by the ambient O_2 and CO_2 levels. Tissue discoloration (e.g., browning or pinking) resulting from the oxidation of accumulated stress-induced phenolic compounds by a group of enzymes called polyphenol oxidases (e.g., polyphenol oxidase, PPO) is lessened under low O_2 atmospheres. The oxidation of phenolic compounds by PPO produces highly reactive quinones that polymerize to cause the browning of cut or bruised fruit. For example, flushing packages of fresh-cut apples to produce atmospheres with $<1\%$ O_2 can quickly inhibit PPO and reduce browning.

It must be emphasized that it is not the external concentration of gases that are responsible for their biological effect on the commodity, but the concentration of gases that exist within the cellular matrix. Barriers to gas diffusion, differences in gas solubility, and the metabolic activity of intervening tissue can all significantly alter internal concentrations of gases from those imposed externally. Experiments that simply alter the external concentration of gases without accounting for these other factors may produce results of dubious validity.

Although low O_2 and elevated CO_2 concentrations (in relation to air) have a pronounced effect on many aspects of plant growth and development, this chapter will focus on the effects of altered gas atmospheres on harvested horticultural commodities. The beneficial effect of low O_2 and elevated CO_2 on suppressing microbial growth is substantial, but they will not be extensively discussed in this chapter.

Basis for low oxygen CA

Reduces respiration: Rates of CO_2 production in 3% O_2 are 10%–46% lower at 0°C and 20%–60% lower at 10°C or 20°C than those that occur in air. The K_m for O_2 equals 3% or 40 mM O_2. The concentration of O_2 in air saturated water is around 270 μM or 8.5 mg O_2/L. The decreased rate of respiration at low O_2 concentrations does not appear to be the result of decreased COX activity, since its K_m is around 0.1 μM O_2 (equivalent to approximately 0.011% O_2 in the surrounding air), but rather through an adaptive reduction and the suppression of other oxidases. Polyphenol oxidase, ascorbic acid oxidase, and glycolic acid oxidase have much lower O_2 affinities than COX. Reductions in O_2 to concentrations affecting COX could induce anaerobic (i.e., fermentative) respiration.

Inhibits ethylene production: Oxygen concentrations around 6% can suppress C_2H_4 production in apples. The K_m for C_2H_4 production by system II in ripening fruit is 3% or 40 mM.

Inhibits ethylene action: Oxygen concentrations around 3% can suppress C_2H_4 action in apples. The K_m for O_2 is 40 μM in solution, which equates with 2.8% in the intercellular gas phase.

Induces glycolysis: Low O_2-induced anoxia promotes glycolysis because of increased synthesis of glycolytic enzymes and increased activators. Anaerobic respiration can result in the formation of lactic acid by the activity of LDH. An increased lactic acid level lowers the intracellular pH and activates PDC while inactivating LDH. Pyruvate is then no longer oxidized, but is decarboxylated to form acetaldehyde, CO_2, and finally ethanol. The main purpose of increased glycolysis during anaerobiosis is to maintain sufficient ATP levels for maintaining the energy status of the cell.

Although the O_2 concentration at which anaerobic respiration occurs can be as low as 0.2% in atmosphere immediately surrounding the plant cell, the gradient of O_2 from atmosphere external to the stored plant organ often requires an O_2 concentration of 1%–3% to maintain aerobic respiration inside the commodity. The actual external O_2 concentration required depends on the rate of respiration and gas diffusion characteristic of the commodity.

Basis for elevated carbon dioxide CA

Reduces ethylene action: Around 5% CO_2 and 1%–3% O_2 is generally an optimum mixture for the CA storage of fruits. The biological activity of ethylene is lowered in the presence of CO_2. Burg studied the interaction of ethylene and CO_2 and found that CO_2 is a competitive inhibitor of ethylene action (Burg, 1990). The affinity for ethylene and CO_2 is about 100,000 to 1.

The K_m for C_2H_4 equals 0.5 nM in solution or 0.1 ppm in the gas phase, while the K_i for CO_2 equals 0.6 mM in solution or 1.5% in the gas phase. Kinetic studies with O_2 showed that C_2H_4 could not bind to the enzyme unless O_2 has already attached to the active site. Inhibition of C_2H_4 action increases when the CO_2 concentration increases, or when the C_2H_4 concentration decreases.

When the CO_2 concentration is lower than 0.15% (i.e., much less than K_i), then CO_2 exerts very little effect on C_2H_4 action. When the CO_2 concentration equals K_i (i.e., 1.5%), then twice the concentration of C_2H_4 is needed to have the same effect as it would in the absence of CO_2. This relationship is shown by the following equation:

$$\frac{[S_i]}{[S_i]} = 1 + \frac{[i]}{K_i}$$

In this equation $[S_i]/[S]$ represents the ratio of substrate (i.e., C_2H_4) concentrations that are required for any given velocity in the presence $[S_i]$ or absence $[S]$ of the inhibitor (i.e., CO_2). For example, when the CO_2 concentration $[i]$ equals three times K_i, then $[S_i]$ equals four times $[S]$.

When the substrate concentration $[S]$ is less than the K_m, then the inhibitor concentration required for 50% inhibition $[i]_{0.5}$ equals K_i. This relationship is shown in the following equation:

$$[i]_{0.5} = 1 + \frac{[S]}{K_m} K_i$$

When the C_2H_4 concentration equals K_m then $[i]_{0.5}$ equals twice K_i or 3.0% CO_2.

Reduces respiration: Elevated CO_2 concentrations also reduce respiration, but levels above 20% CO_2 can stimulate anaerobic respiration with accumulation of ethanol and acetaldehyde. These high levels of CO_2 can change the activity of specific respiratory enzymes and may have an effect on uncoupling oxidative phosphorylation. High CO_2 levels result in the accumulation of tricarboxylic acid cycle intermediates (e.g., succinic acid, citric acid, fumarate, etc.).

CA and MA storage

The terms "controlled atmosphere" (CA) and "modified atmosphere" (MA) refer to atmospheres in which the gas composition is different from air (i.e., 78.08% N_2, 20.95% O_2, 0.93% argon, and 0.04% CO_2). Under one atmosphere of pressure, these concentrations can be converted to partial pressures (1 atm = 101 kPa) (see following section on Gas Laws). Air at one atmosphere has 77.3 kPa of N_2, 20.7 kPa of O_2, 0.92 kPa of argon, and 0.04 kPa of CO_2. At high RH (near 100%), water vapor can occupy 0.6%, 1.2%, 2.2%, and 3.8% of the volume of air at 0°C, 10°C, 20°C, and 30°C, respectively. While other gases can be of interest in CA and MA (e.g., C_2H_4, SO_2, CO, H_2O), the effect of O_2 and CO_2 will be the primarily focus of this chapter.

CA and MA usually involve reducing O_2 levels below 5% and/or elevating CO_2 levels above 3%. Low pressure (hypobaric) storage (LPS) is another method used to establish a CA atmosphere (Burg, 1990). In LPS, the commodity is held under reduced pressure (e.g., 10 kPa) which can produce partial pressures of gases similar to those found in a CA at normal atmospheric pressure. For example, a CA with 2% O_2 at normal pressure (101 kPa) will have nearly the same partial pressure of O_2 (2.0 kPa) as an atmosphere of air at 10 kPa.

MA differs from CA only in the degree of precision in controlling partial pressures of O_2 and CO_2; CA is usually more exact than MA because CA entails periodic measurement of the gas concentrations and adjustments to bring the concentrations into the desired range (Kader, 2016). Other terms used to refer to certain CA storage conditions include reduced (or low) O_2 storage, elevated (or high) CO_2 storage, and nitrogen storage. The last term is meaningless since N_2 is an inert gas with no physiological effects. Manipulation of O_2 and CO_2 levels was described as "gas storage" in the 1920s; however, this term had unpleasant connotations after World War I and was soon replaced by the presently used terms (Smock, 1979).

CA and MA are supplements to and not substitutes for maintaining the optimum temperature and relative humidity. Both CA and MA can be used during transport, temporary storage, and/or long-term storage of horticultural commodities to either prolong storage life or to maintain higher quality.

Gas laws

General laws

Gas molecules can move as the result of diffusion (driven by a difference in concentration across a distance) or by mass flow (driven by a difference in pressure across a distance). Gas movement into and out of plant organs and in all CA, MA, and LPS systems is driven by diffusion from regions of high concentration to regions of lower concentrations. It is therefore important to understand the general physical laws governing the activity of gases in biological systems. Following is a brief introduction to the most important physical laws governing the movement of gases.

Avogadro's Law states that one mole of any compound contains 6.0228×10^{23} molecules, and that this quantity of gas at STP (i.e., Standard Temperature and Pressure; 0°C (273 K) and 1 atm (101 kPa)) will occupy 22.414 L. Equal volumes of gases at the same pressure and temperature contain the same number of molecules. At 20°C, one mole of gas occupies 24.06 L. The relationship among pressure, volume, and temperature of a gas is given by the Ideal Gas Law.

The Ideal Gas Law states that PV/T is a constant, and that for one mole of gas it becomes the universal molar gas constant usually symbolized by "R." For n moles of an ideal gas, PV/T equals nR, or written in a more standard form:

$$PV = nRT$$

where P, pressure in atmospheres; V, volume in liters; n, moles of gas; R, universal molar gas constant (0.082 L atm/degree mole); T, absolute temperature (°C + 273).

Dalton's Law of Partial Pressures states that the total pressure of a mixture of gases is the sum of the partial pressures of the component gases. It follows from this and from the Ideal Gas Law that the partial pressure of any gas in a mixture is in the same ratio to the total pressure as its quantity in moles is related to the total number of moles of all the gases in the mixture.

$$P_{\text{Total}} = PO_2 + PCO_2 + PN_2 + \text{partial pressure of any other gases}$$

It is preferable to use partial pressures of gases instead of percentages when talking about their biological activity. At a constant pressure, the partial pressure of a gas is directly proportional to its concentration in a mixture, but at different pressures, the same concentration will have different partial pressures.

Fick's First Law of diffusion describes the rate of transfer of a substance from a region of high concentration to a region of lower concentration. It is summarized in the following equation:

$$F = -D \times A \times 1/L \times (C_i - C_o)$$

where F, flux, that is, the rate of transfer; D, the diffusion coefficient (the negative sign indicates the substance is moving in the direction of decreasing concentration); A, the area of the barrier to diffusion (the total area times the fraction through which gas is actually diffusing); L, the length of the diffusion path; C_i, the initial, or inside concentration; C_o, the later, or outside concentration.

Since the normal temperature range encountered during postharvest handling rarely exceeds 0°C to 30°C, the maximum influence of temperature on the rate of diffusive flux is only 10% (i.e., 273–303 K) and would usually be $<4\%$.

Graham's Law of Diffusion states that the rate of diffusion of a gas is inversely proportional to the square root of its density. Since the density of a gas at a fixed temperature and pressure is directly proportional to its molecular weight, the rate of diffusion should be inversely proportional to the square root of the molecular weight of the gas.

$$\text{Flux} \approx 1/\sqrt{\text{MW}} \approx 1/\sqrt{\text{Density}}$$

The mean free path of diffusion, and therefore the rate of diffusion, can be increased by replacing the 78% N_2 in air with other gases having a smaller kinetic diameter. The kinetic diameter expresses the likelihood that a molecule in a gas will collide with another molecule, while the mean free path is the distance between collisions. For example, the molecular weight of H_2, He, N_2, and Ar are 2, 4, 28, and 40 g/mol, respectively, and their kinetic diameters are 0.27, 0.22, 0.37, and 0.36 nm, respectively. When pure, their mean free path of diffusion at 25 °C and 1 atm are 125, 193, 66, and 70 nm, respectively. Replacing N_2 that has a MW of 28 with He that has a MW of 4 would increase diffusivity by 2.6-fold ($\sqrt{28}/\sqrt{4}$) as calculated from molecular weights, or 2.9-fold (193/66), as calculated from the mean free paths.

Another way to increase diffusivity is to reduce pressure. For example, an 80% reduction in pressure would produce the same partial pressure of O_2 in an atmosphere of air as would a 2% O_2 CA. The lower pressure would also have the added benefit of greatly enhanced diffusion of O_2 into, and CO_2 and C_2H_4 out of

TABLE 1 Diffusion of O_2, CO_2, and C_2H_4 in cm^2/s through air, hydrogen, and water at 0°C

Gas	Oxygen	Carbon dioxide	Ethylene
Air	0.18	0.14	0.12
Hydrogen	0.69	0.55	0.49
Water	2200	1500	1700

the tissue. An 80% reduction in pressure from 101 to 20 kPa (i.e., 1.0–0.2 atm., or 760–152 mmHg) would result in a corresponding 80% decrease in the density of the medium and a more rapid rate of diffusion since the mean free path would be increased about fivefold.

Even a thin film of water represents a significant barrier to gas diffusion. The diffusion of some gases through various mediums is given in Table 1. Notice that gas movement through other gases is 10,000 times faster than through an equal thickness of water. Dipping tissue in aqueous solutions, cutting tissue or solute leakage during ripening or senescence can all produce a film of water and significantly alter gas diffusion to and from the tissue.

Water represents a special case for diffusion since it exists as both a liquid and a gas in plant tissue at normal biological temperatures. For example, around 95% of the gas exchange in a tomato fruit occurs through the stem scar, while only around 65% of the water loss is through the stem scar. Liquid water can penetrate the cuticle and migrate by mass flow to the surface where it evaporates.

Gas solubility in water

The solubility of gases in liquids can be calculated using the Bunsen absorption coefficient for the gas. The Bunsen absorption coefficient is defined as the volume of gas in liters at STP that saturates 1 L of liquid. A number of Bunsen absorption coefficients for some biologically important gases are listed in Table 2.

The mechanism by which gases dissolve in water is different for various gas species. For example, nonpolar gases such as CO, H_2, N_2, and O_2 dissolve in aqueous solution as gas molecules, while polar gases such as CO_2, SO_2^-, NH_3^+ react with water to form H_2CO_3, H_2SO_4, and NaOH, respectively. This reactivity of the gas greatly increases the amount of gas that will dissolve in a given volume of water. At 0°C and 101 kPa partial pressures, the solubility of N_2, O_2, and C_2H_4 in water are 0.03, 0.07, and 0.28 g/kg, respectively. In comparison, the solubility of CO_2, SO_2^-, NH_3^+ in water is 3.4, 230, and 900 g/kg, respectively.

The solubility of CO_2 is especially important in CA and MA since it accumulates as a product of respiration, is often added to the atmosphere to reduce

TABLE 2 The Bunsen absorption coefficient for some biologically important gases in water

	Temperature (°C)			
Gas	0	10	20	30
Carbon dioxide	1713[a]	1194	878	665
Carbon monoxide	35.4	28.2	23.2	20.0
Ethylene	256	178	135	105
Hydrogen	21.5	19.5	18.2	17.0
Nitrogen	23.5	18.6	15.5	13.4
Oxygen	48.9	38.0	31.0	26.1

[a]*Milliliters (STP) of gas per liter of water at 1.0 atm.*

respiration and the biological activity of C_2H_4, and as a fungistat. Carbonic acid (H_2CO_3) is formed when CO_2 dissolves in water and it can dissociate into ions according to the following equation:

$$CO_2 + H_2O \rightleftharpoons H_2CO_3 \rightleftharpoons HCO_3{}^- + H^+ \rightleftharpoons CO_3{}^{-2} + 2H^+$$

Thus CO_2 in an aqueous solution exists not only as molecules of CO_2, but also as H_2CO_3, and HCO_3^- and CO_3^{-2} ions. This accounts for the much greater solubility of CO_2 in water that the nonpolar gases listed in Table 2. The H^+ produced acidifies the cytoplasm which causes changes in metabolism (i.e., induces PDC activity and alcoholic fermentation) and cellular damage (e.g., loss of tonoplast integrity).

The solubility of a gas is not only affected by the reactivity of the gas itself and the solvent temperature, but also by the concentration of solutes in the solvent. The solubility of gases in water usually decreases when other solutes, particularly electrolytes, are added to the solution. This "salting out" effect is used in the separation of proteins where the progressive addition of salts precipitates different macromolecules. The added electrolytes impose structure on water thereby effectively reducing the amount of water available for molecules to dissolve into. For example, at 0°C the solubility of O_2 is 21% less in seawater than in fresh water. Addition of salts (e.g., NaCl, $(NH_4)_2SO_4$) to air saturated water will degas the solution as evidenced by the formation of gas bubbles in the solution. This degassing is dramatically illustrated by the eruption of a fountain of solution upon the addition of salt to a bottle of carbonated beverage. Consequently, under the same partial pressure in the gas phase, the concentration of a gas dissolved in a subcellular portion of a tissue (and therefore its biological activity in that region) may differ significantly depending of the solute concentration within that region.

Henry's Law states that the mass of any gas that will dissolve in a given volume of liquid is directly proportional to the pressure of the gas above that liquid, and that the various components of a gas mixture behave independently of each other. For example, the volume of CO_2 that will dissolve in 100 mL of water at 10°C is proportional to the partial pressure of CO_2 in the gas mixture above the water. If the gas mixture is 10% or 50% CO_2, then the amount of CO_2 that will dissolve in the 100 mL of water will be 10% or 50% of what it would have been if the atmosphere was 100% CO_2. Since 100 mL of water can contain 119.4 mL of CO_2 at 10 °C (see earlier), the 100 mL would contain about 1.2 or 60.0 mL of CO_2 when in equilibrium with a gas mixture containing 10% or 50% CO_2.

External versus internal

Dermal resistance to diffusion

The diffusion gradient driving the movement of gases into or out of a commodity depends on the commodity's surface-to-volume ratio, resistance to diffusion (a function of the number and size of stomata, lenticels, breaks in the epidermis, and cuticle structure), the rate of respiration by the commodity, and the atmospheric composition surrounding the commodity. In an actively respiring commodity, such as apple or tomato, a concentration gradient develops within the commodity as the O_2 and CO_2 diffusing across the skin is consumed or produced by the respiring tissue.

The change in gas concentration may not be linear with distance into the commodity since the difference in tissue density, diffusivity, and metabolic activity can vary significantly. Internal oxygen levels can vary from around 3% to near zero in permeable versus dense bulky plant tissues. In general though, the concentration of CO_2 increases and the concentration of O_2 decreases from the skin to the center of a bulky commodity. In contrast, there may be little difference in the concentration of gases within commodities that are thin and have bilateral symmetry, for example, leaves.

Changing the resistance to gas diffusion by applying an impervious barrier to the commodity will first reduce evolution of gas from the tissue. The internal concentration will increase (in the case of CO_2) or decrease (in the case of O_2) until the gradient is sufficient to reestablish a rate of diffusion from the tissue that is equal to the rate of production or consumption, that is, until equilibrium is reestablished (see earlier, Fick's First Law of diffusion). This of course assumes that the change in the internal concentration of a gas does not affect respiration.

Since the resistance to gas diffusion can be defined as $r = L/(A \times D)$, and L and A are constant for a given fruit, it follows that r should vary inversely with D and be similar for different gases within a fruit. This relationship is presented in Table 3. Notice how the resistance (here presented as the ratio between the internal concentration in ppm and the rate of evolution) is the same for both CO_2 and C_2H_4 among the fruit even though these gases have very different physical properties.

TABLE 3 Ratio between the internal concentration of CO_2 or C_2H_4 in ppm and the production of these gases in μL/kg hr by a number of fruit

Fruit	Carbon dioxide	Ethylene
Apple	3.5	4.7
Cantaloupe	3.9	3.2
Avocado	1.4	1.8
Passion fruit	6.6	9.0
Orange	3.7	4.0
Tomato	2.5	3.8

Given a constant rate of production and a constant resistance to diffusion in a tissue, the internal concentration and the rate of loss will also be constant. Under these conditions alteration in the external concentration of a gas (i.e., establishment of a CA or MA) will be reflected in changes in the internal concentration. We can therefore calculate what the external gas concentration must be to produce a desired internal concentration. The additive resistances to gas diffusion of barriers surrounding a commodity in CA are shown in Fig. 2. This illustrates how manipulation of the environment can occur in an MA system. In this model, the three levels of barriers are as follows:

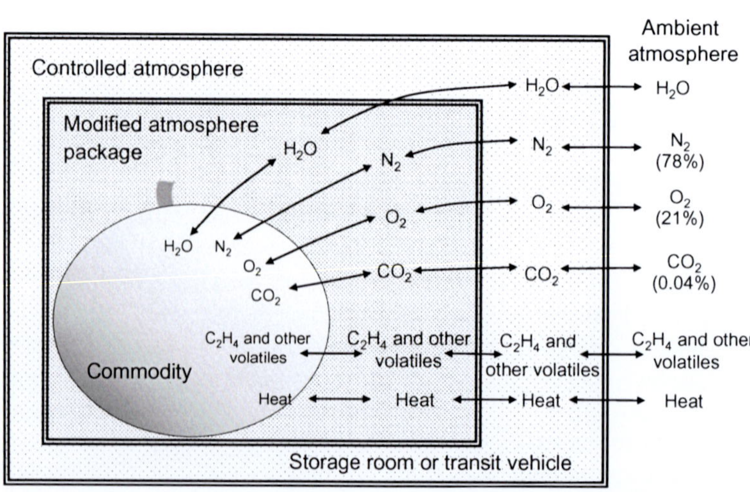

FIG. 2 Model of a commodity within a controlled atmosphere environment, illustrating gas exchange across three barriers: the commodity, the package, and the storage room or transit vehicle. *(Modified from Kader, A. A., 2002. Modified atmospheres during transport and storage. In: Postharvest Technology of Horticultural Crops. University of California, Oakland, pp. 135–144.)*

1. *The commodity*: natural dermal system (epidermis, cuticle, stomata, lenticels, etc.) and added barriers (wax coatings, film wraps, etc.).
2. *The package*: permeability of packaging materials, use of plastic liners in package, ventilation openings, and so on.
3. *The storage room or transit vehicle wall*: degree of gas tightness, ventilation openings for introduction of outside fresh air, and so on.

Additional barriers may include consumer packages inside the master container and plastic pallet covers enclosing several containers on a pallet.

Surface-to-volume ratio

The rate at which gases diffuse into and out of a commodity is a function of its surface area. As a commodity increases in size, the rate at which the volume of respiring tissue increases is greater (a cubic function; e.g., r^3) than the rate at which the surface area increases (a square function; e.g., r^2). This means that there is less surface area for the gases to diffuse through per unit volume (or weight) in a large commodity than there is in a smaller version of the same commodity. Under similar conditions, the larger commodity will have higher internal levels of CO_2 and lower levels of O_2 than the smaller commodity. A CA that is well tolerated by small versions of a commodity may not be tolerated by larger versions of the same commodity because the internal concentration of CO_2 and O_2 would have risen, or fallen to harmful levels, respectively.

Simple formulae for calculating surface area and volume (Table 4) of commodities reveal the major differences in the surface-to-volume ratio between large and small commodities (Table 5).

Gas movement around the commodity

The flow of gases around a commodity influences the atmospheric composition surrounding the commodity and the rate of diffusion into and out of the commodity. In still air, gas diffusion is driven by the concentration gradient between the commodity and the surrounding atmosphere. As gas diffuses into or out of the commodity, the gas concentration in the surrounding atmosphere will

TABLE 4 Equations for the calculation of the area and volume for a sphere (e.g., a orange), a cylinder with spherical ends (e.g., a cucumber), and a right circular cone with a hemispherical end (e.g., a carrot)

	Sphere	Cylinder	Cone
Area	$4\ \pi r^2$	$2\ \pi Lr+4\ \pi r^2$	$\pi r\ [r+(r^2+L^2)^{0.5}]+2\ \pi r^2$
Volume	$4/3\ \pi r^3$	$\pi Lr^2+4/3\ \pi r^3$	$1/3\ \pi Lr^2+2/3\ \pi r^3$

r, radius, L, length of the cylinder or cone, and π, 3.14159.

TABLE 5 Dimensions (length, width, area, volume, and the ratio of area/volume) of some common perishable commodities

Commodity	Radius (cm)	Length (cm)	Area (cm^2)	Volume (cm^3 or mL)	Area/volume
Tomato (small)	3		113	113	1.00
Tomato (large)	6		452	905	0.50
Cucumber (small)	3	8	264	339	0.78
Cucumber (large)	6	16	1056	2714	0.39
Carrot (small)	2	8	89	50	1.78
Carrot (large)	4	16	358	402	0.89

change and the gradients will decrease. Depending on the surface characteristics of the commodity (e.g., fuzzy or smooth), this shell of gases can be thick or thin. Moving air sweeps away these shells and reestablishes the original gradient which promotes increased diffusion which can reduce CO_2 and increase O_2 concentrations within the tissue. Adequate ventilation around the stored commodities is not only needed to maintain the desired atmospheric composition around the commodity, but it is also needed to remove evolved heat and moisture.

Effects of CA on commodities

Potential benefits or injuries of CA or MA are dependent upon the commodity, variety, physiological age, atmospheric composition, temperature, and duration of exposure. This helps explain the wide variability in results among published reports for a given commodity. In most cases, the difference between beneficial and harmful MA combinations is relatively small. MA combinations that are necessary to control decay and/or insects, for example, cannot be tolerated by most commodities and result in faster deterioration. The following benefits translate into overall reduction in quantitative and qualitative losses during handling (Kader, 2016; Wang, 2006).

Potential beneficial effects of reduced O_2 and/or elevated CO_2 include the following:

1. Retardation of senescence (ripening) and associated biochemical and physiological change; for example, reduced respiration, ethylene production, softening, and compositional changes such as reduced chlorophyll loss.

2. Reduction of fruit sensitivity to ethylene action.
3. Alleviation of certain physiological disorders (e.g., chilling injury in avocado, bitter flavor in carrots, and discoloration in lettuce).
4. Inhibition of growth and breaking of dormancy in potatoes.
5. MA can have a direct or indirect effect on postharvest pathogens. Decay incidence and susceptibility to decay can be reduced when ripening and senescence are delayed. Oxygen levels below 1% and CO_2 above 10% can suppress fungal growth, but not all commodities can tolerate these conditions.
6. MA can be a useful tool for insect control in some commodities.

Potential harmful effects of reduced O_2 and/or elevated CO_2 include the following:

1. Initiation and/or aggravation of certain physiological disorders (e.g., internal and external browning, surface pitting).
2. Irregular ripening.
3. Under O_2 concentration below 1% and/or CO_2 levels above 20% there can be a shift from predominantly aerobic to predominantly anaerobic respiration with the development of off-flavor and off-odors. Such a shift is characterized by glycolysis replacing the citric acid cycle as the main source of energy production.
4. Increased susceptibility to decay.
5. Stimulation of sprouting and retardation of periderm development in "root" vegetables.
6. Induced physiological disorders (e.g., brown stain in lettuce).
7. Increased acidity of tissue (Accumulation of succinate, citrate, $CO_2 + H_2O \rightarrow H^+ + HCO_3^-$).
8. Reduced periderm formation during wound healing.

Fruit and vegetable tolerances

Although the respiratory pathways are similar among most harvested horticultural commodities, and (although more variable) the diffusivity does not vary that much, there is a surprisingly large differences in the tolerance of various commodities to low O_2 and elevated CO_2 (Fig. 3–5).

The limits of tolerance of fruits and vegetables to reduced O_2 and elevated CO_2 levels are shown in Figs. 4 and 5 (Kader and Morris, 1977). These tolerance values should not be construed as recommended MA or CA atmospheres, but rather as levels of O_2 below which, and CO_2 above which physiological damage would be expected. In using these limits, the following points should be considered:

These limits of tolerance can be different at temperatures above or below the recommended temperatures for each commodity. Also, a given commodity may tolerate higher levels of CO_2 or lower levels of O_2 than those indicated for short

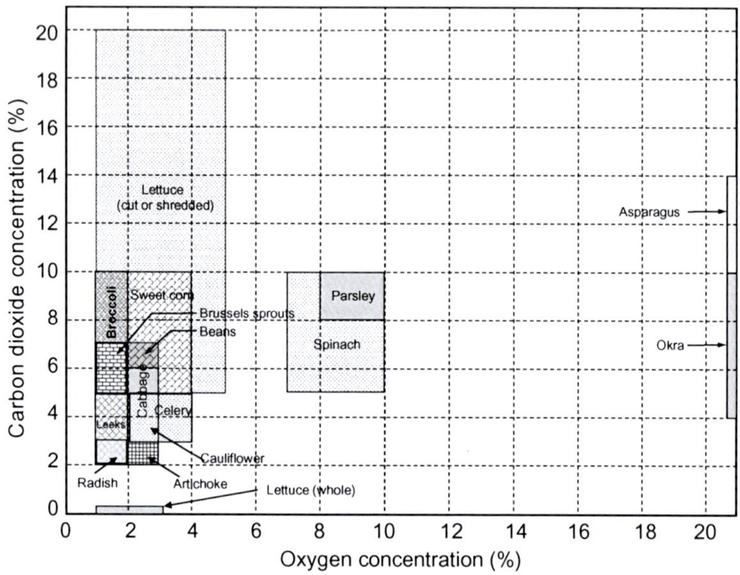

FIG. 3 Limits of O_2 and CO_2 concentrations for some vegetable crops. *(Modified from Saltveit, M.E., 1993. A summary of CA and MA requirements and recommendations for the storage of harvested vegetables. In: Sixth International Controlled Atmosphere Research Conference, Cornell University, Northeast Regional Agricultural Engineering Series 71, pp. 800–818.)*

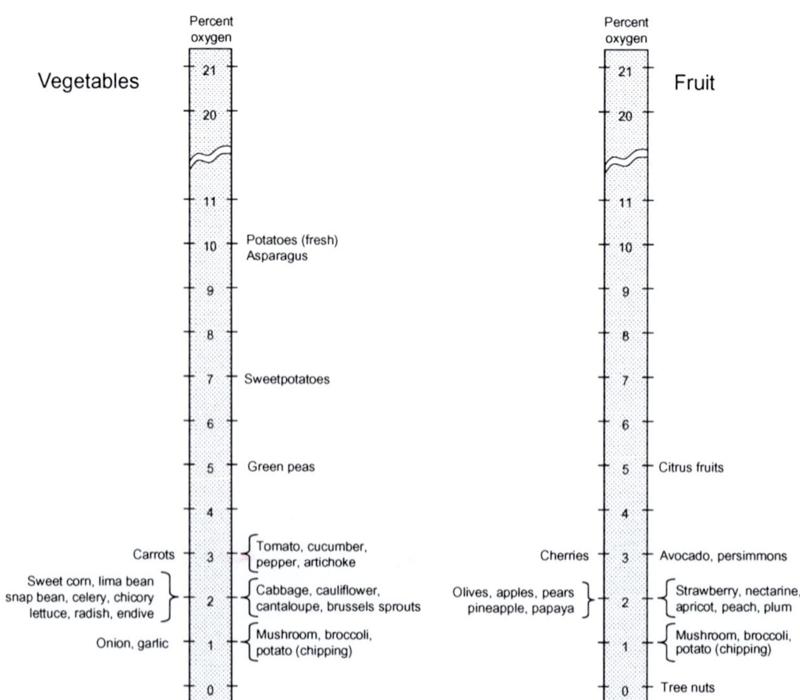

FIG. 4 Relative tolerance of fruits and vegetables to reduced O_2 levels at recommended storage temperatures. *(Modified from Kader, A. A., Morris, L. L. 1977. Relative tolerance of fruits and vegetables to elevated CO2 and reduced O2. levels. Controlled Atmospheres for the Storage and transport of Perishable Agricultural commodities, Michigan State University Horticultural Report No. 28, pp. 260–265.)*

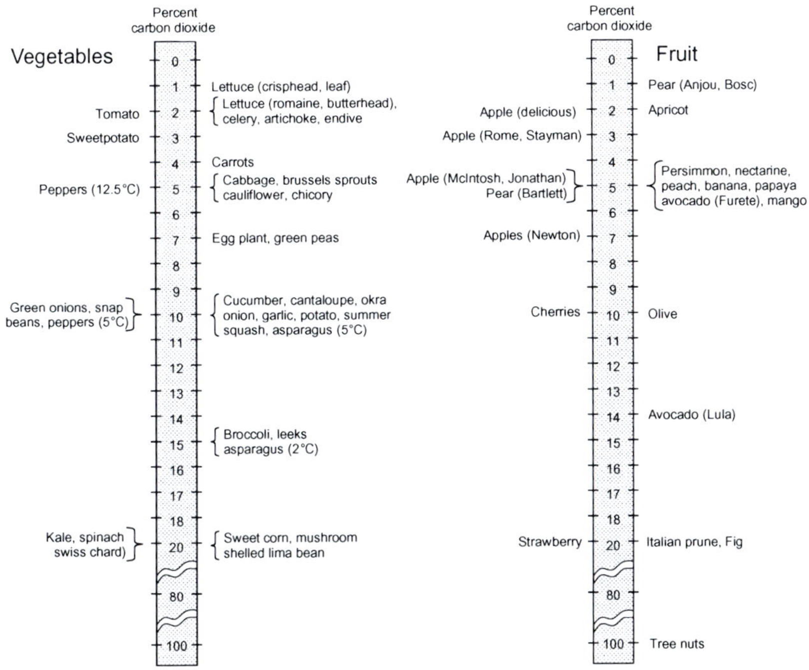

FIG. 5 Relative tolerance of fruit and vegetables to elevated CO_2 levels at recommended storage temperatures. *(Modified from Kader, A. A., Morris, L. L. 1977. Relative tolerance of fruits and vegetables to elevated CO2 and reduced O2 levels. Controlled Atmospheres for the Storage and transport of Perishable Agricultural commodities, Michigan State University Horticultural Report No. 28, pp. 260–265.)*

durations. The limit of tolerance to low O_2 would be at a higher concentration as storage temperature and/or duration increases since O_2 requirements for aerobic respiration of the tissue increase with higher temperatures. Depending on the commodity, damage associated with CO_2 may either increase or decrease with an increase in temperature. CO_2 production increases with temperature but its solubility decreases. Thus CO_2 in the tissue can be increased or decreased by an increase in temperature. Further, the physiological effect of CO_2 could be temperature dependent.

The limits of tolerance to either elevated CO_2 or reduced O_2 (shown in Figs. 4 and 5) are based on the assumption that the other component is near its normal (air) concentration. Tolerance limits to elevated CO_2 decrease with the decrease in O_2 level, and similarly the tolerance limits to reduced O_2 would increase with the increase in CO_2 level. In fact, a high concentration of CO_2 can induce fermentation in some commodities at O_2 concentration similar to those found in air.

The possible effects of supplemental gases (CO, SO_2, C_2H_4, etc.) on modifying the tolerance limits of O_2 and CO_2 should be considered.

In some commodities, their physiological age at harvest may influence their susceptibility to elevated CO_2 and/or reduced O_2 damage. For example, ripe fruits may tolerate higher levels of CO_2 than mature-green fruits.

The form in which a given commodity is handled can affect its tolerance to elevated CO_2 or reduced O_2. Segmented lettuce tolerates much higher levels of CO_2 and/or lower levels of O_2 than intact heads of lettuce. This is also true for slices of apples or potatoes versus intact organs.

The lower limit tolerated by most fruits and vegetables is about 2% O_2 (Figs. 4 and 5). Below this O_2 concentration, anaerobic respiration may result in the development of off-flavor and off-odors. Tolerance limits for elevated CO_2 are more variable. Most tree fruits have a limit of about 5% CO_2. Vegetables with a wide variety of botanical structures differ greatly in their CO_2 tolerance. For example, lettuce is damaged by 1%–2% CO_2, while spinach tolerates 20% CO_2, and broccoli tolerates 15% CO_2, while cauliflower is damaged by 5% CO_2.

Differences in susceptibility to elevated CO_2 and/or reduced O_2 levels among commodities or among cultivars of a given commodity may be due to structural (anatomical) differences rather than metabolic differences. The role of natural barriers and differences in solubility and diffusion coefficients of CO_2 should be considered in studying the relationship between internal and external CO_2 and O_2 concentrations.

Supplemental treatments

Prestorage treatments with elevated CO_2. Tests at several Experiment Stations indicated that treating apples for 2 weeks or pears for 2 to 4 weeks with 12% CO_2 (at 0–5°C) before CA storage delayed fruit softening. However, this treatment caused varying amounts of internal and external CO_2 injury depending on variety, season, and production area. Its commercial application is currently limited to "Golden Delicious" in the Northwest. Elevated CO_2 treatments have also been shown to alleviate chilling injury symptoms on some subtropical and tropical fruits, but this treatment is not used commercially.

Ethylene removal from CA storage. Most research workers have assumed that ethylene removal from MA storage rooms is not necessary because ethylene effects on fruit ripening at 0–5°C and under MA conditions are negligible. However, there is some evidence from recent studies to indicate that the presence of ethylene, at concentrations likely to occur in MA and CA rooms, can enhance fruit softening. Thus C_2H_4 removal is recommended in long-term CA storage of apples and pears, and any storage of kiwifruits.

Use of carbon monoxide (CO). CO is a colorless, tasteless, odorless, and flammable gas with explosive limits between 12.5% and 74.2% (by volume) in air. It is extremely toxic to humans and adequate safety precautions must be followed when used. When added to plant tissues, CO is oxidized to CO_2 that is metabolized into organic acids and other constituents.

Beneficial effects of CO include the following. CO (1% to 5%) added to <5% O_2-atmospheres inhibits discoloration of lettuce butts and mechanically damaged tissue. Similar effects occur on other commodities. This inhibition of discoloration is lost upon removal of the commodity from MA to air during destination marketing.

CO (5% to 10%) added to MA has been found to inhibit growth of several important postharvest pathogens and to prevent decay development on several fruits and vegetables. The fungistatic effects of CO are maximized at O_2 levels below 5%.

Harmful effects of CO include the following: CO may aggravate certain physiological disorders. For example, if CO used in a situation where CO_2 accumulated above 2% during transit of lettuce, it will increase the severity of brown stain (a CO_2-induced disorder).

CO mimics ethylene effects such as enhancing fruit ripening and induction of certain physiological disorders. However, when CO is used in combination with reduced O_2 and/or elevated CO_2, such effects are minimized and become insignificant except for those commodities that are extremely C_2H_4 sensitive such as kiwifruits.

Because of its extreme toxicity to humans and flammability, the commercial use of CO has been very limited. It is used as a component of MA during transport of whole and cut lettuce and other commodities especially for export shipments.

References

Banti, V., Giuntoli, B., Gonzali, S., Loreti, E., Magneschi, L., Novi, G., Paparelli, E., Parlanti, S., Pucciariello, C., Santaniello, A., Perata, P., 2013. Low oxygen response mechanisms in green organisms. Int. J. Mol. Sci. 14, 4734–4761.

Burg, S.P. 1990. Theory and practice of hypobaric storage. In: Food Preservation by Modified Atmospheres, M. Calderon and R. Barkai-Golan (eds.). CRC Press, Boca Raton, FL. pp. 353–372.

Davies, D.D., Grego, S., Kenworthy, P., 1974. The control of the production of lactate and ethanol by higher plants. Planta 118, 297–310.

Drew, M.C., 1997. Oxygen deficiency and root metabolism: Injury and acclimation under hypoxia and anoxia. Annu. Rev. Plant Physiol. Plant Mol. Biol. 48, 223–250.

Geigenberger, P., 2003. Response of plant metabolism to too little oxygen. Curr. Opin. Plant Biol. 6, 247–256.

Kader, A.A., 2016. Controlled atmosphere Storage. In: Gross, K.C., Wang, C.Y., Saltveit, M.E. (Eds.), USDA Handbook 66—The Commercial Storage of Fruits, Vegetables, and Florist and Nursery Crops. United States Department of Agriculture, pp. 22–25. http://www.ba.ars.usda.gov/hb66/contents.html.

Kader A.A. 2002. Modified atmospheres during transport and storage. In: Postharvest Technology of Horticultural Crops, Oakland: University of California, pp 135–144.

Kader, A.A., Morris, L.L. 1977. Relative tolerance of fruits and vegetables to elevated CO_2 and reduced O_2. levels. Controlled Atmospheres for the Storage and transport of Perishable Agricultural commodities, Michigan State University Horticultural Report No. 28. 260–265.

Kosmacz, M., Weits, D.A., 2014. Oxygen perception in plants. In: van Donger, J.T., Licausi, F. (Eds.), Low-Oxygen Stress in Plants. Oxygen Sensing and Adaptive Responses to Hypoxia. Springer, pp. 3–18.

Saltveit, M.E., 1993. A summary of CA and MA requirements and recommendations for the storage of harvested vegetables. In: Sixth International Controlled Atmosphere Research Conference, Cornell University, Northeast Regional Agricultural Engineering Series 71, pp. 800–818.

Saltveit, M.E., 2016a. Ethylene effects. In: Gross, K.C., Wang, C.Y., Saltveit, M.E. (Eds.), USDA Handbook 66—The Commercial Storage of Fruits, Vegetables, and Florist and Nursery Crops. United States Department of Agriculture, pp. 76–82. http://www.ba.ars.usda.gov/hb66/contents.html.

Saltveit, M.E., 2016b. Respiratory metabolism. In: Gross, K.C., Wang, C.Y., Saltveit, M.E. (Eds.), USDA Handbook 66—The Commercial Storage of Fruits, Vegetables, and Florist and Nursery Crops. United States Department of Agriculture, pp. 68–75. http://www.ba.ars.usda.gov/hb66/contents.html.

Smock, R.M., 1979. Controlled atmosphere storage of fruit. Hortic. Rev. 1, 301–336.

Toro, G., Pinto, M., 2015. Plant respiration under low oxygen. Chilean J. Agric. Sci. 75, 57–70.

Wang, C.Y., 2006. Biochemical basis of the effects of modified and controlled atmospheres. Stewart Postharvest Rev. 5, 1–4.

Zabalza, A., van Dongen, J.T., Froehlich, A., Oliver, S.N., Faix, B., Gupta, K.J., Schmälzlin, E., Igal, M., Orcaray, L., Royuela, M., Geigenberger, P., 2009. Regulation of respiration and fermentation to control the plant internal oxygen concentration. Plant Physiol. 149, 1087–1098.

Chapter 2

Cultivar effects on CA/MA requirements of fruits and vegetables

Christopher B. Watkins
School of Integrative Plant Science, Horticulture Section, Cornell University, Ithaca, NY, United States

Introduction

The tolerances of fruits and vegetables, whether whole or fresh cut, to low O_2 and high CO_2 concentrations, have been well established. These tolerances have formed the basis of the identification, usually empirically, of optimum controlled atmosphere (CA) and modified atmosphere (MA) storage regimes, in combination with appropriate storage temperatures, which result in maximum storage periods for each product type. Although an extensive literature and summary of optimal CA/MA conditions is available for most fruits and vegetables (Kupferman, 1997; Saltveit, 1997), commercial application of these technologies to whole products is limited to relatively few of them, most notably apples, pears, kiwifruit, cabbage, and sweet onions, with more restricted use, often during transportation, for asparagus, avocados, bananas, blueberries, broccoli, cherries, figs, lettuce, mangoes, nectarines, peaches, pears, plums, raspberries, strawberries, persimmons, pomegranates, tomatoes, nuts, and dried fruits. Reasons for limited use of CA/MA on most products include the cost of CA technology either as land built facilities or fitted for containers, sometimes limited increase of storage life provided by CA/MA conditions, logistical concerns, and ultimately the cost/benefit ratio of CA/MA application for a given product. Rapid commercial growth of MA has been through application of MA packaging (MAP) to maintain quality of fresh-cut fruits and vegetables through the marketing chain. Skin coatings are also often used for fruit such as apple and citrus to improve appearance and reduce water loss, and these coatings can result in modified internal O_2 and CO_2 concentrations in the fruit (Bai et al., 2002a; Bai et al., 2003; Baldwin et al., 1995).

Controlled and Modified Atmospheres for Fresh and Fresh-Cut Produce
https://doi.org/10.1016/B978-0-12-804599-2.00003-X

This chapter is concerned about the responses of different fruit and vegetable genotypes (cultivars), both whole and fresh, to CA/MA storage. However, most products in the marketplace are not identified by cultivar, the most notable exceptions being apples and pears. The literature on fruits and vegetables under CA/MA storage conditions reflects the commercial situation. For apple, the large body of information about cultivar responses reflects its importance as a fruit grown in many temperate growing regions around the world, export to markets worldwide, long-term storage, and its susceptibility to a number of senescent, storage atmosphere and temperature-related physiological disorders. However, in general, while cultivar differences for storage atmospheres have been described for several fruits and vegetables, information about cultivar effects is relatively limited (Gross et al., 2016). When many of these products are stored under CA/MA conditions, cultivar comparisons are often limited to identification of responses under the same atmospheres; a "safe" atmosphere is selected that will maintain product quality without necessarily being optimal for all cultivars in a CA storage room or MA package. This is also the case for fresh-cut products where generic atmospheres are used and cultivar identification is less important. Here, the effects of genotype on product responses are outlined by reviewing the physiological and biochemical underpinning of differing responses of cultivars to CA/MA, and then the apple, pear, strawberry, and fresh-cut fruits and vegetables for which most information is known.

Physiological and biochemical underpinning of differing responses of cultivars to CA/MA

Variations of physical structure and chemical composition as well as shape and size are large given the range of fruit and vegetable types available in the marketplace. The biological basis of differences in response to CA/MA of these products remains understudied even though it is well known that different cultivars of the same product can vary widely in storage potential, composition, and responses to biotic and/or abiotic stresses (Gil et al., 2002; Stommel et al., 2016). Interestingly, crop response to storage conditions has never been subjected to direct natural selection, and only very recently to human selection. Even today, most breeding programs are directed primarily toward appearance, sensory properties, pest and disease resistance rather than storage potential. Therefore one could postulate that responses to postharvest treatments such as CA/MA are more variable and less heritable than other traits of horticultural significance (Watkins and Pritts, 2001). Our understanding of the inheritance and expression of crop response to CA/MA storage conditions may be constrained by the plethora of responses existing in the cultivated germplasm, and little predictability in response of cultivars to these conditions exist.

Two factors that have received most attention are first, physiology of the product as it affects gas exchange between the product and the environment and thereby O_2 availability for primary and secondary metabolism, and second, the metabolic responses of products to low O_2 and high CO_2 under stressful CA/MA conditions.

A gradient of O_2 concentrations exists between the ambient environment and the internal tissues as a result of respiratory O_2 consumption and CO_2 production. While the O_2 gradient is from high in the external environment to low in the interior tissues, the reverse is true for CO_2. Gas exchange is especially important for bulky organs such as fruits and roots exposed to CA/MA conditions because anoxia can lead to insufficient O_2 concentrations in the organ resulting in cell death. The gradient is a function of several factors including the respiration rate, which is a function of inherent rates of metabolism and temperature, as well as the surface area-to-volume ratio and the nature of the surface (e.g., cuticle thickness, stomata, lenticels). The extent of these gradients in turn further affects the respiration rate of the product (Kader and Saltveit, 2003b). Resistance to gas exchange within bulky tissues appears to be relatively small, the cell walls and membranes having relatively little effect on gas exchange. The resistance of apple skin, for example, is 10- to 20-fold higher than that of the flesh depending on cultivar (Solomos, 1987). However, microstructure is important in understanding O_2 responses of plant organs as intercellular spaces are the main diffusion route for O_2, while cells are an additional route for CO_2 transport (Ho et al., 2011; Verboven et al., 2008).

Resistance to gas exchange is a function of cultivar but changes in the gas diffusion during fruit development and storage are also relevant. As reviewed by Park et al. (1993), increases in resistance to gas exchange may be associated with dehydration of the epidermal cells leading to closure of lenticels, shrinkage of fruit volume at low RH, water logging of tissue caused by cellular leakage, reduction of the intercellular void space and restriction of gaseous package resulting from high cell turgor, formation of a water barrier during fruit ripening, and continuous deposition of wax. Mechanisms that may be associated with decreases in resistance are reduction in cell turgor, fruit expansion, continuous development of open lenticel structure, and widening of skin pores as observed in bananas.

The second factor to consider in understanding of genotype effects in response to CA/MA conditions is the metabolic responses to reduced O_2 and elevated CO_2. Low O_2 and high CO_2 concentrations alone, additively or synergistically, can reduce respiration rates, inhibit ethylene production and action, and retard compositional changes (Kader and Saltveit, 2003a), but the threshold of concentrations of these gases at which aerobic respiration changes to anaerobic respiration can vary. The atmosphere at which minimum respiration rates occurs is a sum of aerobic and anaerobic respiration, known as the anaerobic compensation point (ACP) or lower oxygen limit (LOL), below which fermentation occurs (Prange et al., 2003).

The hypothesis that anaerobic respiration produces less energy for cellular functions, such as membrane maintenance, cytoplasmic acidosis, and results in cell death, has been supported by a number of studies (Franck et al., 2007;

Ho et al., 2013; Lum et al., 2016b; Peppelenbos and Oosterhaven, 1998; Saquet et al., 2000, 2003). Further insight has been gained by omic studies that have also supported this hypothesis: low O_2 stress is associated with genomic, proteomic, and metabolomic changes that demonstrate alterations of primary and secondary metabolism, including energy-related (linked to the tricarboxylic acid cycle and the electron transport chain), oxidative stress and lipid-related genes related to membrane alterations and fatty acid oxidation (Bekele et al., 2015, 2016; Brizzolara et al., 2017; Cukrov et al., 2016; Pedreschi et al., 2009; Pedreschi et al., 2007).

Elevated CO_2 concentrations are frequently used in addition to low CO_2 concentrations in CA and MA storage conditions, and can also be used as pretreatments before storage. The mechanisms of action of CO_2 on metabolism are thought to include cellular acidification as a result of generation of carbonic acid. The associated decrease of cellular pH could affect activity of many enzymes including cytochrome *c* oxidase (GonzalezMeler et al., 1996), TCA dehydrogenases (Mathooko, 1996), specifically succinate dehydrogenase (Hulme, 1956; Shipway and Bramlage, 1973), and those involved in glycolysis (Kerbel et al., 1988).

Genotype (cultivar) responses to CA/MA for selected fruits and vegetables

Apple

A summary of CA recommendations for selected apple cultivars is provided in Chapter 15. More extensive summaries are found for cultivars grown around the world found in resources such as proceedings of the International CA meetings (Kupferman, 1997). While these summaries describe large differences in O_2 and CO_2 recommendations for individual cultivars, they do not necessarily reflect physiological tolerances of these cultivars to different atmosphere regimens. Rather, recommendations can reflect factors such as desired storage times, industry sophistication such as room quality (air tightness), which affects atmospheres that can be achieved, and the availability of automated gas analysis that allows safe maintenance of lower O_2 concentrations than possible with manual systems. CA technology has progressed from controlled ventilation (6%–18% O_2; 3%–15% CO_2) to conventional CA (2%–5% O_2, 2%–5% CO_2) to low O_2 (LO; 1.5%–2% O_2; 1%–3% CO_2) to ultralow O_2 (ULO; 0.8%–1.2% O_2; 0.5%–2% CO_2) storage (Hoehn et al., 2009). Rapid CA, where O_2 in the storage atmosphere is reduced to 3%–5% within a few days of room closing is now common. In general, increased investment in CA facilities around the world appears to be resulting in less regional differences for CA storage recommendations. That said, standard CA recommendations for cultivars are affected by physiological responses to low O_2 and high CO_2 as discussed later, and differences of cultivars in benefits of these atmospheres on apple quality (Bai et al., 2005; Jobling and

McGlasson, 1995). The CA storage period required to demonstrate a benefit over that achievable with air storage also varies by cultivar; for example, soft apple cultivars such as "McIntosh", "Empire", "Jonagold", and "Golden Delicious" can show more substantial benefits under short-term CA storage than hard apple cultivars such as "Delicious", "Granny Smith", and "Fuji" (Drake, 1993).

The importance of different physiological responses associated with cultivar has been further accentuated by the development of dynamic CA (DCA) storage, in which very low O_2 concentrations are used. DCA is based on the principle of lowering the O_2 concentrations to the lowest tolerated by the fruit, without causing excessive anaerobic respiration. Maintaining low O_2, typically $<0.8\%$ and CO_2 is $<1.5\%$, can result in further maintenance of quality over that achieved even by ULO. Fruit responses to low O_2 can now be "sensed" in real time as the ACP is reached, so that it is possible to alter the O_2 concentrations in response to changes in fruit physiology that affect the ACP in storage over time, hence the term "dynamic". Available technologies are DCA-chlorophyll fluorescence (DCA-CF; HarvestWatch^{TM}) (Prange et al., 2013), ethanol dynamic control system (DCS) (Schouten et al., 1997) and DCA respiratory quotient (DCA-RQ) (Bessemans et al., 2016; Gasser et al., 2008). Regardless of the sensor system used, the O_2 concentration is typically increased to about 0.2% above the critical threshold to avoid damage to the fruit. Firmness and color retention is often greater for DCA-stored fruits than ULO or standard CA-stored fruits (Bessemans et al., 2016; Both et al., 2017; DeLong et al., 2004; Gabioud Rebeaud and Gasser, 2015; Veltman et al., 2003; Zanella and Rossi, 2015), although only small benefits are found for some cultivars (Gasser and von Arx, 2015; Zanella and Rossi, 2015). A major driver for the use of DCA is as a nonchemical alternative instead of 1-MCP to replace the antioxidant diphenylamine (DPA) for control of superficial scald (Bessemans et al., 2016; DeLong et al., 2004; Zanella and Sturz, 2013). Control of scald is usually very good but may be affected by cultivar given that relatively poor control is sometimes reported (DeLong et al., 2007).

Optimal O_2 thresholds can vary by cultivar and can be affected by growing season. De Long et al. (2004) found that the low O_2 threshold was 0.4% for "Delicious" and "Honeycrisp"; 0.5% for "Cortland", "Golden Delicious", and "Jonagold"; and 0.8% for "McIntosh" based on DCA-CF. CO_2 concentrations also have to be defined in DCA storage; Zanella and Sturz, 2013 empirically defined the appropriate CO_2 for each O_2 concentration from 0.4% to 1%, for several cultivars, not solely on the basis of fruit responses but also on the economic costs of maintaining low CO_2 concentrations in the storage environment.

Loss of fruit quality, especially firmness, after fruit are removed from CA storage is similarly affected by cultivar. Soft cultivars are especially susceptible to these poststorage losses. Softening during the shelf life period is reduced in DCA-stored fruit, although less consistently than in fruit treated with the inhibitor of ethylene perception 1-methylcyclopropene (1-MCP) (Zanella and Rossi, 2015).

Apples are susceptible to a range of physiological disorders; several of these are related to either low O_2 concentrations and/or high CO_2 concentrations in the storage environment, as well as low temperature-related disorders associated with extended storage periods afforded by use of CA technology (Lee et al., 2012; Lurie and Watkins, 2012; Volz et al., 1998; Watkins and Nock, 2012a; Wilkinson and Fidler, 1973). Susceptibility of fruit to these disorders is affected by cultivar, especially those that are susceptible to injury from high CO_2 concentrations in the storage atmosphere such as "Braeburn", "Fuji", "McIntosh", "Cortland", "Empire", and "Honeycrisp". Susceptibility of the fruit to CO_2 injury is at its highest during early CA storage, and the risk of injury can be reduced by maintaining low levels of CO_2 in storage atmospheres during the early storage period of storage. Management practices are incorporated into CA recommendations, whether standard, ULO or DCA, to avoid disorder development where necessary. Examples include:

1. "McIntosh", where 2%–3% CO_2 (in 2%–3% O_2) atmospheres are maintained for the first 4–6 weeks of storage before it is increased to 5% to reduce risk of CO_2 injury while obtaining the benefit of the high CO_2 concentration of flesh firmness (Blanpied, 1977).
2. "Honeycrisp", where fruit are conditioned at warmer temperatures before CA storage to reduce susceptibility of fruit to CO_2 injury (Contreras et al., 2014; Lum et al., 2016a).
3. "Braeburn", where CA application is delayed for 2 weeks and then CO_2 concentrations are maintained at $<1\%$ (Elgar et al., 1998; Zanella and Rossi, 2015).

Additional factors that affect cultivar responses, such as growing regions, have been identified. For example, "Delicious" apples, which can be stored routinely in 0.7% O_2 CA storage in British Columbia and Washington, result in unacceptable low O_2 injury in fruit from other North American-growing locations (Lau et al., 1998). Also, high CO_2 pretreatments (12%–17% for 4–5 days) were used commercially to maintain firmness of "Golden Delicious" apples in Washington State for several years. This treatment could be applied without injury in that region, but resulted in unacceptable CO_2 injury incidences of fruit and/or poorer quality responses in fruit grown in the US northeast (Tietjen and Hudson, 1984). Results for other cultivars such as "McIntosh" were similarly disappointing because of fruit sensitivity to CO_2 injury (Bramlage et al., 1977).

1-MCP treatments of fruit after harvest is used routinely by apple industries worldwide to maintain quality of fruit, especially during the marketing period after their removal from CA storage conditions (Mattheis, 2008; Watkins, 2008; Watkins, 2015). Recommendations for O_2 and CO_2 concentrations in CA to maintain quality factors such as firmness, soluble solids concentrations, and titratable acidity for individual cultivars that have been treated with 1-MCP have largely remained unchanged. However, significant changes for some cultivars have occurred because of treatment effects on susceptibility of fruit to storage disorders. Examples include:

1. Watercore, where 1-MCP can delay loss of watercore during storage, and can increase risk of internal browning disorder development, especially under CA storage conditions. Protocols to manage the effects of 1-MCP on watercore have included using air instead of CA storage, delaying imposition of low O_2 conditions, and delaying 1-MCP application (Mattheis, 2008).
2. External and internal CO_2 injury, because 1-MCP maintains sensitivity of fruit to CO_2 (Argenta et al., 2002b; Fawbush et al., 2008). Protocols to manage this increased risk include ensuring that low CO_2 concentrations in the storage environment and maintained and/or delaying application of CA storage conditions (Watkins and Nock, 2012b).

Edible coatings as wax are often applied to apples after storage to improve appearance, and the effects of these coatings on quality and internal gas atmospheres can also be greatly affected by cultivar. When "Delicious", "Fuji", "Braeburn", and "Granny Smith" fruit were treated with five experimental coatings with different resistances to gas exchange the internal CO_2 and O_2 concentrations ranged from 1% to 25% and 1% to 20%, respectively (Bai et al., 2002b). Treating "Delicious", "Fuji", and "Braeburn" apples with carnauba–shellac mixture and candelilla coatings, which have intermediate gas resistance, resulted in intermediate values of CO_2 and O_2, and the highest concentrations of butyl acetate and 2-methylbutyl acetate. Shellac and shellac-protein coatings, which have the highest gas resistance, resulted in high CO_2 and low O_2 and anaerobic fermentation for "Braeburn" and "Granny Smith" apples. In a related study, Bai et al. (2003) found that shellac coatings resulted in maximum fruit gloss, lowest internal O_2 and highest CO_2, and least loss of flesh firmness for all of the cultivars. "Granny Smith" had $<2\%$ O_2 in both freshly harvested and 5-month stored apples, while treatment of freshly harvested "Braeburn" resulted in 25% CO_2. This excessive modification of internal gas induced an abrupt rise of the RQ and ethanol accumulation in both "Braeburn" and "Granny Smith", and flesh browning in "Braeburn". Shellac coatings also resulted in ethanol accumulation in freshly harvested and 5-month stored "Fuji". Candelilla and carnauba–shellac coatings maintained more optimal internal O_2 and CO_2 and better quality for "Fuji", "Braeburn", and "Granny Smith" apples.

Physiological and biochemical responses of apple fruit to low O_2 in CA/MA storage

Most research has been focused on differences in gas diffusion of cultivars. Low O_2 limits based on fermentation thresholds were found to vary from 0.7% for "Delicious" and 0.8% for "Law Rome", to 1.9% for "McIntosh" (Gran and

Beaudry, 1993). Tissues of "Fuji" with watercore have a lower intercellular air space volume, reduced permeance to gas diffusion, and increased internal CO_2 concentrations, and accumulation of ethanol and acetaldehyde can occur in affected fruit even in air (Argenta et al., 2002a). "Braeburn" apples have greater resistance to gas diffusion than "Jonica" apples, which may be related to the higher O_2 and low CO_2 concentrations in the storage environment required to minimize development of storage disorders (Schotsmans et al., 2002). The lower diffusivity for both O_2 and CO_2 of "Braeburn" apples compared with "Jonica" apples appears to be also related with very dense outer cortex tissue (smaller cell size and intercellular space volume) of "Braeburn". Similar differences were described by Ho et al. (2010) for "Braeburn" compared with "Jonagold", a cultivar that is stored in ULO with minimal problems, and "Kanzi". Ho et al. (2013) used a gas transport model that incorporated respiration kinetics and prediction of ATP production to investigate the lower O_2 beyond which there is cell death using "Kanzi", "Jonagold", and "Braeburn" apples. The critical O_2 concentration of the tissue was similar in all cultivars because it was independent of the maximal respiration and fermentation rate of the fruit. However, differences in resistance to gas diffusion of each cultivar meant that the O_2 concentrations translated into different minimums. These minimums are consistent with different CA storage requirements for "Braeburn" compared with the other cultivars. Other factors that may affect responses of cultivars to low O_2 include fruit maturity (Ho et al., 2013) and plant growth regulators (Brackmann et al., 2014) but these are not well characterized.

Variations in tolerances can also exist within strains of cultivars. The Marshall strain of "McIntosh" has a higher susceptibility to low O_2 injury than other strains of the cultivar, and this difference is related to its greater resistance to gas diffusion through the skin (Park et al., 1993). The Marshall strain accumulates tenfold more ethanol than the Rogers strain at the same fruit internal O_2 (Fig. 1), also indicating different thresholds for anaerobic metabolism, a finding that is consistent with more recent modeling studies with other cultivars (Ho et al., 2013).

The tolerance of apple cultivars to low O_2 concentrations may be affected by CO_2 concentrations. The effects of 0% to 8% CO_2 on the ACP and the internal fermentation threshold based either on the RQ or ethanol accumulation were investigated using "Cox's Orange Pippin" and "Braeburn" (Yearsley et al., 1997). The results indicated that ACP remained constant at 0.5% O_2 for "Cox's Orange Pippin" and 0.8–1.0% O_2 for "Braeburn" apples over the CO_2 range tested for both cultivars and at both 0 and 20°C. However, no consistent effect of CO_2 on the fermentation threshold (RQ or ethanol-based) was detected at 20°C. The fermentation threshold was 0.2%–0.8% O_2 higher at 8% CO_2 than at 0% CO_2 (with the exception of the threshold based on RQ for "Cox's Orange Pippin"). A small decrease in O_2 uptake (estimated from the difference in external and internal O_2 atmospheres) was observed between 2 and 8% CO_2 at 20°C. Elevated CO_2 slightly lowered the RQ, estimated from the ratio of differences between external and internal concentrations of CO_2 and O_2 of "Cox's Orange

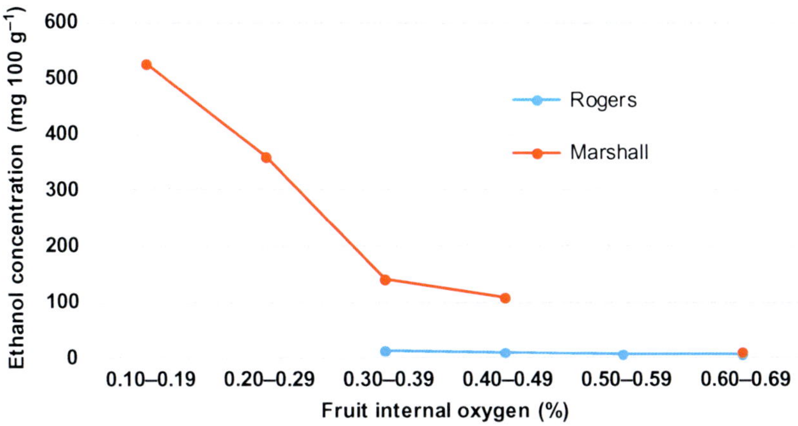

FIG. 1 Ethanol concentrations in relation to internal oxygen concentration of Rogers and Marshall "McIntosh" fruit on removal from storage in 0.8% oxygen and 3% carbon dioxide *(Modified from Park, Y.M., Blan-pied, G.D., Jozwiak, Z., Liu, F.W., 1993. Posthar-vest stud-ies of re-sis-tance to gas dif-fu-sion in McIn-tosh ap-ples. Posthar-vest Biol. Tech-nol. 2, 329–339.)*

Pippin" in 8% CO_2 and "Braeburn" in 2%–8% CO_2 at 20°C, and more markedly in 8% CO_2 at 0°C. The lower RQ of "Braeburn" at 20°C compared with "Cox's Orange Pippin" apples indicated that "Braeburn" had a higher permeance to CO_2 relative to O_2 than "Cox's Orange Pippin". This study indicates the tolerance of "Cox's Orange Pippin" and "Braeburn" apples to low O_2 concentrations may be affected by levels of CO_2.

Pear

Pear fruit also have a wide variation of recommended CA storage regimens (Chapter 14.2), but less is known about the underlying physiology than for apples. "Conference" pears have a lower resistance to gas diffusion than "Doyenne du Comice" (Schotsmans et al., 2002), but "Doyenne du Comice" are less susceptible to storage defects than "Conference" pears. Therefore skin resistance to gas diffusion is not the only important factor in relationship to storage disorders.

Amarante (1998), as described by Bai et al. (2009), found that differences in skin structure of different pear cultivars affected performance of skin coatings. In "Bartlett", "Comice", and "Packham's Triumph", cultivars without lignified cells in the skin, the diffusion of water vapor, CO_2, and O_2 decreases markedly with a small increase of total solids in a carnauba wax. In these pears, the coating covers the entire skin surface including the lenticels. In "Bosc" pears, which have lignified cells in the skin, carnauba wax reduced CO_2 and O_2 diffusion, but that of water vapor was only slightly affected. Although the coating effectively covers the lenticels, it is not effective in covering the lignified cells in the epidermis which is the major pathway for water vapor exchange.

Optimization of surface coatings for pears must account for differences between cultivars and ripening stage at which fruit are coated, as well as the temperature at which the fruit are going to be kept (Amarante et al., 2001). When "Bartlett", "Beurre Bosc", "Doyenne du Comice", and "Packham's Triumph" were treated at harvest or after cold storage with concentrations of 0%, 5%, 10%, 20%, 40%, and 100% (v/v) of a commercial carnauba-based wax emulsion, respiration rates, softening, and color change followed a Michaelis–Menten model when plotted against internal % O_2, while internal CO_2 had little explanatory power for these variables. Plots of internal CO_2 and respiration versus internal O_2 for fruit coated either preclimacteric (at harvest) or entering the climacteric (after cold storage) indicated that "Doyenne du Comice" and "Packham's" were tolerant to hypoxia whereas, with "Bartlett" and "Bosc", tolerance reduced with advancing ripening. "Bosc" pears were the least tolerant of all four cultivars.

Strawberry

Strawberry fruit are often treated with high CO_2 concentrations (10%–15%) in pallet shrouds before transport to markets (Mitcham, 2016). CO_2 treatments increase and maintain flesh firmness (Harker et al., 2000) and reduce decay (Harris and Harvey, 1973), but can result in accumulation of fermentative metabolites depending on concentrations of the gas and exposure time (Ke et al., 1991; Larsen and Watkins, 1995). Effects of cultivar have been described. Four of 25 cultivars did not show a CO_2-related increase of firmness at a single treatment time of 42h (Smith and Skog, 1992), and Watkins et al. (1999) found that the firmness responses of seven cultivars interacted with treatment time. Effects of 100% CO_2 for 4h followed by air storage (5°C) on flesh firmness were markedly influenced by cultivar (Matsumoto et al., 2010). The postharvest life and flavor quality of "Aromas", "Diamante", and "Selva" in air plus 20% CO_2 (compared also with air alone) at 5°C highlighted cultivar differences in postharvest life based on flavor life rather than appearance alone (Pelayo et al., 2003). The postharvest life based on appearance was 9, 11, and 13days for "Aromas", "Diamante", and "Selva" fruit, respectively, but was 2days shorter if based on flavor components (sugars, organic acids, aroma compounds) and fermentative metabolites, as well as the results of sensory evaluations.

Although elevated CO_2 can delay anthocyanin accumulation in strawberry fruit (Gil et al., 1997; Zhang and Watkins, 2005) and increase rate of ascorbic acid loss (Agar et al., 1997), little is known about the effects of cultivar on antioxidant compounds and antioxidant activity. Pelayo et al. (2003) did not find effects of CO_2 on phenolic concentrations of three cultivars. In another study, total flavonoids, phenolics and ascorbic acid concentrations, and total antioxidant activity were lower in CO_2-treated fruit than air-stored fruit but except for differences related to initial levels at harvest, no effect of cultivar on treatment responses was found (Shin et al., 2008).

Accumulations of the fermentative metabolites, acetaldehyde, ethanol, and ethyl acetate in fruit treated with 20% CO_2 were also affected by cultivar (Watkins et al., 1999). Two cultivars classified as tolerant to 20% CO_2 treatment as indicated by little accumulations of fermentative metabolites were shown to have higher activities of pyruvate decarboxylase (PDC) and alcohol dehydrogenase (ADH) in CO_2-treated fruit than in two intolerant cultivars as indicated by high accumulation of fermentative metabolites (Fernandez-Trujillo et al., 1999). Also, while succinate accumulation, usually associated with inhibition of succinate dehydrogenase activity by elevated CO_2, occurred in all cultivars, these were greater in tolerant than nontolerant cultivars. γ-Aminobutyric acid, a metabolite that often accumulates in response to stress (Shelp et al., 1999), increased in CO_2-treated fruit to different degrees in four cultivars, but the maximum concentrations were not related to different accumulations of fermentative metabolites (Deewatthanawong et al., 2010, 2011).

"Jewel" strawberry fruit treated with CO_2 accumulated acetaldehyde and ethanol but little ethyl acetate, while "Cavendish" accumulated little acetaldehyde or ethanol but accumulated ethyl acetate (Ponce-Valadez and Watkins, 2008). In CO_2-treated fruit, PDC activity was positively correlated with ethyl acetate accumulation in "Jewel" but not in "Cavendish," while no differential effect of atmosphere was observed on its gene expression. ADH activity and gene expression were correlated with ethanol accumulation in "Cavendish". In "Jewel", there was a positive correlation between ADH gene expression and enzyme activity; however, this correlation does not explain ethanol accumulation in this cultivar. Ethyl acetate accumulation did not show any correlation with AAT activity and gene expression in any of the cultivars. Succinate concentrations were highest and those of malate lowest in CO_2-treated fruit of both cultivars, but MDH and ME activities were not affected by CO_2. Gene expression of malate dehydrogenase (*MDH*) and malic enzyme (*ME*) was not affected by atmosphere in "Cavendish", although in "Jewel" *MDH* expression was slightly lower in CO_2 than air-treated fruit. The results of this study show that differences in fermentation products and malate accumulation in CO_2-treated strawberry fruit are not consistently correlated with enzyme activities and gene expression.

Differences between the two cultivars were further investigated using a heterologous fruit microarray system (tomato) (Ponce-Valadez et al., 2009). The results indicated that protein synthesis in "Jewel" might be more impaired than in "Cavendish", suggesting that protein synthesis might be more impaired by CO_2 in "Jewel" than in "Cavendish". Some of the cDNAs with altered expression patterns in "Cavendish", but not in "Jewel", were homologous to regulatory genes including several transcription factors. Because these genes likely regulate the expression of other genes, changes in their expression may contribute to the difference in number of cDNAs whose expression was affected by elevated CO_2 between the two cultivars. Expression of genes encoding *MDH* and *PDC* was higher in "Jewel" than in "Cavendish". Overall, "Cavendish" fruit had fewer genes with differential expression affected by CO_2 than "Jewel", perhaps suggesting better homeostatic maintenance of "Cavendish" compared with "Jewel" under elevated CO_2 conditions. No further research on mechanisms of differences in responses of strawberry fruit to CO_2 as affected by cultivar appears available. In addition, while differences in cultivar morphology such as the epidermis may affect softening and disease susceptibility (Nishizawa et al., 2003), no studies on how these may affect tolerance of fruit to CA storage conditions exist.

Fresh-cut products

Recommendations for fresh-cut products are available (Beaulieu and Gorny, 2016; Chapter 19) and typically differ from those for the associated whole product. The reason for such differences is assumed to be because minimal processing reduces the importance of the skin to gas exchange.

Reviews on the physiological and biochemical responses of fresh-cut fruits and vegetables rarely report studies on comparisons between cultivars (Rico et al., 2007; Rojas-Grau et al., 2009). This is not surprising since most studies are carried out with single cultivars and the major factors in quality control are focused on postharvest handling, processing steps, and storage periods. Also, postharvest treatments with antioxidants such as ascorbic acid and calcium salts can be more effective than MAP for preservation of some fresh-cut products (Oms-Oliu et al., 2008). Natural variation in the raw material and its dynamic response to processing and storage may make it difficult to truly optimal storage atmospheres (Saltveit, 2003), a situation likely exacerbated by use of different cultivars. Little evidence exists that modifications to MAPs are made commercially based on cultivar differences. However, cultivar selection can affect maintenance of fresh-cut fruits and vegetables freshness under CA/MA conditions, primarily because of inherent characteristics of any given cultivar such as texture, browning potential, and others that affect sensory perception. Indeed, selection of appropriate cultivars for fresh cut through screening of cultivars and breeding lines is an important part of decisions made by breeders, growers, and processors (Gil, 2016). Therefore while cultivar selection typically is less of a research focus, the storage life of fresh-cut products is strongly affected by cultivar, examples including apples (Calderon-Lopez et al., 2005; Kim et al., 1993), cantaloupe (Beaulieu, 2005), lettuce (Lopez-Galvez et al., 1996; Tudela et al., 2016), pears (Gorny et al., 1999), potatoes (Cantos et al., 2002; Laurila et al., 1998), nectarines and peaches (Gorny et al., 1999), and zucchini (Blanco-Diaz et al., 2016). Differences among cultivars in responses of fresh-cut products to MAP are illustrated by the following examples.

Apples

Kim et al. (1993) compared quality of fresh-cut slices prepared from 12 apple cultivars and were the first to show that cultivar characteristics such as firmness (texture) and browning potential were more important than respiration rates. Cultivar also affects the appropriate processing temperatures (Toivonen and Hampson, 2009), for example, visual quality ratings for "Granny Smith" at 3 weeks was poor for slices processed at 1°C, 5°C, 13°C, and 20°C temperatures, "Ambrosia" slices maintained acceptable quality ratings over all temperature ranges, "Gala" slices had lower quality ratings when processed at warmer temperatures, whereas Aurora Golden "Gala" had increased quality ratings with warmer processing temperatures.

There is a surprising absence of research on cultivar effects on MAP of apple slices, but that existing highlights that atmospheres in MAP are less critical than other products (Rossle et al., 2011). Of 10 cultivars, "Gloster", "Alwa", "Idared", and "Jonica" were selected on the basis of antioxidant capacity and quality factors. Slices from different cultivars varied greatly in CO_2 production and O_2 depletion in the same MAPs. However, the effects of increasing CO_2 concentrations up to 60% on acetaldehyde and ethanol accumulations in slices produced from "McIntosh", "Empire", and "Delicious" were affected by cultivar and storage period (Amissah et al., 2006).

Pears

When "Bartlett", "Bosc", "Anjou", and "Red Anjou" pears were compared (Gorny et al., 2000), "Bartlett" was the best suited cultivar for fresh-cut processing because of its longest postcutting shelf life, even though it had the highest respiration and ethylene production rates. However, only "Bartlett" was tested further under CA conditions (Gorny et al., 2002). Arias et al. (2008) found that "Conference" pears were more suitable than "Williams" and "Passacrassana" for fresh-cut slices because of lower browning and sensory qualities.

Peaches and nectarines

Fresh slices made from 13 peach and eight nectarine cultivars varied between 2 and 12 days at 0°C (Gorny et al., 1999). Atmospheres of 0.25% O_2 and/or 10% or 20% CO_2 extended the shelf life at 10°C of "O'Henry" or "Elegant Lady" peach slices by 1–2 days beyond the air control.

Lettuce

The quality of fresh-cut lettuce is affected greatly by mechanical wounding during processing that can result in physiological responses such as tissue browning (Heimdal et al., 1995). Browning can be inhibited by low O_2 concentrations in

CA/MA (Ballantyne et al., 1988), but these concentrations in combination with elevated CO_2 concentrations can result in development of off-flavors (Martinez-Sanchez et al., 2011). The quality of iceberg lettuce was maintained by CA to a greater extent than romaine, green leaf, and red leaf, while that of butter lettuce was lowered as a result of CO_2 injury (Lopez-Galvez et al., 1996). The extent of accumulation of off-flavors in fresh-cut iceberg lettuce was affected by cultivar as well as maturity stage and seasonality (Tudela et al., 2013). The effect of cultivar was further investigated by Tuleda et al. (2016) who compared different storage conditions (air, CA, passive and active MAP) to evaluate responses of 18 iceberg and 7 romaine lettuce cultivars. The cultivars had a range of respiration rates resulting in differences in O_2 consumption and CO_2 production that impacted the equilibrium gas concentrations achieved. Therefore a single MAP provided a means of screening of the cultivars for browning and off-flavor as well as respiration rates.

Zucchini

Blanco-Diaz et al. (2016) compared the responses of fresh-cut zucchini prepared from whole fruit of four cultivars with a range of metabolic rates and stored in MAP. Physiochemical, nutritional, and sensory quality of the fresh-cut zucchini was affected by cultivar as well as maturity and film selection, with one ("Cronos") performing much better than the others, and one ("Parador") performing poorly as a result of more active metabolism and more rapid depletion of energy reserves.

Conclusions

The optimum CA and MA storage regimes, which result in maximum storage periods for most fruits and vegetables are well established. However, comparison of cultivar effects on CA/MA information is absent for most products, reflecting that commercial conditions are established with atmospheres that encompass wide "safety" limits for O_2 and CO_2 that maximize benefits without incurring risk of anaerobic conditions. In other words, the product has to fit the technology, rather than the other way around, and cultivars of any product that do not perform are discarded. Therefore the focus is on the breeder to select cultivars with desirable responses. The important exception to this rule are apples, and to a lesser extent pears, because the number of cultivars is so large and the need to maximize storage periods over many months in order to market the fruit is so critical. The application of new technologies such as DCA, which push the boundaries of lower O_2 limits, is ongoing. Most information about the underlying physiological and biochemical principles underlying the responses of cultivars to CA/MA storage conditions is focused on apple, and extension of such investigations across a broader range of products is needed.

References

Agar, I.T., Streif, J., Bangerth, F., 1997. Effect of high CO_2 and controlled atmosphere (CA) on the ascorbic and dehydroascorbic acid content of some berry fruits. Postharvest Biol. Technol. 11, 47–55.

Amarante, C.V.T. (1998). Gas exchange, ripening behaviour and postharvest quality of coated pears. Ph.D. Dissertation, Massey University, Palmerston North, New Zealand, 292 pp.

Amarante, C., Banks, N.H., Ganesh, S., 2001. Characterising ripening behaviour of coated pears in relation to fruit internal atmosphere. Postharvest Biol. Technol. 23, 51–59.

Amissah, J.G.N., Hotchkiss, J.H., Watkins, C.B., 2006. Diphenylamine and pre-slicing storage effects on the responses of apple slices to elevated CO_2 atmospheres. Postharvest Biol. Technol. 39, 178–184.

Argenta, L., Fan, X.T., Mattheis, J., 2002a. Impact of watercore on gas permeance and incidence of internal disorders in 'Fuji' apples. Postharvest Biol. Technol. 24, 113–122.

Argenta, L.C., Fan, X.T., Mattheis, J.P., 2002b. Responses of 'Fuji' apples to short and long duration exposure to elevated CO_2 concentration. Postharvest Biol. Technol. 24, 13–24.

Arias, E., Gonzalez, J., Lopez-Buesa, P., Oria, R., 2008. Optimization of processing of fresh-cut pear. J. Sci. Food Agric. 88, 1755–1763.

Bai, J.H., Baldwin, E.A., Hagenmaier, R.H., 2002a. Alternatives to shellac coatings provide comparable gloss, internal gas modification, and quality for 'Delicious' apple fruit. HortScience 37, 559–563.

Bai, J.H., Hagenmaier, R.D., Baldwin, E.A., 2002b. Volatile response of four apple varieties with different coatings during marketing at room temperature. J. Agric. Food Chem. 50, 7660–7668.

Bai, J.H., Hagenmaier, R.D., Baldwin, E.A., 2003. Coating selection for 'Delicious' and other apples. Postharvest Biol. Technol. 28, 381–390.

Bai, J.H., Baldwin, E.A., Goodner, K.L., Mattheis, J.P., Brecht, J.K., 2005. Response of four apple cultivars to 1-methylcyclopropene treatment and controlled atmosphere storage. HortScience 40, 1534–1538.

Bai, J.H., Prange, R.K., Toivonen, P.M., 2009. Pome fruits. In: Yahia, E.M. (Ed.), Modified and Controlled Atmospheres for the Storage, Transportation, and Packaging of Horticultural Commodities. CRC Press, Baca Raton, pp. 267–285.

Baldwin, E.A., Nisperoscarriedo, M., Shaw, P.E., Burns, J.K., 1995. Effect of coatings and prolonged storage-conditions on fresh orange flavor volatiles, degrees brix, and ascorbic acid levels. J. Agric. Food Chem. 43, 1321–1331.

Ballantyne, A., Stark, R., Selman, J.D., 1988. Modified atmosphere packaging of shredded lettuce. Int. J. Food Sci. Technol. 23, 267–274.

Beaulieu, J.C., 2005. Within-season volatile and quality differences in stored fresh-cut cantaloupe cultivars. J. Agric. Food Chem. 53, 8679–8687.

Beaulieu, J.C., Gorny, J.R., 2016. Fresh-cut fruits. In: Gross, K.C., Wang, C.Y., Saltveit, M.E. (Eds.), The Commercial Storage of Fruits, Vegetables, and Florist and Nursery Stocks, USDA-HB66. United States Department of Agriculture.

Bekele, E.A., Beshir, W.F., Hertog, M., Nicolai, B.M., Geeraerd, A.H., 2015. Metabolic profiling reveals ethylene mediated metabolic changes and a coordinated adaptive mechanism of 'Jonagold' apple to low oxygen stress. Physiol. Plant. 155, 232–247.

Bekele, E.A., Ampofo-Asiama, J., Alis, R.R., Hertog, M., Nicolai, B.M., Geeraerd, A.H., 2016. Dynamics of metabolic adaptation during initiation of controlled atmosphere storage of 'Jonagold' apple: effects of storage gas concentrations and conditioning. Postharvest Biol. Technol. 117, 9–20.

Bessemans, N., Verboven, P., Verlinden, B.E., Nicolai, B.M., 2016. A novel type of dynamic controlled atmosphere storage based on the respiratory quotient (RQ-DCA). Postharvest Biol. Technol. 115, 91–102.

Blanco-Diaz, M.T., Perez-Vicente, A., Font, R., 2016. Quality of fresh cut zucchini as affected by cultivar, maturity at processing and packaging. Packag. Technol. Sci. 29, 365–382.

Blanpied, G.D., 1977. Requirements and recommendations for eastern and midwestern apples. In: Controlled Atmospheres for the Storage and Transport of Perishable Agricultural Commodities. Michigan State University, East Lansing, pp. 225–230 [see FSTA (1979) 11 8J1407].

Both, V., Thewes, F.R., Brackmann, A., Anese, R.D., Ferreira, D.D., Wagner, R., 2017. Effects of dynamic controlled atmosphere by respiratory quotient on some quality parameters and volatile profile of 'Royal Gala' apple after long-term storage. Food Chem. 215, 483–492.

Brackmann, A., Thewes, F.R., Anese, R.D., Both, V., 2014. Effect of growth regulators on 'Brookfield' apple gas diffusion and metabolism under controlled atmosphere storage. Pesqui. Agropecu. Bras. 49, 323–329.

Bramlage, W.J., Bareford, P.H., Blanpied, G.D., Dewey, D.H., Taylor, S., Porritt, S.W., Lougheed, E.C., Smith, W.H., McNicholas, F.S., 1977. Carbon dioxide treatments for McIntosh apples before CA storage. J. Am. Soc. Hortic Sci. 102, 658–662.

Brizzolara, S., Santucci, C., Tenori, L., Hertog, M., Nicolai, B., Stürz, S., Zanella, A., Tonutti, P., 2017. A metabolomics approach to elucidate apple fruit responses to static and dynamic controlled atmosphere storage. Postharvest Biol. Technol. 127, 76–87.

Calderon-Lopez, B., Bartsch, J.A., Lee, C.Y., Watkins, C.B., 2005. Cultivar effects on quality of fresh cut apple slices from 1-methylcyclopropene (1-MCP)-treated apple fruit. J. Food Sci. 70, S221–S227.

Cantos, E., Tudela, J.A., Gil, M.I., Espin, J.C., 2002. Phenolic compounds and related enzymes are not rate-limiting in browning development of fresh-cut potatoes. J. Agric. Food Chem. 50, 3015–3023.

Contreras, C., Alsmairat, N., Beaudry, R., 2014. Prestorage conditioning and diphenylamine improve resistance to controlled atmosphere related injury in 'Honeycrisp' apples. HortScience 49, 76–81.

Cukrov, D., Zermiani, M., Brizzolara, S., Cestaro, A., Licausi, F., Luchinat, C., Santucci, C., Tenori, L., Van Veen, H., Zuccolo, A., Ruperti, B., Tonutti, P., 2016. Extreme hypoxic conditions induce selective molecular responses and metabolic reset in detached apple fruit. Front. Plant Sci. 7, 146. https://doi.org/10.3389/fpls.2016.00146.

DeLong, J.M., Prange, R.K., Leyte, J.C., Harrison, P.A., 2004. A new technology that determines low-oxygen thresholds in controlled-atmosphere-stored apples. HortTechnology 14, 262–266.

DeLong, J.M., Prange, R.K., Harrison, P.A., 2007. Chlorophyll fluorescence-based low O_2 CA storage of organic 'Cortland' and 'Delicious' apples. Acta Hortic. 737, 31–37.

Deewatthanawong, R., Nock, J.F., Watkins, C.B., 2010. γ-Aminobutyric acid (GABA) accumulation in four strawberry cultivars in response to elevated CO_2 storage. Postharvest Biol. Technol. 57, 92–96.

Deewatthanawong, R., Nock, J.F., Watkins, C.B., 2011. γ-Aminobutyric acid (GABA) accumulation in four strawberry cultivars in response to elevated CO_2 storage. Vol. 57, p. 92, 2010Postharvest Biol. Technol. 60, 173.

Drake, S.R., 1993. Short-term controlled atmosphere storage improved quality of several apple cultivars. J. Am. Soc. Hortic Sci. 118, 486–489.

Elgar, H.J., Burmeister, D.M., Watkins, C.B., 1998. Storage and handling effects on a CO_2-related internal browning disorder of 'Braeburn' apples. HortScience 33, 719–722.

Fawbush, F., Nock, J.F., Watkins, C.B., 2008. External carbon dioxide injury and 1-methylcyclopropene. Postharvest Biol. Technol. 48, 92–98.

Fernandez-Trujillo, J.P., Nock, J.F., Watkins, C.B., 1999. Fermentative metabolism and organic acid concentrations in fruit of selected strawberry cultivars with different tolerances to carbon dioxide. J. Am. Soc. Hortic Sci. 124, 696–701.

Franck, C., Lammertyn, J., Ho, Q.T., Verboven, P., Verlinden, B., Nicolai, B.A., 2007. Browning disorders in pear fruit. Postharvest Biol. Technol. 43, 1–13.

Gabioud Rebeaud, S., Gasser, F., 2015. Fruit quality as affected by 1-MCP treatment and DCA storage - a comparison of the two methods. Eur. J. Hortic. Sci. 80, 18–24.

Gasser, F., von Arx, K., 2015. Dynamic CA storage of organic apple cultivars. Acta Hortic. 1071, 527–532.

Gasser, F., Eppler, T., Naunheim, W., Gabioud, S., Hoehn, E., 2008. Control of the critical oxygen level during dynamic CA storage of apples. Agrarforschung 15, 98–103.

Gil, M.I., 2016. Preharvest factors and fresh-cut quality of leafy vegetables. Acta Hortic. 1141, 57–64.

Gil, M.I., Holcroft, D.M., Kader, A.A., 1997. Changes in strawberry anthocyanins and other polyphenols in response to carbon dioxide treatments. J. Agric. Food Chem. 45, 1662–1667.

Gil, M.I., Tomas-Barberan, F.A., Hess-Pierce, B., Kader, A.A., 2002. Antioxidant capacities, phenolic compounds, carotenoids, and vitamin C contents of nectarine, peach, and plum cultivars from California. J. Agric. Food Chem. 50, 4976–4982.

GonzalezMeler, M.A., Ribas-Carbo, M., Siedow, J.N., Drake, B.G., 1996. Direct inhibition of plant mitochondrial respiration by elevated CO_2. Plant Physiol. 112, 1349–1355.

Gorny, J.R., Hess-Pierce, B., Kader, A.A., 1999. Quality changes in fresh-cut peach and nectarine slices as affected by cultivar, storage atmosphere and chemical treatments. J. Food Sci. 64, 429–432.

Gorny, J.R., Cifuentes, R.A., Hess-Pierce, B., Kader, A.A., 2000. Quality changes in fresh-cut pear slices as affected by cultivar, ripeness stage, fruit size, and storage regime. J. Food Sci. 65, 541–544.

Gorny, J.R., Hess-Pierce, B., Cifuentes, R.A., Kader, A.A., 2002. Quality changes in fresh-cut pear slices as affected by controlled atmospheres and chemical preservatives. Postharvest Biol. Technol. 24, 271–278.

Gran, C.D., Beaudry, R.M., 1993. Determination of the low oxygen limit for several commercial apple cultivars by respiratory quotient breakpoint. Postharvest Biol. Technol. 3, 259–267.

Gross, K.C., Wang, C.Y., Saltveit, M.E. (Eds.), 2016. The Commercial Storage of Fruits, Vegetables, and Florist and Nursery Stocks, USDA-HB66. United States of Agriculture. https://www.ars.usda.gov/is/np/CommercialStorage/CommercialStorage.pdf.

Harker, F.R., Elgar, H.J., Watkins, C.B., Jackson, P.J., Hallett, I.C., 2000. Physical and mechanical changes in strawberry fruit after high carbon dioxide treatments. Postharvest Biol. Technol. 19, 139–146.

Harris, C.M., Harvey, J.M., 1973. Quality and decay of California strawberries stored in CO_2-enriched atmospheres. Plant Disease Reporter 57, 44–46.

Heimdal, H., Kuhn, B.F., Poll, L., Larsen, L.M., 1995. Biochemical changes and sensory quality of shredded and MA-packaged iceberg lettuce. J. Food Sci. 60, 1265–1268.

Ho, Q.T., Verboven, P., Verlinden, B.E., Schenk, A., Delele, M.A., Rolletschek, H., Vercammen, J., Nicolai, B.M., 2010. Genotype effects on internal gas gradients in apple fruit. J. Exp. Bot. 61, 2745–2755.

Ho, Q.T., Verboven, P., Verlinden, B.E., Herremans, E., Wevers, M., Carmeliet, J., Nicolai, B.M., 2011. A three-dimensional multiscale model for gas exchange in fruit. Plant Physiol. 155, 1158–1168.

Ho, Q.T., Verboven, P., Verlinden, B.E., Schenk, A., Nicolai, B.M., 2013. Controlled atmosphere storage may lead to local ATP deficiency in apple. Postharvest Biol. Technol. 78, 103–112.

Hoehn, E., Prange, R.K., Vigneault, C., 2009. Storage technology and applications. In: Yahia, E.M. (Ed.), Modified and Controlled Atmospheres for the Storage, Transportation, and Packaging of Horticultural Commodities. CRC Press, Baca Raton, pp. 17–50.

Hulme, A.C., 1956. Carbon dioxide injury and the presence of succinic acid in apples. Nature 178, 218–219.

Jobling, J.J., McGlasson, W.B., 1995. A comparison of ethylene production, maturity and controlled-atmosphere storage life of Gala, Fuji and Lady Williams apples (*Malus domestica*, Borkh). Postharvest Biol. Technol. 6, 209–218.

Kader, A.A., Saltveit, M.E., 2003a. Atmosphere modification. In: Bartz, J.A., Brecht, J.K. (Eds.), Postharvest Physiology and Pathology of Vegetables. Marcel Dekker, New York, pp. 229–246.

Kader, A.A., Saltveit, M.E., 2003b. Respiration and gas exchange. In: Bartz, J.A., Brecht, J.K. (Eds.), Postharvest Physiology and Pathology of Vegetables. Marcel Dekker, New York, pp. 7–30.

Ke, D.Y., Goldstein, L., Omahony, M., Kader, A.A., 1991. Effects of short-term exposure to low O_2 and high CO_2 atmospheres on quality attributes of strawberries. J. Food Sci. 56, 50–54.

Kerbel, E.L., Kader, A.A., Romani, R.J., 1988. Effects of elevated CO_2 concentrations on glycolysis in intact bartlett pear fruit. Plant Physiol. 86, 1205–1209.

Kim, D.M., Smith, N.L., Lee, C.Y., 1993. Quality of minimally processed apple slices from selected cultivars. J. Food Sci. 58, 1115–1117.

Kupferman, E. (1997). Controlled atmosphere storage of apples and Pears. Seventh International Controlled Atmosphere Research Conference, University of California, Davis, Vol. 2, Postharvest Horticulture Series 17, 1–30.

Larsen, M., Watkins, C.B., 1995. Firmness and aroma composition of strawberries following short term high carbon dioxide treatments. HortScience 30, 303–305.

Lau, O.L., Barden, C.L., Blankenship, S.M., Chen, P.M., Curry, E.A., DeEll, J.R., Lehman-Salada, L., Mitcham, E.J., Prange, R.K., Watkins, C.B., 1998. A North American cooperative survey of 'Starkrimson Delicious' apple responses to 0.7% O_2 storage on superficial scald and other disorders. Postharvest Biol. Technol. 13, 19–26.

Laurila, E.K., Hurme, E.U., Ahvenainen, R.T., 1998. Shelf life of sliced raw potatoes of various cultivar varieties—substitution of bisulfites. J. Food Prot. 61, 1363–1371.

Lee, J., Rudell, D.R., Davies, P.J., Watkins, C.B., 2012. Metabolic changes in 1-methylcyclopropene (1-MCP)-treated 'Empire' apple fruit during storage. Metabolomics 8, 742–753.

Lopez-Galvez, G., Saltveit, M.E., Cantwell, M., 1996. The visual quality of minimally processed lettuces stored in air or controlled atmosphere with emphasis on romaine and iceberg types. Postharvest Biol. Technol. 8, 179–190.

Lum, G.B., Brikis, C.J., Deyman, K.L., Subedi, S., DeEll, J.R., Shelp, B.J., Bozzo, G.G., 2016a. Pre-storage conditioning ameliorates the negative impact of 1-methylcyclopropene on physiological injury and modifies the response of antioxidants and γ-aminobutyrate inHoneycrisp' apples exposed to controlled-atmosphere conditions. Postharvest Biol. Technol. 116, 115–128.

Lum, G.B., Shelp, B.J., DeEll, J.R., Bozzo, G.G., 2016b. Oxidative metabolism is associated with physiological disorders in fruits stored under multiple environmental stresses. Plant Sci. 245, 143–152.

Lurie, S., Watkins, C.B., 2012. Superficial scald, its etiology and control. Postharvest Biol. Technol. 65, 44–60.

Martinez-Sanchez, A., Tudela, J.A., Luna, C., Allende, A., Gil, M.I., 2011. Low oxygen levels and light exposure affect quality of fresh-cut Romaine lettuce. Postharvest Biol. Technol. 59, 34–42.

Mathooko, F.M., 1996. Regulation of respiratory metabolism in fruits and vegetables by carbon dioxide. Postharvest Biol. Technol. 9, 247–264.

Matsumoto, K., Hwang, Y.S., Lee, C.H., Huber, D.J., 2010. Changes in firmness and pectic polysaccharide solubility in three cultivars of strawberry fruit following short-term exposure to high pCO_2. J. Food Qual. 33, 312–328.

Mattheis, J.P., 2008. How 1-methylcyclopropene has altered the Washington state apple industry. HortScience 43, 99–101.

Mitcham, E.J., 2016. Strawberries. In: Gross, K.C., Wang, C.Y., Saltveit, M.E. (Eds.), The Commercial Storage of Fruits, Vegetables, and Florist and Nursery Stocks. United States Department of Agriculture.

Nishizawa, T., Kondo, Y., Nagasawa, S., Retamales, J.B., Lavin, A., Motomura, Y., 2003. Storability and its relationship with the anatomy of epidermal tissues of ripe Chilean strawberry (*Fragaria chiloensis*) fruit. Acta Hortic. 626, 347–352.

Oms-Oliu, G., Martinez, R., Soliva-Fortuny, R., Martin-Belloso, O., 2008. Effect of superatmospheric and low oxygen modified atmospheres on shelf-life extension of fresh-cut melon. Food Control 19, 191–199.

Park, Y.M., Blanpied, G.D., Jozwiak, Z., Liu, F.W., 1993. Postharvest studies of resistance to gas diffusion in McIntosh apples. Postharvest Biol. Technol. 2, 329–339.

Pedreschi, R., Vanstreels, E., Carpentier, S., Hertog, M., Lammertyn, J., Robben, J., Noben, J.-P., Swennen, R., Vanderleyden, J., Nicolaie, B.M., 2007. Proteomic analysis of core breakdown disorder in Conference pears (*Pyrus communis* L.). Proteomics 7, 2083–2099.

Pedreschi, R., Franck, C., Lammertyn, J., Erban, A., Kopka, J., Hertog, M., Verlinden, B., Nicolai, B., 2009. Metabolic profiling of 'Conference' pears under low oxygen stress. Postharvest Biol. Technol. 51, 123–130.

Pelayo, C., Ebeler, S.E., Kader, A.A., 2003. Postharvest life and flavor quality of three strawberry cultivars kept at 5°C in air or air+20 kPa CO_2. Postharvest Biol. Technol. 27, 171–183.

Peppelenbos, H.W., Oosterhaven, J., 1998. A theoretical approach on the role of fermentation in harvested plant products. Acta Hortic. 464, 381–386.

Ponce-Valadez, M., Watkins, C.B., 2008. Fermentation and malate metabolism in response to elevated CO(2) concentrations in two strawberry cultivars. Physiol. Plant. 134, 121–133.

Ponce-Valadez, M., Fellman, S.M., Giovannoni, J., Gan, S.S., Watkins, C.B., 2009. Differential fruit gene expression in two strawberry cultivars in response to elevated CO_2 during storage revealed by a heterologous fruit microarray approach. Postharvest Biol. Technol. 51, 131–140.

Prange, R.K., DeLong, J.M., Harrison, P.A., Leyte, J.C., McLean, S.D., 2003. Oxygen concentration affects chlorophyll fluorescence in chlorophyll-containing fruit and vegetables. J. Am. Soc. Hortic Sci. 128, 603–607.

Prange, R.K., Wright, A.H., DeLong, J.M., Zanella, A., 2013. History, current situation and future prospects for dynamic controlled atmosphere (DCA) storage of fruits and vegetables, using chlorophyll fluorescence. Acta Hortic. 1012, 905–915.

Rico, D., Martin-Diana, A.B., Barat, J.M., Barry-Ryan, C., 2007. Extending and measuring the quality of fresh-cut fruit and vegetables: a review. Trends Food Sci. Technol. 18, 373–386.

Rojas-Grau, M.A., Oms-Oliu, G., Soliva-Fortuny, R., Martin-Belloso, O., 2009. The use of packaging techniques to maintain freshness in fresh-cut fruits and vegetables: a review. Int. J. Food Sci. Technol. 44, 875–889.

Rossle, C., Gormley, T.R., Brunton, N., Butler, F., 2011. Quality and antioxidant properties of fresh cut apple wedges from 10 cultivars during modified atmosphere packaging storage. Food Sci. Technol. Int. 17, 267–276.

Saltveit, M.E., 1997. A summary of CA and MA requirments and recommendations for the controlled and modified atmosphere storage of harvested vegetables. Seventh International

Controlled Atmosphere Research Conference. Vegetables and Ornamentals. University of California, Davis, Vol. 4, Postharvest Horticulture Series 18, 98–117.

Saltveit, M.E., 2003. Is it possible to find an optimal controlled atmosphere? Postharvest Biol. Technol. 27, 3–13.

Saquet, A.A., Streif, J., Bangerth, F., 2000. Changes in ATP, ADP and pyridine nucleotide levels related to the incidence of physiological disorders in 'Conference' pears and 'Jonagold' apples during controlled atmosphere storage. J. Hortic Sci. Biotech. 75, 243–249.

Saquet, A.A., Streif, J., Bangerth, F., 2003. Energy metabolism and membrane lipid alterations in relation to brown heart development in 'Conference' pears during delayed controlled atmosphere storage. Postharvest Biol. Technol. 30, 123–132.

Schotsmans, W., Verlinden, B.E., Lammertyn, J., Peirs, A., Jancsok, P.T., Scheerlinck, N., Nicolai, B.M., 2002. Factors affecting skin resistance measurements in pipfruit. Postharvest Biol. Technol. 25, 169–179.

Schouten, S.P., Prange, R.K., Verschoor, J., Lammers, T.R., Oosterhaven, J., 1997. Improvement of Quality of Elstar Apples by Dynamic Control of ULO Conditions. Postharvest Horticulture Series. Department of Pomology, University of California, pp. 71–78.

Shelp, B.J., Bown, A.W., McLean, M.D., 1999. Metabolism and functions of γ-aminobutyric acid. Trends Plant Sci. 4, 446–452.

Shin, Y., Ryu, J.A., Liu, R.H., Nock, J.F., Polar-Cabrera, K., Watkins, C.B., 2008. Fruit quality, antioxidant contents and activity, and anti proliferative activity of strawberry fruit stored in elevated CO_2 atmospheres. J. Food Sci. 73, S339–S344.

Shipway, M.R., Bramlage, W.J., 1973. Effects of carbon dioxide on activity of apple mitochondria. Plant Physiol. 51, 1095–1098.

Smith, R.B., Skog, L.J., 1992. Postharvest carbon dioxide treatment enhances firmness of several cultivars of strawberry. HortScience 27, 420–421.

Solomos, T., 1987. Principles of gas exchange in bulky plant tissues. HortScience 22, 766–771.

Stommel, J.R., Camp, M.J., Luo, Y.G., Welten-Schoevaars, A.M., 2016. Genetic diversity provides opportunities for improvement of fresh-cut pepper quality. Plant Genet. Resour. 14, 112–120.

Tietjen, W.H., Hudson, D.E., 1984. Market quality of eastern-grown Golden Delicious apples after prestorage CO_2 treatment and controlled atmosphere storage. HortScience 19, 427–429.

Toivonen, P.M.A., Hampson, C.R., 2009. Apple cultivar and temperature at cutting affect quality of fresh slices. HortTechnology 19, 108–112.

Tudela, J.A., Marin, A., Martinez-Sanchez, A., Luna, M.C., Gil, M.I., 2013. Preharvest and postharvest factors related to off-odours of fresh-cut iceberg lettuce. Postharvest Biol. Technol. 86, 463–471.

Tudela, J.A., Hernandez, N., Perez-Vicente, A., Gil, M.I., 2016. Comprehensive evaluation of different storage conditions for the varietal screening of lettuce for fresh-cut performance. Postharvest Biol. Technol. 120, 36–44.

Veltman, R.H., Verschoor, J.A., van Dugteren, J.H.R., 2003. Dynamic control system (DCS) for apples (*Malus domestica* Borkh. cv 'Elstar'): optimal quality through storage based on product response. Postharvest Biol. Tech. 27, 79–86.

Verboven, P., Kerckhofs, G., Mebatsion, H.K., Ho, Q.T., Temst, K., Wevers, M., Cloetens, P., Nicolai, B.M., 2008. Three-dimensional gas exchange pathways in pome fruit characterized by synchrotron x-ray computed tomography. Plant Physiol. 147, 518–527.

Volz, R.K., Biasi, W.V., Grant, J.A., Mitcham, E.J., 1998. Prediction of controlled atmosphere-induced flesh browning in 'Fuji' apple. Postharvest Biol. Technol. 13, 97–107.

Watkins, C.B., 2008. Overview of 1-methylcyclopropene trials and uses for edible horticultural crops. HortScience 43, 86–94.

Watkins, C.B., 2015. Advances in the use of 1-MCP. In: Wills, R.B.H., Golding, J. (Eds.), Advances in Postharvest Fruit and Vegetable Technology. CRC Press, pp. 117–146.

Watkins, C.B., Nock, J.F., 2012a. Controlled atmosphere storage of 'Honeycrisp' apples. HortScience 47, 886–892.

Watkins, C.B., Nock, J.F., 2012b. Rapid 1-methylcyclopropene (1-MCP) treatment and delayed controlled atmosphere storage of apples. Postharvest Biol. Technol. 69, 24–31.

Watkins, C.B., Pritts, M.P., 2001. The influence of cultivar on postharvest performance of fruits and vegetables. Acta Hortic. (553), 59–63.

Watkins, C.B., Manzano-Mendez, J.E., Nock, J.F., Zhang, J.Z., Maloney, K.E., 1999. Cultivar variation in response of strawberry fruit to high carbon dioxide treatments. J. Sci. Food Agric. 79, 886–890.

Wilkinson, B.G., Fidler, J.C., 1973. Physiological disorders. In: Fidler, J.C., Wilkinson, B.G., Edney, K.L., Sharples, R.O. (Eds.), The Biology of Apple and Pear Storage. Commonwealth Agricultural Bureau, East Malling, Kent, pp. 67–80.

Yearsley, C.W., Banks, N.H., Ganesh, S., 1997. Effect of carbon dioxide on the internal lower oxygen limits of apple fruit. Postharvest Biol. Technol. 12, 1–13.

Zanella, A., Rossi, O., 2015. Post-harvest retention of apple fruit firmness by 1-methylcyclopropene (1-MCP) treatment or dynamic CA storage with chlorophyll fluorescence (DCA-CF). Eur. J. Hortic. Sci. 80, 11–17.

Zanella, A., Sturz, S., 2013. Replacing DPA postharvest treatment by strategical application of novel storage technologies controls scald in 1/10th of EU's apples producing area. Acta Hortic. 1012, 419–426.

Zhang, J.Z.J., Watkins, C.B., 2005. Fruit quality, fermentation products, and activities of associated enzymes during elevated CO_2 treatment of strawberry fruit at high and low temperatures. J. Am. Soc. Hortic Sci. 130, 124–130.

Chapter 3

Effect of maturity on CA/MA responses

Peter M.A. Toivonen
Agriculture and Agri-Food Canada, Summerland Research and Development Centre, Summerland, BC, Canada

Introduction

Maturity (physiological age) of a fruit or vegetable at harvest has a significant influence on its postharvest physiology and consequently will have a significant influence on storage life, poststorage quality, and response to postharvest procedures and treatments (Toivonen and Beveridge, 2005; Toivonen, 2008). Controlled atmosphere or modified atmosphere storage has been shown to significantly improve quality retention of climacteric fruits; however, the effect of atmospheric modification on the attenuation of tissue deterioration of intact nonclimacteric fruits or vegetables can be very limited and often responses are associated with control of decay or insect infestations (Yahia, 2009). The reason for the greater response for climacteric fruit is that low oxygen and high carbon dioxide can significantly attenuate the dramatic rise in respiration and metabolism that occurs during this type of ripening (Fig. 1) and it is this attenuation that is the physiological basis for the successful application modified or controlled atmospheres (Kader, 1997). Generally, respiration and metabolism decline after harvest for nonclimacteric fruits or vegetables (Kader, 1997) and therefore the impact of atmospheric modification has less direct impact on quality retention. An exception to this would be fresh-cut fruits or vegetables, where the produce incurs significant wounding or injury that leads to increases in respiration, ethylene production, and in various metabolic pathways associated with wounding and defense (Toivonen and DeEll, 2002). Much of the discussion in this chapter will be focused on the importance of the harvest maturity of climacteric fruits and implications to consequent response to modified and controlled atmosphere storage.

Controlled and Modified Atmospheres for Fresh and Fresh-Cut Produce
https://doi.org/10.1016/B978-0-12-804599-2.00004-1

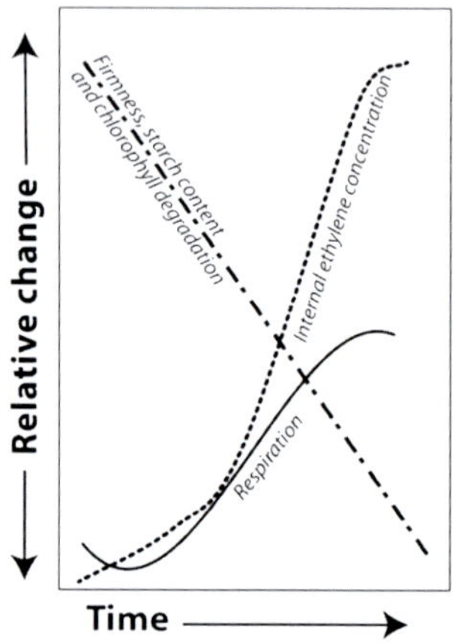

FIG. 1 Relative changes in firmness, starch content, peel chlorophyll degradation, respiration, and autocatalytic ethylene production over time as climacteric fruit matures and then begins to ripen. This pattern is consistent with what has been labeled as a climacteric pattern of ripening (Kader, 1997). Note that climacteric respiratory rise and autocatalytic ethylene production are initiated at a similar timing as the decline in firmness, starch content, and peel chlorophyll degradation.

Implications of climacteric fruit maturity and/or ripening stage on response to controlled atmospheres

Fruit having a typical climacteric ripening pattern display significant rise in respiration and ethylene production rates at the onset of the ripening process (Fig. 1). In immature climacteric fruit, exposure to ethylene results in feedback inhibition of ethylene production (Saltveit, 1999). Once climacteric fruit advances in maturity to the early phase of ripening, this negative feedback is lost and the fruit then begin to show a positive feedback response, when exposure to ethylene leads to consequent autocatalytic production of ethylene that accelerates the ripening process and final progress toward senescence (Saltveit, 1999). It is this switch in primary response to ethylene exposure that defines harvest maturity for many climacteric fruit and has tremendous implications on response to postharvest handling and treatments. For example, where exogenous ethylene is applied to ripen bananas, maturity of the fruit must be sufficiently advanced such that it has switched from the mode of ethylene synthesis feedback inhibition to the mode of feedback enhancement of autocatalytic ethylene synthesis (Saltveit, 1999). If the bananas are harvested before the change from the mode of feedback inhibition, response to exogenous ethylene will not have the desired outcome of accelerating degreening of the peel and flesh ripening.

In general, the goal of maturity management for whole fruits and fruit-vegetables directed to modified or controlled atmosphere storage or shipping is to ensure that they are at a late enough physiological stage that will ensure good eating quality after storage or shipping, but also early enough to ensure good quality retention. As fruits mature, they undergo significant shifts in metabolism that can be followed by monitoring changes in metabolites and/or protein and gene expression (Toivonen, 2007) and some of these changes are differentially expressed in different cultivars of a particular fruit. For example, Wiersma et al. (2007) found that the gene expression responsible for the ethylene feedback inhibition of ethylene production was significant in "Golden Delicious" apple and hardly detectable in "Sunrise" apple. These results may explain why "Golden Delicious," which has a distinct preclimacteric stage, can be managed to maximize long-term controlled atmosphere storage potential, whereas "Sunrise," which does not have a distinct preclimacteric stage, cannot be managed for longer term storage. Similar observations have been made for apples and other fruits, for which some cultivars have a classic climacteric ripening pattern and others do not (Watkins et al., 1989; Paul et al., 2012).

Harvest maturity in nonclimacteric fruits and vegetables is largely based on size, density, shape, and or color (Reid, 2002). Maturity in immature plant parts such as stems, leaves, or leafy heads is due to the fact that the fruit or vegetable is in or near the end of the vegetative phase of growth (Reid, 2002) and as a consequence there are no or few shifts in physiological or metabolic events associated with the onset of ripening. Cabbage is one of the few nonclimacteric fruit or vegetable products that are commercially stored in controlled atmospheres and harvest maturity decisions are based on density of the head (Prange, 2016). Undermature heads of cabbage are subject to excessive water loss and overmature heads are subject to splitting and decay, but neither under- or overmaturity is directly related to response to controlled atmosphere condition. However, controlled atmospheres will inhibit yellowing in the long term, likely due to the effects of elevated carbon dioxide in the atmosphere inhibiting ethylene response to ethylene accumulations that occur due to endogenous production of ethylene by the cabbage in storage. Carbon dioxide also inhibits the growth of decay organisms on the heads (Prange, 2016).

Sweet cherries are nonclimacteric fruit which are commonly shipped long distances in ocean containers (Wang et al., 2015). Modified atmosphere packaging (MAP) will minimize color change (darkening) and maintain green stem color and control decay over longer storage durations for sweet cherries. Despite this, the level of benefit is determined in part by fruit maturity at harvest. Recent evidence indicates that titratable acidity and soluble solids contents increase with maturity up to a dark mahogany maturity stage and those differences are maintained over long storage/shipping durations (Toivonen et al., 2004; Toivonen, 2015a). Susceptibility to many defects decreases with greater maturity (i.e., darker colored cherries) and therefore more advanced maturity will in fact lead to better outcomes from long-term MAP shipping, provided temperature is managed well (Toivonen, 2014, 2015a). However, shipping of later harvest maturities (dark mahogany to black color) can lead to greater susceptibility to stem browning when fruit are packaged in MAP (Toivonen, 2015a).

Strawberries are another fruit which are routinely shipped under MAP conditions (Riad and Brecht, 2005). They are nonclimacteric fruit which respond better to MAP when they are harvested at an earlier maturity stage (Nunes et al., 2002) which is in contrast to sweet cherry, where later harvests lead to better responses to MAP. A part of this difference in response to maturity may be linked to the fact that in strawberries firmness and titratable acidity decline with more advanced maturity (Nunes et al., 2002), whereas those two attributes continue to increase with advancing maturity in sweet cherries (Toivonen et al., 2004; Toivonen, 2015a).

Kiwifruit, which has a climacteric pattern of ripening (Richardson et al., 2011) shows a better response, much like sweet cherry, when the fruit are harvested later and have higher soluble solids and dry matter content (Lallu et al., 2003). It should be noted that while kiwifruit has a climacteric ripening pattern, softening and other maturity-associated quality changes occur well before the onset of autocatalytic ethylene production in that fruit (Fig. 2) which is unlike many apple cultivars where autocatalytic ethylene production is associated closely with softening (Blanpied and Silsby, 1992).

Avocado is climacteric fruit with third distinct ripening characteristic pattern (as compared with apple and kiwifruit), where autocatalytic ethylene production and softening do not occur until the fruit is harvested from the tree (Rhodes, 1970). Onset of ripening occurs days after harvest from the tree (Ben-Yehoshua et al., 1963).

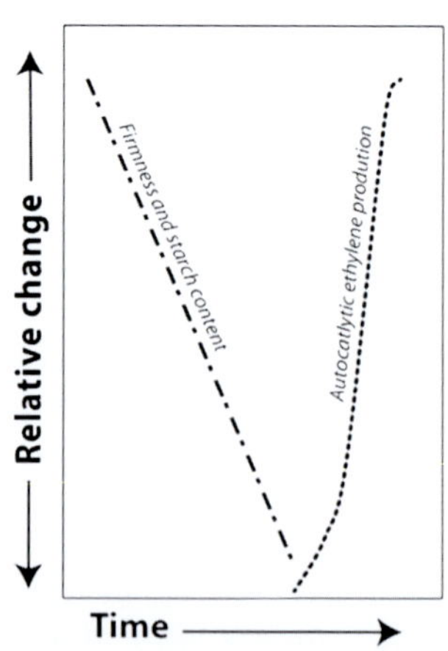

FIG. 2 Relative changes in firmness, starch content, and autocatalytic ethylene production in kiwifruit over time as the fruit matures and then begins to ripen. Note that autocatalytic ethylene production is not initiated until most of the changes in loss of firmness and starch content has been completed. *(Graph based on discussion and data from Richardson, A.C., Boldingh, H.L., McAtee, P.A., Gunaseelan, K., Luo, Z., Atkinson, R.G., David, K.M., Burdon, J.N., Schaffer, R.K., 2011. Fruit development of the diploid kiwifruit, Actinidia chinensis 'Hort16A'. BMC Plant Biol. 11, 182, http://www.biomedcentral.com/1471-2229/11/182.)*

Effects of maturity in climacteric on response fruits to storage atmospheres

In general, if climacteric fruit are harvested at a preclimacteric stage they will have greater storage potential as compared with fruit harvested after the climacteric respiratory rise has been initiated (Kader, 2016). However, while, harvested in early stages of physiological maturation may show better response to low oxygen and high carbon dioxide atmospheres (Kader, 2016) they will have poorer flavor quality and lower titratable acidity levels after controlled atmosphere storage (Fellman et al., 2003). In the case of apples, firmness after controlled atmosphere storage can be influenced by maturity at harvest for some cultivars (e.g., "Gala"), whereas this may not be the case for other cultivars (e.g., "Fuji" or "Lady Williams") (Jobling and McGlasson, 1995). This difference in poststorage response to harvest maturity has been attributed to differing ripening patterns in apple cultivars in regards to the onset of climacteric ethylene production (Jobling and McGlasson, 1995). In the case of kiwifruit and avocado (where the climacteric respiration is delayed past maturation), the harvest date has the greatest effect on quality retention in storage, where late- or mid-season fruit, respectively, show the best quality retention (Mitchell et al., 1991; Dixon et al., 2003). A part of this improved response relates directly to the importance of extending the time on the vine or tree to allow more accumulation of dry matter in the fruit (Mitchell et al., 1991; Dixon et al., 2003). However, the response to controlled atmospheres during storage does not appear to be influenced by the part of the season in which the fruit is harvested (Lallu et al., 2003; Burdon et al., 2010).

In addition to maturity influencing poststorage organoleptic quality, storage-associated defects can occur in apples stored for long durations in controlled atmospheres under cold storage conditions. Occurrence and severity of superficial scald, while influenced by many preharvest factors, is generally considered to be a chilling-related problem that is more prominent when fruit are harvested at a less mature stage (Emonger et al., 1994; DeLong et al., 2014). Corky flesh browning in "McIntosh" apples has been found to occur in apples harvested at a more advanced stage of maturity before placement in controlled atmosphere storage and is considered to be, at least, partly associated with senescence processes in the fruit (Lau et al., 1987). Internal CO_2-related flesh browning in "Braeburn" apples in controlled atmosphere storage has been associated with fruit harvested at more advanced maturity (Elgar et al., 1999; Streif et al., 2003). Internal browning of "Fuji" apples stored in controlled atmospheres increases in severity with more advanced maturity and higher carbon dioxide levels in the atmosphere enhance this response (Kweon et al., 2013). Clearly, at least in apples, maturity has an influence on severity and incidence of peel and/or flesh browning; however, it must be kept in mind that there are other factors that can also be involved (Lau et al., 1987; Emonger et al., 1994; Elgar et al., 1999; Streif et al., 2003; Kweon et al., 2013). Hence maturity management is only a part of the solution in terms of minimizing disorders occurring in controlled atmosphere storage.

The mechanism as to how maturity may influence response to controlled atmosphere can be linked to the physiology of autocatalytic ethylene production in climacteric fruit (Burg and Burg, 1967). It has been demonstrated that low oxygen and high carbon dioxide levels in the storage atmosphere will directly inhibit the autocatalytic ethylene production through delaying and suppressing expression of 1-aminocyclopropane-1-carboxylic acid synthase at the transcriptional level and by reducing the abundance of 1-aminocyclopropane-1-carboxylic acid oxidase protein (Gorny and Kader, 1996). Since these two enzymes are important in the signal transduction for upregulation of ethylene biosynthesis at the onset of climacteric ripening (Burg and Burg, 1967; Gorny and Kader, 1996), the application of controlled atmospheres (high carbon dioxide and low oxygen) for fruit harvested before or at the early climacteric stage will result in a greater response to quality retention in the fruit than if fruit are harvested after the onset of the autocatalytic ethylene biosynthesis.

Dynamic controlled atmosphere storage has become a commercially feasible approach to enhance storage potential in apples (Bai et al., 2009) and avocados (Burdon et al., 2010). Dynamic controlled atmosphere achieved by monitoring the response of the fruit to the oxygen level as this level approaches the respiratory compensation point and adjusting the oxygen in storage to a point slightly above this point (Watkins, 2008). The concern with such a technology is that perhaps fruit from all orchards, if they are harvested at differing maturities, may not have the same respiratory compensation point (Gasser et al., 2003), and if so then if many orchards are stored together in DCA, that there might be a risk of injury to sensitive lots. However, research to date suggests that this not an issue with commercially grown apples of a particular cultivar, within a single harvest season (Watkins, 2008).

Effects of maturity in fresh-cut fruits and vegetables to modified atmosphere storage

One of the major factors in fresh-cut fruits and vegetables is that the cutting and/or peeling process can lead to transient and/or persistent increases in respiration and ethylene production which will result in shortened shelf life and reduction in quality (Toivonen and DeEll, 2002). Mencarelli et al. (1998) demonstrated that low oxygen levels common in modified atmospheres were effective at modulating wound-induced increases in respiration and ethylene production. Maturity of a climacteric fruit can influence the level of respiratory and/or ethylene rise, which in cantaloupe, for example, results in lower respiration and ethylene production if the fruit is cut at an earlier maturity as compared with the fully mature stage (Madrid and Cantwell, 1993). Therefore fresh-cut fruit made from less mature cantaloupe will show a better response to MAP than cut fruit from a fully mature stage. In contrast for the nonclimacteric bell pepper, fruit cut at a fully mature colored stage shows better response to modified atmospheres than immature green peppers (López-Gálvez et al., 1997). A third example is that of fresh-cut apple, which browns in response to cutting. Fresh-cut slices made from immature "Granny Smith" apples similarly have a greater propensity to develop browning in response to cutting than do slices made from more mature apples (Toivonen, 2008). The browning of the apple slices is so severe that neither MAP nor antibrowning dips will prevent it until fruit having a more advanced maturity are used to make fresh-cut slices.

Flavor is an important attribute for fresh-cut fruits packaged in modified atmosphere packages and maturity of the fruit at the time of cutting can have a significant influence on subsequent flavor (Beaulieu and Baldwin, 2002). In papaya, fruit must be of intermediate maturity (55%–80% yellow) in order to maintain quality and have good flavor after shelf holding in modified atmospheres (Paull and Chen, 1997). Pieces cut from less mature fruit will have the largest increases in ethylene production and respiration, but the cut pieces will be too firm to be acceptable for consumption, whereas pieces from fully mature fruit show very little increase in respiration and ethylene production, but will be too fragile to handle and package. Similar examples of choosing an intermediate maturity stage to ensure quality and good flavor for fresh-cut packaged product have been demonstrated for peaches, nectarines, pears, and cantaloupe (Gorny et al., 1998, 1999; Beaulieu and Grimm, 2001).

It is clear from the previous discussion that there are two major issues for which maturity selection is a compromise in fresh-cut fruits and vegetables that are packaged in modified atmospheres: (1) if the fruit softens with advanced maturity, an earlier stage will provide fruit with firmness required to withstand cutting and handling injuries and (2) fruit must be advanced in maturity to ensure good organoleptic quality to the consumer (i.e., it must have the capacity to properly ripen) (Beaulieu and Baldwin, 2002). In the case of bell peppers, the fruit does not soften as it transforms from green to colored stage and the quality of the fresh-cut fruit is improved with advanced maturity purely on the basis of flavor quality (López-Gálvez et al., 1997). The decline in browning observed with slices made from "Granny Smith" apples of increased maturity may be associated with reduced polyphenol oxidase activity and/or decreased soluble phenolics content with advancing apple maturity (Holderbaum et al., 2010).

Indicators for maturity

A challenge to manage climacteric fruits or fruit-vegetables maturity is to find a reliable method to measure the stage of maturity. Many different measures have been proposed or used as indicators of maturity in fruits, including firmness, starch clearing, ethylene production, and background peel color (Reid, 2002). The literature suggests that ethylene coordinates most, if not all, maturation changes that occur in many climacteric fruit (Liu et al., 2015). However, ethylene-coordinated ripening may not always proceed in a synchronous manner since it has been shown that processes associated with ripening (i.e., starch clearing, flavor volatile production, softening, and peel background color change) are not all similarly sensitive or dependent on ethylene (Pech et al., 2008; Johnston et al., 2009). Starch clearing, which is used as an index of maturity for apples is, for example, very sensitive to exposure to ethylene but its progress is not dependent on ethylene production (i.e., it is not tightly linked to the onset of respiratory climacteric). In addition, not all cultivars of apples show the classic climacteric pattern of synchronous ethylene and respiration rise at the onset of ripening as shown by "Delicious" (Fig. 3). For those cultivars of apples, the respiratory climacteric rise may not be apparent as the apple matures and ripens and the rise in ethylene production is much attenuated and delayed well beyond the optimal harvest date as determined by accepted maturity indices such as starch clearing which are not tightly linked to the onset of climacteric ripening (e.g., Salish in Fig. 3). The consequence is that measures of maturity such as a climacteric rise in ethylene production may not be useful to monitor harvest maturity for all apple cultivars (Watkins et al., 1989).

In practice, apples are probably the most studied fruit and vegetable in regards to response to controlled atmosphere storage. Maturity in apples has been measured using several different indices, including peel color (background green color), seed color, flesh firmness, soluble solids content, starch content, starch stain clearing, titratable acidity, respiration, ethylene production or internal ethylene

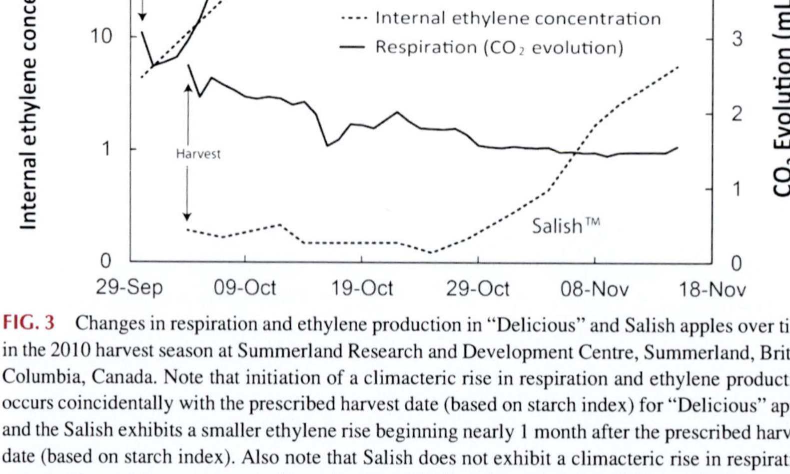

FIG. 3 Changes in respiration and ethylene production in "Delicious" and Salish apples over time in the 2010 harvest season at Summerland Research and Development Centre, Summerland, British Columbia, Canada. Note that initiation of a climacteric rise in respiration and ethylene production occurs coincidentally with the prescribed harvest date (based on starch index) for "Delicious" apple and the Salish exhibits a smaller ethylene rise beginning nearly 1 month after the prescribed harvest date (based on starch index). Also note that Salish does not exhibit a climacteric rise in respiration over the time of the experiment. *(Data from unpublished data (Toivonen P.M.A., 2018).)*

concentration, production of odor-active molecules (Blanpied and Silsby, 1992; Mitcham and Mitchell, 2002). In many older apple cultivars the use of internal ethylene or ethylene production has been a useful indicator of attainment of physiological maturity (Blanpied and Silsby, 1992; Luton, 1996); however, this approach is not useful for other cultivars of apples where initiation of climacteric respiration and ethylene production occurs much later, at a time when optimal harvest dates occur before the onset of the climacteric (Fig. 3). However, it must be kept in mind that the main goal of harvest maturity targeting is to allow the fruit to approach physiological maturity to ensure good eating quality and also to ensure that the fruit will be firm and crisp after longer term storage (i.e., harvest at the onset or before the onset of the climacteric autocatalytic ethylene production).

Recently, work has been published evaluating the physiological linkage of various maturity indicators to ethylene (Johnston et al., 2009). In that work apples were transformed to knock out ethylene production capability. Dependency and sensitivity to ethylene for different indicators of maturity were assessed in Gala apples. It is clear from this work that there are early and late-occurring changes in apple maturation and these separate in regards to ethylene dependency and sensitivity, with the late events being less sensitive but highly dependent on ethylene (Johnston et al., 2009) and these findings are parallel to what has been found with melon ripening (Liu et al., 2015). In apple, starch clearing is often used as a primary approach to assessing maturity in apples (Blanpied and Silsby, 1992); however, it has been shown that this index is not consistently reliable to predict harvest for firm fruit after storage, at least in some growing regions for some apple cultivars (Toivonen, 2015b). The reason for this may be that starch index values are not closely linked to flesh softening in regards to relative dependence or sensitivity to ethylene (Johnston et al., 2009). Background color and flavor volatile synthesis are more closely linked to firmness in regards to dependence and sensitivity to ethylene (Johnston et al., 2009). Flavor volatile analyses using electronic nose technology have been tested for maturity measures (Pathange et al., 2006), but no commercial development has ensued. In general, the use of background (green to yellow) color of many fruit is considered to be a good indicator of maturity (Kupferman, 1994); however, the background color can be masked by a red anthocyanin blush color making the assessment of maturity difficult using this criterion (personal observation). There have been many reports of measuring background color of apples spectrophotometrically (Knee, 1980; Merzlyak et al., 1997, 1999, 2002; Zude and Herold, 2002; Zude-Sasse et al., 2002; Solovchenko et al., 2005). Recently, simplified and portable spectrometric tools have been developed that allow for measurement of chlorophyll in fruit peel to assess maturity (Zude and Herold, 2002; Zude-Sasse et al., 2002; Rutkowski et al., 2008; DeLong et al., 2014; Toivonen, 2015b; Spadoni et al., 2016). If the spectral bands for this analysis are selected carefully (Toivonen, 2015b), the presence of anthocyanin (responsible for the red blush color in apples) or carotenoids (responsible for peel background yellow color) is not of concern since these pigments do not absorb above 650 nm (Sims and Gamon, 2002; Merzlyak et al., 2003).

One of the issues often not discussed is the possibility that there are potential artifact errors associated with the any maturity index measure that can result in inaccurate interpretation of maturity. While there have been great strides in understanding biological variations in age in populations of fruit on a tree (i.e., biological shift factor) (Tijskens et al., 2005, 2009), there are very real issues around the validity or variation of measures due to factors other than maturity that have received very little research focus. Fruit does not grow in a uniform or benign environment and there can be many environmental influences which can modulate levels of components that are used to assess maturity even within a tree (Nilsson and Gustavsson, 2007). The environmental physiology and biochemistry associated with changes in maturity measures and indices must be fully understood if there is to be a real improvement in protocol development and reliable harvest maturity interpretations. The fact that there will be a distribution of maturity classes on any one tree (Nilsson and Gustavsson, 2007) must also be understood in the design of a sampling strategy that will provide the most useful maturity index for that population of fruit.

The concept of starch degradation into sugars is well accepted for apples as an indicator of maturity (Blanpied and Silsby, 1992); however, what is often glossed over is that initial starch accumulation, before the apple begins to mature and ripen, is likely dependent on both photosynthesis and transport of photosynthate into the fruit (Toivonen, 2015b). In addition, nighttime foliar respiration is a process that can influence the amount of photosynthate that is available to be translocated to the fruit (Richardson et al., 2004; Fisher et al., 2012) and hence the amount of starch that accumulates. In addition, starch conversion to sugars is induced by cool temperatures (Smith et al., 1979) and cool temperatures have also been shown to enhance gene expression associated with ethylene metabolism (Bagnaresi et al., 2008), which may explain why starch clearing is often associated with maturation and ripening in apple. Starch clearing is not ethylene dependent, but induction of clearing is very ethylene sensitive (Johnston et al., 2009), which supports the hypothesis in regard to the association of chill-induced ethylene and starch clearing. Luton (1996) indicated that the starch clearing was of limited predictive power for maturity in Cox's Pippin apple when comparing orchards from different growing areas, but could yield relative maturation within a local growing area, which suggests that starch clearing is very dependent on local climatic profiles. In addition, it has been demonstrated that starch clearing as a maturity index may not be consistently useful in different growing seasons in some areas (Lau, 1998; Toivonen, 2015b).

The weakness of the starch clearing index between growing regions has been recognized and attempts have been made to improve on predictive power by including other measures associated with maturation and ripening in apples. The Streif index incorporates data for firmness and soluble solids content (Streif, 1996) and has been found to improve the reliability for decisions made regarding harvest dates in some cases (De Jager and Roelofs, 1996; Streif, 1996; DeLong et al., 1999) but not others (Baumann, 1996; Alegre et al., 2006; Lötze and Bergh, 2012). It is clear, looking at the outcomes of this research that while refinements can be made in the starch clearing indexing, that it may be that indices developed in one part of the world may not be as useful elsewhere.

Nondestructive harvest maturity estimation using internal quality measures such as dry matter, soluble solids, or firmness, has been suggested as an another approach (Peirs et al., 2000; Nicolaï et al., 2007). While theoretically the internal measures are possible using nondestructive tools, there needs to be some consideration for the fact that fruit constituents may not be uniformly distributed in all parts of the fruit (Peiris et al., 1999) and hence measured values will not be similar at all measurement sites in that piece of fruit (Fig. 4). Bobelyn et al. (2010) also determined that there are variations in internal light absorption

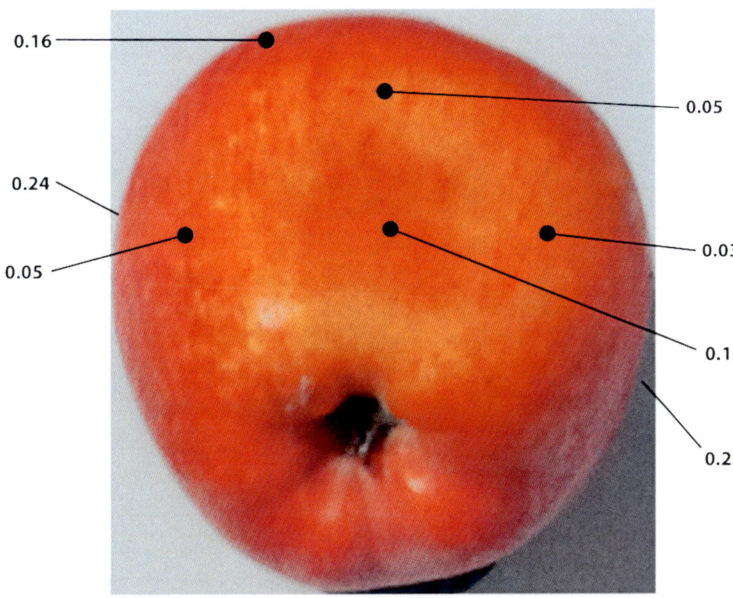

FIG. 4 Differences in I_{AD} values generated with a DA meter (Sintéleia, Bologna, Italy) on a sunscalded "Ambrosia" apple. Note that the value in the severely burned and darkened area of the sunscald is higher than the values in the "Halo" area surrounding the sunburn (Felicetti and Schrader, 2009). As the measurements were taken further at the edge of the "Halo," the I_{AD} value increased to 0.16, but only reliable and consistent I_{AD} values were obtained by measuring on the other side of the fruit which was not affected by sunburn (i.e., 0.24 and 0.23). Similar patterns are seen even if the sun exposure does not produce actual sunburn symptoms (Toivonen et al., 2016). These observations indicate that reliable I_{AD} values that will represent apple maturity, and not include artifacts associated with sun injury, can only be made when measuring an apple surface that is not exposed to direct sunlight. Note: The term DA meter represents "difference in absorbance meter" since this meter measures absorbance at the chlorophyll "a" peak value of 670nm and subtracts from this the absorbance at 720, which is a baseline portion of the absorbance spectrum for apple peel. The resultant value from this subtraction is termed I_{AD}, a dimensionless value which refers to the "index of the absorbance difference." In principle the instrument provides an indication of maturity because as the fruit ripens, chlorophyll levels decline in the peel and consequently the measured I_{AD} values decline. However, as this example shows, there are artifacts of measurement that can be caused by sun exposure and consequent chlorophyll bleaching or development of browning of the peel.

spectra in fruits based on cultivar, orchard, postharvest handling differences, and season and this biological variation must be accounted for in making robust models for routine use in estimating internal apple quality/maturity. In order to implement nondestructive instruments for measuring maturation and ripening in fruit in the orchard, there must be intensive evaluation of the variances in measures dependent on numerous factors that must be considered and appropriate protocols developed to standardize the measurements and to build robust models that will give accurate nondestructive estimates using visible/near-infrared spectrometers. It also must be noted that in regard to uniformity of constituents at different locations of the fruit, that much of this can be a consequence of stress response in the orchard and hence variance of fruit maturity might be reduced with improved production strategies (Schrader et al., 2009). In the end, it is important to develop a protocol where a selected location is used for a standardized measure to ensure that maturity measures between fruit, trees, and orchards can be compared as valid equivalent readings (Toivonen et al., 2016).

Nondestructive determination of maturity has been investigated in fruit other than apples. Much research has been devoted to nondestructive prediction of dry matter or oil content in avocados since these constituents are associated with poststorage quality (Wedding et al., 2011; Blakey, 2016; Olarewaju et al., 2016). There has also been significant research to develop nondestructive measures of kiwifruit maturity measured as soluble solids, sugar contents, dry matter, or flesh color (Slaughter and Crisosto, 1998; Schaare and Fraser, 2000). In mango, near-infrared spectroscopy has been evaluated for maturity assessment based on multiple internal composition components (Subedi et al., 2007; Jha et al., 2014; Seifert et al., 2014) and is currently implemented as a commercial evaluation method for mango in Australia (Walsh, 2016). There are also some studies showing promise of this approach for citrus (Kim et al., 2015), tomatoes (Slaughter et al., 1996; Seifert et al., 2014), and apricots (de Oliveira et al., 2014).

Conclusions and future trends

Maturity of a fruit or vegetable is an important determinant of the successful application of either modified atmosphere or controlled atmosphere storage. In vegetables, the indicators or indices for maturity are largely visual or tactile in nature, whereas the indices for maturity in fruits are largely based on physiological or compositional measures. In the past, destructive measures have been employed to assess maturity, with varying results and often an approach being useful in a specific growing region and less useful in others. Such limitations may be attributed to artifacts that influence the index component being measured. Recent advances in nondestructive instruments, primary spectroscopic tools, have opened up new possibilities for rapid, real time, and reliable measures of fruit constituents that are associated with maturation and ripening.

Nondestructive tools allow the possibility to build models that are multicomponent in nature and thus provide an index which has a complex of constituents associated with maturation and ripening (Peirs et al., 2000; Jha et al., 2014). Using such multicomponent indices might lead to improved decision making than has been possible in the past. However, it must be kept in mind that there is a significant biological variance for maturation in the orchard (Tijskens et al., 2009; Bobelyn et al., 2010) and that sampling protocols must take this into account. Also, it must be understood that there are measurement artifacts associated with this approach and that the concentration of any one component in a piece fruit varies according to position (Peiris et al., 1999; Nilsson and Gustavsson, 2007) or light exposure (Klein et al., 2001; Schrader et al., 2009; Li et al., 2013; Toivonen et al., 2016). Therefore a sampling strategy must be established to avoid these artifacts so that maturity becomes the major determinant of differences in comparable measures between fruit, trees, and orchards. Year-to-year variation must also be considered in building models so that the maturity model is robust and generally useable regardless of the growing season (Peirs et al., 2000; Alegre et al., 2006; Bobelyn et al., 2010). Finally, maturity measures should be integrated into a spatial analysis using global positioning technologies and real-time mapping on the progress of maturity, allowing maturity assessment to become a truly predictive tool (Zude, 2008) and assuring increased reliability of outcomes when the fruit is placed into modified or controlled atmosphere storage or transport.

References

Alegre, S., Molina, D.P., Recasens, I., Casals, M., Bonany, J., Carbó, J., Casero, T., Iglesias, I., 2006. Seasonal trends in harvest indices for 'Golden Smoothee'® apples in Spain. J. Fruit Ornamental Plant Res. 14, 65–75. Suppl. 2.

Bagnaresi, P., Moschella, A., Beretta, O., Vitulli, F., Ranalli, P., Perata, P., 2008. Heterologous microarray experiments allow the identification of the early events associated with potato tuber cold sweetening. BMC Genomics 9, 176.

Bai, J., Prange, R., Toivonen, P.M.A., 2009. Pome fruits. In: Yahia, E. (Ed.), Modified and Controlled Atmospheres for Storage, Transportation and Packaging of Horticultural Commodities. Taylor and Francis Group, Boca Raton, FL, pp. 267–285.

Baumann, H., 1996. Optimum harvest date of some apple varieties in the 'Rheneland' from 1989 to 1992. In: de Jager, A., Johnson, D., Hohn, E. (Eds.), Determination and Prediction of Optimum Harvest Date of Apples and Pears. Cost94: Proceedings of a meeting of the Working Group on optimum harvest date held on 9 and 10 June in Lofthus, Norway, pp. 33–38.

Beaulieu, J.C., Baldwin, E.A., 2002. Flavor and aroma of fresh-cut fruits and vegetables. In: Lamikanra, O. (Ed.), Fresh-Cut Fruits and Vegetables. Science, Technology, and Market. CRC Press, Boca Raton, FL, pp. 391–425.

Beaulieu, J.C., Grimm, C.C., 2001. Identification of volatile compounds in cantaloupe at various developmental stages using solid phase microextraction. J. Agric. Food Chem. 49, 1345–1352.

Ben-Yehoshua, S., Robertson, R.N., Biale, J.B., 1963. Respiration and internal atmosphere of avocado fruit. Plant Physiol. 38, 194–201.

Blakey, R.J., 2016. Evaluation of avocado fruit maturity with a portable near-infrared spectrometer. Postharvest Biol. Technol. 121, 101–105.

Blanpied, G.D., Silsby, K.J., 1992. Predicting harvest date windows for apples. In: Cornell Cooperative Extension Information Bulletin. vol. 221. 12 pp.

Bobelyn, E., Serban, A.-S., Nicu, M., Lammertyn, J., Nicolai, B.M., Saeys, W., 2010. Postharvest quality of apple predicted by NIR-spectroscopy: study of the effect of biological variability on spectra and model performance. Postharvest Biol. Technol. 55, 133–143.

Burdon, J., Lallu, N., Haynes, G., Pidakala, P., Billing, D., McDermott, K., 2010. Dynamic controlled atmosphere storage of New Zealand-grown 'Hass' avocado fruit. Acta Hortic. (876), 47–54.

Burg, S.P., Burg, E.A., 1967. Molecular requirements for the biological activity of ethylene. Plant Physiol. 42, 144–152.

De Jager, A., Roelofs, F.P.M.M., 1996. Prediction of optimum harvest date of Jonagold. In: De Jager, A., Johnson, D., Hohn, E. (Eds.), Determination and Prediction of Optimum Harvest Date of Apples and Pears. COST 94—The Postharvest Treatment of Fruit and Vegetables. European Commission, Luxembourg, pp. 21–31.

de Oliveira, G.A., Bureau, S., Marie-Geneviève, C., Renard, C., Pereira-Netto, A.B., de Castilhos, F., 2014. Comparison of NIRS approach for prediction of internal quality traits in three fruit species. Food Chem. 143, 223–230.

DeLong, J., Prange, R.K., Harrison, P.A., Schofield, R.A., DeEll, J.R., 1999. Using the Streif index as a final harvest window for controlled-atmosphere storage of apples. Hortscience 34, 1251–1255.

DeLong, J., Prange, R., Harrison, P., Nichols, D., Wright, H., 2014. Determination of optimal harvest boundaries for Honeycrisp™ fruit using a new chlorophyll meter. Can. J. Plant Sci. 94, 361–369.

Dixon, J., Pak, H.A., Smith, D.B., Elmsly, T.A., Cutting, J.G.M., 2003. New Zealand avocado fruit quality: the impact of storage temperature and maturity. In: New Zealand Growers' Ann. Res. Rep. vol. 3. pp. 48–53.

Elgar, H.J., Watkins, C.B., Lallu, N., 1999. Harvest date and crop load effects on a carbon dioxide–related storage injury of 'Braeburn' apple. Hortscience 34, 305–309.

Emonger, V.E., Murr, D.P., Lougheed, E.C., 1994. Preharvest factors that predispose apples to superficial scald. Postharvest Biol. Technol. 4, 289–300.

Felicetti, D.A., Schrader, L.E., 2009. Changes in pigment concentrations associated with sunburn browning of five apple cultivars. I. Chlorophylls and carotenoids. Plant Sci. 176, 78–83.

Fellman, J.K., Rudell, D.R., Mattinson, D.S., Mattheis, J.P., 2003. Relationship of harvest maturity to flavor regeneration after CA storage of 'Delicious' apples. Postharvest Biol. Technol. 27, 39–51.

Fisher, G., Almanza-Merchán, P.J., Ramírez, F., 2012. Relación fuente-vertedero en especies frutales. Una revisión. Revista Colombiana de Ciencias Hortícolas 6, 238–253. http://www.scielo.org.co/scielo.php?script=sci_arttext&pid=S2011-21732012000200011&lng=en&tlng=es (accessed 2016.11.16).

Gasser, F., Dätwyler, D., Schneider, K., Naunheim, W., Hoehn, E., 2003. Effects of decreasing oxygen levels in the storage atmosphere on the respiration of Idared apples. Acta Hortic. (600), 189–192.

Gorny, J.R., Kader, A.A., 1996. Controlled-atmosphere suppression of ACC synthase and ACC oxidase in 'Golden Delicious' apples during long-term cold storage. J. Am. Soc. Hortic. Sci. 121, 751–755.

Gorny, J.R., Gil, M.I., Kader, A.A., 1998. Postharvest physiology and quality maintenance of fresh-cut pears. Acta Hortic. (464), 231–236.

Gorny, J.R., Hess-Pierce, B., Kader, A.A., 1999. Postharvest physiology and quality maintenance of fresh-cut nectarines and peaches. Acta Hortic. 485, 173–179.

Holderbaum, D.F., Kon, T., Kudo, T., Guerra, M.P., 2010. Enzymatic browning, polyphenol oxidase activity, and polyphenols in four apple cultivars: dynamics during fruit development. Hortscience 45, 1150–1154.

Jha, S.N., Narsaiah, K., Jaiswal, P., Bhardwaj, R., Gupta, M., Kumar, R., Sharma, R., 2014. Non-destructive prediction of maturity of mango using near infrared spectroscopy. J. Food Eng. 124, 152–157.

Jobling, J.J., McGlasson, W.B., 1995. A comparison of ethylene production, maturity and controlled atmosphere storage life of Gala, Fuji and Lady Williams apples (*Malus domestica*, Borkh.). Postharvest Biol. Technol. 6, 209–218.

Johnston, J.W., Gunaseelan, K., Pidakala, P., Wang, M., Schaffer, R., 2009. Co-ordination of early and late ripening events in apples is regulated through differential sensitivities to ethylene. J. Exp. Bot. 60, 2689–2699.

Kader, A.A., 1997. Biological bases of O_2 and CO_2 effects on postharvest-life of horticultural products. In: Saltveit, M.E. (Ed.), CA'97, Proceedings Volume 4: Vegetables and Ornamentals. Department of Pomology, University of California, Davis, CA, USA, pp. 160–163.

Kader, A.A., 2016. Controlled atmosphere storage. In: Gross, K.C., Wang, C.Y., Saltveit, M.E. (Eds.), The Commercial Storage of Fruits, Vegetables, and Florist and Nursery Stocks, USDA Agricultural Research Service Agriculture Handbook Number 66. pp. 22–25 Washington, D.C.

Kim, S.W., Ahn, M.S., Kwon, Y.K., Song, S.Y., Kim, J.K., Ha, S.-H., Kim, I.-J., Liu, J.R., 2015. Monthly metabolic changes and PLS prediction of carotenoid content of citrus fruit by combined Fourier transform infrared spectroscopy and quantitative HPLC analysis. Plant Biotechnol. Rep. 9, 247–258.

Klein, J.D., Dong, L., Zhou, H.-W., Lurie, S., 2001. Ripeness of shaded and sun-exposed apples (*Malus domestica*). Acta Hortic. (553), 95–98.

Knee, M., 1980. Methods of measuring green colour and chlorophyll content of apple fruit. J. Food Technol. 15, 493–500.

Kupferman, E., 1994. Maturity and storage of Gala, Fuji, and Braeburn apples. Washington State University, Tree Fruit Research and Extension Center, Postharvest Information Network. http://postharvest.tfrec.wsu.edu/pages/J5I3A.

Kweon, H.-J., Kang, I.-K., Kim, M.-J., Lee, J., Moon, Y.-S., Choi, C., Choi, D.G., Watkins, C.B., 2013. Fruit maturity, controlled atmosphere delays and storage temperature affect fruit quality and incidence of storage disorders of 'Fuji' apples. Sci. Hortic. 157, 60–64.

Lallu, N., Burdon, J., Yearsley, C.W., Billing, D., 2003. Commercial practices used for controlled atmosphere storage of 'Hayward' kiwifruit. Acta Hortic. (610), 245–251.

Lau, O.L., 1998. Influence of climate, harvest maturity, waxing, O_2 and CO_2 on browning disorders of 'Braeburn' apples. Postharvest Biol. Technol. 14, 131–141.

Lau, O.L., Yastremski, R., Meheriuk, M., 1987. Influence of maturity, storage procedure, temperature, and oxygen concentration on quality and disorders of 'McIntosh' apples. J. Am. Soc. Hortic. Sci. 112, 93–99.

Li, P., Ma, F., Cheng, L., 2013. Primary and secondary metabolism in the sun-exposed peel and the shaded peel of apple fruit. Physiol. Plant. 148, 9–24.

Liu, M., Pirrello, J., Chervin, C., Roustan, J.-P., Bouzayen, M., 2015. Ethylene control of fruit ripening: revisiting the complex network of transcriptional regulation. Plant Physiol. 169, 2380–2390.

López-Gálvez, G., El-Bassuoni, R., Nie, X., Cantwell, M., 1997. Quality of red and green fresh-cut peppers stored in controlled atmospheres. In: Proc. 7th Int. CA Res. Conference. vol. 5. pp. 152–157.

Lötze, E., Bergh, O., 2012. Evaluating the Streif index against commercial subjective predictions to determine the harvest date of apples in South Africa. S. Afr. J. Plant Soil 29, 53–56.

Luton, M.T., 1996. Ten years of optimum harvest date data for the variety 'Cox's Orange Pippin'. In: de Jager, A., Johnson, D., Hohn, E. (Eds.), Determination and Prediction of Optimum Harvest Date of Apples and Pears.Cost94: Proceedings of a meeting of the Working Group on optimum harvest date held on 9 and 10 June in Lofthus, Norway, pp. 39–48.

Madrid, M., Cantwell, M., 1993. Use of high CO_2 atmospheres to maintain quality of intact and fresh-cut melon. In: Proc. 6th Int. Controlled Atmosphere Research Conf. NRAES-71, Ithaca, NY, pp. 736–745.

Mencarelli, F., Botondi, F.G.R., Tonutti, P., 1998. Postharvest physiology of peach and nectarine slices. Acta Hortic. 465, 463–470.

Merzlyak, M.N., Gitelson, A.A., Pogosyan, S.I., Chivkunova, O.B., Lehimena, L., Garson, M., Buzulikova, N.P., Shevyreva, V.V., Rumyantseva, V.B., 1997. Reflectance spectra of leaves and fruits during their development and senescence under stress. Russ. J. Plant Physiol. 44, 707–716.

Merzlyak, M.N., Gitelson, A.A., Chivkunova, O.B., Rakitin, V.Y., 1999. Non-destructive optical detection of pigment changes during leaf senescence and fruit ripening. Physiol. Plant. 106, 135–141.

Merzlyak, M.N., Solovchenko, A.E., Chivkunova, O.B., 2002. Patterns of pigment changes in apple fruits during adaption to high sunlight and sunscald development. Plant Biochem. Physiol. 40, 679–684.

Merzlyak, M.N., Solovchenko, A.E., Gitelson, A.A., 2003. Reflectance spectral features and non-destructive estimation of chlorophyll, carotenoid and anthocyanin content in apple fruit. Postharvest Biol. Technol. 27, 88–103.

Mitcham, E.J., Mitchell, F.G., 2002. Postharvest handling systems: pome fruits. In: Kader, A.A. (Ed.), Postharvest Technology of Horticultural Crops, third ed. University of California, Agriculture and Natural Resources Publication, Oakland, CA, pp. 333–344. 3311.

Mitchell, F.G., Mayer, G., Biasi, W., 1991. Effect of harvest maturity on storage performance of 'Hayward' kiwifruit. Acta Hortic. 297, 617–625.

Nicolaï, B.M., Beullens, K., Bobelyn, E., Peirs, A., Saeys, W., Theron, K.I., Lammertyn, J., 2007. Nondestructive measurement of fruit and vegetable quality by means of NIR spectroscopy: a review. Postharvest Biol. Technol. 46, 99–118.

Nilsson, T., Gustavsson, K.-E., 2007. Postharvest physiology of 'aroma' apples in relation to position on the tree. Postharvest Biol. Technol. 43, 36–46.

Nunes, M.C.N., Morais, A.M.M.B., Brecht, J.K., Sargent, S.A., 2002. Fruit maturity and storage temperature influence response of strawberries to controlled atmospheres. J. Am. Soc. Hortic. Sci. 127, 836–842.

Olarewaju, O.O., Bertling, I., Magwaza, L.S., 2016. Non-destructive evaluation of avocado fruit maturity using near infrared spectroscopy and PLS regression models. Sci. Hortic. 199, 229–236.

Pathange, L.P., Mallikarjunan, P., Marini, R.P., O'Keefe, S., Vaughan, D., 2006. Non-destructive evaluation of apple maturity using an electronic nose system. J. Food Eng. 77, 1018–1023.

Paul, V., Pandey, R., Srivastava, G.C., 2012. The fading distinctions between classical patterns of ripening in climacteric and non-climacteric fruit and the ubiquity of ethylene—an overview. J. Food Sci. Technol. 49, 1–21.

Paull, R.E., Chen, W., 1997. Minimal processing of papaya (*Carica papaya* L.) and the physiology of halved fruit. Postharvest Biol. Technol. 12, 93–99.

Pech, J.C., Bouzayen, M., Latché, A., 2008. Climacteric fruit ripening: ethylene-dependent and independent regulation of ripening pathways in melon fruit. Plant Sci. 175, 114–120.

Peiris, K.H.S., Dull, G.G., Leffler, R.G., Kays, S.J., 1999. Spatial variability of soluble solids or dry-matter content within individual fruits, bulbs, or tubers: Implications for the development and use of NIR spectrometric techniques. Hortscience 34, 114–118.

Peirs, A., Lammertyn, J., Ooms, K., Nicolaï, B.M., 2000. Prediction of the optimal picking date of different apple cultivars by means of VIS/NIR-spectroscopy. Postharvest Biol. Technol. 21, 189–199.

Prange, R.K., 2016. Cabbage. In: Gross, K.C., Wang, C.Y., Saltveit, M.E. (Eds.), The Commercial Storage of Fruits, Vegetables, and Florist and Nursery Stocks, USDA Agricultural Research Service Agriculture Handbook Number. In: vol. 66. pp. 255–260.

Reid, M.S., 2002. Maturation and maturity indices. In: Kader, A.A. (Ed.), Postharvest Technology of Horticultural Crops. third ed. University of California, Agriculture and Natural Resources Publication, Oakland, CA, pp. 55–62 3311.

Rhodes, M.J.C., 1970. The climacteric and ripening of fruits. In: Hulme, A.C. (Ed.), The Biochemistry of Fruits and their Products. In: vol. 1. Academic Press, London, pp. 521–533.

Riad, G.S., Brecht, J.K., 2005. Simulated long-distance transport of strawberries in a passive modified atmosphere marine container. Proc. Fla. State Hortic. Soc. 118, 396–399.

Richardson, A.C., Marsh, K.B., Boldingh, H.L., Pickering, A.H., Bulley, S.M., Frearson, N.J., Ferguson, A.R., Thornber, S.E., Bolitho, K.M., MaCrae, E.A., 2004. High growing temperatures reduce fruit carbohydrate and vitamin C in kiwifruit. Plant Cell Environ. 27, 423–435.

Richardson, A.C., Boldingh, H.L., McAtee, P.A., Gunaseelan, K., Luo, Z., Atkinson, R.G., David, K.M., Burdon, J.N., Schaffer, R.J., 2011. Fruit development of the diploid kiwifruit, *Actinidia chinensis* 'Hort16A'. BMC Plant Biol. 11, 182. http://www.biomedcentral.com/1471-2229/11/182.

Rutkowski, K.P., Michalczuk, B., Konopacki, P., 2008. Nondestructive determination of 'Golden Delicious' apple quality and harvest maturity. J. Fruit Ornamental Plant Res. 16, 39–52.

Saltveit, M.E., 1999. Effect of ethylene on quality of fresh fruits and vegetables. Postharvest Biol. Technol. 15, 279–292.

Schaare, P.N., Fraser, D.G., 2000. Comparison of reflectance, interactance and transmission modes of visible-near infrared spectroscopy for measuring internal properties of kiwifruit (*Actinidia chinensis*). Postharvest Biol. Technol. 20, 175–184.

Schrader, L.E., Zhang, J., Sun, J., Xu, J., Elfving, D.C., Kahn, C., 2009. Postharvest changes in internal fruit quality in apples with sunburn browning. J. Am. Soc. Hortic. Sci. 134, 148–155.

Seifert, B., Pflanz, M., Zude, M., 2014. Spectral shift as advanced index for fruit chlorophyll breakdown. Food Bioprocess Technol. 7, 2050–2059.

Sims, D.A., Gamon, J.A., 2002. Relationships between leaf pigment content and spectral reflectance across a wide range of species, leaf structures and developmental stages. Remote Sens. Environ. 81, 337–354.

Slaughter, D.C., Crisosto, C.H., 1998. Nondestructive internal quality assessment of kiwifruit using near-infrared spectroscopy. Semin. Food Anal. 3, 131–140.

Slaughter, D.C., Barrett, D., Boersig, M., 1996. Nondestructive determination of soluble solids in tomatoes using near infrared spectroscopy. J. Food Sci. 61, 695–697.

Smith, R.B., Lougheed, E.C., Franklin, E.W., McMillan, I., 1979. The starch iodine test for determining stage of maturation in apples. Can. J. Plant Sci. 59, 725–735.

Solovchenko, A.E., Chivkunova, O.B., Merzlyak, M.N., Gudkovsky, V.A., 2005. Relationships between chlorophyll and carotenoid pigments during on- and off-tree ripening of apple fruit as revealed non-destructively with reflectance spectroscopy. Postharvest Biol. Technol. 38, 9–17.

Spadoni, A., Cameldi, I., Noferini, M., Bonora, E., Costa, G., Mari, M., 2016. An innovative use of DA-meter for peach fruit postharvest management. Sci. Hortic. 201, 140–144.

Streif, J., 1996. Optimum harvest date for different apple cultivars in the ‘Bodensee’ area. In: de Jager, A., Johnson, D., Hohn, E. (Eds.), Determination and Prediction of Optimum Harvest Date of Apples and Pears.Cost94: Proceedings of a meeting of the Working Group on optimum harvest date held on 9 and 10 June in Lofthus, Norway, pp. 15–20.

Streif, J., Saquet, A.A., Xuan, H., 2003. CA-related disorders of apples and pears. Acta Hortic. 600, 223–230.

Subedi, P.P., Walsh, K.B., Owens, G., 2007. Prediction of mango eating quality at harvest using short-wave near infrared spectrometry. Postharvest Biol. Technol. 43, 326–334.

Tijskens, L.M.M., Heuvelink, E., Schouten, R.E., Lana, M.M., van Kooten, O., 2005. The biological shift factor: biological age as a tool for modelling in pre- and postharvest horticulture. Acta Hortic. 687, 39–46.

Tijskens, L.M.M., Unuk, T., Tojnko, S., Hribarc, J., Simčič, M., 2009. Biological variation in the colour development of Golden Delicious apples in the orchard. J. Sci. Food Agric. 89, 2045–2051.

Toivonen, P.M.A., 2007. Fruit maturation and ripening and their relationship to quality. Stewart Postharvest Rev. 3 (2), 1–5.

Toivonen, P.M.A., 2008. Influence of harvest maturity on cut-edge browning of ‘Granny Smith’ fresh apple slices treated with anti-browning solution after cutting. LWT- Food Sci. Technol. 41, 1607–1609.

Toivonen, P.M.A., 2014. Relationship of typical core temperatures after hydrocooling on retention of different quality components in sweet cherry. HortTechnology 24, 457–462.

Toivonen, P.M.A., 2015a. Integrated analysis for improving export of sweet cherries and how a small industry can compete by focusing on premium quality. Acta Hortic. (1079), 71–82.

Toivonen, P.M.A., 2015b. Comparison of I_{AD} and starch-iodine indices at harvest and how they relate to post-storage firmness retention in Ambrosia™ apples over three growing seasons. Can. J. Plant Sci. 95, 1177–1180.

Toivonen, P.M.A., Beveridge, T.H.J., 2005. Maturity, ripening and quality relationships. In: Lamikanra, O., Imam, S.H., Ukuku, D.O. (Eds.), Produce Degradation: Reaction Pathways and Their Prevention. CRC Press, Boca Raton, FL, pp. 55–77.

Toivonen, P.M.A., DeEll, J.R., 2002. Physiology of fresh-cut fruits and vegetables. In: Lamikanra, O. (Ed.), Fresh-Cut Fruits and Vegetables: Science, Technology, and Market. CRC Press, Boca Raton, FL, pp. 91–123.

Toivonen, P.M.A., Kappel, F., Stan, S., McKenzie, D.-L., Hocking, R., 2004. Firmness, respiration and weight loss of ‘Bing’, ‘Lapins’, and ‘Sweetheart’ cherries in relation to fruit maturity and susceptibility to surface pitting. Hortscience 39, 1066–1069.

Toivonen, P.M.A., Lu, C., Lannard, B., 2016. Visible spectroscopy I_{AD} measures in apple: impact of stress and shading on maturity indexing. Acta Hortic. 1119, 243–249.

Walsh, K., 2016. Dry Matter Matters. http://www.industry.mangoes.net.au/resource-collection/2016/3/6/dry-matter-matters (accessed 2016.11.17).

Wang, Y., Bai, J., Long, L.E., 2015. Quality and physiological responses of two late-season sweet cherry cultivars ‘Lapins’ and ‘Skeena’ to modified atmosphere packaging(MAP) during simulated long distance ocean shipping. Postharvest Biol. Technol. 110, 1–8.

Watkins, C.B., 2008. Dynamic controlled atmosphere storage—a new technology for the New York storage industry? New York Fruit Quar. 16, 23–26.

Watkins, C.B., Bowen, J.H., Walker, V.J., 1989. Assessment of ethylene production by apple cultivars in relation to commercial harvest dates. N. Z. J. Crop. Hortic. Sci. 17, 327–331.

Wedding, B.B., White, R.D., Grauf, S., Wright, C., Tilse, B., Hofman, P., Gadek, P.A., 2011. Non-destructive prediction of 'Hass' avocado dry matter via FT-NIR spectroscopy. J. Sci. Food Agric. 91, 233–238.

Wiersma, P.A., Zhang, H., Lu, C., Quail, A., Toivonen, P.M.A., 2007. Survey of the expression of genes for ethylene synthesis and perception during maturation and ripening of 'Sunrise' and 'Golden Delicious' apple fruit. Postharvest Biol. Technol. 44, 204–211.

Yahia, E.M. (Ed.), 2009. Modified and Controlled Atmospheres for the Storage, Transportation, and Packaging of Horticultural Commodities. Taylor & Francis Group, CRC Press, Boca Raton, FL.

Zude, M., 2008. Optical Monitoring of Fresh and Processed Agricultural Crops. CRC Press, Boca Raton, FL.

Zude, M., Herold, B., 2002. Optimum harvest date determination for apples using spectral analysis. Eur. J. Hortic. Sci. 67, 199–204.

Zude-Sasse, M., Truppel, I., Herold, B., 2002. An approach to non-destructive apple fruit chlorophyll determination. Postharvest Biol. Technol. 25, 123–133.

Chapter 4

CA requirements for postharvest pest control

Yong-Biao Liu
USDA-ARS, Salinas, CA, United States

Introduction

Research on CA treatment for postharvest pest control on fresh fruit and vegetables has a history of about 50 years. However, the number of studies is very limited considering the large numbers of pests that need to be treated on a wide variety of fresh fruit and vegetables. Most studies did not yield a successful treatment that warrants commercial development. This is in part due to higher cost of CA treatment as compared with other available treatment methods but largely due to their negative impact on postharvest quality of the treated fresh products. These outcomes highlight the importance of selecting right combinations of susceptible pests with tolerant fresh products as the primary requirement for any successful CA treatment. Typically, CA treatments for pest control use combinations of low O_2 and high CO_2. Most fresh fruit and vegetables, however, cannot tolerate high CO_2 and also need to be stored at low temperatures to maintain postharvest quality and, therefore, CA treatment just with ultralow oxygen (ULO treatment) may need to be used. All these factors put great constraints on CA treatment conditions for pest control. In addition, plant respiration and physiological states also affect tolerance of fresh products to CA treatments. All of these highlight the challenges in developing safe and effective CA treatments for pest control.

As methyl bromide production is being phased out globally, its exempted and limited usage for preshipment fumigation treatment may not be sustainable in the future. Phosphine fumigation as its main alternative for postharvest pest control cannot meet postharvest pest control demands due to its long treatment times and ineffectiveness against many pests (Bell, 2000; Hole et al., 1976; Liu, 2011a; Liu et al., 2014). A newly discovered fumigant nitric oxide is effective against a wide variety of pests on fresh and stored products (Liu, 2013a; Liu, 2016). However, its usage will also be limited to selected products due to its requirement for ultralow environment and will not be available for several

Controlled and Modified Atmospheres for Fresh and Fresh-Cut Produce
https://doi.org/10.1016/B978-0-12-804599-2.00005-3

years as it needs to be registered for commercial use as a pesticide. CA treatment as an environmentally friendly alternative has not been adequately explored and should be studied to develop safe and effective treatment for postharvest pest control on some specific fresh products.

Studies on CA treatments for pest control have been well reviewed (Carpenter and Potter, 1994; Fleurat-Lessard, 1990; Liu, 2010; Mitcham, 2003; Mitcham et al., 2001) and results were summarized (Ke and Kader, 1992; Liu, 2010; Mitcham et al., 2001). In this paper, CA requirements for pest control are discussed. Relevant literatures are also reviewed to emphasize the importance of various requirements of CA treatment including variations in pest susceptibility and product tolerance to CA treatment; combinations of low O_2, temperature, and treatment time; as well as physiological activities and metabolism. Equipment requirements and operation for CA treatments for pest control are also discussed.

Major factors for CA treatment success

Several major factors determine success or failure of a CA treatment such as selection of susceptible target pests; selection of tolerant fresh products; and selection of optimal combinations of O_2, CO_2, temperature, and time. Additionally, proper maintenance of atmosphere composition and reduction of physiological activities of fresh products may also be important factors when sensitive products are subjected to long CA treatments.

Not all pests can be controlled with CA treatment within an acceptable time frame under conditions that can be tolerated by fresh commodities. In fact, only small fraction of pest species can be expected to be controlled on fresh commodities based on past research and our knowledge on tolerance of fresh products to CA conditions. There are large variations among pest groups, species, and different life stages in susceptibility to CA treatment (Liu, 2010; Mitcham et al., 2001).

Different fresh products vary greatly in tolerance to CA treatment conditions such as low O_2 or elevated CO_2 for the duration of the treatment at a higher than the storage temperature. This is illustrated clearly in CA storage conditions for various fresh products and tolerance to low O_2 (Beaudry, 1999; Beaudry, 2000; Watkins and Nock, 2012). Some products can tolerate high CO_2 such as green onion and broccoli while others can be extremely sensitive to CO_2 such as lettuce (Beaudry, 1999; Lipton et al., 1972; Liu and Liu, 2014). These variations are results of the diversity in physical, chemical, and physiological characteristics of different fresh products. Some products such as bulbs, tubers, and rootstocks are valued by their viability rather than cosmetic quality and therefore are more tolerant to CA treatment than products which are evaluated for their cosmetic appearance (Liu et al., 2010).

It is, therefore, critically important to understand susceptibility of different pests and tolerance of fresh products to CA treatment factors such as

O_2, CO_2, and temperature to identify a good match of a susceptible pest with a tolerant fresh product for CA treatment. It also needs to be realistic on expectations for using CA treatment to control a particular pest on a particular product and chances are that, in majority of cases, a CA treatment for the specific pest on the particular fresh product may not be feasible.

Fresh fruit and vegetables vary greatly in physiology and metabolic activities such as O_2 consumption and CO_2 productions (Watkins and Nock, 2012). These variations in large part determine the intrinsic susceptibility or tolerance of fresh products to CA treatment. Low O_2 conditions in CA treatments may present harsh stress to products with active metabolism that requires high consumption of O_2. Accumulation of metabolites may also pose a negative impact on product quality.

For fresh vegetables and nonclimacteric fruit, physiological activities slow down greatly over time after harvest as they are chilled and stored at low temperatures and use up stored metabolic substrates. This process may also have significant impact on tolerance to CA treatment for pest control and may be required to prevent injuries to some sensitive products against some pests as demonstrated in the ULO treatment for control of western flower thrips on lettuce (Liu, 2008b; Liu, 2011b). Preconditioning also has significant effects in lessening negative impact of CA treatment on some fresh products. Postharvest physiological process can also be modulated with chemical treatments such as 1-methylcyclopropene (1-MCP) and nitric oxide to extend storage and shelf life (Blankenship and Dole, 2003; Freiman et al., 2012; Manjunatha et al., 2010; Saadatian et al., 2012; Soegiarto and Wills, 2004; Watkins and Nock, 2012). These external factors may also enhance tolerance of fresh products to CA treatment and should be explored in the development of CA treatment in the future.

Selection of suitable target pests and fresh products

The success of CA or ULO treatments depends on types of pests and commodities. In general, mobile life stages of pests have higher metabolic rates than immobile life stages such as egg and pupa and are more susceptible to low O_2 stress of CA treatments and are better targets for CA treatments. There are also large variations in susceptibility to low O_2 and elevated CO_2 among different taxonomic groups of pests. A special example is black widow spider which is perhaps the most susceptible pest to low O_2 stress. Complete control of black widow spiders can be achieved in 24 h under 0.5% O_2 at 2°C (Liu et al., 2008). This is due to the fact that spiders have book lungs and open blood circulation systems and O_2 is transported through circulating blood by a complex hemocyanin protein which is only about 5% efficient as compared with hemoglobin in mammals (Anderson and Prestwich, 1985; Gillespie and Spagna, 2003). In comparison, most insects have well-developed tracheal systems for gas exchange. Oxygen enters insect bodies via spiracles and diffuses through branching tracheae and tracheoles to reach organs and muscles

(Harrison, 2003). The efficient gas exchange in insects renders them less susceptible to low O_2 stress as compared with spiders in CA treatments.

There are also great variations among different taxonomic groups of insects in susceptibility to low O_2 stress or combinations of low O_2 and elevated CO_2 levels (Held et al., 2001; Liu, 2003; Liu, 2008a; Liu, 2008c; Liu, 2010; Liu, 2013b; Mitcham et al., 2001). Lettuce aphid (*Nasonovia ribisnigri*) and western flower thrips (*Frankliniella occidentalis*) are relatively susceptible to ULO treatment and can be controlled in 1–2 days without injuries to lettuce. Pupae of California pea leafminer (*Liriomyza langei*) which also occur on lettuce could not be successfully controlled on lettuce (Liu, 2008c). Others such as grape mealybug (*Pseudococcus maritimus*) require as long as 14 days under ULO of 30-ppm O_2 at 2°C to achieve complete control of all life stage and the treatment is safe to grape quality (Liu, 2013b). In this case, even though mealybugs take long time to be controlled with CA treatment, table grapes have very high tolerance to ULO conditions to make the CA treatment successful.

In addition, external and internal feeding insects may also differ significantly to CA treatment. Under ULO conditions, as products consume O_2 and are under low O_2 stress, internal plant tissues may experience more extreme low O_2 stress as compared with plant surfaces. Therefore internal feeding insects may experience greater low O_2 stress as compared with external feeding insects. Assuming external and internal feeding insects have same level of susceptibility to low O_2 stress, it is conceivable that a ULO treatment may be more effective against the internal feeder than against the external feeder, even though there is a lack of experimental data for comparisons between external and internal feeding insects of related species.

Fresh fruit and vegetables vary greatly in CA storage requirements (Beaudry, 1999; Beaudry, 2000) and, therefore, vary in tolerance to CA treatments for pest control. ULO treatment for control of western flower thrips can damage lettuce but is safe on table grapes (Liu, 2008a; Liu, 2012). Even in the same species, there is a great variation among different cultivars in tolerance to low O_2 stress as demonstrated with iceberg lettuce to ULO treatment (Liu, 2008a; Liu and Liu, 2014). Because of the great variation and high sensitivity of many fresh products to low O_2, elevated CO_2, and temperature, selection of suitable products for specific CA treatment is critical. In general, products with low physiological activities and long storage life are more tolerant to CA treatment. Plant materials at dormant stages are more tolerant than physiologically active plant materials. Propagating materials such benchgrafts, roots, tubers, stalks, and shoots that are valued for germination and growth are likely to be more tolerant to CA treatment for pest control than fresh products valued for cosmetic appearance such as most fresh fruits and vegetables.

It is critical to identify reasonably susceptible pests to match products which will likely tolerate the CA treatment for the identified pest to increase the chance of safe and effective control of pests on the selected fresh products. For example, complete control of green peach aphid (*Myzus persicae*), sweetpotato whitefly (*Bemisia* sp.), twopotted spider mite (*Tetranychus urticae*), and western flower thrips is achieved with 18-h exposure to >99% N_2 and CO_2 at 20°C. However, the treatment is safe to begonia seedlings but not to seedlings of chrysanthemum and geranium varieties which only tolerate for <12-h exposure to the treatment conditions (Held et al., 2001). Vine mealybug (*Planococcus ficus*) is a quarantined pest on grape nursery stocks in California as well as in export markets of US table grapes. ULO treatments at 15°C and 25°C were developed successfully for control of vine mealybug on dormant grape benchgrafts without any negative impact on germination or growth of the benchgrafts (Liu et al., 2010). These cases illustrate the great variations in both susceptibility of different pest species and tolerance of different plant species or varieties to CA treatment.

Optimal combinations of low O_2, temperature, and time

A target pest can be controlled effectively in CA treatments with different combinations of low O_2, elevated CO_2, temperature, and treatment time. Typically, CA or ULO treatment with lower O_2 or higher temperature takes shorter time to control a pest than a CA or ULO treatment with higher O_2 or lower temperature. But different CA or ULO treatments with same efficacy against a pest can differ significantly in their impact on product quality. This is well illustrated by ULO treatment for control of western flower thrips on broccoli. Both 5-day ULO treatment with 0.003% O_2 and 10-day ULO treatment with 0.03% O_2 successfully control thrips on ice-packed broccoli at 1°C. However, the 10-day ULO treatment with higher O_2 level results in detectable production of off-odor mainly methanethiol at the end of the treatment while the 5-day ULO treatment does not have detectable off-odor (Liu, 2007). Therefore the shorter and intense treatment is safe than the 10-day longer and milder treatment to broccoli quality. These results show that shorter and more intensive ULO treatments tend to achieve effective control of target pests and better tolerance by the fresh products. There are other examples where brief intensive CA treatments are successful in controlling pests without significant impact on product quality. Combined heat and CA treatment is a special example of high-temperature CA for postharvest pest control and is effective against codling moth in sweet cherries without negative impact on product quality (Neven, 2005; Neven and Mitcham, 1996).

Maintain proper atmospheric compositions in CA treatment

Atmospheric composition needs to be managed to ensure efficacy against the target pest and safety to product quality. Consumption of O_2 and release of CO_2 and other metabolites such as ethylene during a CA treatment may significantly alter atmospheric composition over time and cause injuries to the treated products especially to sensitive products in a long CA treatment. CO_2 can be scrubbed using soda lime and various ethylene scrubbing products are also available commercially. Oxygen may also need to be replenished to maintain in a desired range to prevent low O_2 injuries to the products. As most fresh products need to be stored at low temperature to maintain long storage life, CA treatment is also best to be conducted at low temperature to minimize impact on product quality if it is effective against the target pest. However, when temperature is too low, CO_2 released by plant tissues may not dissipate from plants tissues rather dissolved in plant tissues and lead to injuries as exemplified by lettuce response to CA treatments (Brecht et al., 1973).

Enhance tolerance to CA treatment

There are large variations in plant physiological status at harvest and after harvest. For the same plant species or variety, postharvest handling, plant maturity, microclimate, harvest time during the day, weather, soil condition, and management practice may affect tolerance of fresh products to a CA treatment for pest control. In most CA studies, fruits and vegetables were subjected to CA treatments with little or no cold storage after harvest and potential effects of postharvest storage history were not considered. Plants' physiological activity undergoes rapid slow down after harvest as metabolic substrates are being used up and plants are chilled and cold stored. A CA treatment with ultralow O_2 levels (ULO treatment) immediately after harvest represents greater stress because of the high demand for O_2 as compared with plants that have experienced a period of cold storage and have very low respiration rate. For example, a ULO treatment for control of western flower thrips caused brown stains to newly harvested lettuce but not to lettuce that has been stored for 1 week at low temperature (Liu, 2008b). In a later study, 3-day pretreatment cold storage was found to be adequate for preventing injury to heart leaves by a 2-day ULO treatment for control of western flower thrips (Liu, 2011b).

Mature plants are physiologically less active and have lower respiration rates than young plants and therefore may have better tolerance to low O_2 stress than young plants. Therefore ULO treatment more likely causes low O_2 injuries to young plants at active growth stage than to mature plants. For example, ULO treatments tend to injure heart leaves of young iceberg lettuce and this is likely due to greater low O_2 stress to the heart leaves which are growth points due to higher demand for O_2 as well as restriction of O_2 dispersion to the center of lettuce heads due to layers of lettuce leaves and perforated plastic wrapper. On the other hand, some mature plants such as iceberg lettuce heads are larger and denser than young ones and air diffusion in larger and denser lettuce heads is more restricted than in smaller and looser lettuce heads. This may lead to CO_2 injuries to lettuce heart leaves especially at low temperatures when CO_2 is more dissolved in water and trapped in plant tissues and causes injuries. In hermetic storage of iceberg lettuce at a low temperature, there is a negative correlation between head weight and lettuce internal quality due to increased

CO_2-induced brown stains to heart leaves of heavy lettuce heads (Liu and Liu, 2014). Understanding these principles will help to make a better selection of fresh products for ULO treatment to minimize the risk of injuries to the products.

Equipment and processes of CA treatment for pest control

Unlike CA storage which can have a relatively broader range of O_2 above the tolerance limit of fresh products, CA treatment for pest control often need O_2 level to be at or below certain threshold to maintain its efficacy against a specific target pest and the O_2 threshold may be close to the low limit which a treated fresh product can tolerate. Therefore most CA treatments for pest control require more stringent control of O_2 levels than CA for postharvest storage. This includes the ability to accurately measure O_2 at very low level and the ability to maintain precise O_2 level in the CA treatment. Oxygen analyzers with zirconia sensors offer long use life and high sensitivity at ppm levels and are suitable for CA treatments for pest control. Tight seal of the CA treatment chambers or rooms is also needed to maintain the extreme low O_2 levels.

In CA storage, a nitrogen–air mix with a certain level of O_2 can be released into CA rooms or chambers to establish a low O_2 condition for CA storage. For CA treatment for pest control, it is recommended that a feedback control mechanism be used to release nitrogen to establish ULO conditions and to maintain the ULO conditions. Adding N_2 or O_2 to a CA treatment room/chamber is controlled through an O_2 analyzer based on continuous measurements of O_2 levels in the CA treatment.

Under ULO conditions in CA treatment, O_2 consumption by fresh products may cause extreme local O_2 deficiency and low O_2 injuries to the products. Therefore O_2 may need to be replenished and adequate air circulation is required to maintain a uniform O_2 level to ensure the desired efficacy in pest control and prevent low O_2 injuries to the treated products. This is especially important for products with high respiration rates and products which are packaged with restricted ventilation.

As CA storage typically uses much higher O_2 concentrations than CA treatments for pest control, membrane-type nitrogen generators may be used at CA storage facilities and they typically produce N_2 at a low purity of about 95%. This type of N_2 generator will not be suitable for CA treatment for pest control. For laboratory studies and small-scale CA treatments, N_2 in compressed cylinders can be used. Alternatively, vaporizing liquid N_2 can also be used to conduct CA treatments for pest control at a lower cost as compared with compressed cylinders. For large-scale commercial CA treatments, a PSA (pressure swing adsorption) type N_2 generator will be required and it can produce nitrogen at a much higher purity of >99.99%.

To ensure efficacy of CA treatment against a target pest, it may be necessary to circulate the air from the CA treatment chamber through a series of small airtight containers containing the target pest located outside of the CA treatment chamber during the treatment as described by Liu (Liu, 2005). The containers can be disconnected to check mortality of the pest before ending the treatment to ensure that the effective control is achieved.

Conclusions

Even though past research on CA treatment for pest control on fresh products yielded very few satisfactory results or practical pest control solutions, research on CA for pest control is very limited and the potential of CA treatment is far from being adequately explored. Given the fact that only limited pests are susceptible enough to low O_2 stress to be controlled in an acceptable time frame and only limited fresh products can withstand a CA treatment for control of a particular pest, selecting right target pest on a suitable fresh product is critically important to ensure successful development of CA treatment for pest control. It is also important to apply our understanding of postharvest physiology of fresh products into development of CA treatment to enhance product tolerance to and prevent or minimize injuries by CA treatment for pest control. Judging from recent progresses on ULO treatment development, CA treatments for practical control of selected pests on selected fresh products are achievable. Given the environmentally friendly nature of CA treatment and the acute lack of methyl bromide alternative fumigation treatments, more research and developmental efforts are warranted to explore and develop practical CA treatments for postharvest pest control on perishable agricultural products.

References

Anderson, J.F., Prestwich, K.N., 1985. The physiology of exercise at and above maximal aerobic capacity in a theraphosid (tarantula) spider, *Brachypelma smithi* (F.O. Pickard-Cambridge). J. Comp. Physiol. B. 155, 529–539.

Beaudry, R.M., 1999. Effect of O_2 and CO_2 partial pressure on selected phenomena affecting fruit and vegetable quality. Postharvest Biol. Technol. 15, 293–303.

Beaudry, R.M., 2000. Responses of horticultural commodities to low oxygen: limits to the expanded use of modified atmosphere packaging. HortTechnology 10, 491–500.

Bell, C.H., 2000. Fumigation in the 21st century. Crop Prot. 19, 563–569.

Blankenship, S.M., Dole, J.M., 2003. 1-Methylcyclopropene: a review. Postharvest Biol. Technol. 28, 1–25.

Brecht, P.E., Kader, A.A., Morris, L.L., 1973. Influence of postharvest temperature on brown stain of lettuce. J. Am. Soc. Hortic. Sci. 98, 399–402.

Carpenter, A., Potter, M.A., 1994. Controlled atmospheres. In: Sharp, J.L., Hallman, G.J. (Eds.), Quarantine Treatments for Pests of Food Plants. Westview Press, Boulder, CO, pp. 171–198.

Fleurat-Lessard, F., 1990. In: Calderon, M., Barkai-Golan, R. (Eds.), Effect of modified atmospheres on insects and mites infesting stored products. CRC, Boca Eaton, FL, pp. 21–38.

Freiman, Z.E., Rodov, V., Yablovitz, Z., Horev, B., Flaishman, M.A., 2012. Preharvest application of 1-methylcyclopropene inhibits ripening and improves keeping quality of 'Brown Turkey' figs (*Ficus carica* L.). Sci. Hortic. 138, 266–272.

Gillespie, R.G., Spagna, J.C., 2003. Spiders. In: Resh, V.H., Carde, R.T. (Eds.), Encyclopedia of Insects. Academic Press, New York, NY, pp. 1060–1073.

Harrison, J.F., 2003. Respiration system. In: Resh, V.H., Carde, R.T. (Eds.), Encyclopedia of Insects. Academic Press, New York, NY, pp. 1002–1007.

Held, D.W., Potter, D.A., Gates, R.S., Anderson, R.G., 2001. Modified atmosphere treatments as a potential disinfestations technique for arthropod pests in greenhouses. J. Econ. Entomol. 94, 430–438.

Hole, B.D., Bell, C.H., Mills, K.A., Goodship, G., 1976. The toxicity of phosphine to all developmental stages of thirteen species of stored product beetles. J. Stored Prod. Res. 12, 235–244.

Ke, D., Kader, A.A., 1992. Potential of controlled atmospheres for postharvest insect disinfestations of fruits and vegetables. Postharvest News Inf. 3, 31N–37N.

Lipton, W.J., Stewart, J.K., Whitaker, T.W., 1972. An illustrated guide to the identification of some market disorders of head lettuce. In: USDA, ARS, Marketing Research Report No. 950, p. 7, pl. 19.

Liu, Y.-B., 2003. Effects of vacuum and controlled atmosphere treatments on insect mortality and lettuce quality. J. Econ. Entomol. 96, 1100–1107.

Liu, Y.-B., 2005. Ultralow oxygen treatment for postharvest control of *Nasonovia ribisnigri* (Homoptera: Aphididae) on iceberg lettuce. J. Econ. Entomol. 98, 1899–1904.

Liu, Y.-B., 2007. Ultralow oxygen treatment for postharvest control of western flower thrips, *Frankliniella occidentalis*, on broccoli. J. Econ. Entomol. 100, 717–722.

Liu, Y.-B., 2008a. Ultralow oxygen treatment for postharvest control of western flower thrips, *Frankliniella occidentalis* (Thysanoptera: Thripidae), on iceberg lettuce: I. Effects of temperature, time, and oxygen level on insect mortality and lettuce quality. Postharvest Biol. Technol. 49, 129–134.

Liu, Y.-B., 2008b. Ultralow oxygen treatment for postharvest control of western flower thrips, *Frankliniella occidentalis* (Thysanoptera: Thripidae), on iceberg lettuce: II. Effects of pre-treatment storage on lettuce quality. Postharvest Biol. Technol. 49, 135–139.

Liu, Y.-B., 2008c. Advances in postharvest pest control on perishable commodities using ultralow oxygen treatment and low temperature phosphine fumigation. In: Proc. 8th Intl. Conf. CA and Fumigation in Stored Products, Sept. 21–26, 2008, Chengdu, China.

Liu, Y.-B., 2010. Recent advances in development of ultralow oxygen treatment for postharvest pest control on perishable commodities. Steward Postharvest Rev. 3:9. https://doi.org/10.2212/spr.2010.3.9.

Liu, Y.-B., 2011a. Oxygen enhances phosphine toxicity for postharvest pest control. J. Econ. Entomol. 104, 1455–1461.

Liu, Y.-B., 2011b. Semi-commercial ultralow oxygen treatment for control of western flower thrips, *Frankliniella occidentalis* (Thysanoptera: Thripidae), on harvested iceberg lettuce. Postharvest Biol. Technol. 59, 138–142.

Liu, Y.-B., 2012. Ultralow oxygen treatment for control of western flower thrips, *Frankliniella occidentalis* (Thysanoptera: Thripidae), on harvested table grapes. J. Asia Pac. Entomol. 15, 269–271.

Liu, Y.-B., 2013a. Nitric oxide as a potent fumigant for postharvest pest control. J. Econ. Entomol. 106, 2267–2274.

Liu, Y.-B., 2013b. Controlled atmosphere treatment for control of grape mealybug, *Pseudococcus maritimus* (Ehrhorn) (Hemiptera: Pseudococcidae), on harvested table grapes. Postharvest Biol. Technol. 86, 113–117.

Liu, Y.-B., 2016. Nitric oxide fumigation for control of western flower thrips and its safety to postharvest quality of fresh fruit and vegetables. J. Asia Pac. Entomol. 19, 1191–1195.

Liu, S.S., Liu, Y.-B., 2014. Reducing CO_2 accumulation and its phytotoxicity to lettuce with absorbent in hermetic storage as a simulation of long-term fumigation. J. Asia Pac. Entomol. 17, 435–439.

Liu, Y.-B., Daane, K.M., Tebbets, J.S., Bettiga, L.J., 2008. Ultralow oxygen treatment for control of *Latrodectus hesperus* (Araneae: Theridiidae) on harvested table grapes. J. Econ. Entomol. 101, 1515–1518.

Liu, Y.-B., Bettiga, L.J., Daane, K.M., 2010. Ultralow oxygen treatment for control of *Planococcus ficus* (Hemiptera: Pseudococcidae) on grape benchgrafts. J. Econ. Entomol. 103, 272–276.

Liu, S.S., Liu, Y.-B., Simmons, G., 2014. Oxygenated phosphine fumigation for control of *Epiphyas postvittana* (Lepidoptera: Tortricidae) eggs on lettuce. J. Econ. Entomol. 107, 1370–1376.

Manjunatha, G., Lokesh, V., Neelwarne, B., 2010. Nitric oxide in fruit ripening: trends and opportunities. Biotechnol. Adv. 28, 489–499.

Mitcham, E.J., 2003. Controlled atmospheres for insect and mite control in perishable commodities. Acta Hortic. 600, 137–142.

Mitcham, E.J., Zhou, S., Kader, A.A., 2001. Potential of CA for postharvest insect control in fresh horticultural perishables: an update of summary tables compiled by Ke and Kader 1992. Dept. of Pomology, Univ. of California, Davis, CA.

Neven, L.G., 2005. Combined heat and controlled atmosphere quarantine treatments for control of codling moth in sweet cherries. J. Econ. Entomol. 98, 709–715.

Neven, L.G., Mitcham, E.J., 1996. CATTS: controlled atmosphere temperature treatment system, a novel approach to the development of quarantine treatments. Am. Entomol. 42, 56–59.

Saadatian, M., Ahmadiyan, S., Akbari, M., Balouchi, Z., 2012. Effects of pretreatment with nitric oxide on kiwifruit storage at low temperature. Adv. Environ. Biol. 6, 1902–1908.

Soegiarto, L.R., Wills, B.H., 2004. Short term fumigation with nitric oxide gas in air to extend the postharvest life of broccoli, green bean, and bok choy. HortTechnology 14, 538–540.

Watkins, C.B., Nock, J.F., 2012. Production guide for storage of organic fruits and vegetables, NYS IPM publication no. 10, Cornell Univ. Cooperative Extension, Ithaca, NY.

Chapter 5

CA/MA requirements for spoilage microorganisms and human pathogens

Hidemi Izumi
Faculty of Biology-Oriented Science and Technology, Kindai University, Kinokawa, Japan

Introduction

Commercial use of controlled atmosphere (CA) usually targets O_2 levels of 1%–5% and/or CO_2 levels of 1%–5%. These atmospheres are beneficial for storage of apples and pears (Kupferman, 1997, 2003; Meheriuk, 1993; Richardson and Kupferman, 1997). The reduced O_2 levels reduce ethylene perception and oxidative browning and suppress respiratory and associated metabolic activity, while the elevated CO_2 levels reduce ethylene perception and chlorophyll degradation and suppress decay caused by fungal sporulation and/or growth (Beaudry, 2010). With respect to decay control, the conventional CA and modified atmosphere (MA) with low O_2 and moderate CO_2 maintain the resistance of the host produce to microorganisms by retarding advanced stages of senescence of the host tissue rather than by affecting the metabolism of the microorganisms. Levels of O_2 of 1% or less are required to obtain appreciable reduction of growth in numerous fungal decay organisms, and CO_2 of 10% and greater can effectively slow or stop the growth of many postharvest fungi (El-Goorani and Sommer, 1981). However, the ultralow O_2 atmospheres that reduce microbial growth will induce production of off-flavors resulting from fermentative metabolism of the plant tissues. Several researchers have reported the fungistatic effects of high CO_2 ranging from 10% to the highest level tolerated by strawberries (Chambroy et al., 1993; Ke et al., 1991; Nunes et al., 2002; Wszelaki and Mitcham, 2000), blueberries (Alsmairat et al., 2011; Beaudry et al., 1998; Cantín et al., 2012; Schotsmans et al., 2007), blackberries (Perkins-Veazie and Collins, 2002), raspberries (Agar and Streif, 1996; Haffner et al., 2002), table grapes (Crisosto et al., 2002; Teles et al., 2014), and sweet cherries (De Vries-Paterson and Jones, 1991; Tian et al., 2001, 2004). Thus MA pallet and CA

Controlled and Modified Atmospheres for Fresh and Fresh-Cut Produce
https://doi.org/10.1016/B978-0-12-804599-2.00006-5

container shipments with 10%–20% CO_2 are used commercially on strawberry, blueberry, raspberry, table grapes, and cherry for decay suppression.

Passive or active modified atmosphere packaging (MAP) is applied commercially to fresh and fresh-cut products. The passive MAP can be employed to establish the low O_2 and elevated CO_2 conditions due to the natural interaction between respiratory O_2 consumption and CO_2 evolution and transfer of gasses through the package film. The active MAP can actively attain the predicted O_2 and CO_2 equilibrium concentration by flushing the package with a desired mixture of gasses. The optimum atmosphere for fresh-cut produce is different from that of whole produce (Gorny, 2003) because fresh-cut products typically can tolerate more extreme levels of O_2 and CO_2 due to the decreased distance for gas to diffuse from the external to internal tissue as compared to the whole commodities (Watada and Qi, 1999). Therefore an elevated CO_2 of 10%–15% within active MAP is commercially used for fresh-cut lettuce to slow the browning reaction by inhibiting induction of phenylalanine ammonia-lyase (PAL) that synthesizes enzymatic browning reactive substrates (Mateos et al., 1993). The high CO_2 MAP also can be predicted to retard the growth of microorganisms on fresh-cut lettuce (Barriga et al., 1991; Carrasco et al., 2008).

In addition to high CO_2 levels of >10%, superatmospheric O_2 levels of >50% have been investigated for keeping quality and safety of fresh and fresh-cut products as an alternative to conventional CA/MAP with low O_2 and moderate CO_2 (Artés et al., 2009; Kader and Ben-Yehoshua, 2000). These reviews suggested that exposure to superatmospheric O_2 levels may have either a desirable effect, no effect, or detrimental effect on physiological and microbial quality of produce. Consequently, there are still gaps in our understanding on the use of high O_2 in MAP and further evaluations are needed.

It is crucial to control not only spoilage microorganisms but also human foodborne pathogens (human pathogens) of fresh and fresh-cut products. Human pathogens could be present in fresh and fresh-cut products due to direct or indirect cross-contamination between produce and feces or other pathogen sources along the farm-to-table food chain (Anonymous, 1999; Beuchat, 1996; Beuchat and Ryu, 1997). Since the early 1990s, a significant increase in produce-related outbreaks has been observed worldwide and the pathogens of great concern were bacteria (e.g., *Salmonella*, *Escherichia coli* O157:H7, *Listeria*, and *Shigella*), viruses (e.g., Norovirus and Hepatitis A), and parasites (e.g., *Cryptosporidium* and *Cyclospora*) (Dewaal et al., 2006; Harris et al., 2003; Olaimat and Holley, 2012; Ramos et al., 2013). The latest data in the United States reported that the produce category was responsible for 643 outbreaks (19% of total solved outbreaks) and 20,456 illnesses (24% of total illnesses) between 2004 and 2013 (Fischer et al., 2015). When explaining the composition of a fungicidal atmosphere, the effects of high CO_2 and high O_2 atmospheres on the activity of both spoilage microorganisms and human pathogens must be taken into consideration.

The benefits of CA storage for fresh produce and MAP for fresh-cut produce are similar and so there is possible use of MAP for maintaining quality similar to

CA storage without the investment in infrastructure and instrumentation. Thus the effects of CA and MA are combined for evaluation as described in this chapter. This chapter focuses on the influence of CA/MA on growth of spoilage microorganisms and human pathogens for fresh and fresh-cut produce and suggests specific CA and MAP technologies and strategies to assure microbial quality and safety of the products.

Origin of microflora of produce

Fresh vegetables

Approximately 80% of the total isolates are bacteria in vegetables, and the remainder are yeasts and molds. Bacterial counts on vegetables vary widely among produce types. The degree of microbial contamination is dependent on the production farm type (e.g., organic, semiorganic, and conventional farms) (Mukherjee et al., 2006), the production season (Ailes et al., 2008; Kaneko et al., 1999), the environmental conditions during growing (Morris and Lucotte, 1993), the morphological stage (Bartz and Wei, 2003), and the physiological and physicochemical condition (Zagory, 1999) of the product. Aerobic bacterial counts of 6.9–9.2 log CFU/g were found in sprouts, 5.6–9.5 log CFU/g in lettuce, 6.6–8.3 log CFU/g in celery, 4.7–8.3 log CFU/g in cauliflower, and 3.4–6.9 log CFU/g in broccoli at retail markets in the United States (Thunberg et al., 2002), and 2.9–6.3 log CFU/g in cabbage, 4.4–7.2 log CFU/g in lettuce, 2.2–6.5 log CFU/g in onion, 6.5–7.3 log CFU/g in spinach, 5.2–6.0 log CFU/g in celery, 5.8–6.8 log CFU/g in cucumber, 2.4–6.1 log CFU/g in Japanese radish, and 2.9–6.3 log CFU/g in carrots at food factories in Japan (Kaneko et al., 1999). Kaneko et al. (1999) and Johnston et al. (2005) also showed that the counts of most vegetables were 1–2 logs lower for coliforms than aerobes. We reported that aerobic bacteria and coliform groups ranged from 4.5 to 6.5 log CFU/g and 3.3 to 4.3 log CFU/g, respectively, in cucumber, lettuce, spinach, and carrots, whereas the bacterial count was below the limit of detection level (2.4 log CFU/g) in tomatoes, onions, and garlic (Izumi et al., 2004). Probably, low pH in tomatoes, low water activity in onions due to curing, and pungent volatile components in garlic resulted in the inhibition of bacterial growth. Bacterial count varies within a given vegetable and was generally higher on the exterior tissues than on interior tissues, because the exterior tissues are exposed more to environmental contamination than inner tissues, as noted with cucumber, lettuce, and carrots (Izumi et al., 2004), cabbage (Kaneko et al., 1999), Japanese radish (Kaneko et al., 1999), and endive (Morris and Lucotte, 1993).

The bacterial flora of vegetables comprises primarily phytopathogenic bacteria such as the genera *Agrobacterium* and *Pseudomonas* and Enterobacteriaceae such as the genera *Enterobacter* and *Pantoea* (Watada et al., 2005). The most frequently identified bacteria are *Pantoea agglomerans* and *Pseudomonas fluorescens*, which produce pectinolytic enzymes. The pectinolytic bacteria are suspected to influence textural changes by degrading the cell wall and cause soft rots in produce (Nguyen-the and Carlin, 1994; Ragaert et al., 2007). Other Gram-negative rods such as the genera *Stenotrophomonas* and *Xanthomonas* and Gram-positive bacteria such as the genera *Bacillus* and *Staphylococcus*, and lactic acid bacteria such as the genera *Enterococcus* and *Leuconostoc* are present in various fresh vegetables. The lactic acid bacteria induce spoilage of vegetables through fermentative production of lactic and acetic acid, ethanol, and CO_2 (Ragaert et al., 2007; Watada et al., 2005). These bacteria isolated from vegetables are found frequently in soil, agricultural water, and pesticide solution and do not usually represent a public health concern (Bartz and Wei, 2003).

Fresh fruits

The microbial count of fresh fruits is generally lower than that of fresh vegetables, except for cantaloupe (Johnston et al., 2005) and berries (Mukherjee et al., 2006). These fruits are grown on bare soil and contaminated from soil more than tree fruits, and the surface topography may favor microbial attachment. We reported that the microbial population in several fresh fruits in Japan (e.g., persimmon, satsuma mandarin, lemon, Japanese apricot, and Japanese plum) is below the detection level (2.4 log CFU/g for bacteria and 3.0 log CFU/g for fungi) except for the peel of a few fruits, in which the counts of aerobic bacteria and fungi range from 2.9 to 3.6 log CFU/g and 3.3 to 4.0 log CFU/g, respectively (Izumi et al., 2007). External barriers such as peel and the interior low pH in flesh prevent microorganisms from entering and growing, respectively.

The frequency of bacteria and yeasts is few and molds comprise approximately 80% of the total isolates in fresh fruits, probably due to low pH and high sugar content in the flesh. The mold flora consists of phytopathogenic organisms such as the genera *Alternaria*, *Diaporthe*, *Fusarium*, *Penicillium*, and *Pestalotia* and soilborne organisms such as the genera *Cercophora*, *Cladosporium*, and *Ochroconis* (Izumi et al., 2007, 2008a,b; Murakami et al., 2009, Watanabe et al., 2011). Some pectinolytic molds such as *Mucor* sp. and *Sclerotinia* sp. and yeasts such as *Trichosporon* sp. and *Kluyveromyces* sp. were reported to be responsible for soft rots by production of pectinolytic enzymes, mainly endo-polygalacturonases (Blanco et al., 1999; Ragaert et al., 2007). We found that human pathogens were not detected in any of the fruits sampled in Japan.

Fresh-cut produce

Microbial counts on fresh-cut produce in retail outlets are affected by storage time and temperature and vary widely among produce. Counts of aerobic bacteria and coliform groups ranged from 3.5 to 9.0 log CFU/g and 1.0 to 7.2 log CFU/g, respectively, in raw salad packs (Watada et al., 2005) and from 2.1 to 4.0 log CFU/g and 0.2 to 0.8 log CFU/g, respectively, in cooked salad (Kaneko et al., 1999). Levels of molds and yeasts are usually less than that of total bacteria in fresh-cut vegetables, while molds and yeasts comprise most of total isolates in fresh-cut fruits.

The characteristic microflora of produce during its growth can persist after harvest and are therefore commonly found on fresh-cut produce. These epiphytic bacteria and fungi isolated from fresh-cut products are mostly phytopathogenic and soilborne organisms that are not harmful to humans similar to those found in fresh fruits and vegetables. However, the microflora of fresh-cut produce is not necessarily the same as that of fresh produce. We showed that the diversity of bacterial flora in cucumber was less in the fresh-cut product than in the fresh product in a commercial plant (Watada et al., 2005). The genera *Staphylococcus*, *Pseudomonas*, and *Stenotrophomonas* were present in both fresh and fresh-cut cucumber, while the genera *Flavimonas*, *Brevundimonas*, and *Enterobacter* were isolated only from fresh-cut cucumber. The epiphytic microflora of the fresh cucumber contained bacteria associated with growing and harvest equipment, whereas fresh-cut cucumber contained bacteria transferred from the equipment and machinery used in preparing it. Similar results have been noted with fresh-cut lettuce (Magnuson et al., 1990) and celery (Robbs et al., 1996).

Interaction between spoilage microorganisms and human pathogens

As mentioned previously, native microorganisms of fresh fruits and vegetables are typically spoilage organisms living in a plant–soil environment, which are nonpathogenic for humans. However, human pathogens can be present, often due to fecal contamination from insects, wild and farm animals, soil, polluted water, animal manure, and human carriers throughout production and packaging (Beuchat, 1996; Beuchat and Ryu, 1997). The opportunities for cross-contamination of fresh-cut produce also exist during processing in food plants, because any processing step where equipment, machinery, or humans comes in contact with produce becomes a potential source of contamination (Anonymous, 1999).

Fresh and fresh-cut produce are linked to the greatest number of outbreaks among food categories in the United States in the last 10 years from 2004 to 2013 (Fischer et al., 2015). Among food outbreaks in the United States, 53% were caused by bacteria, 42.5% by viruses, and 4.5% by parasites from about 1100 produce-reported outbreaks (Ramos et al., 2013). The pathogenic bacteria and viruses most frequently linked to produce-related outbreaks include two enteric bacteria, *Salmonella* and *E. coli* O157:H7, and Norovirus, respectively. In addition to these pathogens, psychrotrophic *Listeria monocytogenes* is recognized by the US Food and Drug Administration (FDA) and the US Department of Agriculture (USDA) as an important cause of foodborne illness due to its wide distribution and long survival in the agricultural and plant environments (Anonymous, 2003). Other pathogens associated with produce outbreaks include bacteria such as *Shigella*, *Bacillus*, *Clostridium*, and *Campylobacter*, viruses such as Hepatitis A, and parasites such as *Cryptosporidium* and *Cyclospora* (Harris et al., 2003; Ramos et al., 2013).

Natural microflora present on fresh produce have been shown to compete with human pathogens. For example, native bacteria associated with spinach (Babic et al., 1997) and lettuce and spinach (Johnston et al., 2009) restricted the growth of *L. monocytogenes* and *E. coli* O157:H7, respectively, in coculture experiments. Indigenous bacteria on chicory endive (Bennik et al., 1996) and baby carrots (Liao, 2007) inhibited the growth of *L. monocytogenes* on chicory endive and *L. monocytogenes*, *Yersinia enterocolitica*, *Salmonella enterica*, and *E. coli* O157:H7 on pepper, respectively. Since competitive microorganisms for human pathogens could reside in mixed populations recovered from fresh produce, several in vitro and in vivo studies have been conducted to find specific organisms among the natural microflora for control of pathogens. An in vitro study showed that a strain of *P. fluorescens* (Babic et al., 1997), *Enterobacter agglomerans* (synonymous with *P. agglomerans*) (Bennik et al., 1996), and *Enterobacter cloacae* (Francis and O'Beirne, 1998) had the strong inhibitory effect against *L. monocytogenes* or *Listeria innocua*, a surrogate for *L. monocytogenes*. Liao and Fett (2001) reported that a strain of *P. fluorescens*, *Pseudomonas aeruginosa*, *Bacillus pumilus*, *Bacillus megaterium*, and *Bacillus mojavensis* was capable of inhibiting the growth of *L. monocytogenes* or *Salmonella* Chester both in vitro and in vivo. Within vivo study, Cooley et al. (2006) reported that *Enterobacter asburiae* reduced the *E. coli* O157:H7 survival on lettuce, and Liao (2009) found that a strain of *P. fluorescens* inhibited the growth of *L. monocytogenes*, *Y. enterocolitica*, *S. enterica*, and *E. coli* O157: H7 on bell pepper. *Pseudomonas graminis* was also effective antagonist against *E. coli* O157:H7 on fresh-cut apple and peach (Alegre et al., 2013) and *S. enterica* and *E. coli* O157:H7 on fresh-cut lettuce (Oliveira et al., 2015).

The degree of antagonistic inhibition was dependent on both the initial population of the competitive microorganisms and the initial population of contaminating pathogens (Babic et al., 1997; Francis and O'Beirne, 1998; Liao, 2009, 2007). Although the antagonistic properties of lactic acid bacteria have been reviewed as producing bacteriocin, generating hydrogen peroxide, and producing organic acid and therefore lowering the pH (Nettles and Barefoot, 1993; Nishie et al., 2012), lactic acid bacteria, while naturally present on vegetables, are usually at a low population. The mechanism of competition between other epiphytic bacteria and human pathogens has not been fully explored, but the control of pathogenic bacteria could be affected by the antimicrobial substances produced by antagonists (Cheng et al., 1995; Johnston et al., 2009) and/or the competition for nutrients between epiphytes and pathogens (Cooley et al., 2006; McKellar, 2007).

Some of the native microflora on produce are considered deleterious because of inducing spoilage, but some of them are considered beneficial by leading to the inhibition of growth of contaminating human pathogens on products. Therefore when evaluating the effect of CA/MAP on human pathogens, the presence of effective antagonists in indigenous bacteria of produce should be considered.

Effect of conventional low O_2/moderate CO_2 atmospheres on microbial growth

The conventional CA and MA with low O_2 (1%–5%) and moderate CO_2 (5%–10%) delay ripening or senescence of host produce, thereby maintaining host resistance to microorganisms, but it does not directly affect the metabolism of the microorganisms. Several investigators showed that the conventional atmospheres had little or no inhibitory effect on both spoilage and pathogenic microorganisms. El-Goorani and Sommer (1981) summarized that low O_2/moderate CO_2 had little effect on growth of fruit-rotting fungi such as *Penicillium expansum* on pome fruits; *P. expansum*, *Botrytis cinerea*, and *Monilinia fructicola* on stone fruits; *B. cinerea* and *Rhizopus stolonifer* on berries; and *Penicillium digitatum* and *Penicillium italicum* on citrus fruits. A low O_2 atmosphere of 3% had no effect on growth of aerobic mesophilic and psychrotrophic bacteria on fresh-cut lettuce stored at 5°C and 10°C (Beuchat and Brackett, 1990a); carrots stored at 5°C and 15°C (Beuchat and Brackett, 1990b); and fresh-cut lettuce, cucumber, and carrots stored at 12°C and 21°C (Abdul-Raouf et al., 1993) when compared with aerobic atmosphere. An atmosphere of low O_2 of 3%–8% with moderate CO_2 of 4%–10% had no apparent effect on populations of mesophiles and psychrotrophs on fresh-cut lettuce stored at 5°C and 25°C (Oliveira et al., 2010), mesophiles and lactic acid bacteria on cherry tomatoes stored at 7°C and 22°C (Daş et al., 2006), and lactic acid bacteria and yeasts on strawberries and raspberries stored at 7°C (Van der Steen et al., 2002) as compared to aerobic condition.

With human pathogens, previous studies on *L. monocytogenes* have shown that conventional CA or active MAP containing 2%–5% O_2 and/or 3%–10% CO_2 had no significant effect on growth and survival of the pathogen on lettuce stored at 4°C or 5°C and 10°C (Beuchat and Brackett, 1990a; Kakiomenou et al., 1998); carrots stored at 4°C or 5°C and 15°C (Beuchat and Brackett, 1990a; Kakiomenou et al., 1998); asparagus, broccoli, and cauliflower stored at 4°C and 15°C (Berrang et al., 1989); and strawberry and raspberry stored at 7°C (Siro et al., 2006). Similar phenomena in the CA or active MAP have been reported for *Salmonella* on fresh-cut lettuce stored at 8°C and 20°C (Horev et al., 2012); cherry tomatoes stored at 7°C and 22°C (Daş et al., 2006); *E. coli* O157:H7 on fresh-cut lettuce, cucumber, and carrots stored at 12°C and 21°C (Abdul-Raouf et al., 1993); and *Shigella* on fresh-cut lettuce stored at 7°C and 12°C (Bagamboula et al., 2002). Survival of *Listeria*, *Salmonella*, or *E. coli* O157:H7 have also been found on fresh-cut apples (Alegre et al., 2010a), pears (Alegre et al., 2010b), and melon and pineapple (Abadias et al., 2012), when held at 5°C in passive MAP equilibrated to 5%–20% O_2 and 3%–15% CO_2. The growth of human pathogens may have resulted from inhibition of the native microbial population in the low O_2 and high CO_2 atmospheres, as shown on fresh-cut lettuce (Horev et al., 2012), apples (Alegre et al., 2010a), and raspberry (Siro et al., 2006). An exception to these results was that *Shigella* on fresh-cut carrots and bell peppers died off during active MAP (3% O_2 and 5%–10% CO_2) at 7°C and 12°C due to a rapid growth of lactic acid bacteria, resulting in a pH drop of the produce (Bagamboula et al., 2002).

Effect of innovative high CO_2/high O_2 atmospheres on microbial growth

Antimicrobial action

High CO_2 levels of >10% have been reported in vitro studies to inhibit microbial growth of spoilage microorganisms and human pathogens, with the inhibition being CO_2 level, temperature, and type of microorganism dependent (Amanatidou et al., 1999; Bennik et al., 1995; Eklund and Jarmund, 1983; Kimura et al., 1999; Molin, 1983). CO_2 is much more soluble in plant sap than O_2 and it will diffuse extremely through plant cells (Burton, 1974). In general, high CO_2 results in an extension of lag phase of microbial growth and a decrease in microbial growth rate during the logarithmic phase. The antimicrobial action of high CO_2 involves its dissolution into the aqueous phase of the produce and the microorganism. This is proposed to cause a decrease of the intercellular pH, which inhibits enzymatically catalyzed reactions and enzyme synthesis, interacts with the cell membrane, and changes physicochemical properties of proteins (Farber, 1991; Molin, 2000).

High O_2 levels of >50% or the combination of high O_2 and high CO_2 levels are also expected to inhibit growth of spoilage and pathogenic microorganisms as noted in vitro study (Amanatidou et al., 1999; Geysen et al., 2006; Jacxsens et al., 2001), although results on the sensitivity of microorganisms to superatmospheric O_2 concentration have been conflicting in the literature. Inhibition of microorganisms by high O_2 levels may result from intercellular generation of reactive oxygen species, which damage vital cellular macromolecules when the oxidative stress overwhelms the defense systems of cells that synthesize the enzyme superoxide dismutase to remove these radicals (Kader and Ben-Yehoshua, 2000; Purvis, 1997).

In vitro studies may enable prediction of the probability of microbial growth and survival in specific produce; however, the competitive relationship among the microbial complex on produce and organoleptic deterioration of produce is generally ignored in these experiments. Therefore in vivo studies involved in the interaction of various microorganisms under high CO_2 and/or high O_2 atmospheres and quality responses of produce to elevated CO_2 and/or O_2 levels need to be conducted to determine the optimum ranges of O_2 and CO_2 acceptable in CA and MAP. The following sections indicate the influence of a high CO_2/high O_2 atmosphere on microbial growth on fresh and fresh-cut produce.

Spoilage microorganisms

In terms of atmospheric effects on postharvest diseases, the high CO_2 levels directly suppressed decay primarily caused by *B. cinerea* on berries (Cantín et al., 2012; Chambroy et al., 1993; Crisosto et al., 2002; Teles et al., 2014; Wszelaki and Mitcham, 2000) and *M. fructicola* on cherries (De Vries-Paterson and Jones, 1991). Most of these studies indicated that the combination of high CO_2 levels from 8% to 30% with low O_2 levels from 1% to 10% was effective in decay control without adversely affecting quality. The effect of atmosphere was the result of the high CO_2 level rather than low O_2 level.

Growth of spoilage microorganisms in high CO_2 atmospheres of >10% varies depending on type of microorganism and produce, when compared to growth in air. Table 1 includes a summary of the literature on the influence of a high CO_2 atmosphere on growth of spoilage microorganisms on produce stored at a low temperature, except for mango at 13°C to avoid injury caused by chilling. High CO_2 of >10%, either alone or in combination with low O_2 of 4%–10% delayed or inhibited growth of aerobes such as mesophilic and psychrotrophic bacteria on several fresh-cut vegetables (Allende et al., 2004; Carlin et al., 1996; Izumi et al., 2016b) and fruits (Aguayo et al., 2007; Budu et al., 2007; Izumi et al., 2015, 2016b; Martínez-Ferrer et al., 2002; Murakami et al., 2012; Poubol and Izumi, 2005a) when compared with an aerobic atmosphere. In comparison, growth of facultative anaerobes such as coliforms, Enterobacteriaceae, and lactic acid bacteria was delayed on fresh-cut spinach and Japanese radish (Allende et al., 2004; Izumi et al., 2016b) or stimulated on fresh-cut carrots, cabbage, and cucumber (Carlin et al., 1990; Izumi et al., 2016b) relative to an aerobic condition. Stimulated growth of lactic acid bacteria was particularly noted under a high CO_2 atmosphere. Previous studies have shown that increases in CO_2 concentration of >20% and 40% induced proliferation of lactic acid bacteria and extensive spoilage in fresh-cut cabbage (Izumi, 2005) and carrots (Carlin et al., 1990), respectively. In contrast, a CA of >15% CO_2 inhibited coliforms and lactic acid bacteria on fresh-cut persimmons, because the predominant microflora of persimmon fruit is yeasts and molds due to the low pH and high sugar content of the fruit, which may discourage high bacterial populations from being established (Izumi et al., 2015; Murakami et al., 2012). The high CO_2 of >10% delayed or inhibited growth of fungi on several fresh-cut fruits (Aguayo et al., 2007; Budu et al., 2007; Izumi et al., 2015; Martínez-Ferrer et al., 2002; Murakami et al., 2012) and stimulated the growth of yeasts on fresh-cut carrots as compared to growth in air (Carlin et al., 1990).

The effect of high O_2 atmospheres varies among microorganisms, products, and the CO_2 concentrations in the atmosphere. Table 2 summarizes the reported influence of a high O_2 atmosphere on spoilage microorganisms on produce held at the chilling temperature except for mango at 13°C. In most products, growth of aerobic bacteria including mesophiles and psychrotrophs and facultative anaerobic bacteria such as Enterobacteriaceae and lactic acid bacteria on

TABLE 1 Influence of a high CO_2 atmosphere on growth of spoilage microorganisms on produce

Target microorganisms	Produce	Storage style	Storage temp. (°C)	Storage atmosphere		Microbial growth relative to aerobic atmosphere	References
				O_2 (%)	*CO_2 (%)*		
Mesophiles	Cut endive	Active MAP	3	10	30	Delayed	Carlin et al. (1996)
				10	50	Delayed	
			10	10	30	Delayed	
				10	50	Delayed	
	Shredded Japanese radish	CA	10	21	15	Delayed	Izumi et al. (2016b)
	Baby spinach	Passive MAP	5	21–0	0–15	Delayed	Allende et al. (2004)
	Sliced melon	CA	5	4	15	Delayed	Aguayo et al. (2007)
				21	15	Delayed	
	Mango cubes	CA	5	21	10	Inhibited	Poubol and Izumi (2005a)
			13	21	10	Inhibited	
	Mango cubes	Active MAP	5	4	10–18	Delayed	Martínez-Ferrer et al. (2002)
	Pineapple cubes	Active MAP	5	4	10	Delayed	Martínez-Ferrer et al. (2002)
	Sliced pineapple	CA	4.5	5	15	Delayed	Budu et al. (2007)
	Persimmon slices	CA	10	21	20	Delayed	Murakami et al. (2012)
	Baby persimmons	CA	10	21	15	Delayed	Izumi et al. (2015)
				21	20	Delayed	

Psychrotrophs	Sliced melon	CA	5	4	15	Delayed	Aguayo et al. (2007)
				21	15	Delayed	
Coliforms	Shredded cabbage	CA	10	21	20	Stimulated	Izumi et al. (2016b)
	Shredded Japanese radish	CA	10	21	15	Delayed	Izumi et al. (2016b)
	Trimmed spinach leaves	CA	10	21	15	Delayed	Izumi et al. (2016b)
	Persimmon slices	CA	10	21	20	Delayed	Murakami et al. (2012)
	Baby persimmons	CA	10	21	15	Inhibited	Izumi et al. (2015)
				21	20	Inhibited	
Enterobacteriaceae	Baby spinach	Passive MAP	5	21–0	0–15	Delayed	Allende et al. (2004)
Lactic acid bacteria	Grated carrots	CA	10	2	25	Stimulated	Carlin et al. (1990)
				2	40	Stimulated	
				10	25	Stimulated	
				10	40	Stimulated	
	Shredded carrots	CA	10	21	15	Stimulated	Izumi et al. (2016b)
	Sliced cucumber	CA	10	21	20	Stimulated	Izumi et al. (2016b)
	Baby persimmons	CA	10	21	15	Inhibited	Izumi et al. (2015)
				21	20	Inhibited	
Yeasts	Grated carrots	CA	10	2	25	Stimulated	Carlin et al. (1990)

Continued

TABLE 1 Influence of a high CO_2 atmosphere on growth of spoilage microorganisms on produce—cont'd

Target microorganisms	Produce	Storage style	Storage temp. (°C)	Storage atmosphere		Microbial growth relative to aerobic atmosphere	References
				O_2 (%)	CO_2 (%)		
				2	40	Stimulated	
				10	25	Stimulated	
				10	40	Stimulated	
	Baby spinach	Passive MAP	5	21–0	0–15	Delayed	Allende et al. (2004)
Mold	Blueberry	CA	1	3	12	Delayed	Cantín et al. (2012)
				3	24	Delayed	
Fungi	Sliced melon	CA	5	4	15	Inhibited	Aguayo et al. (2007)
				21	15	Inhibited	
	Mango cubes	Active MAP	5	4	10–18	Delayed	Martínez-Ferrer et al. (2002)
	Pineapple cubes	Active MAP	5	4	10	Delayed	Martínez-Ferrer et al. (2002)
	Sliced pineapple	CA	4.5	5	15	Inhibited	Budu et al. (2007)
	Persimmon slices	CA	10	21	20	Delayed	Murakami et al. (2012)
	Baby persimmons	CA	10	21	15	Inhibited	Izumi et al. (2015)
				21	20	Inhibited	

TABLE 2 Influence of a high O_2 atmosphere on growth of spoilage microorganisms on produce

Target microorganisms	Produce	Storage style	Storage temp. (°C)	Storage atmosphere		Microbial growth relative to aerobic atmosphere	References
				O_2 (%)	CO_2 (%)		
Mesophiles	Bell pepper cubes	CA	5	100	0	Unaffected	Conesa et al. (2007)
				80	15	Delayed	
				50	15	Delayed	
	Sliced carrots	CA	8	50	30	Unaffected	Amanatidou et al. (2000)
				90	10	Unaffected	
	Cut lettuce	CA	7	75	0	Delayed	Escalona et al. (2007)
				75	15	Delayed	
	Baby spinach	Active MAP	5	100–80	0–18	Delayed	Allende et al. (2004)
	Mango cubes	CA	13	60	0	Stimulated	Poubol and Izumi (2005b)
	Mango cubes	Active MAP	5	60–15	0–10	Delayed	Martínez-Ferrer et al. (2002)
	Pineapple cubes	Active MAP	5	60–15	0–65	Delayed	Martínez-Ferrer et al. (2002)
	Sliced pineapple	CA	4.5	80	15	Delayed	Budu et al. (2007)
Psychrotrophs	Bell pepper cubes	CA	5	100	0	Unaffected	Conesa et al. (2007)
				80	15	Delayed	
				50	15	Delayed	

Continued

TABLE 2 Influence of a high O_2 atmosphere on growth of spoilage microorganisms on produce—cont'd

Target microorganisms	Produce	Storage style	Storage temp. (°C)	Storage atmosphere		Microbial growth relative to aerobic atmosphere	References
				O_2 (%)	CO_2 (%)		
Enterobacteriaceae	Bell pepper cubes	CA	5	100	0	Unaffected	Conesa et al. (2007)
				80	15	Delayed	
				50	15	Delayed	
	Sliced carrots	CA	8	50	30	Delayed	Amanatidou et al. (2000)
				90	10	Stimulated	
	Baby spinach	Active MAP	5	100–80	0–18	Unaffected	Allende et al. (2004)
Lactic acid bacteria	Sliced carrots	CA	8	50	30	Unaffected	Amanatidou et al. (2000)
				90	10	Unaffected	
	Cut lettuce	CA	7	75	0	Unaffected	Escalona et al. (2007)
				75	15	Unaffected	
Yeasts	Bell pepper cubes	CA	5	100	0	Unaffected	Conesa et al. (2007)
				80	15	Delayed	
				50	15	Delayed	
	Grated celeriac	Active MAP	4	95–20	0–46	Delayed	Jacxsens et al. (2001)
	Shredded chicory endive	Active MAP	4	95–12	0–32	Delayed	Jacxsens et al. (2001)

	Cut lettuce	CA	7	75	0	Unaffected	Escalona et al. (2007)
				75	15	Unaffected	
	Baby spinach	Active MAP	5	100–80	0–18	Unaffected	Allende et al. (2004)
	Mushroom slices	Active MAP	4	95–3	0–76	Unaffected	Jacxsens et al. (2001)
	Mango cubes	CA	13	60	0	Stimulated	Poubol and Izumi (2005b)
	Raspberry	Active MAP	7	95–14	0–73	Unaffected	Van der Steen et al. (2002)
	Strawberry	Active MAP	7	95–12	0–75	Unaffected	Van der Steen et al. (2002)
Fungi	Mango cubes	Active MAP	5	60–15	0–10	Delayed	Martínez-Ferrer et al. (2002)
	Pineapple cubes	Active MAP	5	60–15	0–65	Delayed	Martínez-Ferrer et al. (2002)
	Sliced pineapple	CA	4.5	80	15	Inhibited	Budu et al. (2007)

fresh-cut produce showed a similar pattern in both high O_2 atmospheres of 50%–100% and air atmosphere (Allende et al., 2004; Amanatidou et al., 2000; Conesa et al., 2007; Escalona et al., 2007). Retardation of microbial growth on fresh-cut bell peppers was observed only when high O_2 was combined with high CO_2 (Conesa et al., 2007). A concern is that 60% and 90% O_2 stimulated bacterial growth on mango cubes (Poubol and Izumi, 2005b) and sliced carrots (Amanatidou et al., 2000), respectively. With fungi, growth was unaffected or delayed on most fresh-cut produce by addition of high CO_2 with high O_2 (Allende et al., 2004; Escalona et al., 2007; Jacxsens et al., 2001; Martínez-Ferrer et al., 2002; Van der Steen et al., 2002), although yeasts on mango cubes proliferated at 60% O_2 (Poubol and Izumi, 2005b). It has also been reported that a high O_2 level in MAP caused an overall poor sensory quality of shredded lettuce (Heimdal et al., 1995), an increase in the respiration rate of mushroom slices (Jacxsens et al., 2001), an off-odor and a loss of tissue integrity of spinach leaves (Allende et al., 2004), and browning of mango cubes (Poubol and Izumi, 2005b). Therefore the physiology and quality responses of each produce to elevated O_2 need to be investigated to use high O_2 MAP.

Human pathogens

Because human pathogens are not present naturally on produce unless pathogenic contamination occurs, in research settings the pathogens are artificially inoculated on produce to determine the effect of CA/MAP on human pathogens. Results of several studies assessing the growth of inoculated pathogens in high CO_2 atmospheres as compared to the initial population on several products stored at various temperatures ranging from 0°C to 25°C are described in Table 3. The high CO_2 condition of >10% combined with low O_2 levels of 0%–13% mostly stimulated the growth of *Salmonella* (Alegre et al., 2010b; Oliveira et al., 2010; Sant'Ana et al., 2013), *E. coli* O157:H7 (Abadias et al., 2012; Alegre et al., 2010b; Brown et al., 2011; Oliveira et al., 2010), *Listeria* (Alegre et al., 2010b; Bennik et al., 1996; Carlin et al., 1996; Kallander et al., 1991; Oliveira et al., 2010; Sant'Ana et al., 2013), and *Shigella* (Satchell et al., 1990) on fresh-cut fruits and vegetables. The presence of epiphytic microorganisms including effective antagonists should be taken into consideration when evaluating the effect of fungicidal atmosphere on human pathogens. Horev et al. (2012) reported that the active MAP containing 10% O_2 and 10% CO_2 had an antimicrobial effect on indigenous lettuce microflora, but not on inoculated *Salmonella*, and it even favored the survival of the *Salmonella*. Alegre et al. (2010a) found similar results for sliced apples inoculated with *Salmonella*, *E. coli* O157:H7, and *L. innocua*. Carlin et al. (1996) found that increases in CO_2 concentrations in active MAP reduced the growth of the aerobic microflora, while *L. monocytogenes* grew better as the CO_2 concentration was increased. This result was supported by Bennik et al. (1996), who demonstrated that *L. monocytogenes* grew better on disinfected chicory endive leaves with

TABLE 3 Influence of a high CO_2 atmosphere on growth of human pathogenic bacteria on produce

Target pathogen	Produce	Storage style	Storage temp. (°C)	Storage atmosphere		Microbial growth relative to initial population	References
				O_2 (%)	CO_2 (%)		
Salmonella	Cut collard greens	Active MAP	7	5	15	Stimulated	Sant'Ana et al. (2013)
			15	5	15	Stimulated	
	Cut lettuce	Active MAP	7	5	15	Stimulated	Sant'Ana et al. (2013)
			15	5	15	Stimulated	
	Shredded lettuce	Passive MAP	25	1–10	11–13	Stimulated	Oliveira et al. (2010)
	Sliced peach	Passive MAP	25	11–13	15–25	Stimulated	Alegre et al. (2010b)
Escherichia coli O157:H7	Chopped endive	Passive MAP	25	0–5	10–17	Slightly stimulated	Abadias et al. (2012)
	Grated carrot	Passive MAP	25	0–5	20–35	Stimulated	Abadias et al. (2012)
	Shredded lettuce	Passive MAP	25	1–10	11–13	Stimulated	Oliveira et al. (2010)
	Baby spinach	Active MAP	4–7	0	20	Stimulated	Brown et al. (2011)
	Sliced apple	CA	15	1	15	Slightly inhibited	Gunes and Hotchkiss (2002)

Continued

TABLE 3 Influence of a high CO_2 atmosphere on growth of human pathogenic bacteria on produce—cont'd

Target pathogen	Produce	Storage style	Storage temp. (°C)	Storage atmosphere		Microbial growth relative to initial population	References
				O_2 (%)	CO_2 (%)		
				1	30	Slightly inhibited	
			20	21	30	Unaffected	
	Cut melon	Passive MAP	25	0–5	10–40	Stimulated	Abadias et al. (2012)
	Cut pineapple	Passive MAP	25	0–5	10–50	Inhibited	Abadias et al. (2012)
	Sliced peach	Passive MAP	25	11–13	15–25	Stimulated	Alegre et al. (2010b)
Listeria monocytogenes	Shredded cabbage	Active MAP	5	0	70	Slightly stimulated	Kallander et al. (1991)
			25	0	70	Stimulated followed by inhibited	
	Cut collard greens	Active MAP	7	5	15	Stimulated	Sant'Ana et al. (2013)
			15	5	15	Stimulated	
	Cut chicory endive	Active MAP	8	0	20	Stimulated	Bennik et al. (1996)
				1.5	20	Stimulated	
	Cut endive	Active MAP	3	10	30	Slightly stimulated	Carlin et al. (1996)

				10	50	Slightly stimulated	
			10	10	30	Stimulated	
				10	50	Stimulated	
	Cut lettuce	Active MAP	7	5	15	Stimulated	Sant'Ana et al. (2013)
			15	5	15	Stimulated	
	Shredded lettuce	Passive MAP	25	1–10	11–13	Stimulated	Oliveira et al. (2010)
Listeria innocua	Shredded carrot and cabbage	Passive MAP	7	0–2	15–25	Slightly inhibited	Finn and Upton (1997)
	Sliced peach	Passive MAP	25	11–13	15–25	Stimulated	Alegre et al. (2010b)
Shigella sonnei	Shredded cabbage	Active MAP	0–6	0	70	Unaffected	Satchell et al. (1990)
			24	0	70	Stimulated followed by inhibited	
Clostridium perfringens	Shredded carrot and cabbage	Passive MAP	7	0–2	15–25	Slightly inhibited	Finn and Upton (1997)
Clostridium sporogenes	Baby spinach	Active MAP	4–7	0	20	Unaffected	Brown et al. (2011)
Bacillus cereus	Shredded carrot and cabbage	Passive MAP	7	0–2	15–25	Inhibited	Finn and Upton (1997)
Staphylococcus aureus	Shredded carrot and cabbage	Passive MAP	7	0–2	15–25	Slightly inhibited	Finn and Upton (1997)

reduced numbers of epiphytic bacteria than on nondisinfected or water-rinsed produce during active MAP storage. These findings indicate that the effective gas conditions for spoilage bacteria may provide growth opportunities for pathogenic bacteria. On the contrary, a few reports showed that the high CO_2 atmosphere was effective in inhibiting growth of *E. coli* O157:H7 on cut pineapple (Abadias et al., 2012) and sliced apple (Gunes and Hotchkiss, 2002), and inhibiting *L. innocua*, *Clostridium perfringens*, *Bacillus cereus*, and *Staphylococcus aureus* on shredded carrot and cabbage (Finn and Upton, 1997). The inhibition of growth may be due to a rapid growth of antagonists such as lactic acid bacteria (Bagamboula et al., 2002), a high CO_2 sensitivity of bacteria such as *B. cereus* (Finn and Upton, 1997), and the unfavorable chemical profile of the produce, including its pH, acid composition, and the presence of unfermentable fibers as in pineapple (Abadias et al., 2012). Because an abusive storage temperature could result in an undesirable and hazardous rise of the pathogenic population, fresh-cut produce in MAP must be refrigerated.

Studies using high O_2 atmospheres to control human pathogens in produce are listed in Table 4. The high O_2 condition with or without CO_2 either inhibited the growth of *Salmonella* (Siro et al., 2006), stimulated the growth of *E. coli* O157:H7 (Brown et al., 2011), *L. innocua* (Escalona et al., 2007), and *Aeromonas caviae* (Allende et al., 2002), or had no effect against *L. monocytogenes* (Allende et al., 2002; Siro et al., 2006) and *Clostridium sporogenes* (Brown et al., 2011) on fresh-cut produce. This erratic effect of high O_2 was similar to that for spoilage microorganisms. Siro et al. (2006) indicated that increased yeast counts by high O_2 exposure may have induced inactivation of *Salmonella* on raspberry, while Allende et al. (2002) reported that growth of both yeast and *A. caviae* seemed to be stimulated by high O_2. The use of high O_2 MAP to control microbial quality and safety is still in early stages and needs to be supported by more research.

Obligate anaerobes such as *Clostridium* grow only in anaerobic environments, but inoculated cut onion, butternut squash (Austin et al., 1998), and shredded cabbage (Solomon et al., 1990) in a passive MAP stored at 22°C–25°C supported the outgrowth of and toxin production by *Clostridium botulinum* before being unacceptable organoleptically. These results also suggest that studies for toxigenicity of pathogens should be conducted when considering whether detectable spoilage precedes or follows toxigenesis. Since an active MAP with either high CO_2 or O_2 did not affect growth of *C. sporogenes* (Brown et al., 2011), MAP of fresh cuts must be kept refrigerated at all times to prevent botulism. On the other hand, data for foodborne viruses such as Norovirus and Hepatitis A and parasites such as *Cryptosporidium* and *Cyclospora* on fresh produce held in CA/MAP are lacking. Although the survival rate of Hepatitis A virus on cut lettuce was reported during high CO_2 active MAP (Bidawid et al., 2001) and passive MAP (Fraisse et al., 2011), further research is needed to address virus and parasites in CA/MAP environments as well as human pathogenic bacteria.

TABLE 4 Influence of a high O_2 atmosphere on growth of human pathogenic bacteria on produce

Target pathogen	Produce	Storage style	Storage temp. (°C)	Storage atmosphere		Microbial growth relative to initial population	References
				O_2 (%)	CO_2 (%)		
Salmonella	Raspberry	Active MAP	7	95–10	0–5	Inhibited	Siro et al. (2006)
E. coli O157:H7	Baby spinach	Active MAP	4–7	80	20	Slightly stimulated	Brown et al. (2011)
L. monocytogenes	Mixed vegetable salad	Active MAP	4	95–60	0–20	Unaffected	Allende et al. (2002)
	Raspberry	Active MAP	7	95–10	0–5	Unaffected	Siro et al. (2006)
	Strawberry	Active MAP	7	95–10	0–8	Unaffected	Siro et al. (2006)
L. innocua	Cut lettuce	CA	7	75	0	Stimulated	Escalona et al. (2007)
			7	75	15	Stimulated	
C. sporogenes	Baby spinach	Active MAP	4–7	80	20	Unaffected	Brown et al. (2011)
Aeromonas caviae	Mixed vegetable salad	Active MAP	4	95–60	0–20	Stimulated	Allende et al. (2002)

Effect of combined antimicrobial and CA/MAP treatments on microbial quality and safety

The atmosphere of high CO_2 and/or high O_2 can possess a bacteriostatic effect to reduce the rate of microbial growth, but it does not eradicate the microorganisms, even when combined with refrigeration (Artés et al., 2009; Izumi et al., 2016b; Kader and Ben-Yehoshua, 2000; Ragaert et al., 2007; Zagory, 1999). Therefore combining CA/MAP with antimicrobial treatments to reduce the initial level of microorganisms and/or the microbiological activity during storage is being done to preserve the microbiological quality and wholesomeness of the products. These techniques include chemical treatments such as disinfectants and physical treatments such as irradiation (Izumi et al., 2016b; Nguyen-the and Carlin, 1994; Olaimat and Holley, 2012; Ramos et al., 2013; Watada et al., 2005; Zagory, 1999).

For example, the combination of commercial CA and fungicide treatments such as metalaxyl/carbendazim (Colgan and Johnson, 1998), fludioxonil (Errampalli et al., 2007), and sodium bicarbonate (Janisiewicz et al., 2008) or heat treatment (Janisiewicz et al., 2003) is recommended for controlling blue mold caused by *P. expansum*, brown rot caused by *Metschnikowia pulcherrima*, or other fungal rots caused by *B. cinerea* and *Nectria galligena* in apples and gray mold caused by *B. cinerea* and blue mold caused by *P. expansum* in pears. With active MAP following disinfectant treatment, Beltrán et al. (2005a) reported that shredded lettuce treated with ozonated water (10 ppm and 20 ppm ozone) and stored under MAP with an initial 4% O_2 and 12% CO_2 at 4°C was more effective in inhibiting the growth of native mesophilic and coliform bacteria without detrimental quality changes compared to when the disinfectant and MAP were applied individually. Similarly, Waghmare and Annapure (2013) reported that the effect of active MAP of 5% O_2 and 10% CO_2 on inhibiting the growth of mesophiles and fungi of diced papaya was increased while retaining sensory and quality characteristics when used in combination with a dipping treatment with a solution of 1% calcium chloride and 2% citric acid. Similar results on controlling spoilage bacteria were obtained for passive MAP at 4°C following washing treatment by ozone (20 ppm) and peroxyacetic acid (300 ppm) solution on potato strips (Beltrán et al. 2005b) and prewashing uncut carrots with sodium hypochlorite (200 ppm available chlorine) solution (Klaiber et al., 2004). Chun and Song (2013) found that the combination of sanitizer mixture (100 ppm aqueous chlorine dioxide and 0.3% fumaric acid) and 2 $kJ\,m^{-2}$ ultraviolet irradiation and active MAP with initial 95% CO_2 reduced the initial counts of preexisting spoilage microorganisms (mesophiles, coliforms, and fungi) and the initial inoculated counts of human pathogens (*E. coli* O157:H7 and *Salmonella typhimurium*) in buckwheat sprouts stored at 4°C. This finding is in agreement with the result of Jin and Lee (2007) on the inhibitory effect of 100 ppm chlorine dioxide followed by active MAP of 100% CO_2 on the growth of mesophiles,

S. typhimurium, and *L. monocytogenes* in mungbean sprouts at 5°C and that of Lee and Beak (2008) on the effect of a solution of either 100ppm chlorine dioxide or 100ppm sodium hypochlorite with active MAP of 100% CO_2 on the growth of *E. coli* O157:H7 in spinach at 7°C.

Besides the aqueous chemicals, the effectiveness of antimicrobials in the vapor phase has been investigated for controlling microbial growth on fresh produce in packages. Guillén et al. (2007) indicated the addition of essential oils containing eugenol, thymol, and carvacrol (25μL each) in passive MAP enhanced the beneficial effects in reducing the growth of mesophiles and fungi and improving the physicochemical quality of table grapes stored at 1°C as compared to air condition. In a study on aroma compounds by Lanciotti et al. (2003), the combination of 20ppm (E)-2-hexenal and active MAP of 20% CO_2 endowed with the highest inhibitory effect against *Salmonella enteritidis* and *E. coli* inoculated in sliced apples stored at 20°C when compared to other combinations of compounds and atmospheres. Allyl isothiocyanate, a pungent component of black mustard and Japanese horse radish (Sekiyama et al., 1996), also inactivated *S. Montevideo* or *E. coli* O157:H7 on lettuce at 4°C, tomato at 13°C, apple at 4°C (Lin et al., 2000), and sliced and whole tomato at 4°C and 10°C (Obaidat and Frank, 2009) in sealed containers, but not in film packages. The study of allyl isothiocyanate used in MAP is anticipated for commercial application of the method.

Ionizing radiation such as gamma rays (Rezende et al., 2014) and X-rays (Mahmoud, 2010) is recognized as a physical intervention that has a strong inactivation effect on human pathogens such as *Salmonella*, *E. coli* O157:H7, and *L. monocytogenes* inoculated on fresh produce without causing significant changes in the quality. Studies have shown that 0.15–2.0kGy gamma irradiation reduced the inherent microflora counts on chopped lettuce (Hagenmaier and Baker, 1997; Prakash et al., 2000) and shredded carrots (Chervin and Boisseau, 1994; Hagenmaier and Baker, 1998), and kept them lower than non-irradiated controls during MA storage at 2°C–10°C. A similar result was obtained in a similar study on shredded lettuce irradiated with 0.1kGy and 2.0kGy X-ray and stored in a plastic clamshell container wrapped in polyvinyl chloride (PVC) film at 4°C (Mahmoud, 2010). These studies indicated that irradiation was beneficial in reducing the initial microbial load of produce and the residual effect persisted during refrigerated storage although the microbial population gradually increased with storage time. Thus atmosphere composition in the MAP would be important to interact with irradiation and MAP synergistically for controlling microorganisms. Ahn et al. (2005) found that gamma irradiation at 1.0–2.0kGy reduced the mesophiles, coliforms, and lactic acid bacteria of cut salted Chinese cabbage, and, additionally, active MAP of 25% or 100% CO_2 enhanced the reduction of the spoilage microorganisms other than lactic acid bacteria during storage at 4°C without a significant loss of quality when compared to passive MAP flushed with air. Generally, the higher levels of lactic acid bacteria in a higher CO_2 condition would be rather favorable for the salted Chinese cabbage destined to process into the fermented product *Kimuchi*. Niemira et al. (2005) also demonstrated that gamma irradiation at 0.3–0.6 kGy and active MAP (5% O_2 and 5% CO_2 or 10% O_2 and 10% CO_2) prevented the regrowth of inherent mesophiles and inoculated *L. monocytogenes* on cut endive during storage at 4°C. These results suggest that the combined treatment of irradiation and active MAP is useful for inhibiting the growth of spoilage and pathogenic bacteria on fresh-cut produce, thereby improving product quality and safety.

Conclusions and future trends

The conventional CA/MAP with low O_2 (1%–5%) and moderate CO_2 (5%–10%) reduces microbial growth for spoilage due to delay of produce senescence and maintenance of resistance to microorganisms. The atmosphere needed to directly affect the metabolism of the microorganisms is practically high CO_2 levels of >10% and high O_2 levels of >50%, although both positive and negative effects of high O_2 on microbiological quality of fresh and fresh-cut produce have been reported. The high CO_2 atmospheres of >10% are helpful in inhibiting growth of spoilage bacteria mainly due to a decrease of intercellular pH by dissolved CO_2 in the aqueous phase. The exception to this is the induction of facultative anaerobic bacteria such as lactic acid bacteria, which favor growth in high CO_2 atmospheres. The gas conditions employed to inhibit spoilage microorganisms also provide growth opportunities for human pathogens. This fact indicates that the MAP containing moderate to high levels of CO_2 inhibits organisms that might warn consumers of spoilage and may allow or promote the growth of pathogens. Based on technologies described in this chapter, I can propose the following three factors that should be taken into account for successful MAP to restrict human pathogens: (1) MAP produce must be refrigerated to prevent hazardous rise of the pathogenic population, (2) optimal atmospheric composition should be selected for growth of effective bacterial antagonists in natural microflora that compete with human pathogens, (3) combining CA/MAP with complementary technologies such as chemical and physical treatments contributes to a decisive effect in controlling the pathogens by the synergistic interaction.

A remaining challenge for ensuring the highest microbial safety is the presence of sublethally injured pathogens that are not normally detected in selective media, but can be resuscitated in a suitable environment such as nonselective media. Sublethally injured pathogens may either retain virulence or exhibit enhanced virulence after recovering injury (Wesche et al., 2009). It is well recognized that bacteria can be injured sublethally as a result of exposure to stress including heating, freezing, drying, radiation, starvation, and chemicals such as acids, sanitizers, and preservation (Jay, 1992; Ray, 1979; Wesche et al., 2009). We found sanitizer-injured coliform bacteria in the production and harvest environment of vegetables (Inoue et al., 2017; Izumi et al., 2016a). Since the injury was caused in agricultural water mixed with chlorine or fungicide and on harvest equipment sprayed with an alcohol agent, the sanitizer-injured bacteria may transfer from the environment to produce in the field. Little work has been done on effects of atmosphere composition on survival of injured bacteria on produce during storage, except that we found that chlorine-injured coliforms and *E. coli* O157:H7 on fresh-cut cabbage were capable of exhibiting different degrees of injury during storage in a high CO_2 CA and MAP (Izumi and Inoue, 2018a,b). More research on behavior of sublethally injured bacteria on produce during CA/MA storage is needed, because the rejuvenation of injured bacteria can increase the risk of underestimating the microbiological quality and safety of products.

References

Abadias, M., Alegre, I., Oliveira, M., Altisent, R., Viñas, I., 2012. Growth potential of *Escherichia coli* O157:H7 on fresh-cut fruits (melon and pineapple) and vegetables (carrot and escarole) stored under different conditions. Food Control 27, 37–44.

Abdul-Raouf, U.M., Beuchat, L.R., Ammar, M.S., 1993. Survival and growth of *Escherichia coli* O157:H7 on salad vegetables. Appl. Environ. Microbiol. 59, 1999–2006.

Agar, I.T., Streif, J., 1996. Effect of high CO_2 and controlled atmosphere (CA) storage on the fruit quality of raspberry. Gartenbauwissenschaft 61, 261–267.

Aguayo, E., Escalona, V.H., Artés, F., 2007. Quality of minimally processed *Cucumis melo* var. *saccharinus* as improved by controlled atmosphere. Europ. J. Hort. Sci. 72, 39–45.

Ahn, H.-J., Kim, J.-H., Kim, J.-K., Kim, D.-H., Yook, H.-S., Byun, M.-W., 2005. Combined effects of irradiation and modified atmosphere packaging on minimally processed Chinese cabbage (*Brassica rapa* L.). Food Chem. 89, 589–597.

Ailes, E.C., Leon, J.S., Jaykus, L.-A., Johnston, L.M., Clayton, H.A., Blanding, S., Kleinbaum, D.G., Backer, L.C., Moe, C.L., 2008. Microbial concentrations on fresh produce are affected by postharvest processing, importation, and season. J. Food Prot. 71, 2389–2397.

Alegre, I., Abadias, M., Anguera, M., Oliveira, M., Viñas, I., 2010a. Factors affecting growth of foodborne pathogens on minimally processed apples. Food Microbiol. 27, 70–76.

Alegre, I., Abadias, M., Anguera, M., Usall, J., Viñas, I., 2010b. Fate of *Escherichia coli* O157:H7, *Salmonella* and *Listeria innocua* on minimally-processed peaches under different storage conditions. Food Microbiol. 27, 862–868.

Alegre, I., Viñas, I., Usall, J., Teixidó, N., Figge, M.J., Abadias, M., 2013. Control of foodborne pathogens on fresh-cut fruit by a novel strain of *Pseudomonas graminis*. Food Microbiol. 34, 390–399.

Allende, A., Jacxsens, L., Devlieghere, F., Debevere, J., Artés, F., 2002. Effect of superatmospheric oxygen packaging on sensorial quality, spoilage, and *Listeria monocytogenes* and *Aeromonas caviae* growth in fresh processed mixed salads. J. Food Prot. 65, 1565–1573.

Allende, A., Luo, Y., McEvoy, J.L., Artés, F., Wang, C.Y., 2004. Microbial and quality changes in minimally processed baby spinach leaves stored under super atmospheric oxygen and modified atmosphere conditions. Postharvest Biol. Technol. 33, 51–59.

Alsmairat, N., Contreras, C., Hancock, J., Callow, P., Beaudry, R., 2011. Use of combinations of commercially relevant O_2 and CO_2 partial pressures to evaluate the sensitivity of nine highbush blueberry fruit cultivars to controlled atmospheres. Hortscience 46, 74–79.

Amanatidou, A., Smid, E.J., Gorris, L.G.M., 1999. Effect of elevated oxygen and carbon dioxide on the surface growth of vegetable-associated micro-organisms. J. Appl. Microbiol. 86, 429–438.

Amanatidou, A., Slump, R.A., Gorris, L.G.M., Smid, E.J., 2000. High oxygen and high carbon dioxide modified atmospheres for shelf-life extension of minimally processed carrots. J. Food Sci. 65, 61–66.

Anonymous, 1999. Microbiological safety evaluations and recommendations on fresh produce. National Advisory Committee on microbiological criteria for foods. Food Control 10, 117–143.

Anonymous, 2003. Quantitative Assessment of Relative Risk to Public Health From Foodborne *Listeria monocytogenes* among Selected Categories of Ready-to-Eat Foods. Center for Food Safety and Applied Nutrition/FDA and Food Safety and Inspection Service/USDA, Washington, DC.

Artés, F., Gómez, P., Aguayo, E., Escalona, V., Artés-Hernández, F., 2009. Sustainable sanitation techniques for keeping quality and safety of fresh-cut plant commodities. Postharvest Biol. Technol. 51, 287–296.

Austin, J.W., Dodds, K.L., Blanchfield, B., Farber, J.M., 1998. Growth and toxin production by *Clostridium botulinum* on inoculated fresh-cut packaged vegetables. J. Food Prot. 61, 324–328.

Babic, I., Watada, A.E., Buta, J.G., 1997. Growth of *Listeria monocytogenes* restricted by native microorganisms and other properties of fresh-cut spinach. J. Food Prot. 60, 912–917.

Bagamboula, C.F., Uyttendaele, M., Debevere, J., 2002. Growth and survival of *Shigella sonnei* and *S. flexneri* in minimal processed vegetables packed under equilibrium modified atmosphere and stored at 7°C and 12°C. Food Microbiol. 19, 529–536.

Barriga, M.I., Trachy, G., Willemot, C., Simard, R.E., 1991. Microbial changes in shredded iceberg lettuce stored under controlled atmospheres. J. Food Sci. 56, 1586–1588, 1599.

Bartz, J.A., Wei, C.-I., 2003. The influence of bacteria. In: Bartz, J.A., Brecht, J.K. (Eds.), Postharvest Physiology and Pathology of Vegetables. second ed. Marcel Dekker, Inc., New York, pp. 519–541.

Beaudry, R.M., 2010. Future trends and innovations in controlled atmosphere storage and modified atmosphere packaging technologies. Acta Hort. 876, 21–28.

Beaudry, R.M., Moggia, C.E., Retamales, J.B., Hancock, J.F., 1998. Quality of "Ivanhoe" and "Bluecrop" blueberry fruit transported by air and sea from Chile to North America. Hortscience 33, 313–317.

Beltrán, D., Selma, M.V., Marín, A., Gil, M.I., 2005a. Ozonated water extends the shelf life of fresh-cut lettuce. J. Agric. Food Chem. 53, 5654–5663.

Beltrán, D., Selma, M.V., Tudela, J.A., Gil, M.I., 2005b. Effect of different sanitizers on microbial and sensory quality of fresh-cut potato strips stored under modified atmosphere or vacuum packaging. Postharvest Biol. Technol. 37, 37–46.

Bennik, M.H.J., Smid, E.J., Rombouts, F.M., Gorris, L.G.M., 1995. Growth of psychrotrophic foodborne pathogens in a solid surface model system under the influence of carbon dioxide and oxygen. Food Microbiol. 12, 509–519.

Bennik, M.H.J., Peppelenbos, H.W., Nguyen-the, C., Carlin, F., Smid, E.J., Gorris, L.G.M., 1996. Microbiology of minimally processed, modified-atmosphere packaged chicory endive. Postharvest Biol. Technol. 9, 209–221.

Berrang, M.E., Brackett, R.E., Beuchat, L.R., 1989. Growth of *Listeria monocytogenes* on fresh vegetables stored under controlled atmosphere. J. Food Prot. 52, 702–705.

Beuchat, L.R., 1996. Pathogenic microorganisms associated with fresh produce. J. Food Prot. 59, 204–216.

Beuchat, L.R., Brackett, R.E., 1990a. Survival and growth of *Listeria monocytogenes* on lettuce as influenced by shredding, chlorine treatment, modified atmosphere packaging and temperature. J. Food Sci. 55, 755–758, 870.

Beuchat, L.R., Brackett, R.E., 1990b. Inhibitory effects of raw carrots on *Listeria monocytogenes*. Appl. Environ. Microbiol. 56, 1734–1742.

Beuchat, L.R., Ryu, J.-H., 1997. Produce handling and processing practices. Emerg. Infect. Dis. 3, 459–465.

Bidawid, S., Farber, J.M., Sattar, S.A., 2001. Survival of hepatitis A virus on modified atmosphere-packaged (MAP) lettuce. Food Microbiol. 18, 95–102.

Blanco, P., Sieiro, C., Villa, T.G., 1999. Production of pectic enzymes in yeasts. FEMS Microbiol. Lett. 175, 1–9.

Brown, A.L., Brooks, J.C., Karunasena, E., Echeverry, A., Laury, A., Brashears, M.M., 2011. Inhibition of *Escherichia coli* O157:H7 and *Clostridium sporogenes* in spinach packaged in modified atmospheres after treatment combined with chlorine and lactic acid bacteria. J. Food Sci. 76, M427–M432.

Budu, A.S., Joyce, D.C., Terry, L.A., 2007. Quality changes in sliced pineapple under controlled atmosphere storage. J. Hortic Sci. Biotech. 82, 934–940.

Burton, W.G., 1974. Some biophysical principles underlying the controlled atmosphere storage of plant material. Ann. Appl. Biol. 78, 149–168.

Cantín, C.M., Minas, I.S., Goulas, V., Jiménez, M., Manganaris, G.A., Michailides, T.J., Crisosto, C.H., 2012. Sulfur dioxide fumigation alone or in combination with CO_2-enriched atmosphere extends the market life of highbush blueberry fruit. Postharvest Biol. Technol. 67, 84–91.

Carlin, F., Nguyen-the, C., Chambroy, Y., Reich, M., 1990. Effects of controlled atmospheres on microbial spoilage, electrolyte leakage and sugar content of fresh "ready-to-use" grated carrots. Int. J. Food Sci. Technol. 25, 110–119.

Carlin, F., Nguyen-the, C., Abreu Da Silva, A., Cochet, C., 1996. Effects of carbon dioxide on the fate of *Listeria monocytogenes*, of aerobic bacteria and on the development of spoilage in minimally processed fresh endive. Int. J. Food Microbiol. 32, 159–172.

Carrasco, E., Pérez-Rodríguez, F., Valero, A., García-Gimeno, R.M., Zurera, G., 2008. Growth of *Listeria monocytogenes* on shredded, ready-to-eat iceberg lettuce. Food Control 19, 487–494.

Chambroy, Y., Guinebretiere, M.-H., Jacquemin, G., Reich, M., Breuils, L., Souty, M., 1993. Effects of carbon dioxide on shelf-life and postharvest decay of strawberries fruit. Sci. Aliments. 13, 409–423.

Cheng, C.-M., Doyle, M.P., Luchansky, J.B., 1995. Identification of *Pseudomonas fluorescens* strains isolated from raw pork and chicken that produce siderophores antagonistic towards foodborne pathogens. J. Food Prot. 58, 1340–1344.

Chervin, C., Boisseau, P., 1994. Quality maintenance of "ready-to-eat" shredded carrots by gamma irradiation. J. Food Sci. 59, 359–361, 401.

Chun, H.H., Song, K.B., 2013. The combined effects of aqueous chlorine dioxide, fumaric acid, and ultraviolet-C with modified atmosphere packaging enriched in CO_2 for inactivating preexisting microorganisms and *Escherichia coli* O157:H7 and *Salmonella typhimurium* inoculated on buckwheat sprouts. Postharvest Biol. Technol. 86, 118–124.

Colgan, R.J., Johnson, D.S., 1998. The effects of postharvest application of surface sterilizing agents on the incidence of fungal rots in stored apples and pears. J. Hortic Sci. Biotech. 73, 361–366.

Conesa, A., Artés-Hernández, F., Geysen, S., Nicolaï, B., Artés, F., 2007. High oxygen combined with high carbon dioxide improves microbial and sensory quality of fresh-cut peppers. Postharvest Biol. Technol. 43, 230–237.

Cooley, M.B., Chao, D., Mandrell, R.E., 2006. *Escherichia coli* O157:H7 survival and growth on lettuce is altered by the presence of epiphytic bacteria. J. Food Prot. 69, 2329–2335.

Crisosto, C.H., Garner, D., Crisosto, G., 2002. High carbon dioxide atmospheres affect stored "Thompson seedless" table grapes. Hortscience 37, 1074–1078.

Daş, E., Gürakan, G.C., Bayındırlı, A., 2006. Effect of controlled atmosphere storage, modified atmosphere packaging and gaseous ozone treatment on the survival of *Salmonella enteritidis* on cherry tomatoes. Food Microbiol. 23, 430–438.

De Vries-Paterson, R.M., Jones, A.L., 1991. Fungistatic effects of carbon dioxide in a package environment on the decay of Michigan sweet cherries by *Monilinia fructicola*. Plant Dis. 75, 943–946.

Dewaal, C.S., Hicks, G., Barlow, K., Alderton, L., Vegosen, L., 2006. Foods associated with foodborne illness outbreaks from 1990 through 2003. Food Prot. Trends 26, 466–473.

Eklund, T., Jarmund, T., 1983. Microculture model studies on the effect of various gas atmospheres on microbial growth at different temperatures. J. Appl. Bacteriol. 55, 119–125.

El-Goorani, M.A., Sommer, N.F., 1981. Effects of modified atmospheres on postharvest pathogens of fruits and vegetables. Hort. Rev. 3, 412–461.

Errampalli, D., Wainman, L.I., Chu, C.L., 2007. Evaluation of fludioxonil for the control of postharvest gray mould and blue mould in pears in cold and controlled atmosphere storages. Int. J. Pest Manage. 53, 101–109.

Escalona, V.H., Geysen, S., Verlinden, B.E., Nicolaï, B.M., 2007. Microbial quality and browning of fresh-cut butter lettuce under superatmospheric oxygen condition. Europ. J. Hort. Sci. 72, 130–137.

Farber, J.M., 1991. Microbiological aspects of modified-atmosphere packaging technology—a review. J. Food Prot. 54, 58–70.

Finn, M.J., Upton, M.E., 1997. Survival of pathogens on modified-atmosphere-packaged shredded carrot and cabbage. J. Food Prot. 60, 1347–1350.

Fischer, N., Bourne, A., Plunkett, D., 2015. Outbreak Alert! 2015. Center for Science in the Public Interest. https://cspinet.org/resource/outbreak-alert-2015 (Accessed October 1, 2016).

Fraisse, A., Temmam, S., Deboosere, N., Guillier, L., Delobel, A., Maris, P., Vialette, M., Morin, T., Perelle, S., 2011. Comparison of chlorine and peroxyacetic-based disinfectant to inactivate feline calicivirus, murine norovirus and hepatitis A virus on lettuce. Int. J. Food Microbiol. 151, 98–104.

Francis, G.A., O'Beirne, D., 1998. Effects of the indigenous microflora of minimally processed lettuce on the survival and growth of *Listeria innocua*. Int. J. Food Sci. Technol. 33, 477–488.

Geysen, S., Escalona, V.H., Verlinden, B.E., Aertsen, A., Geeraerd, A.H., Michiels, C.W., Van Impe, J.F., Nicolaï, B.M., 2006. Validation of predictive growth models describing superatmospheric oxygen effects on *Pseudomonas fluorescens* and *Listeria innocua* on fresh-cut lettuce. Int. J. Food Microbiol. 111, 48–58.

Gorny, J.R., 2003. A summary of CA and MA requirements and recommendations for fresh-cut (minimally processed) fruits and vegetables. Acta Hort. 600, 609–614.

Guillén, F., Zapata, P.J., Martínez-Romero, D., Castillo, S., Serrano, M., Valero, D., 2007. Improvement of the overall quality of table grapes stored under modified atmosphere packaging in combination with natural antimicrobial compounds. J. Food Sci. 72, S185–S190.

Gunes, G.G., Hotchkiss, J.H., 2002. Growth and survival of *Escherichia coli* O157:H7 on fresh-cut apples in modified atmospheres at abusive temperatures. J. Food Prot. 65, 1641–1645.

Haffner, K., Rosenfeld, H.J., Skrede, G., Wang, L., 2002. Quality of red raspberry *Rubus idaeus* L. cultivars after storage in controlled and normal atmospheres. Postharvest Biol. Technol. 24, 279–289.

Hagenmaier, R.D., Baker, R.A., 1997. Low-dose irradiation of cut iceberg lettuce in modified atmosphere packaging. J. Agric. Food Chem. 45, 2864–2868.

Hagenmaier, R.D., Baker, R.A., 1998. Microbial population of shredded carrot in modified atmosphere packaging as related to irradiation treatment. J. Food Sci. 63, 162–164.

Harris, L.J., Farber, J.N., Beuchat, L.R., Parish, M.E., Suslow, T.V., Garrett, E.H., Busta, F.F., 2003. Outbreaks associated with fresh produce: Incidence, growth, and survival of pathogens in fresh and fresh-cut produce. Compr. Rev. Food Sci. Food Saf. 2 (Supplement), 78–141.

Heimdal, H., Kühn, B.F., Poll, L., Larsen, L.M., 1995. Biochemical changes and sensory quality of shredded and MA-packaged iceberg lettuce. J. Food Sci. 60, 1265–1268, 1276.

Horev, B., Sela, S., Vinokur, Y., Gorbatsevich, E., Pinto, R., Rodov, V., 2012. The effects of active and passive modified atmosphere packaging on the survival of *Salmonella enterica* serotype Typhimurium on washed romaine lettuce leaves. Food Res. Int. 45, 1129–1132.

Inoue, A., Nakata, Y., Izumi, H., 2017. Enumeration and identification of ethanol-injured coliform bacteria found on harvest equipment and the cross-contamination on cabbage. J. Food Nutr. Disor. 6, 1–5.

Izumi, H., 2005. Microbiological quality of fresh-cut produce stored in CA/MAP. In: Kanlayanarat, S., McGlasson, W.B. (Eds.), Proceeding of the APEC Symposium on Assuring Quality and Safety of Fresh Produce, KMUTT, Bangkok, Thailand, pp. 119–124.

Izumi, H., Inoue, A., 2018a. Viability of sublethally injured coliform bacteria on fresh-cut cabbage stored in high CO_2 atmospheres following rinsing with electrolyzed water. Int. J. Food Microbiol. 266, 207–212.

Izumi, H., Inoue, A., 2018b. Viability of chlorine-injured *Escherichia coli* O157:H7 on fresh-cut cabbage during cold storage in high CO_2 atmospheres. Biocontrol Sci. 23, 199–206.

Izumi, H., Nagano, M., Ozaki, Y., 2004. Microbial evaluation of fresh marketed vegetables. Mem. Faculty B. O. S. T. Kinki University 13, 15–22.

Izumi, H., Morimoto, K., Yamawaki, N., Murakami, Y., Kida, K., 2007. Microbiological evaluation of fruits certified as specially grown agricultural products by an accredited certification agency. Mem. Faculty B. O. S. T. Kinki University 20, 1–8.

Izumi, H., Tsukada, Y., Poubol, J., Hisa, K., 2008a. On-farm sources of microbial contamination of persimmon fruit in Japan. J. Food Prot. 71, 52–59.

Izumi, H., Poubol, J., Hisa, K., Sera, K., 2008b. Potential sources of microbial contamination of Satsuma mandarin fruit in Japan, from production through packing shed. J. Food Prot. 71, 530–538.

Izumi, H., Murakami, Y., Ishimaru, M., Ozaki, Y., 2015. Microbiological and visual quality of enzymatically peeled baby persimmons stored in high CO_2 controlled atmospheres. Acta Hort. 1071, 203–209.

Izumi, H., Nakata, Y., Inoue, A., 2016a. Enumeration and identification of coliform bacteria injured by chlorine or fungicide mixed with agricultural water. J. Food Prot. 79, 1789–1793.

Izumi, H., Rodov, V., Bai, J., Wendakoon, S.K., 2016b. Physiology and quality of fresh-cut produce in CA/MA storage. In: Pareek, S. (Ed.), Fresh-Cut Fruits and Vegetables: Technology, Physiology, and Safety. CRC Press, Boca Raton, FL, pp. 253–305.

Jacxsens, L., Devlieghere, F., Van der Steen, C., Debevere, J., 2001. Effect of high oxygen modified atmosphere packaging on microbial growth and sensorial qualities of fresh-cut produce. Int. J. Food Microbiol. 71, 197–210.

Janisiewicz, W.J., Leverentz, B., Conway, W.S., Saftner, R.A., Reed, A.N., Camp, M.J., 2003. Control of bitter rot and blue mold of apples by integrating heat and antagonist treatments on 1-MCP treated fruit stored under controlled atmosphere conditions. Postharvest Biol. Technol. 29, 129–143.

Janisiewicz, W.J., Saftner, R.A., Conway, W.S., Yoder, K.S., 2008. Control of blue mold decay of apple during commercial controlled atmosphere storage with yeast antagonists and sodium bicarbonate. Postharvest Biol. Technol. 49, 374–378.

Jay, J.M., 1992. Modern Food Microbiology, fourth ed. Chapman & Hall, New York.

Jin, H.-H., Lee, S.-Y., 2007. Combined effect of aqueous chlorine dioxide and modified atmosphere packaging on inhibiting *Salmonella typhimurium* and *Listeria monocytogenes* in mungbean sprouts. J. Food Sci. 72, M441–M445.

Johnston, L.M., Jaykus, L.-A., Moll, D., Martinez, M.C., Anciso, J., Mora, B., Moe, C.L., 2005. A field study of the microbiological quality of fresh produce. J. Food Prot. 68, 1840–1847.

Johnston, M.A., Harrison, M.A., Morrow, R.A., 2009. Microbial antagonists of *Escherichia coli* O157:H7 on fresh-cut lettuce and spinach. J. Food Prot. 72, 1569–1575.

Kader, A.A., Ben-Yehoshua, S., 2000. Effects of superatmospheric oxygen levels on postharvest physiology and quality of fresh fruits and vegetables. Postharvest Biol. Technol. 20, 1–13.

Kakiomenou, K., Tassou, C., Nychas, G.-J., 1998. Survival of *Salmonella enteritidis* and *Listeria monocytogenes* on salad vegetables. World J. Microbiol. Biotechnol. 14, 383–387.

Kallander, K.D., Hitchins, A.D., Lancette, G.A., Schmieg, J.A., Garcia, G.R., Solomon, H.M., Sofos, J.N., 1991. Fate of *Listeria monocytogenes* in shredded cabbage stored at 5 and 25°C under a modified atmosphere. J. Food Prot. 54, 302–304.

Kaneko, K.-I., Hayashidani, H., Ohtomo, Y., Kosuge, J., Kato, M., Takahashi, K., Shiraki, Y., Ogawa, M., 1999. Bacterial contamination of ready-to-eat foods and fresh products in retail shops and food factories. J. Food Prot. 62, 644–649.

Ke, D., Goldstein, L., O'Mahony, M., Kader, A.A., 1991. Effects of short-term exposure to low O_2 and high CO_2 atmosphere on quality attributes of strawberries. J. Food Sci. 56, 50–54.

Kimura, B., Yoshiyama, T., Fujii, T., 1999. Carbon dioxide inhibition of *Escherichia coli* and *Staphylococcus aureus* on a pH-adjusted surface in a model system. J. Food Sci. 64, 367–370.

Klaiber, R.G., Baur, S., Magel, L., Hammes, W.P., Carle, R., 2004. Quality of shredded, packaged carrots as affected by different washing treatments. J. Food Sci. 69, SNQ161–SNQ166.

Kupferman, E., 1997. Controlled atmosphere storage of apples. In: Mitcham, E.J. (Ed.), CA'97 Proc. Vol. 2: Apples and Pears. Postharvest Horticulture Series No. 16, University of California, Davis, CA, pp. 1–30.

Kupferman, E., 2003. Controlled atmosphere storage of apples and pears. Acta Hort. (600), 729–735.

Lanciotti, R., Belletti, N., Patrignani, F., Gianotti, A., Gardini, F., Guerzoni, M.E., 2003. Application of hexanal, (E)-2-hexenal, and hexyl acetate to improve the safety of fresh-sliced apples. J. Agric. Food Chem. 51, 2958–2963.

Lee, S.-Y., Baek, S.-Y., 2008. Effect of chemical sanitizer combined with modified atmosphere packaging on inhibiting *Escherichia coli* O157:H7 in commercial spinach. Food Microbiol. 25, 582–587.

Liao, C.-H., 2007. Inhibition of foodborne pathogens by native microflora recovered from fresh peeled baby carrot and propagated in cultures. J. Food Sci. 72, M134–M139.

Liao, C.-H., 2009. Control of foodborne pathogens and soft-rot bacteria on bell pepper by three strains of bacterial antagonists. J. Food Prot. 72, 85–92.

Liao, C.-H., Fett, W.F., 2001. Analysis of native microflora and selection of strains antagonistic to human pathogens on fresh produce. J. Food Prot. 64, 1110–1115.

Lin, C.-M., Kim, J., Du, W.-X., Wei, C.-I., 2000. Bactericidal activity of isothiocyanate against pathogens on fresh produce. J. Food Prot. 63, 25–30.

Magnuson, J.A., King Jr., A.D., Török, T., 1990. Microflora of partially processed lettuce. Appl. Environ. Microbiol. 56, 3851–3854.

Mahmoud, B.S.M., 2010. Effects of X-ray radiation on *Escherichia coli* O157:H7, *Listeria monocytogenes*, *Salmonella enterica* and *Shigella flexneri* inoculated on shredded iceberg lettuce. Food Miclobiol. 27, 109–114.

Martínez-Ferrer, M., Harper, C., Pérez-Muñoz, F., Chaparro, M., 2002. Modified atmosphere packaging of minimally processed mango and pineapple fruits. J. Food Sci. 67, 3365–3371.

Mateos, M., Ke, D., Cantwell, M., Kader, A.A., 1993. Phenolic metabolism and ethanolic fermentation of intact and cut lettuce exposed to CO_2-enriched atmospheres. Postharvest Biol. Technol. 3, 225–233.

McKellar, R.C., 2007. Role of nutrient limitation in the competition between *Pseudomonas fluorescens* and *Escherichia coli* O157:H7. J. Food Prot. 70, 1739–1743.

Meheriuk, M., 1993. CA storage conditions for apples, pears, and nashi. In: Proceeding from the Sixth International Controlled Atmosphere Research Conference NRAES-71. Vol. 2. NRAES, Ithaca, New York, pp. 819–858.

Molin, G., 1983. The resistance to carbon dioxide of some food related bacteria. Eur. J. Appl. Microbiol. Biotechnol. 18, 214–217.

Molin, G., 2000. Modified atmospheres. In: Lund, B.M., Baird-Parker, T.C., Gould, G.W. (Eds.), The Microbiological Safety and Quality of Food. In: Vol. 1. Aspen Publishers, Inc., Gaithersburg, MD, pp. 214–234.

Morris, C.E., Lucotte, T., 1993. Dynamics and variability of bacterial population density on leaves of field-grown endive destined for ready-to-use processing. Int. J. Food Sci. Technol. 28, 201–209.

Mukherjee, A., Speh, D., Jones, A.T., Buesing, K.M., Diez-Gonzalez, F., 2006. Longitudinal microbiological survey of fresh produce grown by farmers in the upper Midwest. J. Food Prot. 69, 1928–1936.

Murakami, Y., Wendakoon, S., Izumi, H., 2009. Microbial contamination of Japanese apricots and its environmental sources during fruit development and harvest. J. Antibact. Antifung. Agents. 37, 803–812 (in Japanese).

Murakami, Y., Ozaki, Y., Izumi, H., 2012. Microbiological quality and shelf life of enzyme-peeled fresh-cut persimmon slices stored in high CO_2 atmospheres. Hortscience 47, 1758–1763.

Nettles, C.G., Barefoot, S.F., 1993. Biochemical and genetic characteristics of bacteriocins of food-associated lactic acid bacteria. J. Food Prot. 56, 338–356.

Nguyen-the, C., Carlin, F., 1994. The microbiology of minimally processed fresh fruits and vegetables. Crit. Rev. Food Sci. Nutr. 34, 371–401.

Niemira, B.A., Fan, X., Sokorai, K.J.B., 2005. Irradiation and modified atmosphere packaging of endive influences survival and regrowth of *Listeria monocytogenes* and product sensory qualities. Radiation Phys. Chem. 72, 41–48.

Nishie, M., Nagao, J.-I., Sonomoto, K., 2012. Antibacterial peptides "Bacteriocins": An overview of their diverse characteristics and applications. Biocontrol Sci. 17, 1–16.

Nunes, M.C.N., Morais, A.M.M.B., Brecht, J.K., Sargent, S.A., 2002. Fruit maturity and storage temperature influence response of strawberries to controlled atmospheres. J. Amer. Soc. Hort. Sci. 127, 836–842.

Obaidat, M.M., Frank, J.F., 2009. Inactivation of *Salmonella* and *Escherichia coli* O157:H7 on sliced and whole tomatoes by allyl isothiocyanate, carvacrol, and cinnamaldehyde in vapor phase. J. Food Prot. 72, 315–324.

Olaimat, A.N., Holley, R.A., 2012. Factors influencing the microbial safety of fresh produce: a review. Food Microbiol. 32, 1–19.

Oliveira, M., Usall, J., Solsona, C., Alegre, I., Viñas, I., Abadias, M., 2010. Effects of packaging type and storage temperature on the growth of foodborne pathogens on shredded "Romaine" lettuce. Food Microbiol. 27, 375–380.

Oliveira, M., Abadias, M., Colás-Medà, P., Usall, J., Viñas, I., 2015. Biopreservative methods to control the growth of foodborne pathogens on fresh-cut lettuce. Int. J. Food Microbiol. 214, 4–11.

Perkins-Veazie, P., Collins, J.K., 2002. Quality of erect-type blackberry fruit after short intervals of controlled atmosphere storage. Postharvest Biol. Technol. 25, 235–239.

Poubol, J., Izumi, H., 2005a. Shelf life and microbial quality of fresh-cut mango cubes stored in high CO_2 atmospheres. J. Food Sci. 70, M69–M74.

Poubol, J., Izumi, H., 2005b. Physiology and microbiological quality of fresh-cut mango cubes as affected by high-O_2 controlled atmospheres. J. Food Sci. 70, M286–M291.

Prakash, A., Guner, A.R., Caporaso, F., Foley, D.M., 2000. Effects of low-dose gamma irradiation on the shelf life and quality characteristics of cut romaine lettuce packaged under modified atmosphere. J. Food Sci. 65, 549–553.

Purvis, A.C., 1997. The role of adaptive enzymes in carbohydrate oxidation by stressed and senescing plant tissues. Hortscience 32, 1165–1168.

Ragaert, P., Devlieghere, F., Debevere, J., 2007. Role of microbiological and physiological spoilage mechanisms during storage of minimally processed vegetables. Postharvest Biol. Technol. 44, 185–194.

Ramos, B., Miller, F.A., Brandão, T.R.S., Teixeira, P., Silva, C.L.M., 2013. Fresh fruits and vegetables-an overview on applied methodologies to improve its quality and safety. Innov. Food Sci. Emerg. Technol. 20, 1–15.

Ray, B., 1979. Methods to detect stressed microorganisms. J. Food Prot. 42, 346–355.

Rezende, A.C.B., Igarashi, M.C., Destro, M.T., Franco, B.D.G.M., Landgraf, M., 2014. Effect of gamma radiation on the reduction of *Salmonella* strains, *Listeria monocytogenes*, and Shiga toxin-producing *Escherichia coli* and sensory evaluation of minimally processed spinach (*Tetragonia expansa*). J. Food Prot. 77, 1768–1772.

Richardson, D.G., Kupferman, E., 1997. Controlled atmosphere storage of pears. In: Mitcham, E.J. (Ed.), CA'97 Proc. Vol. 2: Apples and Pears. Postharvest Horticulture Series No. 16, University of California, Davis, CA, pp. 31–35.

Robbs, P.G., Bartz, J.A., Sargent, S.A., McFie, G., Hodge, N.C., 1996. Potential inoculum sources for decay of fresh-cut celery. J. Food Sci. 61, 449–452, 455.

Sant'Ana, A.S., Landgraf, M., Destro, M.T., Franco, B.D.G.M., 2013. Growth potential of *Salmonella* and *Listeria monocytogenes* in ready-to-eat lettuce and collard greens packaged under modified atmosphere and in perforated film. J. Food Prot. 76, 888–891.

Satchell, F.B., Stephenson, P., Andrews, W.H., Estela, L., Allen, G., 1990. The survival of *Shigella sonnei* in shredded cabbage. J. Food Prot. 53, 558–562.

Schotsmans, W., Molan, A., MacKay, B., 2007. Controlled atmosphere storage of rabbiteye blueberries enhances postharvest quality aspects. Postharvest Biol. Technol. 44, 277–285.

Sekiyama, Y., Mizukami, Y., Takada, A., Oosono, M., Nishimura, T., 1996. Effect of mustard extract vapor on fungi and spore-forming bacteria. J. Antibact. Antifung. Agents. 24, 171–178.

Siro, I., Devlieghere, F., Jacxsens, L., Uyttendaele, M., Debevere, J., 2006. The microbial safety of strawberry and raspberry fruits packaged in high-oxygen and equilibrium-modified atmospheres compared to air storage. Int. J. Food Sci. Technol. 41, 93–103.

Solomon, H.M., Kautter, D.A., Lilly, T., Rhodehamel, E.J., 1990. Outgrowth of *Clostridium botulinum* in shredded cabbage at room temperature under a modified atmosphere. J. Food Prot. 53, 831–833, 845.

Teles, C.S., Benedetti, B.C., Gubler, W.D., Crisosto, C.H., 2014. Prestorage application of high carbon dioxide combined with controlled atmosphere storage as a dual approach to control *Botrytis cinerea* in organic "flame seedless" and "crimson seedless" table grapes. Postharvest Biol. Technol. 89, 32–39.

Thunberg, R.L., Tran, T.T., Bennett, R.W., Matthews, R.N., Belay, N., 2002. Microbial evaluation of selected fresh produce obtained at retail markets. J. Food Prot. 65, 677–682.

Tian, S., Fan, Q., Xu, Y., Wang, Y., Jiang, A., 2001. Evaluation of the use of high CO_2 concentrations and cold storage to control *Monilinia fructicola* on sweet cherries. Postharvest Biol. Technol. 22, 53–60.

Tian, S.-P., Jiang, A.-L., Xu, Y., Wang, Y.-S., 2004. Responses of physiology and quality of sweet cherry fruit to different atmospheres in storage. Food Chem. 87, 43–49.

Van der Steen, C., Jacxsens, L., Devlieghere, F., Debevere, J., 2002. Combining high oxygen atmospheres with low oxygen modified atmosphere packaging to improve the keeping quality of strawberries and raspberries. Postharvest Biol. Technol. 26, 49–58.

Waghmare, R.B., Annapure, U.S., 2013. Combined effect of chemical treatment and/or modified atmosphere packaging (MAP) on quality of fresh-cut papaya. Postharvest Biol. Technol. 85, 147–153.

Watada, A.E., Qi, L., 1999. Quality of fresh-cut produce. Postharvest Biol. Technol. 15, 201–205.

Watada, A.E., Izumi, H., Luo, Y., Rodov, V., 2005. Fresh-cut produce. In: Ben-Yehoshua, S. (Ed.), Environmentally Friendly Technologies for Agricultural Produce Quality. CRC Press, Boca Raton, FL, pp. 149–203.

Watanabe, M., Tsutsumi, F., Konuma, R., Lee, K.-I., Kawarada, K., Sugita-Konishi, Y., Kumagai, S., Takatori, K., Konuma, H., Hara-Kudo, Y., 2011. Quantitative analysis of microflora on commercial domestic fruits in Japan. J. Food Prot. 74, 1488–1499.

Wesche, A.M., Gurtler, J.B., Marks, B.P., Ryser, E.T., 2009. Stress, sublethal injury, resuscitation, and virulence of bacterial foodborne pathogens. J. Food Prot. 72, 1121–1138.

Wszelaki, A.L., Mitcham, E.J., 2000. Effects of superatmospheric oxygen on strawberry fruit quality and decay. Postharvest Biol. Technol. 20, 125–133.

Zagory, D., 1999. Effects of post-processing handling and packaging on microbial populations. Postharvest Biol. Technol. 15, 313–321.

Chapter 6

Effect of CA/MA on sensory quality

Anne Plotto, Jinhe Bai and Elizabeth Baldwin
USDA, ARS US Horticulture Research Laboratory, Fort Pierce, FL, United States

Introduction

Modifying storage atmosphere of fresh or fresh-cut fruits and vegetables is a common practice to extend storage. By reducing oxygen (O_2) and increasing carbon dioxide (CO_2) in the atmosphere surrounding the plant tissue, respiration is reduced and all other ensuing metabolic reactions of ripening and senescence are delayed (Wang, 2006). Most importantly, low O_2 and high CO_2 inhibit ethylene synthesis at various levels in the ethylene pathway, reducing its production, and many ripening reactions are further delayed by the lack of ethylene. Those include reactions that would change fruit appearance (action on synthesis of carotenoids, anthocyanins, and reduced loss of chlorophyll), texture (change in cell wall constituents and lipid metabolism), taste (reduced loss in carboxylic acids), and flavor (change in aroma volatile production). For fresh-cut fruits or vegetables, an atmosphere with low O_2 has a direct effect on browning of the cut surface as polyphenol oxidase, an enzyme that requires O_2, is involved in the browning reaction. High CO_2 inhibits the development of spoilage microorganisms on cut fruit surface, therefore preventing formation of off-flavors. However, excessive modification of atmosphere, extremely high CO_2 and/or low O_2, may induce anaerobic respiration with high ethanol and acetaldehyde, inducing off-flavors from fermentation metabolism.

While controlled atmosphere (CA) establishes the target atmosphere in a fast and controlled manner, in passive modified atmosphere packaging (MAP), the atmosphere is modified by the fruit respiration as the fruit cells take up O_2 and release CO_2 (Porat and Fallik, 2008). Active MAP uses gas displacement, gas scavengers or absorbers to establish a desired gas mix within the package (Caleb et al., 2013a). Due to the uncontrolled nature of passive and even active MAP and CA, modeling techniques have been used to try to predict the gas levels based on storage time, commodity respiration, and porosity and temperature (Beaudry, 2000; Beaudry et al., 1996; Ho et al., 2016; Lakakul

Controlled and Modified Atmospheres for Fresh and Fresh-Cut Produce
https://doi.org/10.1016/B978-0-12-804599-2.00007-7

et al., 1999) to try to assess final effects on quality and flavor. This is important, not only for intact fruit, but also fresh-cut fruits and vegetables like cut pear (Gomes et al., 2010) and cantaloupe (Gomes et al., 2012). The resulting atmosphere depends on the commodity respiration, which in turn, is dependent upon storage temperature, and thus is less controlled (Rosenfeld et al., 1999). Surface morphology and anatomy of the commodity, such as presence of trichomes, stomata, lenticels, cuticle thickness and cracks, can determine the permeability of the fruit peel, and varietal differences in surface characteristics can determine tolerance toward reduced O_2 and elevated CO_2 (Paul and Pandey, 2014).

Usually CA or modified atmosphere (MA) is not applied alone and is complemented by low temperature as a tool to extend storage, which is critical to reduce respiration. MAP is also used to reduce water loss in produce in order to preserve a fresh appearance. This chapter will focus on the effect of CA and MA on biochemical processes and resulting effects on perceived texture, taste, and flavor.

Effect of CA/MA on cell membrane and fruit texture

The most obvious effect of reduced O_2 and high CO_2 on stored fruit has been maintenance of firmness and acidity. A study showed that in the mind of consumers, texture of foods is an integral quality of a food product, and unless there is a defect in texture, they will affirm that the most important attribute in a food is flavor (Szczesniak and Kahn, 1971). Multiple examples with apple studies showed that consumers preferred CA-stored apples, which are firmer/crisper than air-stored apples, in spite of lower volatile production and aroma intensity (Cliff et al., 1998; Echeverría et al., 2008; López et al., 2007; Moya-Leon et al., 2007). When shrivel, wilt, or bruises appear on a commodity, for example, an apple which is expected to be crispy and firm is instead "soggy" or "mealy," consumers will judge the fruit or vegetable as not being fresh anymore (Szczesniak and Kahn, 1971). To the sensory scientist, texture is a complex attribute that includes several descriptors, which need to be well defined. Trained panelists analyze, quantify, and describe each step of the mechanical movements made by the mouth, including biting, chewing, and swallowing, and record impressions resulting from each action: "first bite" (with front teeth), "compression" (between the molars), "chewiness" (time to masticate the food until swallowed), "juiciness" (amount of liquid released) are common texture attributes used by trained panelists. "Melting" or "fondant" is defined as the force applied when pressing a piece of fruit between the tongue and the palate (Mehinagic et al., 2003; Ngamchuachit et al., 2015). Typical sensory descriptors for apples are crispness, crunchiness, firmness, chewiness, juiciness, and mealiness (Harker et al., 2002; Mann et al., 2005; Mehinagic et al., 2003). Melting, slipperiness, and fibrousness were descriptors specifically chosen for mangoes (Ngamchuachit et al., 2015), while "bursting energy" and "skin toughness" were used in describing blueberries (Blaker et al., 2014; Saftner et al.,

2008). As many as there are attributes to describe fruit texture by sensory analysis, relating these attributes to instrumental measurements would be useful in practice to sort fruit for quality on a packing line, or predict firmness loss in storage, for example. Instrumental measurements reduce the variation that is inevitable between panelists and provide common ground (units) by which to compare between studies, laboratories, industry, and so on (Abbott, 1999). Ideally, an instrument should be nondestructive, fast, sensitive to small changes in texture, and most importantly, relevant, that is, providing data correlating with relevant sensory attributes (Harker et al., 1997a). The puncture test was developed in the 1920s and readily adopted by the fruit industry to measure the force until the tissue ruptures (Abbott, 1999; Bourne, 1979; Harker et al., 1997a). Puncture instruments or penetrometers such as Magness Taylor (MT), Effigi, and similar devices are now widely used by the fruit industry; they use 7.9 mm or 11.1 mm diameter probes, with convex tip with a well-defined curvature (Abbott, 1999; Harker et al., 1997a). Puncture tests measure a combination of shear and compression, but can result in highly variable data when using handheld devices, and also the response depends on fruit cell structure, whether being fibrous or not. Variability may be reduced by having the probe attached to a supporting device and applying a constant force, which the Instron (Instron, Northwood, MA, USA) or Texture Analyzer (Stable Micro Systems Ltd., Godalming, Surrey, UK) instruments do, and it is important to puncture the fruit in the same direction of cell orientation (Abbott, 1999; Harker et al., 1997a). Puncture tests correlate with sensory firmness, hardness, and crispness and are still being used as a reference for apples, kiwifruit, and pear (Abbott, 1999; Cliff et al., 1998; Harker et al., 2002; Mehinagic et al., 2004). However, puncture tests are not that useful for soft fruit such as cherries, blueberries, citrus, tomatoes, and for these commodities, methods that measure elastic properties at very small deformation are preferred (Abbott, 1999).

The constituents of texture in fruit is a complex combination of structural and biochemical components including cell size, various arrangements of cells (geometry), and the substance that keep cells together, mainly the middle lamella (pectin). When fruit are unripe, or for firm fruit such as crunchy apples, cells rupture upon chewing, while in softer fruit, cells remain intact and are rearranged in an intercellular fluid matrix upon compression (Harker et al., 1997a). This was illustrated by showing images of cell structure of various fruit (carrot, apple, avocado, banana, muskmelon) by low temperature scanning electron microscopy before and after applying tensile tests on fruit tissue (Fig. 1) (Harker et al., 1997b). The softening phenomena involved in fruit ripening and senescence include breakdown of tissue cohesiveness and structure as well as differences in water turgor (e.g., blueberry, soft fruit in general). Most of fleshy fruit are composed of parenchyma cells, which include the vacuole, and are interconnected by the middle lamella, a pectin-rich matrix (Harker et al., 1997a). Cell walls are made up of cellulose and hemicellulose (such as xyloglucans) microfibrils embedded in a matrix of pectin polysaccharides and proteins

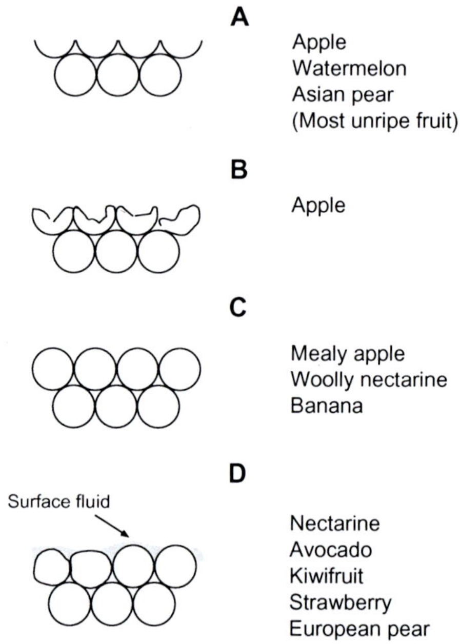

FIG. 1 Pattern of cell failure tensile testing of ripe fruit (unless otherwise indicated). Cells either break open (A and B) or separate from neighboring cells (C and D). Data summarized from Harker and Hallett (1992), Harker and Sutherland (1993), Harker and Hallett (1994), and Harker et al. (1997b). *(Reproduced with permission from Harker, F.R., Redgwell, R.J., Hallett, I.C., Murray, S.H., Carter, G., 1997a. Texture of fresh fruit. In: Janick, J. (Ed.), Horticultural Reviews, vol. 20. John Wiley & Sons, Inc., Oxford, UK, pp. 121–224.)*

(Gapper et al., 2013; Gwanpua et al., 2014; Rose and Bennett, 1999). The degradation and solubilization of hemicellulose and pectin in the middle lamella have been considered the main causes of loss of structure leading to fruit softening during ripening (Lashbrook, 2005). However, pectin degradation involves many enzymes that differ between and among fruit species, as pectin itself is not a simply linear structure, but can be branched, methylated, carboxylated, or acetylated in various places. Enzymes involved in pectin solubilization in fruit include polygalacturonase (PG), pectin esterase (PE) or pectin methyl esterase (PME), β-galactosidase (β-Gal), α-arabinofuranosidase (α-AF), and pectate lyases (PL), all existing in multigene families (Lashbrook, 2005). In fact, it is believed that cell wall metabolism during ripening involves more than 50 related genes (Gapper et al., 2013). Polygalacturonase (PG) has long been known to be a primary enzyme in tomato softening, and antisense RNA was first used to maintain tomato firmness (Sheehy et al., 1988).

Understanding the enzymatic changes occurring during ripening may help manage CA, which has long been shown to maintain firmness in apples (Cliff et al., 1998; Lidster et al., 1983; Siddiqui et al., 1996). Two β-gal enzymes and one α-AF were shown to play important roles in pectin solubilization early in ripening in addition to PG, and their activities were suppressed in CA (Gwanpua et al., 2016, 2014). In contrast, PME seemed to play a role later in ripening. Such knowledge might be considered when managing cold and CA storage timing in order to control better changes in firmness. In peaches, "woolliness" or "mealiness" is a symptom of chilling injury occurring in cold storage, resulting in dry texture as opposed to juicy in fruit ripened normally (Lurie and Crisosto, 2005). "Woolly" fruit have lower PG activity than normally found during ripening, resulting in an imbalance in PG/PME ratio (Ben-Arie and Sonego, 1980; Zhou et al., 2000a). PG and many other enzymes involved in fruit softening are suppressed in cold storage in comparison with normal fruit, and after a brief decline, there is an increase in PME activity with increasing time in cold storage (Brummell et al., 2004). Those biochemical changes in the cell wall result in long chains of deesterified pectin molecules that form a gel retaining intralocular water, providing this feeling of dryness when eating a mealy fruit (Zhou et al., 2000b). Furthermore, the cell wall becomes more elastic, and cells do not break and release juice upon chewing as with fruit ripened normally (Brummell et al., 2004). CA reduced chilling injury symptoms in peaches and nectarines, including woolliness (Lurie, 1992; Lurie and Crisosto, 2005; Zhou et al., 2000a), but the atmosphere regimes are cultivar and maturity dependent.

In strawberries, applications of high CO_2 improved firmness in many cultivars (Smith and Skog, 1992; Watkins et al., 1999). Responses depended on cultivar (Smith and Skog, 1992; Watkins et al., 1999), concentration (Almenar et al., 2006; Harker et al., 2000; Smith, 1992), and exposure time (Watkins et al., 1999). Interestingly, a burst in CO_2 resulted in greater firmness than letting the same level of CO_2 slowly establish as with MAP (Larsen and Watkins, 1995). An increase in cell-to-cell adhesion was measured using the tensile strength test showing an increase in firmness without structural change in strawberry fruit tissue, as observed through electron microscopy (Harker et al., 2000). The increase in firmness from high CO_2 treatment was suggested to be a result of pectin rearrangement due to a change in pH in the apoplast (containing the middle lamella). Opposite results were observed in blueberries stored under MAP: depending on the studies and the cultivars, there were detrimental effects on firmness with CO_2 greater than 6–12 kPa inducing fruit softening (Harb and Streif, 2004; Rodriguez and Zoffoli, 2016).

Effect of CA/MA on sugars and acids

Sugars and acids contribute sweetness and sourness to fruits and vegetables and are generally stored in the vacuole (Etienne et al., 2013). For ripening fruits, acids generally decrease while sugars increase. Acidity of fruits and vegetables is often measured as titratable acidity (TA), pH, or quantification of individual acids, and sugars as soluble solids (SSC) or individual sugars (Bai et al., 2016).

The combination of SSC and TA is very often used as a quality index for many crops including citrus, apples, blueberries, grape, and so on. The SSC/TA index is the most commonly used, although other indicators such as BrimA and sucrose equivalent, which provide better correlation with perceived sensation of sweetness and/or sourness have been used to relate sugars and acids with corresponding taste (Baldwin et al., 1998; Jordan et al., 2001; Koehler and Kays, 1991; Obenland et al., 2009). A very thorough review on fleshy fruit acidity reported that the main organic acids found in fruit are malic and citric acids (Etienne et al., 2013), which contribute to the flavor quality in terms of acidity (Kader, 2008). Apples are an example of fruits where the major acid is malic, and citrus exemplifies fruit where the major organic acid is citric (Etienne et al., 2013). There are exceptions like carambola where the major organic acid is oxalic acid (Wilson et al., 1982), or grape, which contains an equal amount of tartaric and malic acids (Flores et al., 2012). This is important, as sensory research has shown organic acids have their own sensory properties in addition to sourness, including bitterness and astringency (Hartwig and McDaniel, 1995; Rubico and McDaniel, 1992). Sugars in fruits are generally sucrose, and/or glucose and fructose in various ratios. For example, citrus contains more sucrose than glucose and fructose (Baldwin, 1993) while tomatoes contain almost no sucrose, and relatively even amounts of glucose and fructose (Baldwin et al., 2015). Likewise, grapes contain almost no sucrose but similar amounts of fructose and glucose, while mangoes contain mostly sucrose, some fructose, and almost no glucose, with both commodities having similar total sugar levels of ~ 13.0 g/100 g (Li et al., 2002). Like with acids, each sugar has its own sweetening properties, with sucrose and fructose having the most sweetening power in model solutions (Moskowitz, 1970). That relative sweetness varies with the relative concentration of each sugar, and also with the medium of evaluation: for example, fructose was less sweet than sucrose in pear nectar (Pangborn, 1963), but another study showed that this held true only at low concentrations in complex foods (Cardello et al., 1979).

The ripening-related decrease in acids is thought to be due to the conversion between tri- (citrate) and dicarboxylates (malate and others) in the mitochondria (tricarboxylic acid cycle, TCA), the cytosol (citrate catabolism), and the glyoxysome (glyoxylate cycle), and degradation of organic acids through the phosphoenolpyruvate (PEP) pathway in the cytosol (Fig. 2) (Etienne et al., 2013). Decarboxylation of oxaloacetate (OAA) and malate results in phosphoenolpyruvate (PEP) and is related to gluconeogenesis, which creates glucose from PEP during ripening when sugars accumulate (Sweetman et al., 2009). The TCA cycle results in a series of conversions of organic acids including citrus and malate. Malate can be carboxylated by the cytosolic enzyme NADP-dependent malic enzyme (NADP-cytME), an enzyme thought to be responsible for the decrease in malate in some fruit species (Etienne et al., 2013; Sweetman et al., 2009). Fermentative metabolism may also result in the decarboxylation of malate and OAA in ripening fruit if too acidic (Sweetman et al., 2009).

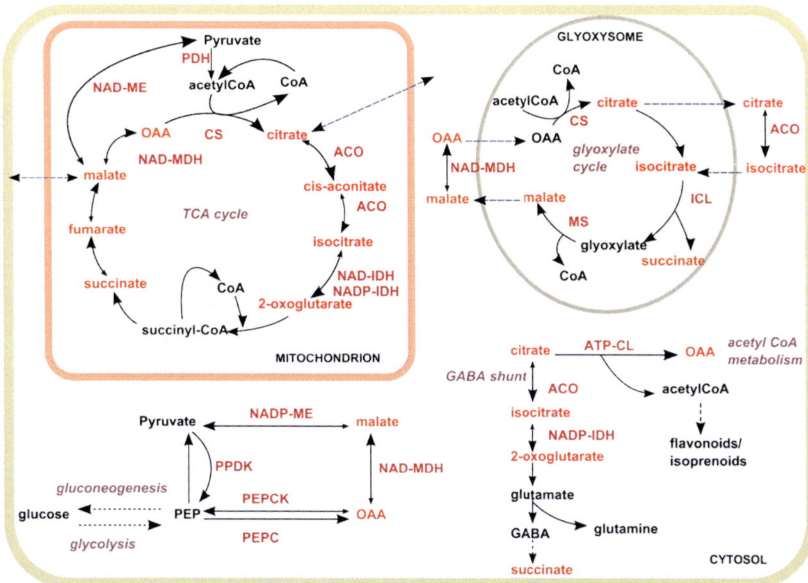

FIG. 2 Citrate and malate metabolic pathways in fruit mesocarp cells. Only the enzymes described in the paper are shown. *ACO*, aconitase; *ATP-CL*, ATP-citrate lyase; *CS*, citrate synthase; *ICL*, isocitrate lyase; *MS*, malate synthase; *NAD-MDH*, NAD-malate dehydrogenase; *NAD-ME*, NAD-malic enzyme; *NAD-IDH*, NAD-isocitrate dehydrogenase; *NADP-ME*, NADP-malic enzyme; *NADP-IDH*, NADP-isocitrate dehydrogenase; *PDH*, pyruvate dehydrogenase; *PEPC*, phosphoenolpyruvate carboxylase; *PEPCK*, phosphoenolpyruvate carboxykinase; *PPDK*, pyruvate orthophosphate dikinase. The probable direction of reversible reactions is indicated by the *large arrow*. *Dashed blue arrows* indicate malate and citrate transport. Names in orange are dicarboxylates and names in red are tricarboxylates. *(Reproduced with permission by Oxford University Press from Etienne, A., Génard, M., Lobit, P., Mbeguié-A-Mbéguié, D., Bugaud, C., 2013. What controls fleshy fruit acidity? A review of malate and citrate accumulation in fruit cells. J. Exp. Bot. 64, 1451–1469.)*

The glyoxylate cycle might be involved in malate accumulation in grape and ripening banana and could provide substrates for gluconeogenesis during banana for sugar accumulation (Surendranathan and Nair, 1976; Terrier et al., 2005). Zhu et al. (2013) reported that CA storage maintained higher sugar contents in apples including sucrose, fructose, and glucose due to enhancement of sucrose synthesis and delay of hydrolysis of sucrose and hexose via increased expression and activity of sucrose phosphate synthase and inhibition of neutral invertase, vacuolar invertase, fructokinase, and hexokinase.

In reality, CA can reduce the loss of acidity in stored apples and pears over the many months of storage (Bai et al., 2005; Thompson, 1998), especially if pretreated with 1-methylcyclopropene (MCP) (Rizzolo et al., 2014; Watkins et al., 2000). CA-stored fruit had higher sourness perceived by trained panels (Cliff et al., 1998; Plotto et al., 1999; Saftner et al., 2002) and were preferred in consumer panels (Echeverría et al., 2008; Saftner et al., 2002). Depending on the studies, sweetness either did not change (Cliff et al., 1998; Saftner et al., 2002) or was perceived much lower than air-stored fruit (Plotto et al., 1999), likely due to a lower sugar/acid ratios and/or less sweet enhancing volatiles produced by the fruit. CA also maintained acidity in kiwifruit (Latocha et al., 2014), but the greatest sensory benefit of CA storage in that study was by preventing bitterness development in comparison to long-term air storage. In "Rich Lady" peaches, CA improved consumer acceptability of cold-stored fruit for 2 weeks, mostly due to increased sweetness and fruity flavor perception, explained by higher SSC, lower TA, and some volatiles (Ortiz et al., 2008, 2009). Lychees stored under 7% O_2 +3% CO_2 had higher SSC/TA than fruit stored at 17% O_2 +6% CO_2, but they were perceived as "too sweet" and had lower sensory scores (Sivakumar and Korsten, 2010). The benefit of CA storage for lychee was very obvious in an atmosphere of 3.5% O_2 +3.5% CO_2, which maintained firmness, color, malic and ascorbic acids, and eating quality in comparison with air cold storage (Mahajan and Goswami, 2004).

MA of other commodities delays loss of acidity in the short term by reducing respiration and ethylene production in order to delay ripening (Porat et al., 2008). For tomato, low O_2 and high CO_2 can lead to anaerobiosis, but MA does not appear to affect tomato fruit SSC. Carbon monoxide (CO) reduced postharvest pathogens and improved tomato total SSC and TA profiles compared to air and reduced botrytis (Beckles, 2012), but is hazardous to use. Increasing pH and decreasing TA in tissues was more prevalent in strawberries stored in high CO_2 atmospheres, which may have affected color as well since pH affects the anthocyanin pigment (Holcroft and Kader, 1999). In that study, taste panels were not performed but it could be speculated that strawberries stored at high CO_2 would be sweeter, as both citric and malic acids, the main acids in strawberry, decreased, and glucose and fructose increased, more so in high CO_2 than air-stored fruit. In a study where blueberries were stored at incremental levels of increasing CO_2 (6, 12, 18, and 24 kPa) with either 2 or 18 kPa O_2, the optimum combination for long-term storage (3–6 weeks) was CO_2 at 6 or 12 kPa with no reduction in O_2 (Harb and Streif, 2004). Levels of CO_2 greater than 12 kPa negatively impacted flavor, firmness, and acidity (Harb and Streif, 2004). However in another study, TA increased continuously in storage, 3 weeks in air and 6 weeks in CA, with CA atmosphere of 2.5 kPa O_2 + 15 kPa CO_2 (Schotsmans et al., 2007), indicating responses to various atmosphere depend on cultivar, and likely on maturity and physiological stage of the fruit at harvest.

Effect of CA/MA on volatiles

Aroma compounds in fruits and vegetables come from diverse chemical pathways, including amino acids, carbohydrates, and fatty acids (Defilippi et al., 2009; Gudenschwager and Defilippi, 2016). These pathways give rise to the major volatile chemical groups that include aldehydes, alcohols, esters, ketones, and terpenes that are commonly found in fruits and vegetables. For fatty acid metabolism, the lipoxygenase (LOX) pathway gives rise to linoleic and linolenic acids and eventually to C-6 and C-9 aldehydes and alcohols via lipoxygenase (LOX), hydroperoxide lyase (HPL), an isomerase, and alcohol dehydrogenase (ADH) (Bai et al., 2011). This is generally thought to happen via tissue maceration, as a wound response (tissue damage) and is regulated at the gene expression level with the LOX gene family having five isoforms (Bai et al., 2011; Chen et al., 2004). The β-oxidation pathway, with successive removal of C2 units via acetyl CoA, eventually gives rise to aldehydes and alcohols and the alcohols can be metabolized by alcohol acetyltransferases to esters. Lactones are also associated with this pathway (Osorio et al., 2010). Amino acids including leucine, isoleucine, valine (Gudenschwager and Defilippi, 2016; Schwab et al., 2008), and phenylalanine give rise to 3-methylbutanal, 3-methylbutanol, 3-methylbutanoic acid (leucine), 2-isobutylthiazole (valine), phenylacetaldehyde, and 2-phenylethanol (phenylalanine), for example (Gudenschwager and Defilippi, 2016; Sanz and Perez, 2010). The contribution of volatile compounds to fruit aroma and flavor depends on the combination of their concentration and odor threshold, usually measured in air, in water, or in the matrix of interest (Buttery et al., 1989; Grosch, 1994; Plotto et al., 2004; Tandon et al., 2000). In general, it can be said that sulfur compounds have extremely low odor thresholds (Engel et al., 1990; Plotto et al., 2017). Among esters, branched-chain esters have much lower thresholds than their straight-chain counterpart (Takeoka et al., 1996, 1995). Some compounds such as ethanol, acetaldehyde, and ethyl acetate, indicators of anaerobic respiration in stored fruit, might only contribute partially to off aroma, as their odor thresholds are rather high in comparison with other aroma-active volatiles (Plotto et al., 2008; Rychlik et al., 1998; van Gemert, 2003). The "fermented/off aroma" odor perceived in fruit stored under anaerobic conditions may instead be due to imbalance in overall volatiles, in addition to high ethanol, ethyl acetate, and acetaldehyde, such as in mandarins (Obenland et al., 2011) or apples (Echeverría et al., 2003). However, in some fruits such as strawberries, ethyl acetate can be present in higher amount than its threshold, and indeed contribute to off-flavor (Berna et al., 2007; Larsen and Watkins, 1995). Maintaining the cold chain is important to avoid anaerobic effects on aroma volatiles; however, chilling injury can take the effect of aroma volatile alteration or reduction before visual symptoms appear. This can be due to genetic expression, transcription of volatile enzymes, and may even involve methylation status of promoters (Bai et al., 2011; Zhang et al., 2016).

Harvest maturity can affect overall fruit aroma during postharvest ripening for climacteric fruit (El Hadi et al., 2013). In some cases fruit harvested early can result in firmer texture, a better sugar/acid ratio, and adequate color during storage at the expense of aroma production (Bangerth et al., 2012). If fruit are not harvested at optimal eating quality due to the need for extended storage or transport, the resulting flavor can be lacking (El Hadi et al., 2013). Treatment of fruit with CA or MAP can delay ripening and aroma synthesis, thus inhibiting the normal development and accumulation of aroma compounds. In general, the aroma profile changes from grassy-notes (aldehydes and alcohols) to fruity-notes (esters in apples, for example) (Fellman et al., 2000; Gudenschwager and Defilippi, 2016) or lactones in peaches (Zhang et al., 2010) and furanones in strawberries (Ménager et al., 2004). For climacteric fruit, ethylene plays a role in ripening and aroma development and these fruit exhibit their maximum aroma development around their climacteric peak (Dixon and Hewett, 2000; Fellman et al., 2000). Use of ethylene inhibitors or suppression of ethylene biosynthesis in transgenic fruits for apple (Dandekar et al., 2004), melon (Bauchot et al., 1998), and tomato (Oeller et al. 1991) has shown that ethylene is needed for aroma biosynthesis (Gudenschwager and Defilippi, 2016), especially for ester production for banana (Golding et al., 1998), plum (Abdi et al., 1998) and apple (Defilippi et al., 2004; Ferenczi et al., 2006). Ester production was reduced more than 90% in apples with suppressed ACC-oxidase or ACC-synthase activity or when treated with the ethylene action inhibitor 1-methylcyclopropene (1-MCP) (Defilippi et al., 2005) due to ethylene regulation of alcohol acyl-CoA transferase (AAT), a key enzyme in ester biosynthesis. Increase of ethyl esters in CA for "Fuji" (Echeverrıa et al., 2004; Lara et al., 2006) and "Pink Lady" (Villatoro et al., 2008) apples as well as in pear (Lara et al., 2003) was associated with the availability of alcohol precursors. A MA of low O_2 and high CO_2 retarded blackcurrant fruit from synthesizing terpenes over 3 weeks of storage and changed the terpene profile (Harb et al., 2008).

There have been many good reviews on the subject of CA/MA effects on fruit and vegetable flavor. Reviews by Thompson (1998), Brecht et al. (2003), Porat et al. (2008), and Beaudry (2010) discuss the complexity of CA/MA effects based on the fruit or vegetable respiration rate, cultivar differences, product maturity or ripeness stage, and the storage atmosphere combined with temperature. While benefits include inhibition of ethylene synthesis and action, browning, chlorophyll degradation, decay, and proliferation of human pathogens (Beaudry, 2010), effects on reduced flavor are well documented. They also discuss the specific off-flavors due to anaerobic respiration that can occur with MAP if the product is stored at abusive temperatures, often due to ethanol, acetaldehyde, and subsequent ester alterations. Also discussed are the effect of CA on general volatile profiles which can be altered, as with the volatile esters, or suppressed (much of this data comes from studies with apples). Bai et al. (2005) and Mattheis et al. (2005) showed that apple fruit stored in CA had reduced production of esters, alcohols, 1-methoxy-4(2-propenyl) benzene, and other volatile compounds, which was clearly perceived by a sensory panel (Plotto et al., 1999). Beaudry et al. (1996) looked at the permeability of packaging to aroma volatiles and modeled their accumulation in a package in conjunction with O_2 and CO_2 in apple. Caleb et al. (2015) reported that CA storage of pomegranate fruit reduced scald and decay, but adversely affected the aroma volatiles, with the accumulation of fermentative metabolites (ethanol, acetaldehyde, and ethyl acetate). Use of MAP for packaging of pomegranate arils exhibited adverse effects if the O_2 was too low, with decreases in volatile ketones, aldehydes, alcohols, and esters, and resulting in shorter aril "flavor life" than shelf life, itself determined by microbial growth (Caleb et al., 2013b, 2015). Another trend in CA storage is O_2 partial pressure monitoring or dynamic controlled atmosphere (DCA), based on the lowest O_2 limit tolerated by a fruit at a specific metabolic stage and thus can be changed throughout storage by measuring ethanol production, chlorophyll fluorescence (CF), or respiratory quotient (RQ) (Both et al., 2017; Rizzolo et al., 2014; Weber et al., 2015). The volatile profile of "Royal Gala" apples stored using DCA-RQ still promoted higher anaerobic metabolism, increased ethanol and ethyl acetate, although below threshold, but had higher flesh firmness and less disorders and resulted in a higher concentrations of volatile compounds including key compounds such as butyl acetate and 2-methylbutyl acetate. Apples stored under DCA-CF, however, showed decreased ester production, especially key volatile 2-methybutyl acetate (Both et al., 2017).

Impairment of aroma production is especially true under long-term CA or ultralow oxygen (ULO) conditions (Song and Forney, 2008). Analysis of "Fuji" apple in air, CA, ULO, or hypoxic conditions showed partial inhibition of LOX activity under hypoxic conditions and decreased pyruvate decarboxylase activity under ULO (Lara et al., 2006) with straight-chain esters higher in air than in CA-stored fruit. Peaches stored under various CA conditions lost flavor under the lower O_2 CA conditions, which was related to certain esters, an aldehyde, an acid, and several lactones (Cano-Salazar et al., 2013). Fruit maturity at harvest impacts the fruit respiratory rate and ethylene production (low respiratory rate and fatty acid levels) (Song and Bangerth, 1993, 1996). Generally, the lower the O_2, the higher the CO_2 and the longer the duration, the more suppressed the aroma volatile production (Fellman et al., 2000). The ULO condition is the most detrimental to aroma and is linked to a lack of fatty acid precursors and disturbed ethylene sensitivity. Fatty acids metabolism is thought to be regulated by ethylene in apple (Defilippi et al., 2005). Ethylene induces accumulation of linoleic and linolenic acids by upregulation of fatty acid desaturase genes producing grassy aroma aldehydes and alcohols (Zhang et al., 2010). Production of straight-chain esters in apples is suppressed by low O_2 which affects LOX activity, but biosynthesis of branched-chain esters is suppressed by high CO_2 as these ester precursors are amino acids from the TCA cycle which is inhibited by high CO_2 (Fellman et al., 2000; Gudenschwager and Defilippi, 2016).

New technologies are emerging including use of solid-state ethylene monitoring, use of 1-methylcyclopropene (1-MCP) to reduce ethylene effects, and breathable patches for MAP (Beaudry, 2010). Use of MAP is more short term than CA and can generate fermented off-flavors including acetaldehyde, ethanol, and ethyl esters, especially ethyl acetate (Almenar et al., 2009; Ke et al., 1994) such as in strawberries (Forney et al., 2000; Ke et al., 1994; Larsen and Watkins, 1995) if the product is not under strict temperature control (Gudenschwager and Defilippi, 2016) designed for that package and that specific fruit or vegetable. Yet the purpose of packaging is to increase shelf life and quality of the product by reducing postharvest decay organisms as well as food-borne pathogens to increase food safety and quality (Kalia and Parshad, 2015; Zhang et al., 2013), using fungicides, microbial control agents, radiation, antimicrobial oils or aromatic compounds, with or without MAP. Exposure of microbes to MAP, however, has sometimes led to survival and growth of food-borne pathogens, while controlling spoilage organisms, thus the produce looks good to eat but may not be safe (Caleb et al., 2013a). Flavor loss of fresh-cut produce can also be traced to diffusional loss or scalping compounds from the package atmosphere. Diffusion is dependent on the flavor compound chemical properties (its partitioning coefficient in the product matrix), concentration gradient between the product and the atmosphere, and the type of packaging (Forney, 2008).

Modification of the internal atmosphere within fruit and associated flavor effects is also true in the case of edible coatings (Dhall, 2013), especially for apples (Bai et al., 2002) and citrus (Baldwin et al., 1995; Cohen et al., 1990; Hagenmaier, 2002), which are often coated and/or stored in CA with resulting alterations in the volatile profile. This is especially true of resin coatings, but other coatings materials are more gas permeable. A carboxymethyl cellulose (CMC)- and guar gum-based silver nanoparticle coating, however, reduced volatile aroma loss in Kinnow citrus fruit stored at low temperature, which was perceived by panelists (Shah et al., 2016). A CMC-chitosan bilayer coating was compared to commercial polyethylene-based coatings on citrus fruit, slightly increasing internal CO_2 and ethanol. The coated fruit were less preferred in sensory tests for mandarins, but there was no difference for orange or grapefruit (Arnon et al., 2014). This has also been found with coated tropical fruits such as mangoes (Baldwin et al., 1999) or mangoes in CA (Bender et al., 2000a,b). Mangoes are particularly intolerant of low O_2 and high CO_2 levels and showed cultivar and maturity differences (Bender et al., 1994, 2000a,b). The coating Semperfresh and an *Aloe vera* gel delayed mango fruit ripening but also reduced aroma development (Dang et al., 2008; El Hadi et al., 2013). Strawberries coated with chitosan with or without lemon essential oil resulted in an altered volatile profile with pure chitosan promoting formation of esters and dimethyl furfural, although not sensorially perceived. Coatings with lemon oil resulted in added terpenes and also modified the normal aroma composition, which was well received by panelists (Perdones et al., 2016). Unfortunately, produce in packaging (or treated with a coating) designed for a specific storage temperature may experience higher (abusive) temperatures during the journey through the transport and marketing chain, leading to anaerobic respiration (Beaudry et al., 1992; Cameron et al., 1994). Products are also affected by fluctuations in temperature that often occurs during retail display (Brecht et al., 2003). Therefore, whether being MAP or coating, maintaining temperature cold chain is key to maintaining produce quality.

Conclusion

Use of CA has allowed some crops such as apples and pears to be available to the consumer all year-round, while it can considerably extend market life of stone fruits such as peaches and nectarines, plums, grapes, and other fruit (Kader, 2002). Yet, within a species, there is much diversity in response to gas composition depending on cultivars and harvest maturity, thus research is continuously needed to adjust combinations of packaging, gas, temperature, and humidity to optimize storage for each commodity. Increasing knowledge of biochemical processes and gene regulations will allow target gene expression to prevent specific storage disorders, or simply to maintain fruit flavor and firmness, as was done with the RIN mutant tomatoes (Giovannoni, 2007; Smith et al., 1990). Even though research is making significant advances in the knowledge of biochemical processes, information about how biochemical changes in stored produce relate to taste and texture together with consumer acceptance is still scant, as evidenced in this chapter. Collaborations between breeders and food technologists have led to more published information relating flavor chemistry and sensory perception for specific produce, probably because sampling from a breeding population allows for a wide range of qualities, making it possible to draw inferences between sensory and chemical data (Bartoshuk and Klee, 2013; Gilbert et al., 2014; Goldenberg et al., 2016, 2015; Schwieterman et al., 2014). Some of this has led to the identification of flavor quality trait loci, leading to marker- or molecular-assisted breeding for flavor (Klee, 2010). There is also the concept of consumer-assisted breeding, where consumer sensory panels drive breeder selections, or at least the understanding of consumer preferences (Bartoshuk and Klee, 2013). The next step is to follow these new cultivars through harvest maturities and postharvest management, including chilling effects, conducting chemical and sensory evaluations along the way. This knowledge will eventually translate into postharvest management for quality of new cultivars.

References

Abbott, J.A., 1999. Quality measurement of fruits and vegetables. Postharvest Biol. Technol. 15, 207–225.

Abdi, N., McGlasson, W.B., Holford, P., Williams, M., Mizrahi, Y., 1998. Responses of climacteric and suppressed-climacteric plums to treatment with propylene and 1-methylcyclopropene. Postharvest Biol. Technol. 14, 29–39.

Almenar, E., Hernández-Muñoz, P., Lagarón, J.M., Catalá, R., Gavara, R., 2006. Controlled atmosphere storage of wild strawberry fruit (*Fragaria vesca* L). J. Agric. Food Chem. 54, 86–91.

Almenar, E., Hernández-Muñoz, P., Gavara, R., 2009. Evolution of selected volatiles in chitosan-coated strawberries (*Fragaria* × *ananassa*) during refrigerated storage. J. Agric. Food Chem. 57, 974–980.

Arnon, H., Zaitsev, Y., Porat, R., Poverenov, E., 2014. Effects of carboxymethyl cellulose and chitosan bilayer edible coating on postharvest quality of citrus fruit. Postharvest Biol. Technol. 87, 21–26.

Bai, J., Hagenmaier, R.D., Baldwin, E.A., 2002. Volatile response of four apple varieties with different coatings during marketing at room temperature. J. Agric. Food Chem. 50, 7660–7668.

Bai, J., Baldwin, E.A., Goodner, K.L., Mattheis, J.P., Brecht, J.K., 2005. Response of four apple cultivars to 1-methylcyclopropene treatment and controlled atmosphere storage. HortScience 40, 1534–1538.

Bai, J., Baldwin, E.A., Imahori, Y., Kostenyuk, I., Burns, J., Brecht, J.K., 2011. Chilling and heating may regulate C6 volatile aroma production by different mechanisms in tomato (*Solanum lycopersicum*) fruit. Postharvest Biol. Technol. 60, 111–120.

Bai, J., Baldwin, E., McCollum, G., Plotto, A., Manthey, J., Widmer, W., Luzio, G., Cameron, R., 2016. Changes in volatile and non-volatile flavor chemicals of "Valencia" orange juice over the harvest seasons. Foods. 5(4). https://doi.org/10.3390/foods5010004.

Baldwin, E.A., 1993. Citrus fruit. In: Seymour, G.B., Taylor, J.E., Tucker, G.A. (Eds.), Biochemistry of Fruit Ripening. Chapman and Hall, New York, pp. 107–149.

Baldwin, E.A., Nisperos-Carriedo, M., Shaw, P.E., Burns, J.K., 1995. Effect of coatings and prolonged storage conditions on fresh orange flavor volatiles, degrees brix, and ascorbic acid levels. J. Agric. Food Chem. 43, 1321–1331.

Baldwin, E.A., Scott, J.W., Einstein, M.A., Malundo, T.M.M., Carr, B.T., Shewfelt, R.L., Tandon, K.S., 1998. Relationship between sensory and instrumental analysis for tomato flavor. J. Am. Soc. Hortic. Sci. 123, 906–915.

Baldwin, E.A., Malundo, T.M.M., Bender, R., Brecht, J.K., 1999. Interactive effects of harvest maturity, controlled atmosphere and surface coating on mango (*Mangifera indica* L.) flavor quality. HortScience 34, 514.

Baldwin, E.A., Scott, J.W., Bai, J., 2015. Sensory and chemical flavor analyses of tomato genotypes grown in Florida during three different growing seasons in multiple years. J. Am. Soc. Hortic. Sci. 140, 490–503.

Bangerth, F.K., Song, J., Streif, J., 2012. Physiological impacts of fruit ripening and storage conditions on aroma volatile formation in apple and strawberry fruit: A review. HortScience 47, 4–10.

Bartoshuk, L.M., Klee, H.J., 2013. Better fruits and vegetables through sensory analysis. Curr. Biol. 23, R374–R378.

Bauchot, A.D., Mottram, D.S., Dodson, A.T., John, P., 1998. Effect of aminocyclopropane-1-carboxylic acid oxidase antisense gene on the formation of volatile esters in cantaloupe Charentais melon (cv. Vedrandais). J. Agric. Food Chem. 46, 4787–4792.

Beaudry, R.M., 2000. Responses of horticultural commodities to low oxygen: Limits to the expanded use of modified atmosphere packaging. HortTechnology 10, 491–500.

Beaudry, R.M., 2010. Future trends and innovations in controlled atmosphere storage and modified atmosphere packaging technologies. Acta Hortic. (876), 21–28.

Beaudry, R.M., Cameron, A.C., Shirazi, A., Dostal-Lange, D.L., 1992. Modified-atmosphere packaging of blueberry fruit: effect of temperature on package O_2 and CO_2. J. Am. Soc. Hortic. Sci. 117, 436–441.

Beaudry, R.M., Deng, W., Lakakul, R., Song, J., 1996. Interaction between the package and the packaged product: Effects on respiratory and aroma volatiles. Int. Postharvest Sci. Conf. Postharvest 96 (464), 327–332.

Beckles, D.M., 2012. Factors affecting the postharvest soluble solids and sugar content of tomato (*Solanum lycopersicum* L.) fruit. Postharvest Biol. Technol. 63, 129–140.

Ben-Arie, R., Sonego, L., 1980. Pectolytic enzyme activity involved in woolly breakdown of stored peaches. Phytochemistry 19, 2553–2555.

Bender, R., Brecht, J.K., Campbell, C.A., 1994. Responses of 'Kent' and 'Tommy Atkins' mangoes to reduced O_2 and elevated CO_2. Proc. Fla. State Hort. Soc. 107, 274–277.

Bender, R., Brecht, J., Baldwin, E., Malundo, T., 2000a. Aroma volatiles of mature-green and tree-ripe 'Tommy Atkins' mangoes after controlled atmosphere vs. air storage. HortScience 35, 684–686.

Bender, R.J., Brecht, J.K., Sargent, S.A., Huber, D.J., 2000b. Mango tolerance to reduced oxygen levels in controlled atmosphere storage. J. Am. Soc. Hortic. Sci. 125, 707–713.

Berna, A.Z., Geysen, S., Li, S., Verlinden, B.E., Lammertyn, J., Nicolaï, B.M., 2007. Headspace fingerprint mass spectrometry to characterize strawberry aroma at super-atmospheric oxygen conditions. Postharvest Biol. Technol. 46, 230–236.

Blaker, K.M., Plotto, A., Baldwin, E.A., Olmstead, J.W., 2014. Correlation between sensory and instrumental measurements of standard and crisp-texture southern highbush blueberries (*Vaccinium corymbosum* L.) interspecific hybrids. J. Sci. Food Agric. 94, 2785–2793.

Both, V., Thewes, F.R., Brackmann, A., de Oliveira Anese, R., de Freitas Ferreira, D., Wagner, R., 2017. Effects of dynamic controlled atmosphere by respiratory quotient on some quality parameters and volatile profile of 'Royal Gala' apple after long-term storage. Food Chem. 215, 483–492.

Bourne, M.C., 1979. Fruit texture—an overview of trends and problems. J. Texture Stud. 10, 83–94.

Brecht, J., Chau, K., Fonseca, S., Oliveira, F., Silva, F., Nunes, M., Bender, R., 2003. Maintaining optimal atmosphere conditions for fruits and vegetables throughout the postharvest handling chain. Postharvest Biol. Technol. 27, 87–101.

Brummell, D.A., Dal Cin, V., Lurie, S., Crisosto, C.H., Labavitch, J.M., 2004. Cell wall metabolism during the development of chilling injury in cold-stored peach fruit: association of mealiness with arrested disassembly of cell wall pectins. J. Exp. Bot. 55, 2041–2052.

Buttery, R.G., Teranishi, R., Flath, R.A., Ling, L.C., 1989. Fresh tomato volatiles: composition and sensory studies. In: Teranish, R., Buttery Ron, G., Shahidi, F. (Eds.), Flavor Chemistry. American Chemical Society, Wahington, DC, pp. 213–222.

Caleb, O.J., Mahajan, P.V., Al-Said, F.A.-J., Opara, U.L., 2013a. Modified atmosphere packaging technology of fresh and fresh-cut produce and the microbial consequences—a review. Food Bioprocess Technol. 6, 303–329.

Caleb, O.J., Opara, U.L., Mahajan, P.V., Manley, M., Mokwena, L., Tredoux, A.G., 2013b. Effect of modified atmosphere packaging and storage temperature on volatile composition and postharvest life of minimally-processed pomegranate arils (cvs.'Acco'and 'Herskawitz'). Postharvest Biol. Technol. 79, 54–61.

Caleb, O.J., Fawole, O.A., Mphahlele, R.R., Opara, U.L., 2015. Impact of preharvest and postharvest factors on changes in volatile compounds of pomegranate fruit and minimally processed arils–Review. Sci. Hortic. 188, 106–114.

Cameron, A.C., Beaudry, R.M., Banks, N.H., Yelanich, M.V., 1994. Modified-atmosphere packaging of blueberry fruit: modeling respiration and package oxygen partial pressures as a function of temperature. J. Am. Soc. Hortic. Sci. 119, 534–539.

Cano-Salazar, J., López, M., Echeverría, G., 2013. Relationships between the instrumental and sensory characteristics of four peach and nectarine cultivars stored under air and CA atmospheres. Postharvest Biol. Technol. 75, 58–67.

Cardello, A.V., Hunt, D., Mann, B., 1979. Relative sweetness of fructose and sucrose in model solutions, lemon beverages and white cake. J. Food Sci. 44, 748–751.

Chen, G., Hackett, R., Walker, D., Taylor, A., Lin, Z., Grierson, D., 2004. Identification of a specific isoform of tomato lipoxygenase (TomloxC) involved in the generation of fatty acid-derived flavor compounds. Plant Physiol. 136, 2641–2651.

Cliff, M.A., Lau, O.L., King, M.C., 1998. Sensory characteristics of controlled atmosphere and air-stored 'Gala' apples. J. Food Qual. 21, 239–249.

Cohen, E., Shalom, Y., Rosenberg, I., 1990. Postharvest ethanol buildup and off-flavor in 'Murcott' tangerine fruits. J. Am. Soc. Hortic. Sci. 115, 775–778.

Dandekar, A.M., Teo, G., Defilippi, B.G., Uratsu, S.L., Passey, A.J., Kader, A.A., Stow, J.R., Colgan, R.J., James, D.J., 2004. Effect of down-regulation of ethylene biosynthesis on fruit flavor complex in apple fruit. Transgenic Res. 13, 373–384.

Dang, K.T., Singh, Z., Swinny, E.E., 2008. Edible coatings influence fruit ripening, quality, and aroma biosynthesis in mango fruit. J. Agric. Food Chem. 56, 1361–1370.

Defilippi, B.G., Dandekar, A.M., Kader, A.A., 2004. Impact of suppression of ethylene action or biosynthesis on flavor metabolites in apple (*Malus domestica* Borkh) fruits. J. Agric. Food Chem. 52, 5694–5701.

Defilippi, B.G., Kader, A.A., Dandekar, A.M., 2005. Apple aroma: alcohol acyltransferase, a rate limiting step for ester biosynthesis, is regulated by ethylene. Plant Sci. 168, 1199–1210.

Defilippi, B.G., Manriquez, D., Luengwilai, K., González-Agüero, M., 2009. Aroma volatiles: biosynthesis and mechanisms of modulation during fruit ripening. Adv. Bot. Res. 50, 1–37.

Dhall, R., 2013. Advances in edible coatings for fresh fruits and vegetables: a review. Crit. Rev. Food Sci. Nutr. 53, 435–450.

Dixon, J., Hewett, E.W., 2000. Factors affecting apple aroma/flavour volatile concentration: a review. N. Z. J. Crop. Hortic. Sci. 28, 155–173.

Echeverría, G., Fuentes, M.T., Graell, J., López, M.L., 2003. Relationships between volatile production, fruit quality and sensory evaluation of Fuji apples stored in different atmospheres by means of multivariate analysis. J. Sci. Food Agric. 84, 5–20.

Echeverria, G., Graell, J., López, M., Lara, I., 2004. Volatile production, quality and aroma-related enzyme activities during maturation of 'Fuji'apples. Postharvest Biol. Technol. 31, 217–227.

Echeverría, G., Graell, J., López, M.L., 2008. Physicochemical measurements in 'Mondial Gala®' apples stored at different atmospheres: Influence on consumer acceptability. Postharvest Biol. Technol. 50, 135–144.

El Hadi, M.A.M., Zhang, F.-J., Wu, F.-F., Zhou, C.-H., Tao, J., 2013. Advances in fruit aroma volatile research. Molecules 18, 8200–8229.

Engel, K.H., Heidlas, J., Tressl, R., 1990. The flavour of tropical fruits (Banana, Melon, Pineapple). In: Morton, I.D., Macleod, A.J. (Eds.), Food Flavours. Part C. The Flavour of Fruits. Elsevier, New York, NY, pp. 195–219.

Etienne, A., Génard, M., Lobit, P., Mbeguié-A-Mbéguié, D., Bugaud, C., 2013. What controls fleshy fruit acidity? A review of malate and citrate accumulation in fruit cells. J. Exp. Bot. 64, 1451–1469.

Fellman, J., Miller, T., Mattinson, D., Mattheis, J., 2000. Factors that influence biosynthesis of volatile flavor compounds in apple fruits. HortScience 35, 1026–1033.

Ferenczi, A., Song, J., Tian, M., Vlachonasios, K., Dilley, D., Beaudry, R., 2006. Volatile ester suppression and recovery following 1-methylcyclopropene application to apple fruit. J. Am. Soc. Hortic. Sci. 131, 691–701.

Flores, P., Hellín, P., Fenoll, J., 2012. Determination of organic acids in fruits and vegetables by liquid chromatography with tandem-mass spectrometry. Food Chem. 132, 1049–1054.

Forney, C.F., 2008. Flavour loss during postharvest handling and marketing of fresh-cut produce. Stewart Postharvest Rev. 3 (5), 1–10.

Forney, C.F., Kalt, W., Jordan, M.A., 2000. The composition of strawberry aroma is influenced by cultivar, maturity, and storage. HortScience 35, 1022–1026.

Gapper, N.E., McQuinn, R.P., Giovannoni, J.J., 2013. Molecular and genetic regulation of fruit ripening. Plant Mol. Biol. 82, 575–591.

Gilbert, J.L., Olmstead, J.W., Colquhoun, T.A., Levin, L.A., Clark, D.G., Moskowitz, H.R., 2014. Consumer-assisted selection of blueberry fruit quality traits. Hortscience 49, 864–873.

Giovannoni, J.J., 2007. Fruit ripening mutants yield insights into ripening control. Curr. Opin. Plant Biol. 10, 283–289.

Goldenberg, L., Yaniv, Y., Kaplunov, T., Doron-Faigenboim, A., Carmi, N., Porat, R., 2015. Diversity in sensory quality and determining factors influencing mandarin flavor liking. J. Food Sci. 80, S418–S425.

Goldenberg, L., Yaniv, Y., Doron-Faigenboim, A., Carmi, N., Porat, R., 2016. Diversity among mandarin varieties and natural sub-groups in aroma volatiles compositions. J. Sci. Food Agric. 96, 57–65.

Golding, J., Shearer, D., Wyllie, S., McGlasson, W., 1998. Application of 1-MCP and propylene to identify ethylene-dependent ripening processes in mature banana fruit. Postharvest Biol. Technol. 14, 87–98.

Gomes, M.H., Beaudry, R.M., Almeida, D.P., Malcata, F.X., 2010. Modelling respiration of packaged fresh-cut 'Rocha'pear as affected by oxygen concentration and temperature. J. Food Eng. 96, 74–79.

Gomes, M.H., Beaudry, R.M., Almeida, D.P., 2012. Influence of oxygen and temperature on the respiration rate of fresh-cut cantaloupe and implications for modified atmosphere packaging. HortScience 47, 1113–1116.

Grosch, W., 1994. Determination of potent odorants in foods by aroma extract dilution analysis (AEDA) and calculation of odour activity values (OAVs). Flavour Fragr. J. 9, 147–158.

Gudenschwager, O., Defilippi, B.G., 2016. Fresh fruit aroma: and integrative overview for a complex flavor trait. In: Parkeek, S. (Ed.), Postharvest Ripening Physiology of Crop CRC Press. Taylor and Francis Group, USA, pp. 513–552.

Gwanpua, S.G., Van Buggenhout, S., Verlinden, B.E., Christiaens, S., Shpigelman, A., Vicent, V., Kermani, Z.J., Nicolai, B.M., Hendrickx, M., Geeraerd, A., 2014. Pectin modifications and the role of pectin-degrading enzymes during postharvest softening of Jonagold apples. Food Chem. 158, 283–291.

Gwanpua, S.G., Mellidou, I., Boeckx, J., Kyomugasho, C., Bessemans, N., Verlinden, B.E., Hertog, M.L.A.T.M., Hendrickx, M., Nicolai, B.M., Geeraerd, A.H., 2016. Expression analysis of candidate cell wall-related genes associated with changes in pectin biochemistry during postharvest apple softening. Postharvest Biol. Technol. 112, 176–185.

Hagenmaier, R.D., 2002. The flavor of mandarin hybrids with different coatings. Postharvest Biol. Technol. 24, 79–87.

Harb, J.Y., Streif, J., 2004. Controlled atmosphere storage of highbush blueberries cv. 'Duke'. Eur. J. Hortic. Sci. 69, 66–72.

Harb, J., Bisharat, R., Streif, J., 2008. Changes in volatile constituents of blackcurrants (*Ribes nigrum* L. cv.'Titania') following controlled atmosphere storage. Postharvest Biol. Technol. 47, 271–279.

Harker, F.R., Hallett, I.C., 1992. Physiological changes associated with development of mealiness of apple fruit during cool storage. HortScience 27, 1291–1294.

Harker, F.R., Hallett, I.C., 1994. Physiological and mechanical properties of kiwifruit tissue associated with texture change during cool storage. J. Am. Soc. Hortic. Sci. 119, 987–993.

Harker, F.R., Sutherland, P.W., 1993. Physiological changes associated with fruit ripening and the development of mealy texture during storage of nectarines. Postharvest Biol. Technol. 2, 269–277.

Harker, F.R., Redgwell, R.J., Hallett, I.C., Murray, S.H., Carter, G., 1997a. Texture of fresh fruit. In: Janick, J. (Ed.), Horticultural Reviews. In: vol. 20. John Wiley & Sons Inc, Oxford, UK, pp. 121–224.

Harker, F.R., Stec, M.G.H., Hallett, I.C., Bennett, C.L., 1997b. Texture of parenchymatous plant tissue: a comparison between tensile and other instrumental and sensory measurements of tissue strength and juiciness. Postharvest Biol. Technol. 11, 63–72.

Harker, F.R., Elgar, H.J., Watkins, C.B., Jackson, P.J., Hallett, I.C., 2000. Physical and mechanical changes in strawberry fruit after high carbon dioxide treatments. Postharvest Biol. Technol. 19, 139–146.

Harker, F.R., Maindonald, J., Murray, S.H., Gunson, F.A., Hallett, I.C., Walker, S.B., 2002. Sensory interpretation of instrumental measurements 1: texture of apple fruit. Postharvest Biol. Technol. 24, 225–239.

Hartwig, P.A.M., McDaniel, M.R., 1995. Flavor characteristics of lactic, malic, citric, and acetic acids at various pH levels. J. Food Sci. 60, 384–388.

Ho, Q.T., Rogge, S., Verboven, P., Verlinden, B.E., Nicolaï, B.M., 2016. Stochastic modelling for virtual engineering of controlled atmosphere storage of fruit. J. Food Eng. 176, 77–87.

Holcroft, D.M., Kader, A.A., 1999. Controlled atmosphere-induced changes in pH and organic acid metabolism may affect color of stored strawberry fruit. Postharvest Biol. Technol. 17, 19–32.

Jordan, R., Seelye, R., McGlone, A., 2001. A sensory-based alternative to brix/acid ratio. Food Technol. 55, 36–44.

Kader, A.A., 2002. Modified atmospheres during transport and storage. In: Kader, A.A. (Ed.), Postharvest Technology of Horticultural Crops. University of California Agriculture and Natural Resources, pp. 135–144. Publication 3311.

Kader, A.A., 2008. Flavor quality of fruits and vegetables. J. Sci. Food Agric. 88, 1863–1868.

Kalia, A., Parshad, V.R., 2015. Novel trends to revolutionize preservation and packaging of fruits/fruit products: microbiological and nanotechnological perspectives. Crit. Rev. Food Sci. Nutr. 55, 159–182.

Ke, D., Zhou, L., Kader, A.A., 1994. Mode of oxygen and carbon dioxide action on strawberry ester biosynthesis. J. Am. Soc. Hortic. Sci. 119, 971–975.

Klee, H.J., 2010. Improving the flavor of fresh fruits: genomics, biochemistry, and biotechnology. New Phytol. 187, 44–56.

Koehler, P.E., Kays, S.J., 1991. Sweet potato flavor: quantitative and qualitative assessment of optimum sweetness. J. Food Qual. 14, 241–249.

Lakakul, R., Beaudry, R., Hernandez, R., 1999. Modeling respiration of apple slices in modified-atmosphere packages. J. Food Sci. 64, 105–110.

Lara, I., Miró, R., Fuentes, T., Sayez, G., Graell, J., López, M., 2003. Biosynthesis of volatile aroma compounds in pear fruit stored under long-term controlled-atmosphere conditions. Postharvest Biol. Technol. 29, 29–39.

Lara, I., Graell, J., López, M., Echeverría, G., 2006. Multivariate analysis of modifications in biosynthesis of volatile compounds after CA storage of 'Fuji'apples. Postharvest Biol. Technol. 39, 19–28.

Larsen, M., Watkins, C.B., 1995. Firmness and concentrations of acetaldehyde, ethyl acetate and ethanol in strawberries stored in controlled and modified atmospheres. Postharvest Biol. Technol. 5, 39–50.

Lashbrook, C.C., 2005. New insights into cell wall disassembly during fruit ripening. Stewart Postharvest Rev. 1 (3), 1–18.

Latocha, P., Krupa, T., Jankowski, P., Radzanowska, J., 2014. Changes in postharvest physicochemical and sensory characteristics of hardy kiwifruit (Actinidia arguta and its hybrid) after cold storage under normal versus controlled atmosphere. Postharvest Biol. Technol. 88, 21–33.

Li, B.W., Andrews, K.W., Pehrsson, P.R., 2002. Individual sugars, soluble, and insoluble dietary fiber contents of 70 high consumption foods. J. Food Compost. Anal. 15, 715–723.

Lidster, P.D., Lightfoot, H.J., McRae, K.B., 1983. Fruit quality and respiration of 'McIntosh' apples in response to ethylene, very low oxygen and carbon dioxide storage atmospheres. Sci. Hortic. 20, 71–83.

López, M.L., Villatoro, C., Fuentes, T., Graell, J., Lara, I., Echeverría, G., 2007. Volatile compounds, quality parameters and consumer acceptance of 'Pink Lady®' apples stored in different conditions. Postharvest Biol. Technol. 43, 55–66.

Lurie, S., 1992. Controlled atmosphere storage to decrease physiological disorders in nectarines. Int. J. Food Sci. Technol. 27, 507–514.

Lurie, S., Crisosto, C.H., 2005. Chilling injury in peach and nectarine. Postharvest Biol. Technol. 37, 195–208.

Mahajan, P.V., Goswami, T.K., 2004. Extended storage life of litchi fruit using controlled atmosphere and low temperature. J. Food Process. Preserv. 28, 388–403.

Mann, H., Bedford, D., Luby, J., Vickers, Z., Tong, C., 2005. Relationship of instrumental and sensory texture measurements of fresh and stored apples to cell number and size. HortScience 40, 1815–1820.

Mattheis, J.P., Fan, X., Argenta, L.C., 2005. Interactive responses of 'Gala' apple fruit volatile production to controlled atmosphere storage and chemical inhibition of ethylene action. J. Agric. Food Chem. 53, 4510–4516.

Mehinagic, E., Royer, G., Bertrand, D., Symoneaux, R., Laurens, F., Jourjon, F., 2003. Relationship between sensory analysis, penetrometry and visible-NIR spectroscopy of apples belonging to different cultivars. Food Qual. Prefer. 14, 473–484.

Mehinagic, E., Royer, G., Symoneaux, R., Bertrand, D., Jourjon, F., 2004. Prediction of the sensory quality of apples by physical measurements. Postharvest Biol. Technol. 34, 257–269.

Ménager, I., Jost, M., Aubert, C., 2004. Changes in physicochemical characteristics and volatile constituents of strawberry (Cv. Cigaline) during maturation. J. Agric. Food Chem. 52, 1248–1254.

Moskowitz, H.R., 1970. Ratio scales of sugar sweetness. Percept. Psychophys. 7, 315–320.

Moya-Leon, M.A., Vergara, M., Bravo, C., Pereira, M., Moggia, C., 2007. Development of aroma compounds and sensory quality of 'Royal Gala' apples during storage. J. Hortic. Sci. Biotech. 82, 403–413.

Ngamchuachit, P., Sivertsen, H.K., Mitcham, E.J., Barrett, D.M., 2015. Influence of cultivar and ripeness stage at the time of fresh-cut processing on instrumental and sensory qualities of fresh-cut mangos. Postharvest Biol. Technol. 106, 11–20.

Obenland, D., Collin, S., Mackey, B., Sievert, J., Fjeld, K., Arpaia, M.L., 2009. Determinants of flavor acceptability during the maturation of navel oranges. Postharvest Biol. Technol. 52, 156–163.

Obenland, D., Collin, S., Mackey, B., Sievert, J., Arpaia, M.L., 2011. Storage temperature and time influences sensory quality of mandarins by altering soluble solids, acidity and aroma volatile composition. Postharvest Biol. Technol. 59, 187–193.

Oeller, P.W., Min-Wong, L., Taylor, L.P., Pike, D.A., Theologis, A., 1991. Reversible inhibition of tomato fruit senescence by antisense RNA. Science 254, 437–439.

Ortiz, A., Lara, I., Graell, J., López, M.L., Echeverría, G., 2002. Sensory acceptance of CA-stored peach fruit. Relationship to instrumental quality parameters. In: Streif, J., McCormick, R. (Eds.), Acta Horticulturae 796. ISHS 2008, pp. 225–230.

Ortiz, A., Echeverría, G., Graell, J., Lara, I., 2009. Overall quality of 'Rich Lady' peach fruit after air- or CA storage. The importance of volatile emission. LWT Food Sci. Technol. 42, 1520–1529.

Osorio, S., Muñoz, C., Valpuesta, V., 2010. Physiology and biochemistry of fruit flavors. In: Hui, Y.-H. (Ed.), Handbook of Fruit and Vegetable Flavors. Wiley & Sons, Hoboken, NJ, pp. 25–43.

Pangborn, R.M., 1963. Relative taste intensities of selected sugars and organic acids. J. Food Sci. 28, 726–733.

Paul, V., Pandey, R., 2014. Role of internal atmosphere on fruit ripening and storability—a review. J. Food Sci. Technol. 51, 1223–1250.

Perdones, Á., Escriche, I., Chiralt, A., Vargas, M., 2016. Effect of chitosan–lemon essential oil coatings on volatile profile of strawberries during storage. Food Chem. 197, 979–986.

Plotto, A., McDaniel, M.R., Mattheis, J.P., 1999. Characterization of 'Gala' apple aroma and flavor: differences between controlled atmosphere and air storage. J. Am. Soc. Hortic. Sci. 124, 416–423.

Plotto, A., Margaría, C.A., Goodner, K.L., Goodrich, R., Baldwin, E.A., 2004. Odour and flavour thresholds for key aroma components in an orange juice matrix: terpenes and aldehydes. Flavour Fragr. J. 19, 491–498.

Plotto, A., Margaría, C.A., Goodner, K.L., Baldwin, E.A., 2008. Odour and flavour thresholds for key aroma components in an orange juice matrix: esters and miscellaneous compounds. Flavour Fragr. J. 23, 398–406.

Plotto, A., Bai, J., Baldwin, E., 2017. Fruits. In: Buettner, A. (Ed.), Springer Handbook of Odor. Springer International Publishing, Cham, pp. 27–28.

Porat, R., Fallik, E., 2008. Production of off-flavours in fruit and vegetables under fermentative conditions. In: Brückner, B., Wyllie, S.G. (Eds.), Fruit and Vegetable Flavour: Recent Advances and Future Prospects. Woodhead Publishing Ltd., Cambridge, UK, pp. 150–164.

Rizzolo, A., Grassi, M., Vanoli, M., 2014. 1-Methylcyclopropene application, storage temperature and atmosphere modulate sensory quality changes in shelf-life of 'Abbé Fétel' pears. Postharvest Biol. Technol. 92, 87–97.

Rodriguez, J., Zoffoli, J.P., 2016. Effect of sulfur dioxide and modified atmosphere packaging on blueberry postharvest quality. Postharvest Biol. Technol. 117, 230–238.

Rose, J.K.C., Bennett, A.B., 1999. Cooperative disassembly of the cellulose–xyloglucan network of plant cell walls: parallels between cell expansion and fruit ripening. Trends Plant Sci. 4, 176–183.

Rosenfeld, H.J., Røed Meberg, K., Haffner, K., Sundell, H.A., 1999. MAP of highbush blueberries: sensory quality in relation to storage temperature, film type and initial high oxygen atmosphere. Postharvest Biol. Technol. 16, 27–36.

Rubico, S.M., McDaniel, M.R., 1992. Sensory evaluation of acids by free-choice profiling. Chem. Senses 17, 273–289.

Rychlik, M., Schieberle, P., Grosch, W., 1998. Compilation of odor thresholds, odor qualities and retention indices of key food odorants. Institut für Lebensmittelchemie der Technischen Universität München und Deutsche Forschungsanstalt für Lebensmittelchemie Garching, Germany.

Saftner, R.A., Abbott, J.A., Conway, W.S., Barden, C.L., Vinyard, B.T., 2002. Instrumental and sensory quality characteristics of 'Gala' apples in response to prestorage heat, controlled atmosphere, and air storage. J. Am. Soc. Hortic. Sci. 127, 1006–1012.

Saftner, R.A., Polashock, J.J., Ehlenfeldt, M., Vinyard, B., 2008. Instrumental and sensory quality characteristics of blueberry fruit from twelve cultivars. Postharvest Biol. Technol. 49, 19–26.

Sanz, C., Perez, A.G., 2010. Plant metabolic pathways and flavor biosynthesis. In: Hui, Y.H. (Ed.), Handbook of Fruit and Vegetable Flavors. Wiley and Sons, Hoboken, NJ, pp. 129–155.

Schotsmans, W., Molan, A., MacKay, B., 2007. Controlled atmosphere storage of rabbiteye blueberries enhances postharvest quality aspects. Postharvest Biol. Technol. 44, 277–285.

Schwab, W., Davidovich-Rikanati, R., Lewinsohn, E., 2008. Biosynthesis of plant-derived flavor compounds. Plant J. 54, 712–732.

Schwieterman, M.L., Colquhoun, T.A., Jaworski, E.A., Bartoshuk, L.M., Gilbert, J.L., Tieman, D.M., Odabasi, A.Z., Moskowitz, H.R., Folta, K.M., Klee, H.J., Sims, C.A., Whitaker, V.W., Clark, D.G., 2014. Strawberry flavor: diverse chemical compositions, a seasonal influence, and effects on sensory perception. PLoS One. 9, e88446.

Shah, S.W.A., Qaisar, M., Jahangir, M., Abbasi, K.S., Khan, S.U., Ali, N., Liaquat, M., 2016. Influence of CMC-and guar gum-based silver nanoparticle coatings combined with low temperature on major aroma volatile components and the sensory quality of kinnow (*Citrus reticulata*). Int. J. Food Sci. Technol. 51, 2345–2352.

Sheehy, R.E., Kramer, M., Hiatt, W.R., 1988. Reduction of polygalacturonase activity in tomato fruit by antisense RNA. Proc. Natl. Acad. Sci. U. S. A. 85, 8805–8809.

Siddiqui, S., Brackmann, A., Streif, J., Bangerth, F., 1996. Controlled atmosphere storage of apples: cell wall composition and fruit softening. J. Hortic. Sci. 71, 613–620.

Sivakumar, D., Korsten, L., 2010. Fruit quality and physiological responses of litchi cultivar McLean's Red to 1-methylcyclopropene pre-treatment and controlled atmosphere storage conditions. LWT Food Sci. Technol. 43, 942–948.

Smith, R.B., 1992. Controlled atmosphere storage of 'Redcoat' strawberry fruit. J. Am. Soc. Hortic. Sci. 117, 260–264.

Smith, R.B., Skog, L.J., 1992. Postharvest carbon dioxide treatment enhances firmness of several cultivars of strawberry. HortScience 27, 420–421.

Smith, C.J.S., Watson, C.F., Morris, P.C., Bird, C.R., Seymour, G.B., Gray, J.E., Arnold, C., Tucker, G.A., Schuch, W., Harding, S., Grierson, D., 1990. Inheritance and effect on ripening of antisense polygalacturonase genes in transgenic tomatoes. Plant Mol. Biol. 14, 369–379.

Song, J., Bangerth, F., 1993. Production and development of volatile aroma compounds of apple fruits at different times of maturity. Int. Symp. Postharvest Treat. Horticultural Crops 368, 150–159.

Song, J., Bangerth, F., 1996. The effect of harvest date on aroma compound production from 'Golden Delicious' apple fruit and relationship to respiration and ethylene production. Postharvest Biol. Technol. 8, 259–269.

Song, J., Forney, C.F., 2008. Flavour volatile production and regulation in fruit. Can. J. Plant Sci. 88, 537–550.

Surendranathan, K., Nair, P.M., 1976. Stimulation of the glyoxylate shunt in gamma-irradiated banana. Phytochemistry 15, 371–373.

Sweetman, C., Deluc, L.G., Cramer, G.R., Ford, C.M., Soole, K.L., 2009. Regulation of malate metabolism in grape berry and other developing fruits. Phytochemistry 70, 1329–1344.

Szczesniak, A.S., Kahn, E.L., 1971. Consumer awareness of and attitudes to food texture. I. Adults. J. Texture Stud. 2, 280–295.

Takeoka, G.R., Buttery, R.G., Turnbaugh, J.G., Benson, M., 1995. Odor thresholds of various branched esters. LWT Food Sci. Technol. 28, 153–156.

Takeoka, G., Buttery, R.G., Ling, L., 1996. Odour thresholds of various branched and straight chain acetates. LWT Food Sci. Technol. 29, 677–680.

Tandon, K.S., Baldwin, E.A., Shewfelt, R.L., 2000. Aroma perception of individual volatile compounds in fresh tomatoes (*Lycopersicon esculentum*, Mill.) as affected by the medium of evaluation. Postharvest Biol. Technol. 20, 261–268.

Terrier, N., Glissant, D., Grimplet, J., Barrieu, F., Abbal, P., Couture, C., Ageorges, A., Atanassova, R., Léon, C., Renaudin, J.-P., 2005. Isogene specific oligo arrays reveal multifaceted changes in gene expression during grape berry (Vitis vinifera L.) development. Planta 222, 832–847.

Thompson, A.K., 1998. Controlled Atmosphere Storage of Fruits and Vegetables. CAB International, Wallingford, Oxon, UK.

van Gemert, L.J., 2003. Odour thresholds: Compilations of odour threshold values in air, water and other media. Oliemans Punter & Partners BV, Utrecht, The Netherlands.

Villatoro, C., Altisent, R., Echeverría, G., Graell, J., López, M., Lara, I., 2008. Changes in biosynthesis of aroma volatile compounds during on-tree maturation of 'Pink Lady®' apples. Postharvest Biol. Technol. 47, 286–295.

Wang, C.Y., 2006. Biochemical basis of the effects of modified and controlled atmospheres. Stewart Postharvest Rev. 2, 1–4.

Watkins, C.B., Manzano-Mendez, J.E., Nock, J.F., Zhang, J., Maloney, K.E., 1999. Cultivar variation in response of strawberry fruit to high carbon dioxide treatments. J. Sci. Food Agric. 79, 886–890.

Watkins, C.B., Nock, J.F., Whitaker, B.D., 2000. Responses of early, mid and late season apple cultivars to postharvest application of 1-methylcyclopropene (1-MCP) under air and controlled atmosphere storage conditions. Postharvest Biol. Technol. 19, 17–32.

Weber, A., Brackmann, A., Both, V., Pavanello, E.P., Anese, R.d.O., Thewes, F.R., 2015. Respiratory quotient: innovative method for monitoring Royal Gala apple storage in a dynamic controlled atmosphere. Sci. Agric. 72, 28–33.

Wilson III, C.W., Shaw, P.E., Knight Jr., R.J., 1982. Analysis of oxalic acid in carambola (Averrhoa carambola L.) and spinach by high-performance liquid chromatography. J. Agric. Food Chem. 30, 1106–1108.

Zhang, B., Shen, J.-Y., Wei, W.-W., Xi, W.-P., Xu, C.-J., Ferguson, I., Chen, K., 2010. Expression of genes associated with aroma formation derived from the fatty acid pathway during peach fruit ripening. J. Agric. Food Chem. 58, 6157–6165.

Zhang, B.-Y., Samapundo, S., Pothakos, V., de Baenst, I., Sürengil, G., Noseda, B., Devlieghere, F., 2013. Effect of atmospheres combining high oxygen and carbon dioxide levels on microbial spoilage and sensory quality of fresh-cut pineapple. Postharvest Biol. Technol. 86, 73–84.

Zhang, B., Tieman, D.M., Jiao, C., Xu, Y., Chen, K., Fei, Z., Giovannoni, J.J., Klee, H.J., 2016. Chilling-induced tomato flavor loss is associated with altered volatile synthesis and transient changes in DNA methylation. Proc. Natl. Acad. Sci. U. S. A. 113, 12580–12585.

Zhou, H.-W., Lurie, S., Lers, A., Khatchitski, A., Sonego, L., Ben Arie, R., 2000a. Delayed storage and controlled atmosphere storage of nectarines: two strategies to prevent woolliness. Postharvest Biol. Technol. 18, 133–141.

Zhou, H.-W., Ben-Arie, R., Lurie, S., 2000b. Pectin esterase, polygalacturonase and gel formation in peach pectin fractions. Phytochemistry 55, 191–195.

Zhu, Z., Liu, R., Li, B., Tian, S., 2013. Characterisation of genes encoding key enzymes involved in sugar metabolism of apple fruit in controlled atmosphere storage. Food Chem. 141, 3323–3328.

Chapter 7

CA/MA on bioactive compounds

Maria Isabel Gil[*], Maria Luisa Amodio[†] and Giancarlo Colelli[†]
[*]*Group on Quality and Safety, Department of Food Science and Technology, CEBAS-CSIC, Murcia, Spain,* [†]*Department of Science of Agriculture, Food, and Environment, University of Foggia, Foggia, Italy*

Introduction

Bioactive compounds are essential and nonessential compounds (e.g., vitamins or polyphenols) that occur in fruit and vegetables and are proved to have an effect on human health (Biesalski et al., 2009). Among all, the most extensively studied are those with antioxidant properties that protect cellular systems from oxidative damage through a variety of mechanisms and reduce the risk of chronic diseases including cancer and cardiovascular disease (Kris-Etherton et al., 2002). Bioactive compounds vary widely in chemical structure, being phenolics, phytoestrogens, carotenoids, tocopherols, and glucosinolates the most important groups. Phenolics are a group of compounds that include flavonoids (e.g., flavonols, flavones, flavan-3-ols, anthocyanidins, flavanones, isoflavones, condensed tannins) and nonflavonoids (e.g., phenolic acids, hydroxycinnamates, stilbenes, hydrolyzable tannins) depending on the number and arrangement of their carbon atoms. Phenolic compounds can act as copigments and can play important roles in attracting pollinators, augmenting the plants' chemical defense mechanism against UV radiation and insect attack, and serve as substrates in browning reactions (Harborne and Williams, 2000). Many phenolic compounds have antioxidant properties with favorable effects in reducing the risks of coronary heart disease. Carotenoids are a class of natural fat-soluble pigments that are responsible for the yellow, orange, or red color of many fruit and vegetables. As phenolic compounds, some carotenoids (i.e., provitamin A compounds such as α-carotene, β-carotene, and β-cryptoxanthin) can act as free radical scavengers, which play a beneficial role in human health. Nonprovitamin A carotenoids, such as lutein and lycopene, may also contribute to the health benefits by protecting against degenerative diseases (Tsao et al., 2007). Glucosinolates are a group of sulfur- and nitrogen-containing secondary metabolites present in cruciferous species such as mustard and broccoli, identified by their anticancer activities. Finally, also

Controlled and Modified Atmospheres for Fresh and Fresh-Cut Produce
https://doi.org/10.1016/B978-0-12-804599-2.00008-9

monoterpenes present in citrus, cherries, and herbs have shown anticarcinogenic actions, decreasing the risk of colorectal cancers as well as cardioprotective effects.

The content of bioactive compounds can be affected by storage time and environmental conditions, particularly temperature, relative humidity (RH), and gas composition (O_2, CO_2, and C_2H_4). Most of the studies related to bioactive compounds have been focused on the evaluation of controlled atmosphere (CA) or modified atmosphere (MA) storage conditions in preserving the content of bioactive compounds, and particularly their antioxidant capacity. Storage under CA/MA offers the possibility to retard the respiration rate, maintain bioactive compounds, and extend the shelf life of fruit and vegetables (Li et al., 2012). When fruit and vegetables are stored within the optimum range of low oxygen (O_2) and/or elevated carbon dioxide (CO_2) atmospheres, respiration and ethylene production rates reduce (Kader, 1997). Optimum CA/MA retards biosynthesis of carotenoids and anthocyanins as well as the oxidation of phenolic compounds. Storage under optimum CA/MA results in better retention of ascorbic acid and other vitamins including antioxidant activity (Kader, 2001). After harvest, storage conditions of low O_2 and/or up to 10% CO_2 atmospheres have a beneficial effect preserving vitamin C content although it varies among commodities, CO_2 levels, storage temperatures, and storage time (Weichmann, 1986; Watada, 1987; Lee and Kader, 2000; Kader, 2001). In fresh-cut fruit and vegetables, changes in the phytochemical status with processing and storage under MA have been described in detail for several commodities (Devlieghere et al., 2002; Gil and Kader, 2008; Yang, 2011; Pirovani et al., 2015). Fresh-cut processing increases oxidative stress, quality losses due to microbial contamination, softening, and browning. However, the depletion of bioactive compounds such as phenolics, ascorbic acid, and carotenoids is minimum (Gil et al., 2006).

This chapter is intended to review those scientific articles dealing with the influence of CA/MA conditions during storage to preserve the quality and the content of bioactive compounds.

Bioactive compounds of fruits as affected by CA/MA

Pome fruits

The optimum levels for CA storage on numerous varieties of apples and pears have been extensively reviewed (Chen, 2016; Watkins et al., 2016). The effect of CA during cold storage on apple phenolic content has been reported to be different depending on the cultivar (Kupferman, 2001; DeEll, 2012). Storage conditions such as atmosphere and temperature appear to have more influence on the quality criteria related to taste and appearance than on the health-related parameters (Remberg et al., 2010). Van der Sluis et al. (2003) studied the effect of CA storage conditions for Jonagold, Golden Delicious, and Elstar

(1.2% O_2 + 2.5% CO_2 at 1.5°C) and Cox's Orange apples (1.2% O_2 + 0.7% CO_2 at 4.0°C). They evaluated the content of flavonol, catechin, phloridzin, chlorogenic acid, as well as antioxidant activities during storage. They found that CA and cold air storage do not apparently compromise antioxidant activity; therefore they maintain the potential health benefits of the fruits. A recent study was conducted to investigate the effect of dynamic controlled atmospheres (DCA) (<0.5% O_2 + 1% CO_2) on phytochemicals in the apple peel of "Granny Smith" stored for up to 20 weeks at 0°C (Mditshwa et al., 2017). Fruit stored in DCA were characterized by higher antioxidant capacity, ascorbic acid, and total phenolic concentration. However, lipid peroxidation and ROS were significantly lower in DCA than in air. This study showed that the high fruit quality in DCA correlates with higher antioxidant capacity, total phenolics, and ascorbic acid concentrations in these fruit. On the other hand, flesh browning in "Cripps' Pink" apples, which causes major economic losses to the South African fresh fruit export industry, was studied after storage for 1, 4, or 7 months in CA conditions (1% CO_2 and 1.5% O_2) at −0.5°C, followed by 4 weeks in air plus 7 days at 20°C (Crouch et al., 2015). These authors observed that late harvested fruit had higher levels of diffuse flesh browning (DFB) and developed DFB sooner during CA storage compared with fruit harvested at optimum maturity. Total ascorbate concentrations in late-harvested fruit decreased in the first 4 months of CA storage, while it took from 4 to 7 months to decrease in CA-stored fruit harvested at optimum maturity. This study indicated that harvest maturity and storage duration are critical factors influencing the content of bioactive compounds. Neuwald et al. (2015) studied the development of a physiological skin burning disorder caused by CO_2 injury during CA storage of "Cameo" apples. Authors observed that delayed CA as well as lower CO_2 partial pressures in CA storage (1 kPa O_2 plus <0.7 kPa CO_2) reduced skin burning of fruit stored at 1°C for 7 months, followed by 7 day of shelf life at 20°C. Fruit with higher blush color showed less skin burning after storage and the skin of affected fruit had lower levels of ascorbic acid.

In pear and apple, depletion of ascorbate has previously been associated with development of stress-related flesh browning of intact fruit. Delaporte (1971) and others observed that storing apples in a reduced O_2 atmosphere reduced loss of ascorbic acid. However, Haffner et al. (1997) observed in various apple cultivars that ascorbic acid levels decreased more under ultralow oxygen (ULO) compared to air storage. Regarding CO_2, Bangerth (1977) reported accelerated ascorbic acid losses in apples stored in elevated CO_2 atmospheres. In pears, Veltman et al. (1999) also observed a 60% loss in ascorbic acid content of "Conference" pears after storage in 2% O_2 + 10% CO_2. Felicetti et al. (2011) studied the possible chemical and biochemical causes for CO_2-induced browning in apple by the evaluation of the ascorbic acid dynamics in "Braeburn," a susceptible cultivar, and "Gala," a resistant cultivar, during on-tree development and storage at 0.5°C in air or CA (1 kPa O_2 and 1, 3, or 5 kPa CO_2). "Braeburn" apples had significantly higher ascorbic acid levels than "Gala" during on-tree development and storage and generally ascorbic acid was diffusely distributed throughout the cortex in both cultivars during on-tree development. During storage, ascorbic acid was localized in the periphery of brown tissue in "Braeburn." Cascia et al. (2013) described the changes in ascorbate content in three pear varieties "Williams Bon Chretien," "Doyenne du Comice," and "Beurre Bosc" stored at 0.5°C in air or CA (2 kPa O_2 and 5 kPa CO_2). They observed small differences in ascorbate and suggested that rather than ascorbate content determining the potential to develop internal browning, it may be how ascorbate is utilized. They suggested that a transitory increase in dehydroascorbate (DHA) during postharvest storage may indicate that fruits are at risk of developing the disorder.

Stone fruits

Sweet cherries (*Prunus avium* L.) are among the most valued seasonal fruit. The content of bioactive compounds is a relevant characteristic of sweet cherries that influences consumer acceptability. Aglar et al. (2017) compared the effect of MAP storage, obtained by storing 5 kg of "0900 Ziraat" sweet cherries in commercial plastic bags (Xtend®, Stepac, Israel), also in combination with a preharvest coating named "Parka," on bioactive compounds, during 21 days of storage and after additional 3 days of shelf life at ambient temperature. In both cold storage and shelf life measurements, vitamin C content of MAP-treated fruits was higher than the control and Parka-treated fruits, which lost, respectively, 62% and 55% of vitamin C by the end of storage period. Phenolic acids and monomeric anthocyanins increased over storage, but "Parka," MAP, and "Parka" + MAP treatments showed lower phenolic content than in control fruits. As a consequence, antioxidant activity measured by FRAP was lower than in control while TEAC activity of fruits stored in MAP was comparable to control fruit. González-Gómez et al. (2017) evaluated different postharvest treatments in order to establish the most appropriate storage conditions to preserve the overall quality of "Sweetheart" cherries including the content of anthocyanins, phenolic acids, and flavonoids. They stored the fruit in CA (8% O_2 and 10% CO_2) at 1°C 1-kg bags of cherries (microperforated bags of Pplus®) for 27 days. According to their results, the MAP storage conditions were the most appropriate ones to maintain or even increase the concentration of bioactive compounds.

Small fruits

In strawberries, CA/MA storage can extend the shelf life in terms of appearance, textural quality, and flavor, but the content of bioactive compounds can be negatively affected. Thus CO_2-enriched atmospheres (>20%) used to reduce decay and extend the postharvest life of strawberries induced a remarkable decrease in anthocyanin content of internal tissues compared with the storage in air that increased the anthocyanin levels (Gil et al., 1997). Holcroft and Kader (1999) related the decrease in strawberry color under CO_2 atmosphere, with a decrease in the activity of two enzymes involved in the biosynthesis of anthocyanins, phenylalanine ammonia lyase and glucosyltransferase. In addition, high CO_2 concentrations (10%–30% CO_2) reduced the content of vitamin C in strawberries and blackberries whereas in black currants, red currants, and raspberries the effect was moderate to negligible (Agar et al., 1997). It has been reported that the reduction of ascorbic acid and the relative increase in dehydroascorbic acid could be an indication that high CO_2 stimulates the oxidation of ascorbic acid by ascorbate peroxidase (Agar et al., 1997).

In grapes, long-term storage in MAP and CA of the cultivar "Napoleon" maintained or even enhanced the content of phenolic compounds (Artés-Hernández et al., 2003). These authors studied the impact of CA and MAP during 38 days of storage at 0°C followed by 6 days of shelf life at 15°C in air. They observed that the best results were obtained in MAP with 5 kPa O_2 + 15 kPa CO_2. In CA, the total anthocyanin content, including the major anthocyanin peonidin 3-glucoside, declined after storage while the total stilbenoid content increased up to twofold after shelf life (Artés-Hernández et al., 2003). In addition, the incorporation of antioxidants in an edible coating on grapes can enhance the level of phenolics, ascorbic acid, and total antioxidant activity (Baraiya et al., 2016).

Tropical fruits

In mangos, Sivakumar et al. (2012) showed that 1-MCP and CA (3% O_2 + 8% CO_2) maintained ascorbic acid, carotenoid, total phenolic and flavonoid contents, and antioxidant scavenging activity in hot water-treated mangoes. In longan fruit, Khan et al. (2017) established the tolerance levels to low O_2 and elevated CO_2 at 2°C and found that CA (5% O_2 + 5% CO_2, 5% O_2 + 10% CO_2 and 5% O_2 + 15% CO_2) reduced the polyphenol oxidase (PPO) activity, maintained L^* value, and slowed down the decrease of total phenolic contents.

Subtropical fruits

In persimmon fruit (*Diospyros kaki*), astringency is removed under anaerobiosis in CO_2- or N_2-enriched atmospheres. These treatments result in the production of acetaldehyde, and deastringency is due to the insolubilization of kaki-tannin by reaction with the acetaldehyde (Haslam et al., 1992). In figs, MAP technique during cold storage has been described as a good tool to improve antioxidant activity and the content of bioactive compounds (Guillén et al., 2015). When figs (*Ficus carica*) were stored at 2°C during 28 days under MAP it was observed that total antioxidant activity and total polyphenols increased while anthocyanins were delayed in comparison with unpacked figs. In guava (*Psidium guajava*) fruit, the effects of CA (2.5, 5, 8, and 10 kPa O_2 + 2.5, 5, and 10 kPa CO_2) in three commercially important cultivars were studied during storage for 30 days at 8°C and after a simulated shelf life period in ambient conditions (25–28°C and 60%–70% RH) for ripening (Singh and Pal, 2008). Storage under CA was effective in reducing weight loss and maintaining fruit firmness. The ascorbic acid and total phenol contents were reduced by CA. High CO_2 (>5 kPa) caused a reduction in ascorbic acid levels. In pomegranate, moderated CO_2 atmospheres (10%) prolonged the storage life and maintained the quality and adequate red color intensity of the arils (Holcroft et al., 1998). However, the color of pomegranate arils stored in air was a deeper red compared to initial controls and those stored in CO_2-enriched atmospheres. In grapefruit, Chaudhary et al. (2015) investigated the effect of MA and RH generated in microperforated and macroperforated bags on the levels of phytochemicals of "Star Ruby" cultivar stored for 16 weeks at 10°C. The MAP treatments did not significantly affect ascorbic acid or limonoids and did not impact any fruit quality parameters, suggesting that MAP can be used to maintain the quality of "Star Ruby" grapefruit with no detrimental effect on health-promoting phytochemicals. In date fruits (*Phoenix dactylifera* L.), Al-Redhaiman (2005) studied the effect of storage in CA of 5%, 10%, or 20% CO_2 at 0°C in mature "Barhi" dates. These authors observed that the quality was preserved for 26 weeks when stored in 20% CO_2, and color, firmness, SSC, total sugar content, and total tannins were maintained. They also observed that CO_2 treatments reduced degradation of caffeoylshikimic acid, one of the major phenolic compounds of date fruit.

Fresh-cut fruits

Modified atmosphere packaging (MAP) is a common and highly useful industrial preservation technique for prolonging freshness of fresh-cut fruits. The study by Putnik et al. (2017) described the changes in phenolics and antioxidant capacity of fresh-cut apples packaged in MAP at 4°C for 14 days and showed that phenolic compounds such as chlorogenic and *p*-coumaroylquinic acids, quercetin, epicatechin, and phloridzin were stable during storage in MAP. The reduction in browning potential of fresh-cut apples was also studied by Tardelli et al. (2013) in response to the 1 and 2 weeks to CA treatment of 1.2 kPa O_2 and 1 kPa CO_2 for 4 months at 3°C prior to slicing. Total phenolic content was also significantly lower in slices from the 1 and 2 weeks in CA treatments due probably to a direct effect on reducing phenolic substrate accumulation in the slices and hence providing less substrate for endogenous oxidative enzymes (Tardelli et al., 2013). For fresh-cut peaches, slices stored in 2 kPa O_2 + 10 kPa CO_2 showed the highest total phenolic content (71 mg gallic acid per 100 g) compared to samples stored in 2 kPa O_2 + 20 kPa CO_2 (46.7 mg gallic acid per 100 g), while in air or 2 kPa O_2 intermediate values were observed (57–59 mg 100 g^{-1}) (González-Buesa et al., 2011). These authors also observed a reduction up to 50% in the total phenolic content of different peach cultivars stored at 4°C in passive MA. According to some authors (Mateos et al., 1993), the reduced activity of the enzyme phenylalanine ammonia-lyase (PAL), as a result of a decrease in cytoplasmic pH, may induce a lower total phenolic content biosynthesis. Regarding vitamin C, high CO_2 concentration has been described to cause its degradation, but no reduction was observed on fresh-cut peaches (Colantuono et al., 2015). Concerning carotenoids, in slices of peaches and persimmons, Wright and Kader (1997b) reported that sliced peaches stored in air + 12 kPa CO_2 reduced the content of β-carotene and β-cryptoxanthin. However, they found that the limit of shelf life was reached before major losses of carotenoids occurred. In fresh-cut pumpkin, low changes of carotenoids have been observed after stored for 25 days at 5°C in MAP (Baskaran et al., 2001). In fresh-cut kiwifruit slices, concentrations of 5%, 10%, or 20% CO_2 decreased the content of vitamin C (Agar et al., 1999). In fresh-cut strawberries and persimmons, Wright and Kader (1997a) found no significant losses of vitamin C after 8 days in CA (2% O_2, air + 12% CO_2, or 2% O_2 + 12% CO_2) at 0°C. In fresh-cut oranges, Petrel et al. (1998) found no changes on the carotenoid content after 11 days at 4°C in MAP (19% O_2 + 5% CO_2 and 3% O_2 + 25% CO_2).

Bioactive compounds of vegetables as affected by CA/MA

Fruit vegetables

Storage of sweet pepper for 6 days at 13°C in CO_2-enriched atmospheres resulted in a reduction in ascorbic acid content (Wang, 1977). It has been suggested that high levels of CO_2 increased ascorbic acid losses by conversion into dehydroascorbic acid due to availability of O_2 in sweet peppers (Petersen and Berends, 1993). In sweet peppers stored for 45 days at 8°C in MAP having a headspace composition of 3–5 kPa O_2 + 2–8 kPa CO_2 and in CA storage atmospheres of 2 kPa O_2 + 3 kPa CO_2 (Palliflex™, PL) the content of ascorbic acid declined and β-carotene content increased progressively with storage time (Dogan et al., 2016). Similarly, in jalapeño pepper rings, the conversion of ascorbic acid to dehydroascorbic acid was higher in MAP than in air (Howard et al., 1994; Howard and Hernández-Brenes, 1998).

In tomato, Sozzi et al. (1999) have observed that CA of 3% O_2 and 20% CO_2 either alone or together with ethylene prevented total carotenoid and lycopene biosynthesis on tomato. After transfer of the fruits to air, total carotenoids and lycopene increased but the content was in all cases significantly lower than those held in air. The influence of MAP for delaying postharvest quality loss of tomatoes has been investigated in two cultivars, "Delizia," which has a short storage life, and "Pitenza," which has a long storage life (Domínguez et al., 2016). These authors studied the impact of storage in MAP at 13°C in bags with different gas permeability combined or not with an ethylene absorber on the content of lycopene, total phenols, and ascorbic acid. They observed that the content of these bioactive compounds was affected by the packaging and the storage length in a cultivar-specific manner. They concluded that ethylene removal in combination with MAP led to a higher content of total phenols and ascorbic acid in the short-life tomato cultivar.

Leafy vegetables

Most of the studies on the effect of CA/MA on nutritional composition of leafy vegetables refer to fresh-cut products, since CA/MA is scarcely applied to whole crops. In particular, MAP with low O_2 and high CO_2 is typically applied to control browning of fresh-cut lettuce (*Lactuca sativa*) (Saltveit, 1997). The mechanism of MAP controlling browning of fresh-cut lettuce has been updated recently (Luna et al., 2016). Browning inhibition is linked to the suppressing action of low O_2 on oxidative enzymes (PAL and PPO activities) and to the effect of CO_2 on the biosynthesis of caffeoyl quinic derivatives, which are generally the limiting factor for the onset and progression of browning due to the low initial level of phenolics of these species. In addition, retention of ascorbic acid was also found in fresh-cut lettuce packaged with low O_2 (Barry-Ryan and O'Beirne, 1999). With a similar topic, Baldassarre et al. (2015) observed in fresh-cut "Iceberg" lettuce stored for 9 days under a constant flow of 50% Ar+47% N_2+3% O_2 and 97% N_2+3% O_2 a reduction in PPO activity compared to holding the lettuce in air. Further, the modified atmosphere led to a better retention of vitamin C during storage under the Ar-enriched atmosphere compared to N_2 treatment and air. Conversely, high levels of CO_2 (30%–40%) have been shown to increase ascorbic acid losses in lettuce by conversion into dehydroascorbic acid (Barry-Ryan and O'Beirne, 1999). In addition, in fresh-cut red lettuce, an atmosphere of 2%–3% O_2+12%–14% CO_2 decreased the content of flavonols and anthocyanins compared to air (Gil et al., 1998b).

The content of ascorbic acid and dehydroascorbic acid decreased in fresh-cut Swiss chard stored at 6°C for 8 days in MAP of 7% O_2+10% CO_2 (Gil et al., 1998a) while for fresh-cut spinach in similar storage condition, the content of these compounds was better preserved (Gil et al., 1999). Regarding the content of flavonoids, these authors observed that for both fresh-cut leafy greens the content of flavonoids remained stable in MAP compared to air (Gil et al., 1998a, 1999). Bioactive compounds of two traditional leafy vegetables in Africa (*Amaranthus cruentus* L and *Solanum retroflexum*), which represent a good source of bioactive compounds, were evaluated following storage in MAP with different polypropylene films (Mampholo et al., 2015). Carotenoids, ascorbic acid, total phenolic compounds, antioxidant scavenging activity, and sensory properties were evaluated at 10°C after 14 days. These authors observed that the film that maintained a MAP of 4.3% O_2 and 7.3% CO_2 for *A. cruentus* leaves and 5.6% O_2 and 6.7% CO_2 for *S. reflexum* retained the overall appearance and the content of bioactive compounds and antioxidant activity (Mampholo et al., 2015).

Brassica vegetables

Brassica vegetables, such as cabbage, Brussels sprouts, broccoli, cauliflower, and rocket are an important dietary source for glucosinolates, present as thioglucosides, which can be hydrolyzed by the endogenous plant enzyme myrosinase to thiocyanates, isothiocyanates, or nitriles among others. The glucosinolate content in Brassica vegetables can vary depending on the variety, cultivation conditions, harvest time, and climate (Martínez-Ballesta et al., 2008). In broccoli, the bioactive isothiocyanates found are mainly sulforaphane, allyl isothiocyanate, and indole-3 carbinol apart from flavonols and carotenoids. Broccoli suffers from quality losses during postharvest storage at nonoptimum conditions of high temperature, low RH, and extremely low O_2 and high CO_2 (Nath et al., 2011). The effects of temperature, relative humidity, and storage under CA/MA on glucosinolate content in broccoli heads have been described (Jones et al., 2006). The most relevant postharvest conditions for maintaining broccoli quality are low temperature (<4°C) and high RH. These conditions maintain cellular integrity and the glucosinolate content by preventing the mixing of glucosinolates with myrosinase. Results obtained by Serrano et al. (2006) showed that in broccoli heads packaged in MAP using either microperforated and nonperforated films and stored at 1°C for 28 days, the total antioxidant activity, ascorbic acid, and total phenolic compound remained almost unchanged. These results have been also confirmed for "Parthenon" broccoli florets in which the retention of bioactive compounds in MAP obtained with a microperforated polypropylene plastic was higher than in air (Fernández-León et al., 2013). Packaging broccoli florets with PE film (40 μm thick) without perforations is a simple, economical, and effective method for maintaining the visual quality and the content of glucosinolates at both low (4°C) and high temperature (20°C) (Jia et al., 2009). Storage in CA/MAP cut broccoli florets and intact heads resulted in greater retention of ascorbic acid and shelf life extension in contrast to air (Barth et al., 1993; Paradis et al., 1996). Jaime and Saltveit (2002) found 90% Ar and 2% O_2 did not suppress the loss of chlorophyll from broccoli florets. Regarding carotenoids, the content of α-carotene in broccoli florets increased at the end of CA storage (2% O_2+6% CO_2) and remained stable after returning the samples to ambient conditions for 24 h (Paradis et al., 1996). Storage of fresh-cut broccoli raab florets at 5°C under low oxygen (3% O_2) delayed postcutting deterioration and slowed the loss of chlorophyll and vitamin C for up to 17 days (Cefola et al., 2010). For rocket leaves, ascorbic acid decreased followed by an increase in dehydroascorbic acid when leaves were stored under very low O_2+high CO_2 (Martínez-Sánchez et al., 2006), while it was maintained during storage at 0°C in MA of 8 kPa O_2 and 13 kPa CO_2 reached after 10 days of storage (Mastrandrea et al., 2017a). In addition, glucosinolates were the most affected bioactive compounds of rocket leaves that were reduced by between 4% and 33% when samples were stored in air while the decrease was between 60% and 100% in extremely low

O_2+high CO_2 (Martínez-Sánchez et al., 2006). Isothiocyanates and thiocyanates derivatives are responsible for the typical pungent flavor of rocket leaves. Mastrandrea et al. (2017b) described a slower loss of typical flavor volatiles, including these compounds, when rocket was packaged in polypropylene bags in passive atmosphere and stored 8 days at 5°C.

Other fresh-cut vegetables

Postharvest life of ready-to-use zucchini flowers as affected by the atmosphere composition has been studied (Cefola et al., 2016). These authors observed that vitamin C content and the nutritional quality as well as the visual appearance were preserved. For fresh-cut Tatsoi baby leaves, the effects of a moderate UV-C pretreatment in combination with superatmospheric O_2-enriched MAP were studied during 11 days storage at 5°C (Tomás-Callejas et al., 2012). The UV-C alone or in combination with O_2-enriched MAP did not affect the phenolic content or the total antioxidant capacity. For prepeeled potatoes, the content of ascorbic acid was slightly higher in 20kPa CO_2+80kPa N_2 compared to 5kPa O_2+20kPa CO_2 (Hagg et al., 1998). For carrot sticks, CA or MAP with low O_2 and high CO_2 concentrations maintained the levels of carotenoids and phenolics and prolonged shelf life of carrots (Simoes et al., 2011). Regarding edible coating, their combination with MAP have been proposed to maintain the quality and prolong shelf life of fresh-cut products. A good example is carrot sticks because the development of white discoloration that can be controlled by an edible coating such as chitosan stored in MAP conditions (10kPa O_2+10kPa CO_2 and 2kPa O_2+15–25kPa CO_2) for 12 days at 4°C (Simoes et al., 2009). Chitosan preserved the overall visual quality and reduced surface whiteness during storage but the content of vitamin C and carotenoids decreased. In contrast, the content of total phenolics markedly increased in coated carrot sticks stored under moderate O_2 and CO_2 levels, while it was unaffected under low O_2 and high CO_2 levels (Simoes et al., 2009). Another example is shredded orange and purple carrots packed in air or MAP conditions (5%O_2+5%CO_2 and 95%O_2+5%CO_2) for 13 days (Alasalvar et al., 2005). For orange carrots, no effect of storage conditions was observed on antioxidant activity, the content of carotenoids and phenolics, or sensory quality while purple carrots can only be stored under 5% O_2+5% CO_2 for up to 10 days to maintain the storage quality and the nutraceutical properties.

Conclusions

Collectively, the literature demonstrates there is no clear or consistent impact of the storage atmosphere on preserving bioactive compounds. Rather, it strongly depends on several environmental and biological factors and can sometimes yield contradictory findings. Beside the different responses among different species, several pre- and postharvest factors, such as variety, stage of maturity, and handling operations affect the final content of bioactive compounds. Moreover, reviewed studies generally assayed the total content of the main family of compounds whereas for a better understanding of the mechanism involved in both biosynthesis and degradation processes, a more specific composition of individual compounds would be recommended. Further research should be focused on studying the effect of O_2 and CO_2 concentrations on the changes of phenolic, glucosinolate, and carotenoid compositions. This objective can be achieved using a CA approach with a fixed level of gas separating the effect of each gas on final composition. These results would then need "field verification" in commercial CA and in MAP conditions, including systems in which the atmosphere undergoes modification over time. New multivariate techniques of data analysis may also support to visualize and understand the main trends of variation among subset of data, extracting patterns not readily discerned among different groups of compounds. Finally, all these considerations are even more important for minor species, on which literature references regarding the subject are still scarce.

References

Agar, I.T., Streif, J., Bangerth, F., 1997. Effect of high CO_2 and controlled atmosphere on the ascorbic and dehydroascorbic acid content of some berry fruits. Postharvest Biol. Technol. 11, 47–55.

Agar, I.T., Massantini, R., Hess-Pierce, B., Kader, A.A., 1999. Postharvest CO_2 and ethylene production and quality maintenance of fresh-cut kiwifruit slices. J. Food Sci. 64, 433–440.

Aglar, E., Ozturk, B., Guler, S.K., Karakaya, O., Uzun, S., Saracoglu, O., 2017. Effect of modified atmosphere packaging and 'parka' treatments on fruit quality characteristics of sweet cherry fruits (*Prunus avium* L. '0900 Ziraat') during cold storage and shelf life. Sci. Hortic. 222, 162–168.

Alasalvar, C., Al-Farsi, M., Quantick, P.C., Shahidi, F., Wiktorowicz, R., 2005. Effect of chill storage and modified atmosphere packaging (MAP) on antioxidant activity, anthocyanins, carotenoids, phenolics and sensory quality of ready-to-eat shredded orange and purple carrots. Food Chem. 89, 69–76.

Al-Redhaiman, K.N., 2005. Chemical changes during storage of 'Barhi' dates under controlled atmosphere conditions. HortScience 40, 1413–1415.

Artés-Hernández, F., Artés, F., Tomás-Barberán, F.A., 2003. Quality and enhancement of bioactive phenolics in cv. Napoleon table grapes exposed to different postharvest gaseous treatments. J. Agric. Food Chem. 51, 5290–5295.

Baldassarre, V., Amodio, M.L., De Chiara, M.L., Colelli, G., 2015. Effect of argon-enriched atmospheres on shelf life of fresh-cut 'iceberg' lettuce. Acta Hortic. (1071), 755–762.

Bangerth, F., 1977. The effect of different partial pressures of CO_2, C_2H_4, and O_2 in the storage atmosphere on the ascorbic acid content of fruits and vegetables. Qual. Plant. 27, 125–133.

Baraiya, N.S., Ramana Rao, T.V., Thakkar, V.R., 2016. Composite coating as a carrier of antioxidants improves the postharvest shelf life and quality of table grapes (*Vitis vinifera* l. var. Thompson seedless). J. Agric. Sci. Technol. 18, 93–107.

Barry-Ryan, C., O'Beirne, D., 1999. Ascorbic acid retention in shredded iceberg lettuce as affected by minimal processing. J. Food Sci. 64, 498–500.

Barth, M.M., Kerbel, E.L., Perry, A.K., Schmidt, S.J., 1993. Modified atmosphere packaging affects ascorbic acid, enzyme activity and market quality of broccoli. J. Food Sci. 57, 954–957.

Baskaran, R.H., Prasad, R., Shivaiah, K.M., 2001. Storage behaviour of minimally processed pumpkin (*Cucurbita maxima*) under modified atmosphere packaging conditions. Eur. Food Res. Technol. 212, 165–169.

Biesalski, H.-K., Dragsted, L.O., Elmadfa, I., Grossklaus, R., Müller, M., Schrenk, D., Walter, P., Weber, P., 2009. Bioactive compounds: definition and assessment of activity. Nutrition 25, 1202–1205.

Cascia, G., Bulley, S.M., Punter, M., Bowen, J., Rassam, M., Schotsmans, W.C., Larrigaudière, C., Johnston, J.W., 2013. Investigation of ascorbate metabolism during inducement of storage disorders in pear. Physiol. Plant. 147, 121–134.

Cefola, M., Amodio, M.L., Rinaldi, R., Vanadia, S., Colelli, G., 2010. Effect of atmosphere composition on the quality of ready-to-use broccoli raab (*Brassica rapa* L.). J. Sci. Food Agric. 90, 789–797.

Cefola, M., Amodio, M.L., Colelli, G., 2016. Extending postharvest life of ready-to-use zucchini flowers: effects of the atmosphere composition. Acta Hortic. (1141), 123–130.

Chaudhary, P.R., Jayaprakasha, G.K., Porat, R., Patil, B.S., 2015. Influence of modified atmosphere packaging on 'star ruby' grapefruit phytochemicals. J. Agric. Food Chem. 63, 1020–1028.

Chen, P.M., 2016. Pear. In: Gross, K.C., Wang, C.Y., Saltveit, M.E. (Eds.), The Commercial Storage of Fruits, Vegetables, and Florist and Nursery Stocks. Agriculture Handbook No. 66. U.S. Department of Agriculture, Washington, D.C. http://ucanr.edu/datastoreFiles/234-2927.pdf. (accessed 17.01.18), pp. 471–480.

Colantuono, F., Amodio, M.L., Colelli, G., 2015. Comparison of different gas compositions on fresh-cut peach quality: a preliminary study. Acta Hortic. (1071), 763–770.

Crouch, E.M., Jooste, M., Majoni, T.J., Crouch, I.J., Bergman, H., 2015. Harvest maturity and storage duration influencing flesh browning in south African 'Cripps pink' apples. Acta Hortic. (1079), 121–127.

DeEll, J., 2012. Controlled Atmosphere Storage Guidelines and Recommendations for Apples. Factsheet Order No. 12-045, Agdex 211/66. Ontario Ministry of Agriculture, Food and Rural Affairs, Canada.

Delaporte, N., 1971. Effect of oxygen content of atmosphere on ascorbic acid content of apple during controlled atmosphere storage. Lebens. Wissen. Technol. 4, 106–112.

Devlieghere, F., Gil, M.I., Debevere, J., 2002. Modified atmosphere packaging. In: Henry, C.J.K., Chapman, C. (Eds.), The Nutrition Handbook for Food Processors. Woodhead Publishing Limited, Cambridge, pp. 342–370.

Dogan, A., Selcuk, N., Erkan, M., 2016. Comparison of pesticide-free and conventional production systems on postharvest quality and nutritional parameters of peppers in different storage conditions. Sci. Hortic. 207, 104–116.

Domínguez, I., Lafuente, M.T., Hernández-Muñoz, P., Gavara, R., 2016. Influence of modified atmosphere and ethylene levels on quality attributes of fresh tomatoes (*Lycopersicon esculentum* Mill). Food Chem. 209, 211–219.

Felicetti, E., Mattheis, J.P., Zhu, Y., Fellman, J.K., 2011. Dynamics of ascorbic acid in 'Braeburn' and 'Gala' apples during on-tree development and storage in atmospheres conducive to internal browning development. Postharvest Biol. Technol. 61, 95–102.

Fernández-León, M.F., Fernández-León, A.M., Lozano, M., Ayuso, M.C., Amodio, M.L., Colelli, G., González-Gómez, D., 2013. Retention of quality and functional values of broccoli 'Parthenon' stored in modified atmosphere packaging. Food Control 31, 302–313.

Gil, M.I., Kader, A.A., 2008. Fresh-cut fruit and vegetables. In: Tomás-Barberán, F.A., Gil, M.I. (Eds.), Improving the Health-Promoting Properties of Fruit and Vegetable Products. CRC Press, Woodhead Publishing Ltd., pp. 475–496.

Gil, M.I., Holcroft, D.M., Kader, A.A., 1997. Changes in strawberry anthocyanins in response to carbon dioxide treatments. J. Agric. Food Chem. 45, 1662–1667.

Gil, M.I., Ferreres, F., Tomás-Barberán, F.A., 1998a. Effect of modified atmosphere packaging on the flavonoids and vitamin C content of minimally processed Swiss chard (*Beta vulgaris* subsp. cycla). J. Agric. Food Chem. 46, 2007–2012.

Gil, M.I., Castañer, M., Ferreres, F., Artés, F., Tomás-Barberán, F.A., 1998b. Modified-atmosphere packaging of minimally processed 'Lollo Rosso' (*Lactuca sativa*). Z. Lebensm. Unters. Forsch. 206, 350–354.

Gil, M.I., Ferreres, F., Tomas-Barberan, F.A., 1999. Effect of postharvest storage and processing on antioxidant constituents (flavanoids and vitamin C) of fresh-cut spinach. J. Agic. Food Chem. 47, 2213–2217.

Gil, M.I., Aguayo, E., Kader, A.A., 2006. Quality changes and nutrient retention in fresh-cut versus whole fruits during storage. J. Agric. Food Chem. 54, 4284–4296.

González-Buesa, J., Arias, E., Salvador, M.L., Oria, R., Ferrer-Mairal, A., 2011. Suitability for minimal processing of non-melting clingstone peaches. Food Sci. Technol. 46, 819–826.

González-Gómez, D., Bernalte, M.J., Ayuso, M.C., Fernández-Leon, M.F., Lozano, M., Rato, A.E., Palma, V., Santos, A.C., 2017. Evaluation of different postharvest conditions to preserve the amount of bioactive compounds, physicochemical quality parameters and sensory attributes of 'sweetheart' cherries. Acta Hortic. (1161), 581–586.

Guillén, F., Castillo, S., Valero, D., Zapata, P.J., Martínez-Romero, D., Díaz-Mula, H.M., Serrano, M., 2015. Use of modified atmosphere packaging improves antioxidant activity and bioactive compounds during postharvest storage of 'collar' figs. Acta Hortic. (1071), 263–268.

Haffner, K., Jansrud, W.K., Tengesdal, G., 1997. L-ascorbic acid contents and other quality criteria in apples (*Malus domestica* Borkh.) after storage in cold store and controlled atmosphere. In: Mitcham, E.L. (Ed.), Postharvest Horticultural Series No 16. University of California, Davis.

Hagg, M., Hakkinen, U., Kumpulainen, J., Ahvenainen, R., Hurme, E., 1998. Effects of preparation procedures, packaging and storage on nutrient retention in peeled potatoes. J. Sci. Food Agric. 77, 519–526.

Harborne, J.B., Williams, C.A., 2000. Advances in flavonoid research since 1992. Phytochemistry 55, 481–504.

Haslam, E., Lilley, T.H., Warminki, E., Liao, H., Cai, Y., Martin, R., Gaffney, S.H., Goulding, P.N., Luck, G., 1992. Polyphenol complexation. In: Ho, I.C.T., Lee, C.Y., Hiang, M.T. (Eds.), Phenolic Compounds in Food and Their Effect on Health. American Chemical Society, Washington, DC.

Holcroft, D.M., Kader, A.A., 1999. Carbon dioxide-induced changes in color and anthocyanin synthesis of stored strawberry fruit. HortScience 34, 1244–1248.

Holcroft, D.M., Gil, M.I., Kader, A.A., 1998. Effect of carbon dioxide on anthocyanins, phenylalanine ammonia lyase and glucosyltransferase in the arils of stored pomegranates. J. Am. Soc. Hortic. Sci. 123, 136–140.

Howard, L.R., Hernández-Brenes, C., 1998. Antioxidant content and market quality of jalapeno pepper rings as affected by minimal processing and modified atmosphere packaging. J. Food Qual. 21, 317–327.

Howard, L.R., Smith, R.T., Wagner, A.B., Villalon, B., Burns, E.E., 1994. Provitamin A and ascorbic acid content of fresh pepper cultivars (*Capsicum annuum*) and processed jalapenos. J. Food Sci. 59, 362–365.

Jaime, P., Saltveit, M.E., 2002. Postharvest changes in broccoli and lettuce during storage in argon, helium, and nitrogen atmospheres containing 2% oxygen. Postharvest Biol. Technol. 26, 113–116.

Jia, C.-G., Xu, C.-J., Wei, J., Yuan, J., Yuan, G.-F., Wang, B.-L., Wang, Q.-M., 2009. Effect of modified atmosphere packaging on visual quality and glucosinolates of broccoli florets. Food Chem. 114, 28–37.

Jones, R.B., Faragher, J.D., Winkler, S., 2006. A review of the influence of postharvest treatments on quality and glucosinolate content in broccoli (*Brassica oleracea* var. italica) heads. Postharvest Biol. Technol. 41, 1–8.

Kader, A.A., 1997. A summary of CA requirements and recommendations for fruits other than pomes fruits. In: Kader, A.A. (Ed.), Proceeding of Seventh International Controlled Atmosphere Research Conference. In: Postharvest Horticulture Series No. 17, University of California, Davis, pp. 1–34.

Kader, A.A., 2001. Physiology of CA treated produce. In: 8th International Controlled Atmosphere Research Conference, Rotterdam, Oosterhaven.

Khan, M.R., Suwanamornlert, P., Leelaphiwat, P., Chinsirikul, W., Chonhenchob, V., 2017. Quality and biochemical changes of longan (Dimocarpus longan Lour cv. Daw) fruit under different controlled atmosphere conditions. Int. J. Food Sci. Technol. 52, 2163–2170.

Kris-Etherton, P.M., Hecker, K.D., Bonanome, A., Coval, S.M., Binkoski, A.E., Hilpert, K.F., Griel, A.E., Etherton, T.D., 2002. Bioactive compounds in foods: their role in the prevention of cardiovascular disease and cancer. Am. J. Med. 113, 71–88.

Kupferman, E., 2001. Controlled Atmosphere Storage of Apples and Pears. Postharvest Information Network. Available at http://postharvest.tfrec.wsu.edu/EMK2000C.pdf.

Lee, S.K., Kader, A.A., 2000. Preharvest and postharvest factors influencing vitamin C content of horticultural crops. Postharvest Biol. Technol. 20, 207–220.

Li, H., Tsao, R., Deng, Z., 2012. Factors affecting the antioxidant potential and health benefits of plant foods. Can. J. Plant Sci. 92, 1101–1111.

Luna, M.C., Tudela, J.A., Tomás-Barberán, F.A., Gil, M.I., 2016. Modified atmosphere (MA) prevents browning of fresh-cut romaine lettuce through multi-target effects related to phenolic metabolism. Postharvest Biol. Technol. 119, 84–93.

Mampholo, M.B., Sivakumar, D., Van Rensburg, J., 2015. Variation in bioactive compounds and quality parameters in different modified atmosphere packaging during postharvest storage of traditional leafy vegetables (*Amaranthus cruentus* L and *Solanum retroflexum*). J. Food Quality 38, 1–12.

Martínez-Ballesta, M.C., López-Pérez, L., Hernández, M., López-Berenguer, C., Fernández-García, N., Carvajal, M., 2008. Agricultural practices for enhanced human health. Phytochem. Rev. 7, 251–260.

Martínez-Sánchez, A., Allende, A., Bennett, R.N., Ferreres, F., Gil, M.I., 2006. Microbial, nutritional and sensory quality of rocket leaves as affected by different sanitizers. Postharvest Biol. Technol. 42, 86–97.

Mastrandrea, L., Amodio, M.L., de Chiara, M.L.V., Pati, S., Colelli, G., 2017a. Effect of temperature abuse and improper atmosphere packaging on volatile profile and quality of rocket leaves. Food Packaging Shelf Life 14, 59–65.

Mastrandrea, L., Amodio, M.L., Pati, S., Colelli, G., 2017b. Effect of modified atmosphere packaging and temperature abuse on flavor volatile compounds of rocket leaves (*Diplotaxis tenuifolia* L.). J. Food Sci. Technol. 54, 2433–2442.

Mateos, M., Ke, D., Cantwell, M., Kader, A., 1993. Phenolic metabolism and ethanolic fermentation of intact and cut lettuce exposed to CO_2-enriched atmospheres. Postharvest Biol. Technol. 3, 225–233.

Mditshwa, A., Fawole, O.A., Vries, F., van der Merwe, K., Crouch, E., Opara, U.L., 2017. Impact of dynamic controlled atmospheres on reactive oxygen species, antioxidant capacity and phytochemical properties of apple peel (cv. Granny Smith). Sci. Hortic. 216, 169–176.

Nath, A., Bagchi, B., Misra, L.K., Deka, B.C., 2011. Changes in post-harvest phytochemical qualities of broccoli florets during ambient and refrigerated storage. Food Chem. 127, 1510–1514.

Neuwald, D.A., Streif, J., Kittemann, D., 2015. CO_2 skin burning of 'cameo' apples during CA storage - reasons and possibilities to reduce the disorder. Acta Hortic. (1071), 711–716.

Paradis, C., Castaigne, F., Desrosiers, T., Fortin, J., Rodrigue, N., Willemot, C., 1996. Sensory, nutrient and chlorophyll changes in broccoli florets during controlled atmosphere storage. J. Food Qual. 19, 303–316.

Petersen, M.A., Berends, H., 1993. Ascorbic acid and dehydroascorbic acid content of blanched sweet green pepper during chilled storage in modified atmospheres. Z. Lebensm. Unters. Forsch. 197, 546–549.

Petrel, M.T., Fernández, P.S., Romojaro, F., Martínez, A., 1998. The effect of modified atmosphere packaging on ready-to-eat oranges. Lebensm. Wiss. Technol. 31, 322–328.

Pirovani, M.E., Piagentini, A.M., Van de Velde, F., 2015. Bioactive compounds in fresh-cut fruits: occurrence and impact of processing and cold storage. In: Biotechnology of Bioactive Compounds: Sources and Applications. Wiley Blackwell, pp. 37–62.

Putnik, P., Bursać Kovačević, D., Herceg, K., Pavkov, I., Zorić, Z., Levaj, B., 2017. Effects of modified atmosphere, anti-browning treatments and ultrasound on the polyphenolic stability, antioxidant capacity and microbial growth in fresh-cut apples. J. Food Process Eng. 40. https://doi.org/10.1111/jfpe.12539.

Remberg, S.F., Hagen, S.F., Borge, G.I.A., Sandberg, E., Jeksrud, W.K., Haffner, K., 2010. Phenolic compounds, antioxidant activity and other quality related criteria of 'Summerred' apples as influenced by postharvest storage conditions. Acta Hortic. (875), 169–176.

Saltveit, M.E., 1997. Physical and physiological changes in minimally processed fruits and vegetables. In: Tomás-Barberán, F.A., Robins, R.J. (Eds.), Phytochemistry of Fruit and Vegetables. Oxford.

Serrano, M., Martinez-Romero, D., Guillén, F., Castillo, S., Valero, D., 2006. Maintenance of broccoli quality and functional properties during cold storage as affected by modified atmosphere packaging. Postharvest Biol. Technol. 39, 61–68.

Simoes, A., Tudela, J.A., Allende, A., Puschmann, R., Gil, M.I., 2009. Edible coatings containing chitosan and moderate modified atmosphere maintain quality and enhance phytochemicals of carrots sticks. Postharvest Biol. Technol. 51, 364–370.

Simoes, A., Allende, A., Tudela, J.A., Puschmann, R., Gil, M.I., 2011. Optimum controlled atmospheres minimise respiration rate and quality losses while increase phenolic compounds of baby carrots. LWT - Food Sci. Technol. 44, 277–283.

Singh, S.P., Pal, R.K., 2008. Controlled atmosphere storage of guava (*Psidium guajava* L.) fruit. Postharvest Biol. Technol. 47, 296–306.

Sivakumar, D., Van Deventer, F., Terry, L.A., Polenta, G.A., Korsten, L., 2012. Combination of 1-methylcyclopropene treatment and controlled atmosphere storage retains overall fruit quality and bioactive compounds in mango. J. Sci. Food Agric. 15, 821–830.

Sozzi, G., Trinchero, G.D., Fraschina, A.A., 1999. Controlled-atmosphere storage of tomato fruit: low oxygen or elevated carbon dioxide levels alter galactosidase activity and inhibit exogenous ethylene action. J. Sci. Food Agric. 79, 1065–1070.

Tardelli, F., Guidi, L., Massai, R.c., Toivonen, P.M., 2013. Effects of 1-methylcyclopropene and post-controlled atmosphere air storage treatments on fresh-cut Ambrosia apple slices. J. Sci. Food Agric. 93, 262–270.

Tomás-Callejas, A., Otón, M., Artés, F., Artés-Hernández, F., 2012. Combined effect of UV-C pretreatment and high oxygen packaging for keeping the quality of fresh-cut Tatsoi baby leaves. Innovative Food Sci. Emerg. Technol. 14, 115–121.

Tsao, R., Wang, M., Deng, Z., 2007. Lutein: separation, antioxidant activity, and potential health benefits. In: Shahidi, F., Ho, C.-T. (Eds.), Antioxidant Measurement and Applications. American Chemical Society, Washington, DC, pp. 352–372.

Van der Sluis, A.A., Dekker, M., Jongen, W.M.F., de Jager, A. 2003. Polyphenolic antioxidants in apples. Effect of storage conditions on four cultivars. In: Proceedings of the Eight International Controlled Atmosphere Research Conference. Oosterhaven, J., Peppelenbos, H.W. (Eds.). Acta Hortic., (600), 533–540.

Veltman, R.H., Sanders, M.G., Persijn, S.T., Peppelenbos, H.W., Oosterhaven, J., 1999. Decreased ascorbic acid levels and brown core development in pears (*Pyrus communis* L. cv. Conference). Physiol. Plant. 107, 39–45.

Wang, C.Y., 1977. Effects of CO_2 treatment on storage and shelf life of sweet pepper. J. Am. Soc. Hort. Sci. 102, 808–812.

Watada, A.E., 1987. Vitamins. Marcel Dekker, New York.

Watkins, C.B., Kupferman, E., Rosenberger, D.A., 2016. Apple. In: Gross, K.C., Wang, C.Y., Saltveit, M.E. (Eds.), The Commercial Storage of Fruits, Vegetables, and Florist and Nursery Stocks. Agriculture Handbook No. 66, U.S. Department of Agriculture, Washington, D.C. http://ucanr.edu/datastoreFiles/234-2927.pdf. (accessed 17.01.18), pp. 176–194.

Weichmann, J., 1986. The effect of controlled-atmosphere storage on the sensory and nutritional quality of fruits and vegetables. Hort. Rev. 8, 101–127.

Wright, K.P., Kader, A.A., 1997a. Effect of slicing and controlled-atmosphere storage on the ascorbate content and quality of strawberries and persimmons. Postharv. Biol. Technol. 10, 39–48.

Wright, K.P., Kader, A.A., 1997b. Effect of controlled-atmosphere storage on the quality and carotenoid content of sliced persimmons and peaches. Postharv. Biol. Technol. 10, 89–97.

Yang, J., 2011. Phytochemical changes of fresh-cut fruits and vegetables in controlled and modified atmosphere packaging. In: Brody, A.L., Zhuang, H., Han, J.H. (Eds.), Modified Atmosphere Packaging for Fresh-Cut Fruits and Vegetables. Wiley-Blackwell Ltd, pp. 101–140.

Chapter 8

The impact of temperature on atmosphere requirements and effects: The limits of design and utility for CA/MA/MAP

Jeffrey K. Brecht*, Eleni D. Pliakoni† and Konstantinos Batziakas†
*Horticultural Sciences Department, University of Florida, Gainesville, FL, United States,
†Department of Horticulture and Natural Resources, Kansas State University, Olathe, KS, United States

Introduction

Temperature management is rightly considered the most important action that is available to maintain the quality and extend the shelf life of fresh fruits and vegetables. This is because all life processes occurring in the plant tissues are responsive to temperature, proceeding at lower rates at lower temperatures and faster rates at higher temperatures. Optimum CA/MA has effects on fruit and vegetable physiology and biochemistry that are nearly as comprehensive, thus acting as a supplement to low-temperature maintenance. The rate of the respiratory process, which is the source of the chemical energy (adenosine triphosphate, ATP) that fuels most biochemical reactions, fluctuates in response to temperature in a characteristic way for different fruits and vegetables known as the Q_{10}. The Q_{10} is the ratio of respiration rates at 10°C intervals and varies from around to two to four for most fruits and vegetables within the typical postharvest temperature range. Respiration also responds directly to O_2 and CO_2 in the concentration ranges employed for CA and MA, and reduced respiration under CA/MA indirectly reduces the rates of numerous metabolic reactions that require ATP to proceed. In addition, the rates of metabolic reactions that require O_2 (e.g., ACC oxidase, polyphenol oxidase, and ascorbate oxidase) and expression of genes that are sensitive to O_2 levels (Kanellis et al., 2009) may be directly affected under low O_2 conditions. Similarly, other specific reactions are inhibited in the presence of CO_2 (e.g., succinate dehydrogenase, isocitrate dehydrogenase, *endo*-beta-1,4-glucanase, ACC synthase, and ACC oxidase;

Controlled and Modified Atmospheres for Fresh and Fresh-Cut Produce
https://doi.org/10.1016/B978-0-12-804599-2.00009-0

Kanellis et al., 2009) while inhibition of gene expression has been shown in strawberry and tomato exposed to elevated CO_2 environments (Ponce-Valadez et al., 2009; Rothan et al., 1997). In addition, ethylene synthesis and action are inhibited by reduced O_2 and elevated CO_2. The result is that both metabolic pathways like phenolic metabolism and ethylene biosynthesis as well as physiological processes like ripening and senescence are slowed or inhibited under CA/MA conditions (Thompson, 2010a).

Traditionally, with the goal of CA/MA research being to maximally extend the postharvest life of a particular crop, the process of developing the treatment has involved first selecting the maturity or ripeness stage and the temperature that result in the absolute longest storage potential without appearance of atmosphere injuries, while maintaining minimally acceptable sensory quality for the crop. This developmental stage and temperature combination then becomes the basis for testing various combinations of reduced O_2 and/or elevated CO_2 to determine the crop's tolerance limits to exposure to those gases, resulting in the longest possible storage duration without induction of physiological injury (Brecht et al., 2003a; Thompson, 2010b). This has been a successful approach resulting in commercial CA storage of several crops, most notably, apples, avocados, kiwifruit, sweet onions, and pears. A newer approach has been to develop optimum CA or MA or MAP conditions for particular distribution systems as disparate as international marine transport routes and domestic distribution systems for fresh-cut products. What these systems share in common is that there are fixed durations involved, often long enough to severely challenge the product shelf life, making exploration of CA/MA inviting. Clearly, there is a time factor involved in fruit and vegetable tolerance to extreme O_2 and CO_2 levels. Fruits and vegetables can tolerate and recover from the absence of O_2 and even 100 kPA CO_2 if the exposure time is short, as has been demonstrated in development of insecticidal atmospheres for quarantine purposes (Neven et al., 2009). Therefore, for fixed durations, the research and development goal becomes identifying a beneficial, noninjurious CA/MA to help deliver the best possible quality within that fixed time frame. Besides the optimal CA/MA likely being more extreme than that used for long-term storage, this may very well also allow using product at more advanced maturity or ripeness stages that, although possessing shorter potential postharvest life than the stage best suited for long-term storage, delivers superior quality to the consumer.

An interesting consideration within the "fixed distribution system" scenario is how it applies to crops that are subject to chilling injury. Sensitivity of fruits to chilling injury is generally greater the less mature or ripe the fruit. Using the traditional goal in CA/MA research of achieving maximum postharvest life for a chilling sensitive crop usually involves selecting a maturity stage and temperature combination that unfortunately increases the risk of chilling injury. This may mean that the storage duration is limited by the appearance of visual chilling injury symptoms. However, a very commonly encountered symptom of chilling injury that precedes appearance of visual symptoms is inhibition of aroma volatile synthesis (Maul et al., 2000), resulting in fruit that may look fine but lack aroma, thus possessing at best minimally acceptable sensory quality. As mentioned previously, it may very well be possible in a fixed distribution system to start with fruit at a more advanced maturity or ripeness stage. For chilling sensitive fruit, use of a lower temperature may also be possible because of the reduced chilling sensitivity of the more advanced fruit. This situation can allow attainment of the desired postharvest life with improved quality. In the case of mango, it was found that more advanced fruit were also more tolerant of elevated CO_2 levels, but they were less tolerant of reduced O_2 (Bender et al., 2000a).

Another situation is inadvertent or unavoidable exposure of crops in MAP to nonoptimum temperatures, which can lead to negative consequences (Brecht et al., 2003a). At temperatures below that for which the MAP was designed, the reduced respiration rate may be insufficient to generate an effective MA; the opposite occurs at higher temperatures, with the MA likely to become too extreme and cause injury to the product in the package. Since temperature fluctuations during commercial handling are largely unavoidable—temperatures during transit, at distribution centers, and at retail are rarely the same (Nunes et al., 2014)—this limits the effectiveness of MAP. However, since fruits and vegetables are often able to recover from brief exposures to extreme atmospheres without occurrence of irreversible tissue injury, duration of temperature excursions also plays a role. Thus, for MAP to be effective, the package must be designed with consideration of the entire distribution system, including possibilities for occurrence of temperature excursions and their likely durations, so that the atmosphere can be mostly maintained within a noninjurious, if not always beneficial, range in most all likely scenarios.

Physiological and metabolic bases for temperature effects on CA/MA requirements

Respiration and temperature

There are basic biological reactions, such as respiration, transpiration, and biochemical and compositional changes that continue after harvest of fresh horticultural commodities. Those processes can be affected by different environmental factors like temperature, humidity, and gas composition, among others. Optimum conditions of those factors can result in several advantages for the commodities. However, nonoptimum conditions could result in problems like physiological disorders and decay. Respiration rate reflects the metabolic rate of the product and is inversely related with the shelf life. Temperature is the most important factor that regulates the metabolic processes, including respiration, and lower temperature extends the shelf life of the produce. The creation and maintenance of an optimal atmosphere inside a MA package depends on the respiration of the produce and the permeability of the films to O_2 and CO_2, both of which are affected by temperature. Establishment of an optimum gas equilibrium in MAP is the result of a choice of a particular combination of film for a given produce item at an optimum temperature. However, temperature fluctuation is very often observed during storage and distribution and that can lead to a nonoptimum gas equilibrium. Produce responses to abusive gas concentrations could result in induction of fermentation by accumulation of succinate and or alanine and subsequent reduction of the ATP production.

To fully understand the influence of temperature on the respiration of fruits and vegetables and on the permeability of films, the Q_{10} for respiration (Q_{10}^R) and the Q_{10} for package permeability (Q_{10}^P) values can be compared. The Q_{10} relationship can describe the degree of acceleration of either a biological or chemical process when the temperature changes by 10°C. Over a temperature range of 0–10°C, at which the values of the Q_{10}^R for mushrooms, broccoli, and tomatoes were 3.0, 2.8, and 2.3, respectively, the Q_{10}^P of MA packages designed for 0°C was less than 1.2 (Exama et al., 1993a). This disparity resulted in an accumulation of CO_2 and a decrease in O_2 inside the packages subjected to temperature fluctuations. In another study, tomatoes in MAP subjected to temperature fluctuation leading to less than 2% O_2 were found to undergo loss of color and uneven ripening (Geeson et al., 1985).

How temperature affects different physiological processes

Texture has long been associated with the quality and storage life of fresh fruits and vegetables. More specifically, softening is directly related to the ripening processes through the compositional changes of the cell wall carbohydrates that occur due to the action of cell wall enzymes. Some of the major cell wall enzymes that are related to ripening and are associated with fruit softening are polygalacturonase (PG), pectinesterase, cellulase, and β-galactosidase. The use of CA/MA/MAP as a storage treatment has resulted, for some fruits, in a delay and/or retardation of ripening (Zagory and Kader, 1988; Wang, 1993). In papaya, a combination of MAP and low temperature resulted in decreased loss of firmness and reduced activities of PG, pectinesterase, and β-galactosidase (Lazan et al., 1993). Also for papaya, activities of enzymes in the ethylene biosynthetic pathway were suppressed in MAP at low temperature (Latifah et al., 1997). Exposure of climacteric fruits to ethylene induces the climacteric rise in respiration rate and results in rapid ripening. The reduction of ethylene production and sensitivity that occurs in CA/MA/MAP can therefore lead to extension of shelf life.

How temperature affects specific metabolic processes, biochemical reactions, and gene expression

When produce is stored under optimum conditions, respiration rate remains low and relatively stable during the storage period. However, when storage conditions are not optimum, respiration may increase or decrease and that will have an effect on the physiological status of the produce (Fennir et al., 2003). Additionally, the changes in the composition of the respiratory gases at equilibrium could affect the rates of respiration. Reduction of O_2 concentration and increase of CO_2 have been related to suppuration of various metabolic reactions including respiration (Beaudry, 1999; Kader, 1997). However, high CO_2 concentrations are not always beneficial: exposure of plant tissues from 10 to 20 kPa CO_2 results in the suppression of aerobic metabolic processes (Mathooko, 1996; Watkins, 2000). Also, very low O_2 or lack of O_2 could result in the development of off-flavors due to the change from aerobic to anaerobic (fermentative) metabolism (Beaudry, 2000). Additionally, either high CO_2 or low O_2 concentrations could indirectly induce off-flavors by stimulating the growth of homo- and hetero-fermentative bacteria and yeasts, which produce organic acids, ethanol, and volatile esters (Babic et al., 1992; Carlin et al., 1989; Kakiomenou et al., 1996).

Generally, the lower O_2 limit for most commodities increases as temperature increases (Beaudry et al., 1992) while the upper CO_2 limit may increase or decrease as temperature declines (Bender et al., 2000b; Costa et al., 1994; Watkins, 2000). Plant responses to modified O_2 and CO_2 levels are related to both primary and secondary metabolism (Kader, 1997; Table 1), but the specific tolerance levels vary by commodity, cultivar, temperature, exposure duration, and potentially interactions between responses to O_2 and CO_2.

For the primary metabolic responses, low O_2 leads to a reduction in respiration rate (i.e., O_2 uptake), which is a beneficial response in terms of quality retention and extended shelf life and can be realized by the reduction in starch degradation and sugar consumption. A negative response related to low O_2 is fermentation, as has been mentioned previously, and also the diversion of glycolytic carbon to form acetaldehyde, ethanol, and lactate (Beaudry, 2000). For the secondary metabolic responses, beneficial low O_2 responses include reduction of ethylene synthesis, reduced chlorophyll degradation, reduced cell wall degradation, and reduced phenolic oxidation (Beaudry, 2000). Additionally, secondary metabolic responses to extremely low O_2 can include reduced aroma biosynthesis for fruits, including apple, banana, pear, peach, strawberry, and other crops (Ke et al., 1994; Mattheis and Fellman, 2000; Shamaila et al., 1992), and off-flavor generation (Kays, 1997). The activity of phenylalanine ammonia lyase (PAL), an enzyme that is central to phenolic metabolism, is inhibited by low O_2 but enhanced by high CO_2 concentrations (Beaudry, 2000; Watkins, 2000). Different fruits and vegetables may be either more or less tolerant of extreme CO_2 levels at lower temperatures. For example, Lipton (1968) showed that 10% CO_2 was injurious to asparagus in 6.1°C storage, but was tolerated at 1.7°C, while Costa et al. (1994) found that "Opus" snap beans held in 2% O_2 plus various elevated CO_2 levels tolerated up to 18% CO_2 at 8°C, but only tolerated 8% CO_2 at 1°C. Mitz (1979) suggested that reduced tolerance of plant tissues to elevated CO_2 at lower temperatures may

TABLE 1 General effects of low temperature, reduced O_2, and elevated CO_2 on primary and secondary metabolism of horticultural crops

Metabolic process	General effects[a]		
	Low temperature	*Reduced O_2*	*Elevated CO_2*
Respiratory metabolism			
Aerobic respiration	−	−	−, 0 or +
Anaerobic respiration	−	+	+
Energy (ATP) production	−	−	−
Ethylene metabolism			
ACC synthase	−	−	−
ACC oxidase	−	−	− or +
Ethylene action	−	−	−
Compositional changes			
Chlorophyll degradation	−	−	−
Anthocyanin biosynthesis	−	−	−
Carotenoid biosynthesis	−	−	−
Total phenolics	−	−	−
Phenolic synthesis (PAL)	−	−	+
Browning (PPO)	−	−	−
Pectin depolymerization (PG)	−	−	−
Starch-to-sugar conversion	− or +	−	−
Total acids content	+	+	+
Succinic acid	−	−	+
Malic acid	+	+	−
Characteristic aromas	−	−	−
Off-flavors	− or +	+	+
Pro-vitamin A (β-carotene) loss	−	−	−
Vitamin C (ascorbic acid) loss	−	−	−

− = decrease; 0 = no effect; + = increase or maintenance.
[a]*Specific O_2 and CO_2 levels at which these effects occur vary by commodity, cultivar, temperature, exposure duration, and potentially interactions between responses to O_2 and CO_2.*
Adapted from Kader, A.A., 1997. Biological bases of O_2 and CO_2 effects on postharvest life of horticultural perishables. In: Saltveit, M.E. (Ed.)., Proc. 7th Intl. Controlled Atmosphere Res. Conf. vol. 4. Vegetables and ornamentals. Postharvest Hort. Ser. 18. Univ. Calif., Davis, pp. 160–163.

be a consequence of greater solubility of the gas in the cell sap at lower temperatures, leading to pH reduction since dissolved CO_2 in an aqueous environment forms carbonic acid.

Strategies for determining optimum CA/MA conditions

Historical approaches in the development of CA/MA

The historical purpose for development of CA/MA storage technology for fresh fruits and vegetables has been to extend the marketable and edible life of the products to the greatest extent possible. The earliest applications of CA/MA were long-term stationary storage of domestic produce for a domestic market. Beginning in the 19th century, CA/MA storage development was focused on achieving maximum extension of edible condition during storage for each product being investigated (Thompson, 2010b). The strategy involved first identifying the minimum acceptable maturity stage in order to start with a product that possesses the maximum potential postharvest life. Similarly, the temperature chosen would be the lowest possible temperature to avoid either freezing injury or chilling injury depending on if the crop is temperate or subtropical/tropical. From these starting points, the optimum combination of O_2 and CO_2 must be determined by empirical means, which characteristically means identifying the most extreme levels of each gas that are tolerated by the crop (i.e., those concentrations that do not result in physiological injuries over the course of extended storage durations) until deterioration of some important quality factor determines the limits of storage duration. This entails observing the changes in the various shelf life-limiting quality factors during storage in experiments in which the product is placed in a limited number of combinations of O_2 and CO_2 and the treatment identified that results in the longest retention of minimal acceptability. The inherent difficulties and limited application of this approach led Saltveit (1997) to propose an alternative approach to application of CA/MA: a feed-back approach in which the O_2 and CO_2 levels are varied dynamically in response to physiological indicators produced by the stored commodity.

The internationalization of fruit and vegetable marketing has been at least partially supported by the second purpose of CA/MA technology, which is its application to long-distance marine shipping in CA/MA refrigerated containers. Internationalization has also affected stationary CA storage since domestic products may now compete with imported products, potentially limiting the storage incentive. The third purpose of CA/MA technology is semipermeable packaging to create MAP, which is applied mostly to allow domestic marketing of the most perishable products, most notably fresh-cut vegetables and fruits. In both international marine shipping and use of MAP, it may very well not be necessary to achieve maximum possible postharvest life, meaning that identification of the optimum atmosphere does not necessarily require starting with minimally mature fruits or marginally safe temperatures in terms of chilling injury or other physiological disorders such as scald. Fixed durations of transport or marketing also mean that more extreme atmospheres, which the product could not tolerate if the exposure times were longer or the temperature lower, may potentially be employed to improve effect. CA/MA conditions must be tailored to the temperature that will be used since the product responses to CA/MA, or the optimum CA/MA, are different at different temperatures, as demonstrated for strawberries (Nunes et al., 1995).

In fresh fruit and vegetable postharvest distribution systems, temperatures encountered by the products are very often the result of compromises. For example, transport temperatures may be set higher than optimal for the product being carried due to limitations of the refrigeration system that may result in freezing if lower temperatures are used. When multiple products are transported in the same container or trailer, a compromise temperature must be chosen, usually based on the requirements of the most valuable product or the most chilling or freezing sensitive product. In distribution centers, it is not feasible to provide optimum temperature storage areas for the hundreds of fruit and vegetable products that pass through the facility—at most, there may be three or four temperature zones, each of which is a compromise for the products it contains. At retail, the situation is similar, with display temperatures chosen not only as compromise settings for multiple products, but with marketing goals also dictating which products are adjacent in displays and that some products requiring low temperature to maintain quality should be placed in open nonrefrigerated aisle displays. The result of this is that CA/MA/MAP may be designed for just one, discrete segment of the distribution system. Examples include pallet cover MAP systems like Tectrol®, which is applied during transport and removed at the distribution center, and fresh-cut MAP systems that may be designed for the retail display conditions, resulting in little benefit during the lower temperature transport phase.

The temperature problem for MAP

For MAP, system design is limited by the package permeability response to this fluctuating temperature environment. The issue is that the rates of changes in package permeability and product respiration rate in response to temperature are quite different. Thus a film that produces the desired atmosphere for a given product at the product's optimal storage temperature will result in more extreme and potentially damaging atmospheres at higher temperatures as the increase in respiration rate outstrips the change in package permeability (Beaudry et al., 1992; Cameron et al., 1993, 1994; Exama et al., 1993b; Joles et al., 1994). Ideas for addressing this issue have included inclusion of a temperature responsive valve or film patch that allows more O_2 to enter the package and more CO_2 to escape at high temperatures (Cameron et al., 1993; Exama et al., 1993b) as well as multilevel CA/MA/MAP systems with each level designed for a different temperature likely to be encountered (Brecht et al., 2003a,b; Rodov et al., 2000; Silva et al., 1999). Commercialization of the former concept as the

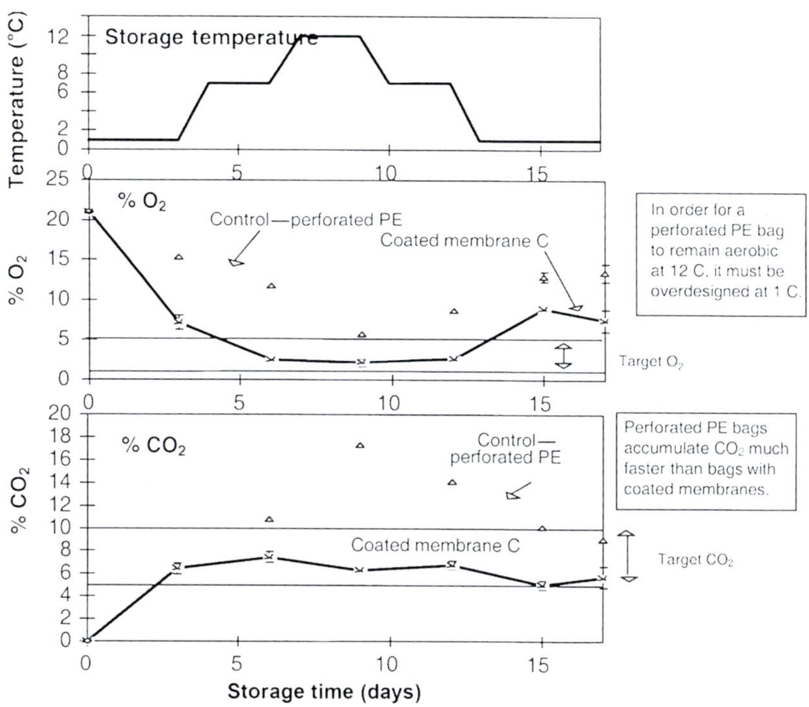

FIG. 1 Atmosphere changes during storage of broccoli with fluctuating temperature in MAP bags of either perforated polyethylene (PE) or with a side chain crystallizable membrane-coated patch over a hole. The storage temperature was 3 days at 1°C, followed by 3 days at 7°C, 3 days at 12°C, and then back down to 7°C for 3 days and 1°C for 5 days. *(With permission Clarke, R., 2001. Temperature switchable membranes for creating and maintaining beneficial package atmospheres for fresh produce. J. Plast. Film Sheet. 17(1), 22–34.)*

Breatheway® MAP technology was made possible by the discovery of side chain crystallizable polymer structures that change permeability via a phase change in response to temperature changes within the range utilized in fresh produce handling (Clarke and DeMoor, 1997). Different semipermeable coatings are placed onto a microporous patch covering an open hole in the package and the coating undergoes a solid–liquid phase change at the chosen temperature, changing its gas permeability, due to selection of the side chains on the polymer backbone. This allows the package to create a new atmosphere that is appropriate for the new temperature (Clarke, 2001; Fig. 1).

Consequences for CA/MA of chilling injury and other temperature-related disorders

Interactions between CA/MA and several physiological disorders have been found to occur with many fruits and vegetables (Schotsmans et al., 2009).

Considering the building consensus that oxidative stress is involved in the mechanisms underlying many physiological disorders (Wismer, 2003), it is tempting to speculate that the reduced O_2 in CA/MA may limit the pool of O_2 available for development of reactive oxygen species, thus reducing the oxidative response of plant tissues to stress and the potential for cellular damage to occur. Most physiological disorders that develop postharvest are similar in that they involve membrane damage and cellular decompartmentalization leading to development of the symptoms of the disorders, often including tissue browning, texture changes, disruption of normal aroma synthesis, and localized necrosis (pitting and scalding). Development of all these symptoms involves to some extent biochemical reactions that require chemical energy to proceed. Thus it should not be surprising that CA/MA conditions, which inhibit respiration and energy production, should alleviate symptom development, whether or not the development of the underlying injury is inhibited.

Chilling injury is a physiological disorder affecting mostly plant species of tropical or subtropical origin that are injured by extended exposure to low, but nonfreezing temperatures. Susceptibility to the disorder in individual plants and plant organs tends to decline as development proceeds. Thus physiologically immature fruits are more susceptible to chilling than mature fruits, and chilling susceptibility declines as fruit ripen. Tolerance to chilling injury is typically expressed in terms of the crop's chilling threshold temperature, which is defined as the lowest temperature at which the crop can be held indefinitely without developing symptoms of chilling injury. To some extent, brief exposure to temperatures below the chilling threshold may be reversible. Differences in chilling susceptibility among crops or developmental stages of the same crop are usually expressed as differences in the chilling threshold temperature or as the ability to tolerate longer or shorter exposures to a given temperature. It seems likely that a technology like CA/MA that extends the potential storage duration of crops by modulating changes in various shelf life-limiting quality aspects could result in an apparent increase in the chilling threshold temperature of susceptible crops. Demonstrating this possibility, however, is complicated by the likelihood that CA/MA is also slowing the rate of the reactions involved in chilling injury and chilling injury symptom development.

Reduced O_2 in CA/MA storage has been found to alleviate chilling injury symptom development in numerous crops (Schotsmans et al., 2009). Elevated CO_2, either as a prestorage treatment or as part of a CA/MA regime, also has been shown to alleviate chilling injury in several crops (Schotsmans et al., 2009). These observations usually involve an apparent lowering of the chilling threshold temperature, supporting the previously stated goal of much CA/MA research to identify conditions that extend the marketable and edible life of the product to the greatest extent possible. As proposed by Pesis (2005), induction of fermentative metabolism in plant tissues by either reduced O_2 or elevated CO_2 may inhibit chilling injury via effects of the resulting acetaldehyde and ethanol produced on membrane function and enzyme activities. This may explain why CA/MA has often been observed to reduce chilling injury at temperatures that normally induce the disorder in air storage. However, the relatively modest modification of the respiratory gases typical of CA/MA storage regimes (usually chosen so as to avoid more than slight fermentation) may inhibit development of symptoms downstream from the site of injury to a greater extent than any alleviation of the injury itself. This would be due to the very low Km value for the terminal respiratory oxidase compared with the Km for other oxidative enzyme reactions. Thus chilling injury may still occur and chilling symptom development (e.g., inhibited aroma development) may proceed upon removal of the crop from CA/MA conditions. It is possible that this is the situation with CA/MA storage of many, if not all, chilling sensitive crops like banana, mango, melons, peach, and tomato, for which the "optimum CA/MA conditions" involve possible lower temperatures and definitely longer storage durations than for air storage.

For fresh fruits and vegetables that are sensitive to chilling injury, the environment that is created within MAP (saturated or near-saturated humidity along with reduction of O_2 and accumulation of CO_2) has been reported to prevent or alleviate the development of chilling injury symptoms. Thus MAP has been reported to be effective in reducing the incidence of CI-related surface pitting and other CI symptoms in papaya, grapefruit, and cucumbers, as well as other fruits (Zagory and Kader, 1988; Latifah et al., 1997). The benefits of high humidity plus modified gas composition with high CO_2 and low O_2 levels surrounding the MAP fruit are beneficial for reducing CI incidence in both climacteric and nonclimacteric fruit (Wang, 1993).

Superficial scald is a type of chilling injury expressed in fruit of several apple and pear cultivars during storage (Lurie and Watkins, 2012; Watkins et al., 1995). Scald development is an example of the interplay that can occur between cultivar, fruit maturity, storage temperature, storage duration, and CA/MA in chilling injury symptom development. Like other chilling injuries, fruit susceptibility to scald decreases with advancing maturity, the injury is more severe at lower temperatures, increases with increasing storage duration, and symptom development is accelerated and usually only visible upon removal from storage and transfer to higher temperatures. The disorder is also thought to be induced by oxidative stress (Whitaker, 2004) and its development can be controlled by maintaining O_2 at the lowest possible levels ("ultralow oxygen") without induction of fermentative metabolism ("dynamic atmosphere control"). Interestingly, ethylene also appears to be involved in scald development as suggested by the effective control of scald achieved by CA/MA atmospheres and chemicals like aminoethoxyvinylglycine that inhibit ethylene biosynthesis (Lurie and Watkins, 2012) and confirmed by the control achieved in both air and CA/MA storage by treatment with 1-methylcyclopropene (1-MCP), which binds almost irreversibly to the ethylene receptors (Zanella, 2003). It is possible that chilling injury/scald could be more effectively controlled by CA/MA applied at storage temperatures higher than those typically employed in air storage.

Inadvertent or unavoidable exposure of crops in MAP to nonoptimum temperature: Negative consequences and limits of concentration range and exposure duration

The quality of MA-packaged fresh produce can be maintained and the shelf life prolonged by following proper handling and distribution practices. This mainly involves sustaining temperature at optimum levels during the storage period, but as previously mentioned this is rarely the case in practice. Produce is usually subjected to a series of temperature alterations throughout the distribution chain. For instance, while the majority of fresh-cut fruits and vegetables are optimally handled at 0°C, they are commonly prepared, shipped, and stored closer to 5°C and have been documented to occasionally reach 10°C (Watada et al., 1996) or in some extreme cases temperatures even above 27°C have been documented (Nunes et al., 2009), if only for brief durations.

Due to the large volumes of produce that are handled, there may be delays in precooling and in packaging, which, due to logistical issues, may exceed 12h (Kim et al., 2005). The temperature abuse after packaging starts at the loading docks, where the temperature control may be limited and continues during transportation. In practice, temperature in commercial refrigerated trucks is often set at 2–4°C, but frequently the load is not uniformly refrigerated. During transportation, the temperature inside the cargo space may vary according to the loading pattern and the volume of the load. In truck trailers, the refrigeration unit is placed on the front ceiling of the cargo space and refrigeration occurs by top-air delivery. Inadequate air pathways in the trailer may not allow proper air circulation and thus uniform temperature. Strawberries loaded against the sidewalls leaving a corridor in the centerline had on average 2°C higher temperature upon arrival after US transcontinental shipments than those loaded along the centerline (Harvey et al., 1980). The relative position of the packaged produce affects its response to temperature changes. The pallet temperature in the outer part of the load changes faster due to the outside environment than in the middle of the load. However, the heat transfer rate is low in the middle of the pallets, which can result in a linear increase of temperature during transport due to respiratory heat generation, especially if the product was not thoroughly cooled before loading (Pelletier et al., 2011; Punt and Huysamer, 2005). The temperature range can be up to 6°C in different parts of the container (Tanner and Amos, 2003) and usually the products placed in the middle and back parts of the truck are the warmest. MAP itself may also contribute to temperature variation during transport by interfering with heat exchange. MAP pallets with strawberries were on average 1 to 4°C warmer than nonpackaged pallets after transcontinental shipments in the United States (Harvey et al., 1980; Macnish et al., 2012). The temperature uniformity and stability is affected as well by the frequency of the truck door openings during the distribution. The temperature discrepancies throughout transportation may reach 13°C (Koseki and Isobe, 2005) while different days of transportation for the same load demonstrate gradients up to 5.5°C (Nunes et al., 2009). The surface temperature of MAP vegetable salads upon arrival at the retailer was found to be 5–7°C above the recommended temperature (Nunes et al., 2009).

The variety of produce handled in the distribution chain requires a range of different temperatures, but usually there are not adequate facilities for storing each commodity at its optimum temperature. The most extended temperature abuse usually occurs at the retail level. In practice, 90% of the products that require storage at 4°C or below are held at higher temperatures at some point during distribution (Paull, 1999). The retail stores have either nonrefrigerated backrooms or, more typically, a cold storage room for produce with temperatures mostly ranging from 0 to 10°C, averaging around 4°C (EcoSure, 2008), while, depending on the retailer, the products may be displayed for sale in refrigerated or nonrefrigerated displays. The refrigerated displays where MAP-packaged produce are stored at the retail stores show significant temperature variation within the display case. The retailer standard temperature setup of the chilled cases is 2–4°C but there are huge discrepancies, depending on the type of display case (open front or full door cabinet) and the position of the bag in the display. Both in open fronted and full door cabinets approximately 95% of the packages with the maximum temperature were found to be located in the front part of the display (Evans et al., 2007). In a survey of retail produce losses, product temperatures in various displays were found to range from −1.2°C (middle shelves) to 19.2°C (front of bottom shelves) inside refrigerated displays, and from 7.6°C (bottom shelves) to 27.7°C (top shelves) inside nonrefrigerated displays (Nunes et al., 2009).

Consumer handling subjects produce in MAP to considerable temperature abuse. Home refrigerators are frequently not set at the proper temperature by the consumers and may also be poorly maintained so as to result in inadequate cooling capability. Consumer studies revealed that 70% of the home refrigerators measured had temperatures above 5°C (Johnson et al., 1998) and 8%–10% were even above 10°C (Bakalis et al., 2003; Taoukis et al., 2005). There is a substantial knowledge gap at the consumer level about proper produce handling during transport from the retailer and during home storage. When asked, approximately 90% of the consumers did not have complete understanding of what the cold chain is and the importance in maintaining it (Ovca and Jevšnik, 2009). A survey showed that only 13% of the consumers are using a cooling bag or box when transferring chilled products from the retailer to home (Evans, 1992) while 7% did not refrigerate chilled products sooner than 90min after purchase (Kennedy et al., 2005).

Time limits for high temperature excursions in terms of package atmosphere modification and product tolerance and injury

MAP is commonly used to supplement temperature control for prolonging the shelf life of fresh produce and, in some cases, it can substitute for refrigeration to a limited extent. A basic limitation of MAP is that the packaging used in the industry is designed for a specific and constant temperature. Thus it can have adverse results when used at other temperatures. As previously mentioned, this implication is related to the difference in responses of the product respiration rate and the film permeability during temperature fluctuation. Respiration rate increases exponentially with temperature while the gas permeability of films does not. For example, for most fruit and vegetable products, the respiration rate increases at several times the rate of Low Density Polyethylene (LDPE) film permeability and 30 times the rate of perforation permeability, when temperature increases from 0 to 15°C (Beaudry, 2000). A MAP that created beneficial low O_2 levels that are not inducing anaerobic respiration when kept at 0°C developed O_2 concentrations that induced fermentation with a 5°C temperature increase (Cameron et al., 1994). This divergence between the respiration rate and the gas permeability of the film results in O_2 depletion and CO_2 accumulation in the MAP, creating injurious gas concentrations.

Storage under too low O_2 concentration and/or too high CO_2 concentration can lead to physiological disorders including discoloration, tissue injury, decay, irregular ripening, and development of off-odor and off-flavor. The tolerance levels in low O_2 and high CO_2 concentrations vary significantly between horticultural commodities. The minimum O_2 concentration tolerated can range from 0.5% to 5% (Kader et al., 1989) and the maximum CO_2 concentration from 1% to 30% (as cited in Watkins, 2000).

Additionally, the rise in temperature itself accelerates the metabolic rate of the produce leading in faster deterioration since every 10°C increase in temperature accelerates deterioration and the rate of loss in nutritional quality by two- to fourfold (Kader, 2002). Generally, the effect of temperature abuse on the quality varies according to the duration and the temperature and is cumulative through the chain. It is also linked to the perishability of the produce with fresh cuts being the most susceptible products handled in the chain. The effect of temperature abuse is pronounced when it occurs at the late shelf life stage. Broccoli packed in MAP demonstrated faster yellowing when subjected to temperature abuse near the end of storage (Ihringer et al., 2010; Fig. 2).

Storing produce in MAP with temperatures different than that for which the package was designed will drastically reduce its quality. Mushrooms, broccoli, and mature green tomatoes in MAP (5 kPa O_2 + 10 kPa CO_2, 3 kPa O_2 + 8 kPa CO_2, and 5 kPa O_2 + 5 kPa CO_2, respectively) when subjected to temperature fluctuations of $\Delta T = 10$°C, demonstrated rapid decrease in O_2 and increase in CO_2, which led to severe quality mitigation (Tano et al., 2007). MAP baby spinach exhibited yellowing, decay, and off-odor development, which resulted in reduced shelf life when stored at 8°C or above (Kou et al., 2014). Sweet cherries packed in passive MAP at 25°C or 20°C had five times shorter shelf life than cherries kept at 0°C, due to increased mold and physiological breakdown (Petracek et al., 2002). Similarly, fresh-cut celery packed in passive MAP at 10°C developed discoloration and loss of crispiness while celery packed at 0°C maintained acceptable quality for 27 days (Viña and Chaves, 2003).

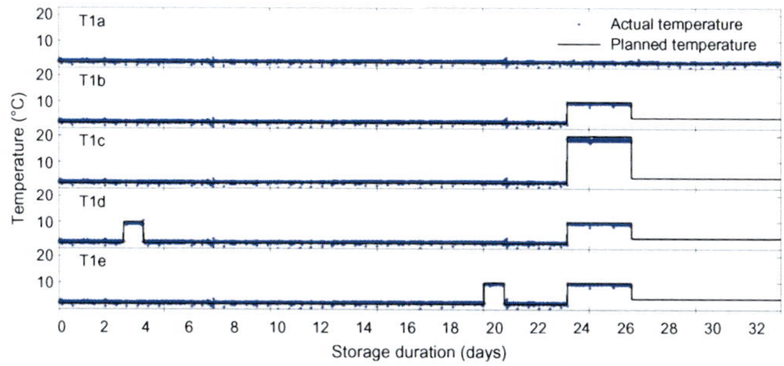

	Hue angle	
	28 days	35 days
T1a	132.4	127.8
T1b	131.8	126.8
T1c	125.0	116.1
T1d	124.3	123.6
T1e	125.4	125.1
s^2	0.8	0.8

FIG. 2 Hue angle for broccoli florets of heads ($n=20$) held under different temperature profiles. Lower hue angle values indicate yellowing of the broccoli florets. *(Adapted from Ihringer, D., Schotsmans, W., Mawson, A.J., MacKay, B., 2010. Effect of temperature abuse during transport of modified atmosphere packaged broccoli. Acta Hortic. 857, 189–196.)*

MAP can be utilized for prolonging the shelf life and maintaining the quality of some commodities stored at ambient temperature by creating beneficial in-package gas composition. MAP (1–2 kPa O_2 + 15–20 kPa CO_2) storage of shredded Galega kale at 20°C extended its shelf life for 2 days more than when stored in air, with no symptoms of CO_2 injury, but off-odor was the limiting factor (Fonseca et al., 2005). At 20°C, broccoli spears in 3 kPa O_2 + 9 kPa CO_2 MAP maintained higher ascorbic acid and chlorophyll contents and increased moisture retention compared with product stored in air without packaging (Barth et al., 1993). Cherry tomatoes packed in MAP using microperforated films that created atmospheres of 15–18 kPa O_2 + 2–4 kPa CO_2, preserved their nutritional, organoleptic, and overall quality when held at ambient temperature (D'Aquino et al., 2016). The quality of litchi fruit packed in permselective MAP was maintained for up to 17, 13, and 9 day at 10, 15, and 20°C, respectively (Mangaraj et al., 2012). Yet, the maximum quality retention and shelf life for those commodities would be achieved when they are at their optimum storage temperature. Development of a MAP for nonoptimum or fluctuating temperature can be a solution when temperature control is not possible or is limited, but it will not result in similar success as MAP applied at optimum temperatures. Also, the optimum atmosphere is not constant across a range of temperatures—at above optimum temperatures, the required minimum O_2 and maximum CO_2 would both tend to be higher than at the optimum temperature.

Conclusions and Future Trends

Determining the best CA/MA conditions for fresh fruits and vegetables in order to maintain the best possible product quality for a given storage or distribution period is a complex undertaking. As we have attempted to show, successful accomplishment of this task requires consideration of several factors in addition to just the best O_2 and CO_2 levels. For most applications of CA/MA besides long-term stationary storage it is necessary to define a discrete time frame during which the goal is to maintain the sensory quality of the product as near to perfect as is possible; however, with competition from offshore production regions increasing all the time, the economic feasibility of traditional, extended, stationary storage must also be reevaluated. Future improvements in CA/MA may involve defining different scenarios for each crop and developing the optimum CA/MA conditions for each specifying higher temperatures and using fruits and vegetables with higher initial quality or quality potential that need be successfully maintained for only a specified, limited time.

For MAP systems, the foremost challenge is overcoming the design limitations currently imposed by the disparity between the behavior of product respiration and packaging permeability when exposed to the fluctuating temperatures that occur during distribution. We may eventually see development of new films or other materials and devices that result in packages for which the QP_{10} can seamlessly match the QR_{10} over a range of temperatures so that the optimum MA is always maintained. However, in this case, it would not be enough for the new MAP to maintain a constant MA since experience indicates the optimum MA would be expected to be different at different temperatures.

References

Babic, I., Hilbert, G., Nguyen-the, C., Guiraud, J., 1992. The yeast flora of stored ready-to-use carrots and their role in spoilage. Int. J. Food Sci. Technol. 27 (4), 473–484.

Bakalis, S., Giannakourou, M.C., Taoukis, P., 2003. Effect of domestic storage and cooking conditions on the risk distribution in ready to cook meat products. In: Presented at: 9th Intl. Congr. Eng. Food, Montpellier, France, 7–11 March 2003, pp. 13–18. http://smas.chemeng.ntua.gr/files/webdoc_15_2_2_2004.pdf.

Barth, M.M., Kerbel, E.L., Broussard, S., Schmidt, S.J., 1993. Modified atmosphere packaging protects market quality in broccoli spears under ambient-temperature storage. J. Food Sci. 58 (5), 1070–1072.

Beaudry, R.M., 1999. Effect of O_2 and CO_2 partial pressure on selected phenomena affecting fruit and vegetable quality. Postharvest Biol. Technol. 15 (3), 293–303.

Beaudry, R.M., 2000. Responses of horticultural commodities to low oxygen: limits to the expanded use of modified atmosphere packaging. HortTechnology 10 (3), 491–500.

Beaudry, R.M., Cameron, A.C., Shirazi, A., Dostal-Lange, D.L., 1992. Modified atmosphere packaging of blueberry fruit: effect of temperature on package O_2 and CO_2. J. Am. Soc. Hort. Sci. 117 (3), 436–441.

Bender, R.J., Brecht, J.K., Baldwin, E.A., Malundo, T.M.M., 2000a. Aroma volatiles of mature-green and tree-ripe 'Tommy Atkins' mangoes after controlled atmosphere vs. air storage. HortScience 35 (4), 684–686.

Bender, R.J., Brecht, J.K., Sargent, S.A., Huber, D.J., 2000b. Low temperature controlled atmosphere storage for tree-ripe mangoes (*Mangifera indica* L.). Acta Hortic. (509), 447–458.

Brecht, J.K., Chau, K.V., Fonseca, S.C., Oliveira, F.A.R., Silva, F.M., Nunes, M.C.N., Bender, R.J., 2003a. Maintaining optimal atmosphere conditions for fruits and vegetables throughout the postharvest handling chain. Postharvest Biol. Technol. 27 (1), 87–101.

Brecht, J.K., Chau, K.V., Fonseca, S.C., Oliveira, F.A.R., 2003b. CA transport of fresh produce in MAP: designing systems for optimal atmosphere conditions throughout the postharvest handling chain. Acta Hortic. (600), 799–802.

Cameron, A.C., Patterson, B.D., Talasila, P.C., Joles, D.W., 1993. Modeling the risk in modified-atmosphere packaging: a case for sense-and-respond packaging. In: Blanpied, G.D., Barstch, J.A., Hicks, J.R. (Eds.). Proc. 6th Intl. Controlled Atm. Res. Conf., vol. 1. Cornell University, Ithaca, NY, NRAES-71, pp. 95–102.

Cameron, A.C., Beaudry, R.M., Banks, N.H., Yelanich, M.V., 1994. Modified-atmosphere packaging of blueberry fruit: modeling respiration and package oxygen partial pressures as a function of temperature. J. Am. Soc. Hortic. Sci. 119 (3), 534–539.

Carlin, F., Nguyen-the, C., Cudennec, P., Reich, M., 1989. Microbiological spoilage of fresh, ready-to-use grated carrots. Sci. Aliment. 9, 371–386.

Clarke, R., 2001. Temperature switchable membranes for creating and maintaining beneficial package atmospheres for fresh produce. J. Plast. Film Sheet. 17 (1), 22–34.

Clarke, R., DeMoor, C.P., 1997. The future in film technology: a tunable packaging system for fresh produce. In: Gorny, J.R. (Ed.). Proc. 7th Intel. Controlled Atm. Res. Conf. Postharvest Horticulture Series 18, vol. 5. University of California, Davis, pp. 68–75.

Costa, M.A.C., Brecht, J.K., Sargent, S.A., Huber, D.J., 1994. Tolerance of snap beans to elevated CO_2 levels. Proc. Fla. State Hort. Soc. 107, 271–273.

D'Aquino, S., Mistriotis, A., Briassoulis, D., Di Lorenzo, M.L., Malinconico, M., Palma, A., 2016. Influence of modified atmosphere packaging on postharvest quality of cherry tomatoes held at 20°C. Postharvest Biol. Technol. 115 (1), 103–112.

EcoSure, 2008. 2007 US Cold Temperature Evaluation. Design and Summary Pages. 10 p. http://foodrisk.org/default/assets/File/EcoSure%202007%20Cold%20Temperature%20Report.pdf.

Evans, J., 1992. Consumer handling of chilled foods: Perception and practice. Int. J. Refrig. 15 (5), 290–298.

Evans, J.A., Scarcelli, S., Swain, M.V.L., 2007. Temperature and energy performance of refrigerated retail display and commercial catering cabinets under test conditions. Int. J. Refrig. 30 (3), 398–408.

Exama, A., Arul, J., Lencki, R., Li, Z., 1993a. Suitability of plastic films for modified atmosphere packaging of fruits and vegetables. J. Food Sci. 58 (6), 1365–1370.

Exama, A., Arul, J., Lencki, R., Li, Z., 1993b. Suitability of various plastic films for modified atmosphere packaging of fruits and vegetables: Gas transfer properties and effect of temperature fluctuation. Acta Hortic. (343), 175–180.

Fennir, M.A., Landry, J.A., Raghavan, G.S.V., 2003. Respiration rate of potatoes (*Solanum tuberosum* L.) measured in a two-bin research scale storage facility, using heat and moisture balance and GC analysis. Can. Biosyst. Eng. 45, 4.1–4.9.

Fonseca, S.C., Oliveira, F.A.R., Brecht, J.K., Chau, K.V., 2005. Influence of low oxygen and high carbon dioxide on shredded Galega kale quality for development of modified atmosphere packages. Postharvest Biol. Technol. 35 (3), 279–292.

Geeson, J.D., Browne, K.M., Maddison, K., Shepherd, J., Guaraldi, F., 1985. Modified atmosphere packaging to extend the shelf life of tomatoes. J. Food Technol. 20 (3), 339–349.

Harvey, J., Harris, C., Tietjen, W., Serio, T., 1980. Quality Maintenance in Truck Shipments of California Strawberries. Advances in agricultural technology (AAT-W-12). U.S. Dept. Agric., Agric. Res. Western Reg., Sci. Educ. Admin. 13 p.

Ihringer, D., Schotsmans, W., Mawson, A.J., MacKay, B., 2010. Effect of temperature abuse during transport of modified atmosphere packaged broccoli. Acta Hortic. (857), 189–196.

Johnson, A.E., Donkin, A., Morgan, K., Lilley, J.M., Neale, R.J., Page, R.M., Silburn, R., 1998. Food safety knowledge and practice among elderly people living at home. J. Epidemiol. Community Health 52 (11), 745–748.

Joles, D.W., Cameron, A.C., Shirazi, A., Petracek, P.D., Beaudry, R.M., 1994. Modified-atmosphere packaging of 'heritage' red raspberry fruit: respiratory response to reduced oxygen, enhanced carbon dioxide, and temperature. J. Am. Soc. Hortic. Sci. 119 (3), 540–545.

Kader, A.A., 1997. Biological bases of O_2 and CO_2 effects on postharvest life of horticultural perishables. In: Saltveit, M.E. (Ed.). Proc. 7th Intl. Controlled Atmosphere Res. Conf. Vegetables and ornamentals. Postharvest horticulture series. 18, vol. 4. University of California, Davis, pp. 160–163.

Kader, A.A., 2002. Pre-and postharvest factors affecting fresh produce quality, nutritional value, and implications for human health. In: Proceedings of the International Congress Food Production and the Quality of Life. vol. 1. 4–8 September, 2000, Sassari, Italy, pp. 109–119.

Kader, A.A., Zagory, D., Kerbel, E.L., 1989. Modified atmosphere packaging of fruits and vegetables. Crit. Rev. Food Sci. Nutr. 28 (1), 1–30.

Kakiomenou, K., Tassou, C., Nychas, G., 1996. Microbiological, physiochemical and organoleptic changes of shredded carrots stored under modified conditions. Int. J. Food Sci. Technol. 31 (4), 359–366.

Kanellis, A.K., Tonutti, P., Perata, P., 2009. Biochemical and molecular aspects of modified and controlled atmospheres. In: Yahia, E.M. (Ed.), Modified and Controlled Atmospheres for the Storage, Transportation, and Packaging of Horticultural Commodities. CRC, Boca Raton, pp. 553–568.

Kays, S.J., 1997. Postharvest Physiology of Perishable Plant Products. Van Nostrand Reinhold, New York.

Ke, D., Zhou, L., Kader, A.A., 1994. Mode of oxygen and carbon dioxide action on strawberry ester biosynthesis. J. Am. Soc. Hortic. Sci. 119 (5), 971–975.

Kennedy, J., Jackson, V., Blair, I., McDowell, D., Cowan, C., Bolton, D., 2005. Food safety knowledge of consumers and the microbiological and temperature status of their refrigerators. J. Food Prot. 68 (7), 1421–1430.

Kim, J.G., Luo, Y., Saftner, R.A., Gross, K.C., 2005. Delayed modified atmosphere packaging of fresh-cut romaine lettuce: effects on quality maintenance and shelf-life. J. Am. Soc. Hortic. Sci. 130 (1), 116–123.

Koseki, S., Isobe, S., 2005. Prediction of pathogen growth on iceberg lettuce under real temperature history during distribution from farm to table. Int. J. Food Microbiol. 104 (3), 239–248.

Kou, L.P., Luo, Y.G., Park, E., Turner, E.R., Barczak, A., Jurick, W.M., 2014. Temperature abuse timing affects the rate of quality deterioration of commercially packaged ready-to-eat baby spinach. Part I: Sensory analysis and selected quality attributes. Postharvest Biol. Technol. 91, 96–103.

Latifah, M.N., Ali, Z.M., Lazan, H., 1997. Effects of modified atmosphere packaging on the quality of Eksotika papaya stored at low temperature. J. Trop. Agric. Food Sci. 25, 95–102.

Lazan, H., Selamat, M.K., Ali, Z.M., 1993. β-Galactosidase, polygalacturonase and pectinesterase in differential softening and cell wall modification during papaya fruit ripening. Physiol. Plant. 95, 106–112.

Lipton, W.J., 1968. Market quality of asparagus: Effects of maturity at harvest and of high CO_2 atmospheres during simulated transit. In: USDA Marketing Research Report No. 817.

Lurie, S., Watkins, C.B., 2012. Superficial scald, its etiology and control. Postharvest Biol. Technol. 65, 44–60.

Macnish, A.J., Padda, M.S., Pupin, F., Tsouvaltzis, P.I., Deltsidis, A.I., Sims, C.A., Brecht, J.K., Mitcham, E.J., 2012. Comparison of pallet cover systems to maintain strawberry fruit quality during transport. HortTechnology 22 (4), 493–501.

Mangaraj, S., Goswami, T.K., Giri, S.K., Tripathi, M.K., 2012. Permselective MA packaging of litchi (cv. Shahi) for preserving quality and extension of shelf-life. Postharvest Biol. Technol. 71, 1–12.

Mathooko, F.M., 1996. Regulation of respiratory metabolism in fruits and vegetables by carbon dioxide. Postharvest Biol. Technol. 7 (1–2), 1–26.

Mattheis, J.P., Fellman, J.K., 2000. Impacts of modified atmosphere packaging and controlled atmospheres on aroma, flavor and quality of horticultural commodities. HortTechnology 10 (3), 507–510.

Maul, F., Sargent, S.A., Sims, C.A., Baldwin, E.A., Balaban, M.O., Huber, D.J., 2000. Tomato flavor and aroma quality as affected by storage temperature. J. Food Sci. 65 (7), 1228–1237.

Mitz, M.A., 1979. CO_2 biodynamics: a new concept of cellular control. J. Theor. Biol. 80 (4), 537–551.

Neven, L.G., Yahia, E.M., Hallman, G.J., 2009. Effects on insects. In: Yahia, E.M. (Ed.), Modified and Controlled Atmospheres for the Storage, Transportation, and Packaging of Horticultural Commodities. CRC, Boca Raton, pp. 233–266.

Nunes, M.C.N., Morais, A.M.B., Brecht, J.K., Sargent, S.A., 1995. Quality of strawberries after storage in controlled atmospheres at above optimum storage temperatures. Proc. Fla. State Hortic. Soc. 108, 273–278.

Nunes, M.C.N., Emond, J.P., Rauth, M., Dea, S., Chau, K.V., 2009. Environmental conditions encountered during typical consumer retail display affect fruit and vegetable quality and waste. Postharvest Biol. Technol. 51 (2), 232–241.

Nunes, M.C.N., Nicometo, M., Emond, J.P., Badia, R.M., Uysal, I., 2014. Improvement in fresh fruit and vegetables logistics quality: Berry logistic field studies. Philos. Trans. A Math. Phys. Eng. Sci. 372 (20130307), 1–19.

Ovca, A., Jevšnik, M., 2009. Maintaining a cold chain from purchase to the home and at home: consumer opinions. Food Control 20 (2), 167–172.

Paull, R., 1999. Effect of temperature and relative humidity on fresh commodity quality. Postharvest Biol. Technol. 15 (3), 263–277.

Pelletier, W., Brecht, J.K., Nunes, M.C.N., Emond, J.P., 2011. Quality of strawberries shipped by truck from California to Florida as influenced by postharvest temperature management practices. HortTechnology 21 (4), 482–493.

Pesis, E., 2005. The role of the anaerobic metabolites, acetaldehyde and ethanol, in fruit ripening, enhancement of fruit quality and fruit deterioration. Postharvest Biol. Technol. 37 (1), 1–19.

Petracek, P.D., Joles, D.W., Shirazi, A., Cameron, A.C., 2002. Modified atmosphere packaging of sweet cherry (*Prunus avium* L., cv. Sams) fruit: Metabolic responses to oxygen, carbon dioxide, and temperature. Postharvest Biol. Technol. 24 (3), 259–270.

Ponce-Valadez, M., Fellman, S.M., Giovannoni, J., Gan, S., Watkins, C.B., 2009. Differential fruit gene expression in two strawberry cultivars in response to elevated CO_2 during storage revealed by a heterologous fruit microarray approach. Postharvest Biol. Technol. 51 (2), 131–140.

Punt, H., Huysamer, M., 2005. Temperature variances in a 12-m integral reefer container carrying plums under a dual temperature shipping regime. Acta Hortic. (687), 289–296.

Rodov, V., Copel, A., Aharoni, N., Aharoni, Y., Wiseblum, A., Horev, B., Vinokur, Y., 2000. Nested modified atmosphere packages maintain quality of trimmed sweetcorn during cold storage and the shelf life period. Postharvest Biol. Technol. 18 (3), 259–266.

Rothan, C., Duret, S., Chevalier, C., Raymond, P., 1997. Suppression of ripening-associated gene expression in tomato fruits subjected to a high CO_2 concentration. Plant Physiol. 114 (1), 255–263.

Saltveit, M.E., 1997. The future of CA & MA recommendations. In: Saltveit, M.E. (Ed.), Proc. 7th Intl. Controlled Atmosphere Res. Conf. Vol. 4. Vegetables and Ornamentals. Postharvest Hort. Ser. 18. University of California, Davis, pp. 2–7.

Schotsmans, W.C., DeLong, J.M., Larrigaudière, C., Prange, R.K., 2009. Effects on physiological disorders. In: Yahia, E.M. (Ed.), Modified and Controlled Atmospheres for the Storage, Transportation, and Packaging of Horticultural Commodities. CRC, Boca Raton, FL, pp. 160–192.

Shamaila, M., Powire, W.D., Skura, B.J., 1992. Analysis of volatile compounds from strawberry fruit stored under modified atmosphere packaging (MAP). J. Food Sci. 57 (5), 1173–1176.

Silva, F.M., Chau, K.V., Brecht, J.K., Sargent, S.A., 1999. Modified atmosphere packaging for mixed loads of horticultural commodities exposed to two postharvest temperatures. Postharvest Biol. Technol. 17 (1), 1–9.

Tanner, D.J., Amos, N.D., 2003. Heat and mass transfer: temperature variability during shipment of fresh produce. Acta Hortic. 599, 193–204.

Tano, K., Oulé, M.K., Doyon, G., Lencki, R.W., Arul, J., 2007. Comparative evaluation of the effect of storage temperature fluctuation on modified atmosphere packages of selected fruit and vegetables. Postharvest Biol. Technol. 46 (3), 212–221.

Taoukis, P., Giannakourou, M., Koutsoumanis, K., Bakalis, S., 2005. Modelling the effect of house hold chilled storage conditions on the risk distribution of meat products. Acta Hortic. (674), 435–439.

Thompson, A.K., 2010a. Effects and Interactions of CA Storage. In: Controlled Atmosphere Storage of Fruits & Vegetables, second ed. CABI, Wallingford, UK, pp. 11–25.

Thompson, A.K., 2010b. Introduction: CA Storage History. In: Controlled Atmosphere Storage of Fruits & Vegetables, second ed. CABI, Wallingford, UK, pp. 1–10.

Viña, S.Z., Chaves, A.R., 2003. Texture changes in fresh cut celery during refrigerated storage. J. Sci. Food Agric. 83 (13), 1308–1314.

Wang, C.Y., 1993. Approaches to reduce chilling injury of fruits and vegetables. Hortic. Rev. 15, 63–95.

Watada, A.E., Ko, N.P., Minott, D.A., 1996. Factors affecting quality of fresh-cut horticultural products. Postharvest Biol. Technol. 9 (2), 115–125.

Watkins, C.B., 2000. Responses of horticultural commodities to high carbon dioxide as related to modified atmosphere packaging. HortTechnology 10 (3), 501–506.

Watkins, C.B., Bramlage, W.J., Cregoe, B.A., 1995. Superficial scald of 'granny smith' apple is expressed as a typical chilling injury. J. Am. Soc. Hortic. Sci. 120 (1), 88–94.

Whitaker, B.D., 2004. Oxidative stress and superficial scald of apple fruit. HortScience 39 (5), 933–937.

Wismer, W., 2003. Low temperature stress as a causative agent of oxidative stress in postharvest crops. In: Hodges, D.M. (Ed.), Postharvest Oxidative Stress in Horticultural Crops. Food Products Press, New York, pp. 55–68.

Zagory, D., Kader, A.A., 1988. Modified atmosphere packaging of fresh produce. Food Technol. 42 (9), 70–74. 76–77.

Zanella, A., 2003. Control of apple scald—a comparison between 1-MCP and DPA postharvest treatments, ILOS and ULO storage. Acta Hortic. (600), 271–275.

Chapter 9

The impact of light on modified atmosphere storage and quality of fresh produce

Maria Isabel Gil and Yolanda Garrido
Group on Quality and Safety, Department of Food Science and Technology, CEBAS-CSIC, Murcia, Spain

Lighting systems for crop production and postharvest storage

D'Souza et al. (2015) reviewed the most common lighting sources in food production and preservation such as high-intensity discharge (HID) lighting, metal halide, xenon fluorescent, and incandescent lamps as well as light-emitting diodes (LEDs). LEDs have great potential as lighting systems for crop production and postharvest storage because of their small size, long operating lifetime, wavelength specificity, and low radiant heat emissions that reduce thermal impacts on the product and the environment, making them suitable for cold storage applications. D'Souza et al. (2015) have described several studies that demonstrated the usefulness of LEDs in the cultivation of different crops. LEDs have proved to be very useful in growing facilities in space because of the greater plant photosynthesis rate (Ilieva et al., 2010), biomass production (Poulet et al., 2014), and nutrition (Shen et al., 2014) per photon generated. It has been also reported that LEDs increase shoot tissue pigments, glucosinolates, and mineral elements of sprouting broccoli seeds after short-duration exposure to blue light (Kopsell and Sams, 2013). Preharvest exposure to lighted conditions can also promote negative effects depending on the exposure time and intensity. Long-term light treatment can result in yield reduction although it increases the antioxidant content in plants (Golan et al., 2006). Bartoli et al. (2009) documented the effects of light quality rather than quantity in the accumulation of ascorbic acid in beans (*Phaseolus vulgaris* L.). They observed that bean growth under low red/far red (R/FR) ratios resulted in much lower leaf AA content, leaf respiration rate, and antioxidant enzyme activities than high (R/FR) ratios. These authors concluded that preharvest light treatments might be applied not only to boost the content of leaf antioxidants, but also to

Controlled and Modified Atmospheres for Fresh and Fresh-Cut Produce
https://doi.org/10.1016/B978-0-12-804599-2.00010-7

maintain or enhance phytochemical level of stored vegetables. Thus a short-term light treatment can improve the content of phytochemicals and improve the postharvest performance. However, the role of antioxidants in preserving plants at harvest and after harvest extending the storage life of fresh produce remains unknown. The control of intensity and duration of light exposure is crucial to prevent possible photo-damage. A useful guide for the selection of suitable light intensity is the light compensation point (LCP) defined as the amount of light required for the rate of photosynthesis to be equivalent to the rate of respiration. During postharvest storage when photon flux is below the light compensation point, senescence of the tissue is accelerated.

The impact of preharvest light on plant tissue health and quality

The impact of light on plants is crucial for photosynthesis, plant growth, and development. However, excess absorbed light is dangerous to plants because it can lead to the enhanced production of reactive oxygen species (ROS) such as hydrogen peroxide (H_2O_2), superoxide, hydroxyl radicals, and singlet oxygen that can damage cellular components including the integrity of membrane lipids (Niyogi, 1999). These ROS are generated in the course of normal oxidative metabolism but their production is often increased under light stress (Whitaker, 2003). To detoxify potentially damaging ROS, plant cells produce antioxidants such as ascorbic acid, glutathione, α-tocopherol (vitamin E), and carotenoids, plus a complement of enzymes such as superoxide dismutase, catalase and glutathione, and ascorbic acid peroxidase (Yu, 1994). Plants grown under high light intensity generally have high vitamin C, and carotenoids and polyphenols compared to plants grown under low light. The amount and intensity of light during the growing season influences the amount of ascorbic acid formed because ascorbic acid is synthetized from sugars supplied through photosynthesis (Lee and Kader, 2000; Dorais and Ehret, 2008). Comparison of senescence parameters such as chlorophyll loss and lipid peroxidation confirmed the visual observation of senescence (Philosoph-Hadas et al., 1994; Whitaker, 2003). These authors showed that the content of soluble reducing compounds had a clear inverse correlation with the rate of senescence.

Light exposure has been proposed to be beneficial for storing seedlings, leaf cuttings, and harvested leafy crops (Kubota and Kozai, 1995). Light in combination with low temperature has been suggested to preserve quality of growing tissues because photosynthetic and regrowth abilities are maintained (Heins et al., 1992; Kubota and Kozai, 1994). In contrast with darkening, light conditions can reduce leaf senescence, by contributing to photosynthesis, increasing the levels of chlorophylls and proteins (Chen et al., 1997; Brouwer et al., 2012). Dark conditions accelerate degradation processes and death through an anticipated damage of the photosynthetic machinery (Keech et al., 2007).

Light is one of the most important factors regulating the biosynthesis and accumulation of phytochemicals as light exposure during growth increases the concentrations of vitamin C, glucosinolates, and polyphenols compared with growth in the dark. This is the case of broccoli sprouts in which light increased the concentration of phytochemicals, mainly during the first 3–5 days of development (Perez-Balibrea et al., 2008). The manipulation of light has been an effective approach toward practical manipulation of plant pigments. This is the case of UV radiation, which has been found to have a positive effect on phytochemicals in a broad range of plant species (Dorais and Ehret, 2008). In general, exterior fruit exposed to maximum sunlight contain higher amounts of antioxidants such as vitamin C (Lee and Kader, 2000), anthocyanins (Gil et al., 1995), and even carotenoids although they do not require induction by light (Ju et al., 1999).

The impact of light in postharvest storage of vegetables

One of the main goals of postharvest technology is to delay senescence using storage conditions that include the combination of adequate temperature, relative humidity, and concentrations of oxygen, carbon dioxide, and ethylene (Kader and Rolle, 2004). Fresh produce is usually stored under dark conditions, even though light radiation is known to retard leaf senescence (Goldthwaite and Laetsch, 1967). The use of light has been explored very little due to the difficulties in the selection of appropriate lighting regimes to extend the postharvest life. Senescence due to light deprivation is the result of reduced photosynthesis, while light can activate photosynthesis and positively affect the levels of carbon metabolites and other metabolic changes in the chloroplast (Costa et al., 2013).

During dark storage, a rapid depletion of phytochemical content often occurs together with a decline in several quality aspects. The use of light for postharvest applications has been managed with the purpose of increasing the content of phytochemicals to deal with oxidative stress defense. This was the case of a short postharvest exposure to pulsed light as a means of stimulating coloration by increasing the accumulation of anthocyanin in some fruits such as figs (*Ficus carica* L.) (Rodov et al., 2012). Another example that illustrates this concept is the enhancement of phytochemicals like stilbenes (mainly resveratrol) after UV-C exposure in grapes (Cantos et al., 2000). The concept of hormesis which is a stimulation by low doses of any potentially harmful agent (Luckey, 1980) establishes that it is possible to obtain a beneficial effect delaying postharvest senescence from the application of a low or sublethal dose of a stress such as UV radiation. Postharvest UV-C treatment has been described as delaying senescence in broccoli (*Brassica oleracea* L.) florets (Costa et al., 2006) and in peppers (*Capsicum annum* L. cv. Zafiro) (Vicente et al., 2005). These authors suggested that UV-C treatments could be a useful nonchemical method to delay chlorophyll degradation, reduce respiration rate, and maintain antioxidant capacity. Light impacts metabolic activity during storage even at low temperature and this can affect the content of phytochemicals, many of which can have an antioxidant effect to balance the unfavorable oxidative status and thus delay senescence. However, it remains unknown if postharvest light treatments have effects apart from stimulating color development and the levels of antioxidants.

Postharvest application of light to vegetables

In the case of leafy green vegetables, postharvest application of light is extremely relevant because the leaves are the main site for photosynthesis and therefore have significant impact on postharvest quality. The influence of light on quality characteristics of fresh and fresh-cut products is controversial, as it has been described as positive, negative, or no effect (Lester et al., 2010; Costa et al., 2013; Gergoff Grozeff et al., 2013; Braidot et al., 2014; Glowacz et al., 2014). This is likely due to the differences in the intensity and duration of light exposure as well as the type of light. In this chapter, we define as low light intensity those treatments $\leq$5–15 $\mu mol\, m^{-2}\, s^{-1}$, medium light intensity those treatments between 15 and 45 $\mu mol\, m^{-2}\, s^{-1}$, and high light intensity those treatments $\geq$45 $\mu mol\, m^{-2}\, s^{-1}$. Clear differences have been observed depending on postharvest light intensity as in the case of romaine lettuce (*Lactuca sativa* L.) in which higher ascorbic acid and total content of polyphenols were observed for a light intensity treatment of 34 $\mu mol\, m^{-2}\, s^{-1}$ while lower ascorbic acid and content of total polyphenols were observed at a lower light intensity of 7 $\mu mol\, m^{-2}\, s^{-1}$ (Zhan et al., 2012a). Regarding the duration of the treatments, short-time light treatments (cycles of 1 or 0.5 h with an average fluency of $1.4 \pm 0.4\, \mu mol^{-2}\, s^{-1}$) in the spectrum of photosynthetically active radiation (PAR) were efficient postharvest light treatments for lamb's lettuce (*Valerianella olitoria* [L.] Pollich) while higher doses were less efficient or even negative (Braidot et al., 2014). Concerning the duration of the light exposure, several studies have described the beneficial effect of continuous low-intensity light as an efficient and low-cost treatment to delay postharvest senescence of broccoli (*B. oleracea* L.) florets (Büchert et al., 2011; Zhan et al., 2012b) or low-intensity light pulses to reduce senescence and improve the content of phytochemicals in basil (*Ocimum basilicum*) (Costa et al., 2013). Regarding the type of light, LED technology has been successfully applied to improve quality of cabbage (*B. oleracea* var. capitate) (Lee et al., 2014). These authors observed that the total chlorophyll content was higher for green followed by white than green and red LED light, while the vitamin C content and total polyphenolic content were higher for blue followed by white than red and green LED light (Lee et al., 2014).

Postharvest light exposure influences the visual quality of some perishables, but as in the case of other quality attributes, its influence depends on the light treatment and type of product (Table 1). Optimizing the light environment during storage at low temperature preserved color, texture, aroma, and overall acceptability in broccoli florets (Zhan et al., 2012b) and diminished browning in lettuce (*Lactuca sativa* L. var. longifolia) (Zhan et al., 2012a) and celery

TABLE 1 Effect of light on visual quality of fresh and fresh-cut vegetables

Vegetable	Effect of light	Observations	References
Spinach (mature leaves)	–	Light exposure promoted yellowing	Toledo et al., 2003 (20–25 μmol m^{-2} s^{-1})
Cauliflower (florets)	–	Light exposure promoted browning of cut edge	Sanz et al., 2007 (35 μmol m^{-2} s^{-1})
Chard (leaves)	–	Light exposure promoted changes in color, texture, and reduced shelf life	Sanz et al., 2008 (35 μmol m^{-2} s^{-1})
Leek (fresh cut)	–	Light exposure promoted changes in color, texture, and reduced shelf life	Ayala et al., 2009 (35 μmol m^{-2} s^{-1})
Asparagus (fresh cut)	–	Light exposure promoted browning of cut edge	Sanz et al., 2009 (35 μmol m^{-2} s^{-1})
Broccoli (florets)	+	Light exposure preserved green color	Büchert et al., 2011 (12 μmol m^{-2} s^{-1})
Lettuce (fresh cut)	+/–	Light exposure increased browning and reduced visual quality and off-odors	Martínez-Sánchez et al., 2011 (6 μmol m^{-2} s^{-1})
Lettuce (fresh cut)	+/–	Low light exposure (7 μmol m^{-2} s^{-1}) increased browning while medium light exposure (34 μmol m^{-2} s^{-1}) reduced browning	Zhan et al., 2012a (7 μmol m^{-2} s^{-1} and 34 μmol m^{-2} s^{-1})
Broccoli (florets)	+	Light exposure preserved color, texture, and acceptability	Zhan et al., 2012b (24 μmol m^{-2} s^{-1} intensity)
Basil (mature leaves)	+	Light exposure reduced darkening	Costa et al., 2013 (30–37 μmol m^{-2} s^{-1})
Celery (mature steaks)	+	Light exposure suppressed browning and improved antioxidant capacity	Zhan et al., 2013a (34 μmol m^{-2} s^{-1})

Continued

TABLE 1 Effect of light on visual quality of fresh and fresh-cut vegetables—cont'd

Vegetable	Effect of light	Observations	References
Spinach (baby leaves)	–	Light exposure promoted yellowing	Glowacz et al., 2014 (30–35 μmol m^{-2} s^{-1} and 130–140 μmol m^{-2} s^{-1})
Endive, chicory (heads)	–	Light exposure promoted greening	Wulfkuehler et al., 2014 (60 μmol m^{-2} s^{-1})
Radish (cotyledonary leaf stage)	–	Light exposure reduced visual quality	Xiao et al., 2014 (30 μmol m^{-2} s^{-1})
Spinach (baby leaves)	–	Light reduced visual quality	Garrido et al., 2016 (5 μmol m^{-2} s^{-1})

(*Apium graveolens* var. duce) (Zhan et al., 2013a, 2014). Light had the beneficial effect of increasing the chlorophyll content in Chinese kale (*B. oleracea* var. alboglabra) (Noichinda et al., 2007), broccoli florets (Zhan et al., 2012b), basil leaves (Costa et al., 2013), mature spinach leaves (*Spinacia oleracea* L.) (Gergoff Grozeff et al., 2013), lettuce (Zhan et al., 2013b), lamb's lettuce (*Valerianella olitoria* [L.] Pollich) (Braidot et al., 2014), chicory (*Cichorium intybus* L. var. foliosum Hegi) (Wulfkuehler et al., 2014), and celery (Zhan et al., 2014). Even though the effect of light was expected to increase the content of chlorophylls, a detrimental influence has also been reported on some products (Table 2). In baby spinach, light shortened the shelf life because it decreased visual quality (Garrido et al., 2016). In addition, high-intensity (130–150 μmol m^{-2} s^{-1}) light has been reported to reduce visual and nutritional quality in rocket (*Eruca sativa* Mill.), chicory (*Cichorium intybus* L.), and Swiss chard (*Beta vulgaris* L.) leaves (Ferrante et al., 2004). One of the negatives effects associated with light is that it stimulates stomatal opening and thereby enhances transpiration and the loss of water vapor from the tissue, leading to greater moisture condensation inside the packaging film in light exposure packages (Glowacz et al., 2014; Lester et al., 2010; Martínez-Sánchez et al., 2011; Noichinda et al., 2007; Olarte et al., 2009; Zhan et al., 2012a,b, 2013b, 2014; Xiao et al., 2014) (Fig. 1). The opening of the stomata is a general response to light leading to weight loss (Braidot et al., 2014; Zhan et al., 2014) (Table 2). However, postharvest short-term LED emitting NIR light has been shown to

TABLE 2 Effect of light on water content, chlorophylls, products of photosynthesis, antioxidant capacity, and bioactive compounds of fresh and fresh-cut vegetables

Vegetable	Water Content	Chlorophylls	Products of photosynthesis	Antioxidant Capacity	Bioactive Compounds	References
Spinach (mature leaves)	–	ND	ND	ND	ND	Toledo et al., 2003 (20–25 $\mu mol\, m^{-2}\, s^{-1}$)
Chinese kale (mature leaves)	–	+	+ levels of starch, fructose and glucose	+	+ Carotenoids, Ascorbic acid	Noichinda et al., 2007 (21.8 $\mu mol\, m^{-2}\, s^{-1}$)
Cauliflower (florets)	–	ND	ND	ND	ND	Sanz et al., 2007 (35 $\mu mol\, m^{-2}\, s^{-1}$)
Chard (leaves)	–	ND	ND	ND	ND	Sanz et al., 2008 (35 $\mu mol\, m^{-2}\, s^{-1}$)
Broccoli and cauliflower (florets)	–	ND	ND	ND	ND	Olarte et al., 2009 (35 $\mu mol\, m^{-2}\, s^{-1}$)
Leek (fresh cut)	–	ND	ND	ND	ND	Ayala et al., 2009 (35 $\mu mol\, m^{-2}\, s^{-1}$)
Spinach (leaves)	–	ND	ND	+	+ Ascorbic acid, folate (Vit B9), phylloquinone (vit K1), R-tocopherol (Vit E) and carotenoids	Lester et al., 2010 (27 $\mu mol\, m^{-2}\, s^{-1}$)

Continued

TABLE 2 Effect of light on water content, chlorophylls, products of photosynthesis, antioxidant capacity, and bioactive compounds of fresh and fresh-cut vegetables—cont'd

Vegetable	Water Content	Chlorophylls	Products of photosynthesis	Antioxidant Capacity	Bioactive Compounds	References
Broccoli	ND	+	+ levels of total and reducing sugars	ND	ND	Büchert et al., 2011 (12 $\mu mol\,m^{-2}\,s^{-1}$)
Lettuce (fresh cut)	–	ND	ND	ND	– Polyphenols	Martínez-Sánchez et al., 2011 (6 $\mu mol\,m^{-2}\,s^{-1}$)
Lettuce (leaves)	–	ND	ND	+ at medium intensity (34 $\mu mol\,m^{-2}\,s^{-1}$) and – at low intensity (7 $\mu mol\,m^{-2}\,s^{-1}$)	+ Ascorbic acid, total phenol (34 $\mu mol\,m^{-2}\,s^{-1}$) and – ascorbic acid, total phenol (7 $\mu mol\,m^{-2}\,s^{-1}$)	Zhan et al., 2012a (7 $\mu mol\,m^{-2}\,s^{-1}$ and 34 $\mu mol\,m^{-2}\,s^{-1}$)
Broccoli (florets)	–	+	ND	+	+ Ascorbic acid, Polyphenols	Zhan et al., 2012b (24 $\mu mol\,m^{-2}\,s^{-1}$)
Basil (mature leaves)	ND	+	+ levels of total and soluble sugars	ND	+ Carotenoids	Costa et al., 2013 (30–37 $\mu mol\,m^{-2}\,s^{-1}$)
Spinach (mature leaves)	ND	+	ND	+	+ Ascorbic acid, Glutathione	Gergoff Grozeff et al., 2013 (30 $\mu mol\,m^{-2}\,s^{-1}$)

Lettuce (fresh cut)	–	+	+ glucose, reducing sugar, and sucrose content	+	Ascorbic acid, soluble sugars	Zhan et al., 2013a ($34\,\mu mol\,m^{-2}\,s^{-1}$)
Celery (mature steaks)	ND	ND	ND	+	Total phenols	Zhan et al., 2013b ($27\,\mu mol\,m^{-2}\,s^{-1}$)
Lamb's lettuce (baby leaves)	ND	+	+ ATP content	–	+ Carotenoids, feofitina	Braidot et al., 2014 ($1.4\,\mu mol\,m^{-2}\,s^{-1}$)
Spinach (baby leaves)	–	–	NS	– at high intensity and NS at medium intensity	– Ascorbic acid, + carotenoids at high intensity and NS at medium intensity	Glowacz et al., 2014 ($30–35\,\mu mol\,m^{-2}\,s^{-1}$ and $130–140\,\mu mol\,m^{-2}\,s^{-1}$)
Endive, chicory (heads)	NS	+	ND	ND	+ Carotenoids	Wulfkuehler et al., 2014 ($60\,\mu mol\,m^{-2}\,s^{-1}$)
Radish (cotyledonary leaf stage)	–	ND	ND	–	+ Ascorbic acid, – Carotenoids (β-carotene, lutein/zeaxanthin)	Xiao et al., 2014 ($30\,\mu mol\,m^{-2}\,s^{-1}$)
Celery (mature stalks)	–	+	+ sucrose, reducing sugars and glucose	+	+ Ascorbic acid	Zhan et al., 2014 ($24\,\mu mol\,m^{-2}\,s^{-1}$)
Spinach (baby leaves)	–	NS	ND	ND	ND	Garrido et al., 2016 ($5\,\mu mol\,m^{-2}\,s^{-1}$)

ND: not described; NS: not significant.

FIG. 1 Visual aspects of the bags of baby spinach stored for 12 d at 7°C under different light conditions: light (A), photoperiod (B), and darkness (C).

reduce water loss during subsequent storage in lettuce (Kozuki et al., 2015). These authors concluded that stimulation of stomatal closure and ROS accumulation by NIR radiation could be involved in the suppression of transpiration rates. Concerning photosynthetic activity, there are also some studies that indirectly assess the ability of the product to perform photosynthesis by analyzing products of photosynthesis such as sugars (Costa et al., 2013; Noichinda et al., 2007; Zhan et al., 2013a, 2014). Other studies describe the effect of light in preventing ascorbic acid degradation and its correlation with a higher sugar content (Noichinda et al., 2007; Zhan et al., 2014) (Table 2). Related to this effect, light exposure during storage increases glucose and fructose levels in Chinese kale (Noichinda et al., 2007), basil (Costa et al., 2013), lettuce, and celery (Zhan et al., 2013a,b). As mentioned before, light is one of the most important factors regulating the biosynthesis and accumulation of phytochemicals (Bian et al., 2015). Exposure to light during storage promotes the increased content of antioxidants and prevents loss of ascorbic acid in Chinese kale (Noichinda et al., 2007), spinach (Lester et al., 2010; Gergoff Grozeff et al., 2013), lettuce (Zhan et al., 2012a,b), broccoli (Zhan et al., 2012b), celery (Zhan et al., 2013a), chicory (Wulfkuehler et al., 2014), radish micro-greens (Xiao et al., 2014), and lamb's lettuce (Braidot et al., 2014) (Table 2).

The impact of light on modified atmosphere

The effect of light on the MA generated during storage has been examined in very few studies. For vegetables, clear differences in the atmospheres generated have been observed between different plant organs [i.e., leaves (lettuce, spinach), inflorescence (cauliflower, broccoli), root (carrot) or leaf base (leek) (Ayala et al., 2009)]. There are also significant differences depending on whether the tissue is white or green. Some authors have reported that light increased respiratory activity, leading to enhanced consumption of O_2 and release of CO_2 in chard and chicory (Sanz et al., 2008; Wulfkuehler et al., 2014). Some of the possible mechanisms of light exposure increasing respiration have been suggested to be due to the differentiation of etioplast into chloroplasts, implying dramatic structural and functional changes of thylakoid membrane components (Wulfkuehler et al., 2014). In contrast with these observations, a lower production of CO_2 has been reported for green fresh products such as broccoli, basil, lamb's lettuce, lettuce, and spinach with exposure to light (Olarte et al., 2009; Martínez-Sánchez et al., 2011; Costa et al., 2013; Braidot et al., 2014; Xiao et al., 2014; Garrido et al., 2016). When white tissue was studied separately from green such as in chard (Sanz et al., 2008) and leek (Ayala et al., 2009), light caused an increase in stomatal aperture and modified the headspace gas composition in a higher extent compared to samples stored in the dark. In green tissues, respiratory activity was compensated for by photosynthesis, and the atmosphere composition remained similar to atmospheric conditions until the end of the storage period. Gil and Garrido conducted some experimental studies exposing some fresh and fresh-cut products to light and dark conditions and the results are presented in this chapter. Green and red baby leaves, fresh-cut iceberg lettuce, and broccoli and cauliflower (*B. oleracea* var. botrytis) florets were stored for 9 days at 7°C. Light exposure of 20 $\mu mol\,m^{-2}\,s^{-1}$ during storage reduced the respiration rate as CO_2 production of green and red baby leaves, fresh-cut iceberg, and broccoli florets compared to darkness (Fig. 2). This was due to the fact that CO_2 production was reduced in fresh-cut iceberg and broccoli florets while in baby leaf lettuce it was compensated by photosynthetic activity under light even at low temperature. These results also agree with those observed for fresh-cut romaine lettuce in which light reduced rate of CO_2 production and reduced the overall rate of O_2 uptake, leading to a higher O_2 concentration in the package and accelerated browning on the cut edges (Martínez-Sánchez et al., 2011). In the case of nongreen products such as cauliflower (Olarte et al., 2009), exposure to light was reported to lead to an increased release of CO_2 from the product resulting in an intensified respiration and accelerated senescence. However, taking into account our experimental data, we did not observe differences in the respiration rate measured as CO_2 emission between cauliflower florets exposed to light (20 $\mu mol\,m^{-2}\,s^{-1}$) and those under darkness (Fig. 2).

The atmosphere composition inside a closed package changes during storage depending on the uptake and release of gases by the tissues as well as on the transmission of gases through the package. When green vegetables are packaged and exposed to light during storage, the atmosphere in the package affected and, the altered atmosphere, in turn, affects metabolic pathways. Increased CO_2 levels or decreased O_2 levels inhibit the decarboxylation reactions of normal respiration (Kader and Saltveit, 2003). The effect of the atmosphere modification depends on CO_2 and O_2 concentrations around the

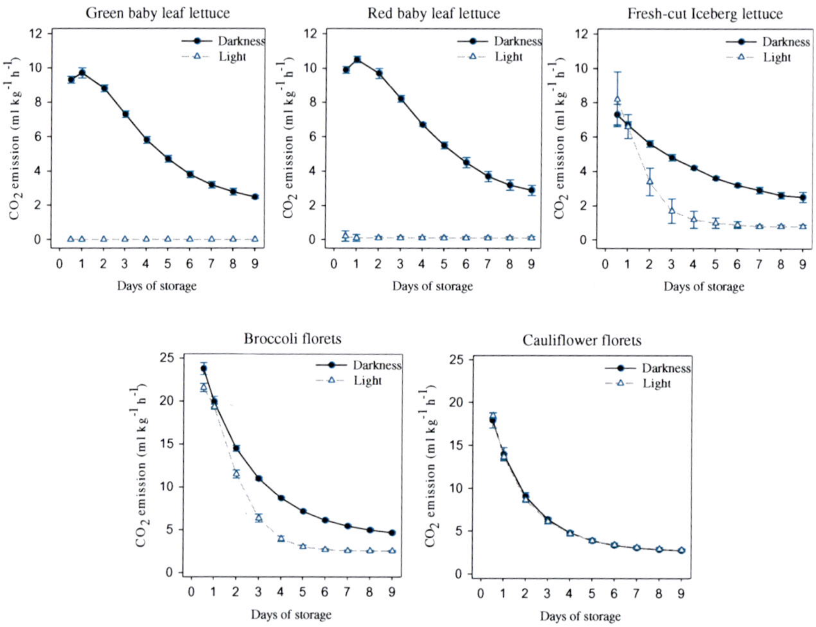

FIG. 2 Effect of light (20 μmol m^{-2} s^{-1}) and darkness on respiration rate of *green and red baby leaf lettuce*, fresh-cut iceberg lettuce, and broccoli and cauliflower florets stored for 9 days at 7°C. Error bars represent standard deviation (SD) of mean values ($n=5$).

commodity, temperature, and duration of light exposure. During storage, low temperature has a dramatic effect in reducing the rates of biological reactions including respiration while light exposure increases the photosynthetic processes and modify the net rate of CO_2 emissions. Under light exposure in storage, carbohydrates assimilated via photosynthesis compensate for the loss of carbohydrates via respiration (Fujiwara, 2008). All these processes influence the atmosphere generated in the headspace of the package, which then impacts the quality characteristics and shelf life of the commodity. When packages of baby spinach were exposed to different light conditions during storage, the atmospheres generated under light were different from those under darkness or photoperiod (Garrido et al., 2016). Olarte et al. (2009) reported that when cauliflower was exposed to light, the CO_2 production was stimulated while in the case of broccoli the CO_2 production was compensated by photosynthetic activity. In radish micro-greens, the large differences in gas compositions observed between light and dark conditions were also probably attributable to the respiratory and photosynthetic activities (Xiao et al., 2014). In the case of fresh-cut romaine lettuce, the same effect was described which caused a higher level of O_2 and a lower level of CO_2 in light-stored packages suggesting active photosynthetic processes during storage (Martínez-Sánchez et al., 2011). Garrido et al. (2016) also reported for baby spinach that photosynthetic processes of baby spinach were engaged by light exposure, leading to a reduction in net O_2 uptake and CO_2 release. Under darkness, a strong MA was generated (0.5 kPa O_2 + 10 kPa CO_2), but under light exposure, only a slight MA was generated (17 kPa O_2 + 2 kPa CO_2). Further, under photoperiod, an intermediate MA was detected with a headspace composition of 7 kPa O_2 + 7 kPa CO_2 at the steady state. The MA generated under the different light conditions also affected the quality characteristics of baby spinach. Under light, MA with high O_2 level promoted the growth of *Pseudomonas* spp., lipid peroxidation, and tissue senescence. Under darkness, MA with high CO_2 promoted intense off-odor development, increase in pH, higher electrolyte leakage, and chlorophyll degradation (Garrido et al., 2016). In our study, when minimally processed green and red baby lettuce, iceberg lettuce, and florets of broccoli and cauliflower were exposed to light (20 μmol m^{-2} s^{-1}) during storage, the MA generated was highly dependent on the light conditions and the commodity. In all cases, a passive MA was generated by the respiration rate of the product and the film permeability of the packages. In the case of green and red baby leaves, MA under light was very different than that generated under darkness (Fig. 3). Under light exposure, photosynthesis compensated for respiration such that CO_2 levels remained nearly unchanged from day 0 to day 9 while there was a sharp increase

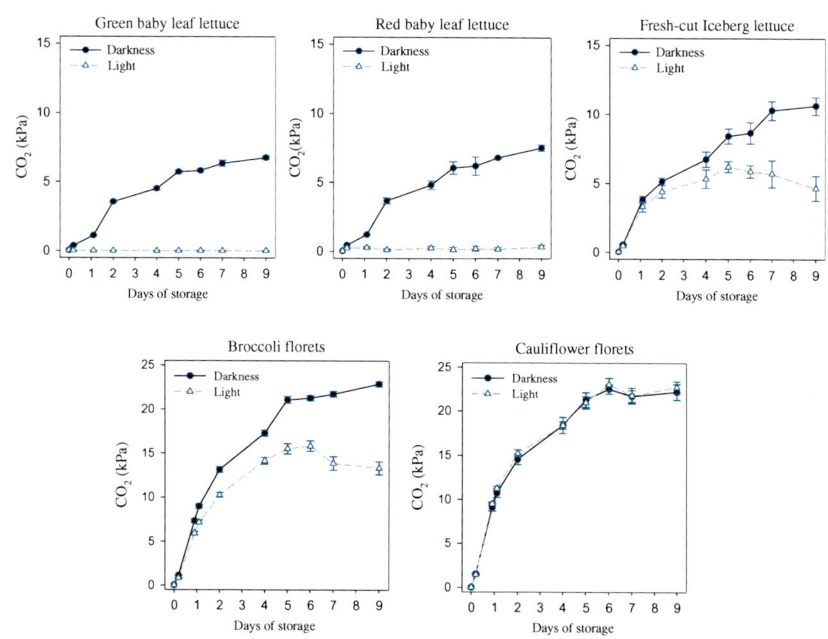

FIG. 3 Effect of light (20 μmol m^{-2} s^{-1}) and darkness on CO_2 concentration in the headspace gas composition inside the packages of *green and red baby leaf lettuce*, fresh-cut iceberg lettuce, and broccoli and cauliflower florets stored for 9 days at 7°C. Error bars represent standard deviation (SD) of mean values (n = 5).

in CO_2 throughout the storage under dark conditions (Fig. 3). In fresh-cut iceberg and broccoli florets, photosynthesis reactions under light exposed were not able to completely compensate for the release of CO_2 due to respiration and thus CO_2 production under light was lower than that under darkness (Fig. 3). This was due to the light exposure (20 $\mu mol\,m^{-2}\,s^{-1}$) being below the light compensation point and thus CO_2 being accumulated during storage, but in much lower concentrations than in darkness. In the case of cauliflower florets, as there was no photosynthesis in response to light exposure, and CO_2 accumulation due to respiration rate was not affected such that there were no significant differences between CO_2 partial pressures in the package headspace in light and in darkness (Fig. 3). The quality characteristics of fresh-cut lettuce were affected negatively by light because light shortened the shelf life because of an increase in cut edge browning due to a higher O_2 level in the headspace in response to photosynthesis (Martínez-Sánchez et al., 2011). Continuous light has been reported to induce negative effects in a number of commodities including increased respiration and accelerated browning of cut edges in cauliflower (Sanz et al., 2007), chard (Sanz et al., 2008), broccoli, and asparagus (Olarte et al., 2009; Sanz et al., 2009). In two studies on packaged baby spinach under darkness, the commodity generated an MA with excessively low O_2 and high CO_2 that was also detrimental because of the intense off-odor development as a consequence of the increased release of ammonium (Tudela et al., 2013; Garrido et al., 2016). It has been shown that ammonium accumulates in detached rice leaves during dark-induced senescence (Chen et al., 1997). Darkness influences senescence, which involves degradation of chlorophyll and proteins and accumulation of ammonium (Chen et al., 1997; Brouwer et al., 2012).

Conclusions

Light intensity and duration during storage have a great impact on maintaining the quality of fresh and fresh-cut vegetables. In general, low-intensity light applications have been shown to have beneficial effects preventing yellowing, increasing the availability of soluble sugars, and contributing to the maintenance of a higher content of bioactive antioxidants. On the other hand, high-intensity light applications can accelerate quality deterioration, oxidative stress, tissue damage, membrane linkage, and weight loss during storage. There is a need for a better understanding of the biological responses to light and its potential application for postharvest quality maintenance on a crop-by-crop basis. The proper combination of temperature, relative humidity, and light intensity and duration is necessary to establish the best storage conditions for green vegetables. This chapter emphasizes the importance of considering light exposure during storage when selecting the most suitable film for packaging, as respiratory activity can be compensated for by photosynthesis.

References

Ayala, F., Echavarri, J.F., Sanz, S., Olarte, C., 2009. Quality characteristics of minimally processed leek packaged using different films and stored in lighting conditions. Int. J. Food Sci. Technol. 44, 1333–1343.

Bartoli, C.G., Tambussi, E.A., Diego, F., Foyer, C.H., 2009. Control of ascorbic acid synthesis and accumulation and glutathione by the incident light red/far red ratio in Phaseolus vulgaris leaves. FEBS Lett. 583, 118–122.

Bian, Z.H., Yang, Q.C., Liu, W.K., 2015. Effects of light quality on the accumulation of phytochemicals in vegetables produced in controlled environments: a review. J. Sci. Food Agric. 95, 869–877.

Braidot, E., Petrussa, E., Peresson, C., Patui, S., Bertolini, A., Tubaro, F., Zancani, M., 2014. Low-intensity light cycles improve the quality of lamb's lettuce (*Valerianella olitoria* [L.] Pollich) during storage at low temperature. Postharvest Biol. Technol. 90, 15–23.

Brouwer, B., Ziolkowska, A., Bagard, M., Keech, O., Gardestrom, P., 2012. The impact of light intensity on shade-induced leaf senescence. Plant Cell Environ. 35, 1084–1098.

Büchert, A.M., Gómez Lobato, M.E., Villarreal, N.M., Civello, P.M., Martínez, G.A., 2011. Effect of visible light treatments on postharvest senescence of broccoli (*Brassica oleracea* L.). J. Sci. Food Agric. 91, 355–361.

Cantos, E., García-Viguera, C., de Pascual-Teresa, S., Tomás-Barberán, F.A., 2000. Effect of postharvest ultraviolet irradiation on resveratrol and other phenolics of cv. Napoleon table grapes. J. Agric. Food Chem. 48, 169–188.

Chen, S.J., Hung, K.T., Kao, C.H., 1997. Ammonium accumulation is associated with senescence of rice leaves. Plant Growth Regul. 21, 195–201.

Costa, L., Vicente, A., Civello, M., Martínez, G., Chaves, A., 2006. UV-C treatments delays postharvest senescence in broccoli florets. Postharvest Biol. Technol. 39, 115–222.

Costa, L., Millan Montano, Y., Carrión, C., Rolny, N., Guiamet, J.J., 2013. Application of low intensity light pulses to delay postharvest senescence of *Ocimum basilicum* leaves. Postharvest Biol. Technol. 86, 181–191.

D'Souza, C., Yuk, H.-G., Khoo, G.H., Zhou, W., 2015. Application of light-emitting diodes in food production, postharvest preservation, and microbial foof safety. Compr. Rev. Food Sci. Food Saf. 14, 719–740.

Dorais, M., Ehret, D.L., 2008. Agronomy and the nutritional quality of fruit. In: Tomás-Barberán, F.A., Gil, M.I. (Eds.), Improving the Health-promoting Properties of Fruit and Vegetable Products. CRC Press, Woodhead Publishing Ltd, pp. 347–391.

Ferrante, A., Incrocci, L., Maggini, R., Serra, G., Tognoni, F., 2004. Colour changes of fresh-cut leafy vegetables during storage. J. Food Agric. Environ. 2, 40–44.

Fujiwara, K., 2008. LED low-light irradiation for quality preservation of green plants during low-temperature storage. Jpn. Soc. Agric. Biol. Environ. Eng. Sci., Shokubutsu Kankyo Kogaku 20, 136–142.

Garrido, Y., Tudela, J.A., Hernández, J.A., Gil, M.I., 2016. Modified atmosphere generated during storage under light conditions is the main factor responsible for the quality changes of baby spinach. Postharvest Biol. Technol. 114, 45–53.

Gergoff Grozeff, G.E., Chaves, A.R., Bartoli, C.G., 2013. Low irradiance pulses improve postharvest quality of spinach leaves (*Spinacia oleraceae* L. cv Bison). Postharvest Biol. Technol. 77, 35–42.

Gil, M.I., García-Viguera, C., Artés, F., Tomás-Barberán, F.A., 1995. Changes in pomegranate juice pigmentation during ripening. J. Sci. Food Agric. 68, 77–81.

Glowacz, M., Mogren, L.M., Reade, J.P., Cobb, A.H., Monaghan, J.M., 2014. High- but not low-intensity light leads to oxidative stress and quality loss of cold-stored baby leaf spinach. J. Sci. Food Agric. 95, 1821–1829.

Golan, T., Müller-Moulé, P., Niyogi, K.K., 2006. Photoprotection mutants of *Arabidopsis thaliana* acclimate to high light by increasing photosynthesis and specific antioxidants. Plant Cell Environ. 29, 879–887.

Goldthwaite, J.J., Laetsch, W.M., 1967. Regulation of senescence in bean leaf discs by light and chemical growth regulators. Plant Physiol. 42, 1757–1762.

Heins, R.D., Lange, N., Wallace Jr., T.F., 1992. Low-temperature storage of bedding-plant plugs. In: Kurata, K., Kozai, T. (Eds.), Transplant Production Systems. Kluwer, Netherlands, pp. 45–64.

Ilieva, I., Ivanova, T., Naydenov, Y., Dandolov, I., Stefanov, D., 2010. Plant experiments with light-emitting diode module in Svet space greenhouse. Adv. Space Res. 46, 840–845.

Ju, Z., Duan, Y., Ju, Z., 1999. Effects of covering the orchard floor with reflecting films on pigment accumulation and fruit coloration in "Fuji" apples. Sci. Hortc. 82, 47–56.

Kader, A.A., Rolle, R.S., 2004. The role of post-harvest management in assuring the quality and safety of horticultural produce. FAO Agricu. Serv. Bull. 65.

Kader, A.A., Saltveit, M.E., 2003. Respiration and gas exchange. In: Bartz, J.A., Brecht, J.K. (Eds.), Postharvest Physiology and Pathology of Vegetables. Marcel Dekker, Inc., New York, Basel, pp. 7–29.

Keech, O., Pesquet, E., Ahad, A., Askne, A., Nordvall, D., Vodnala, S.M., Tuominen, H., Hurry, V., Dizengremel, P., Gardestrom, P., 2007. The different fates of mitochondria and chloroplasts during dark-induced senescence in Arabidopsis leaves. Plant Cell Environ. 30, 1523–1534.

Kopsell, D.A., Sams, C.E., 2013. Increases in shoot tissue pigments, glucosinolates, and mineral elements in sprouting broccoli after exposure to short-duration blue light from light emitting diodes. J. Am. Soc. Hortic. Sci. 138, 31–37.

Kozuki, A., Ishida, Y., Kakibuchi, K., Mishima, T., Sakurai, N., Murata, Y., Nakano, R., Ushijima, K., Kubo, Y., 2015. Effect of postharvest short-term radiation of near infrared light on transpiration of lettuce leaf. Postharvest Biol. Technol. 108, 78–85.

Kubota, C., Kozai, T., 1994. Low-temperature storage for quality preservation and growth suppression of broccoli plantlets cultured photoautotrophically in vitro. Hortscience 29, 1191–1194.

Kubota, C., Kozai, T., 1995. Low-temperature storage of transplants at the light com-pensation point – Air-temperature and light-intensity for growth suppressionand quality preservation. Sci. Hortic. 61, 193–204.

Lee, K.S., Kader, A.A., 2000. Preharvest and postharvest factors influencing vitamin C content of horticultural crops. Postharvest Biol. Technol. 20, 207–220.

Lee, Y., Ha, J., Oh, J., Cho, M., 2014. The effect of LED irradiation on the quality of cabbage stored at a low temperature. Food Sci. Biotechnol. 23, 1087–1093.

Lester, G.E., Makus, D.J., Hodges, D.M., 2010. Relationship between fresh-packaged spinach leaves exposed to continuous light or dark and bioactive contents: effects of cultivar, leaf size, and storage duration. J. Agric. Food Chem. 58, 2980–2987.

Luckey, T.D., 1980. Hormesis with Ionizing Radiation. CRC Press, Boca Raton.

Martínez-Sánchez, A., Tudela, J.A., Luna, C., Allende, A., Gil, M.I., 2011. Low oxygen levels and light exposure affect quality of fresh-cut Romaine lettuce. Postharvest Biol. Technol. 59, 34–42.

Niyogi, K.K., 1999. Photoprotection revisited: genetic and molecular approaches. Annu. Rev. Plant Physiol. Plant Mol. Biol. 50, 333–359.

Noichinda, S., Bodhipadma, K., Mahamontri, C., Narongruk, T., Ketsa, S., 2007. Light during storage prevents loss of ascorbic acid, and increases glucose and fructose levels in Chinese kale (*Brassica oleracea* var. alboglabra). Postharvest Biol. Technol. 44, 312–315.

Olarte, C., Sanz, S., Federico Echávarri, J., Ayala, F., 2009. Effect of plastic permeability and exposure to light during storage on the quality of minimally processed broccoli and cauliflower. LWT- Food Sci. Technol. 42, 402–411.

Perez-Balibrea, S., Moreno, D.A., Garcia-Viguera, C., 2008. Influence of light on health-promoting phytochemicals of broccoli sprouts. J. Sci. Food Agric. 88, 904–910.

Philosoph-Hadas, S., Meir, S., Akiri, B., Kanner, J., 1994. Oxidative defense system in leaves of three edible herb species in relation to their senescence rates. J. Agric. Food Chem. 42, 2376–2381.

Poulet, L., Massa, G.D., Morrow, R.C., Bourget, C.M., Wheeler, R.M., Mitchell, C.A., 2014. Significant reduction in energy for plant-growth lighting in space using targeted LED lighting and spectral manipulation. Life Sci. Space Res. 2, 43–53.

Rodov, V., Vinokur, Y., Horev, B., 2012. Brief postharvest exposure to pulsed lightstimulates coloration and anthocyanin accumulation in fig fruit (*Ficus carica* L.). Postharvest Biol. Technol. 68, 43–46.

Sanz, S., Olarte, C., Echavarri, J., Ayala, F., 2007. Influence of exposure to light on the sensorial quality of minimally processed cauliflower. J. Food Sci. 72, S12–S18.

Sanz, S., Olarte, C., Ayala, F., Echavarri, J.F., 2008. The response to lighting of minimally processed chard: influence on its shelf life. J. Sci. Food Agric. 88, 1622–1631.

Sanz, S., Olarte, C., Ayala, F., Echavarri, J.F., 2009. Evolution of quality characteristics of minimally processed asparagus during storage in different lighting conditions. J. Food Sci. 74, S296–S302.

Shen, Y.Z., Guo, S.S., Ai, W.D., Tang, Y.K., 2014. Effects of illuminants and illumination time on lettuce growth, yield and nutritional quality in a controlled environment. Life Sci. Space Res. 2, 38–42.

Toledo, M.E.A., Ueda, Y., Imahori, Y., Ayaki, M., 2003. L-ascorbic acid metabolism in spinach (*Spinacia oleracea* L.) during postharvest storage in light and dark. Postharvest Biol. Technol. 28, 47–57.

Tudela, J.A., Marín, A., Garrido, Y., Cantwell, M., Medina-Martínez, M.S., Gil, M.I., 2013. Off-odour development in modified atmosphere packaged baby spinach is an unresolved problem. Postharvest Biol. Technol. 75, 75–85.

Vicente, A.R., Pineda, C., Lemoine, L., Civello, P.M., Martínez, G.A., Chaves, A.R., 2005. UV-C treatments reduce decay, keep quality and alleviate chilling injury in peppers. Postharvest Biol. Technol. 35, 69–78.

Whitaker, B.D., 2003. Chemical and Physical changes in membranes. In: Bartz, J.A., Brecht, J.K. (Eds.), Postharvest Physiology and Pathology of Vegetables. Marcel Dekker, Inc., New York, Basel, pp. 79–110.

Wulfkuehler, S., Gras, C., Carle, R., 2014. Influence of light exposure during storage on the content of sesquiterpene lactones and photosynthetic pigments in witloof chicory (*Cichorium intybus* L. var. foliosum Hegi). LWT- Food Sci. Technol. 58, 417–426.

Xiao, Z., Lester, G.E., Luo, Y., Xie, Z.K., Yu, L.L., Wang, Q., 2014. Effect of light exposure on sensorial quality, concentrations of bioactive compounds and antioxidant capacity of radish microgreens during low temperature storage. Food Chem. 151, 472–479.

Yu, B.P., 1994. Cellular defenses against damage from reactive oxygen species. Biol. Rev. 74, 139–162.

Zhan, L., Li, Y., Hu, J., Pang, L., Fan, H., 2012a. Browning inhibition and quality preservation of fresh-cut romaine lettuce exposed to high intensity light. Innovative Food Sci. Emerg. Technol. 14, 70–76.

Zhan, L., Hu, J., Li, Y., Pang, L., 2012b. Combination of light exposure and low temperature in preserving quality and extending shelf-life of fresh-cut broccoli (*Brassica oleracea* L.). Postharvest Biol. Technol. 72, 76–81.

Zhan, L., Hu, J., Lim, L.-T., Pang, L., Li, Y., Shao, J., 2013a. Light exposure inhibiting tissue browning and improving antioxidant capacity of fresh-cut celery (*Apium graveolens* var. dulce). Food Chem. 141, 2473–2478.

Zhan, L., Hu, J., Ai, Z., Pang, L., Li, Y., Zhu, M., 2013b. Light exposure during storage preserving soluble sugar and L-ascorbic acid content of minimally processed romaine lettuce (*Lactuca sativa* L. var. longifolia). Food Chem. 136, 273–278.

Zhan, L., Hu, J., Pang, L., Li, Y., Shao, J., 2014. Effects of light exposure on chlorophyll, sugars and vitamin C content of fresh-cut celery (Apium graveolens var. dulce) petioles. Int. J. Food Sci. Technol. 49, 347–353.

Chapter 10

Packaging design: Functions and materials

Jeffrey S. Brandenburg
The JSB Group, LLC, Greenfield, MA, United States

Relationship between packaging and produce physiology

When designing MAP the convergence of three unique and separate sciences must take place. These are the sciences of produce physiology, polymer engineering, and converting technology (Fig. 1).

Effective modified atmosphere package design can only effectively take place within the intersection of these three disciplines. In many package design scenarios, two of the three take place; when this happens it is impossible to optimize the package for quality and shelf life. Surrounding these scientific disciplines is the impact and requirements of marketing, brand management, and the desires and needs of the consumer. What makes fresh produce packaging unique is the impact of the fresh produce. Unlike almost all other packaging applications, produce MAP involves packaging a living product. Therefore in order to effectively design a package, a quantitative understanding of the physiological properties and requirements of the packaged product is critical.

Modified atmospheres

Most extended shelf life packaging technologies involve the use of oxygen (O_2) and carbon dioxide (CO_2) in quantities (partial pressures) and ratios different from those found in air. The partial pressures, diffusion properties, and permeation rates through polymer films, microperforations, membranes, and so on, of each gas are independent of the other gases present. This simplifies matters since we do not have to be concerned with the gas interactions. Most of the reactions with food involving O_2 are linked to a degradation process involving the oxidative breakdown of foods. Because of this, many packaging strategies seek to exclude oxygen and thus slow these degradative processes. Many spoilage microorganisms such as pseudomonas, yeasts, and molds require O_2 and will grow and cause off-odors in the presence of sufficient O_2. Oxygen is also

Controlled and Modified Atmospheres for Fresh and Fresh-Cut Produce
https://doi.org/10.1016/B978-0-12-804599-2.00011-9

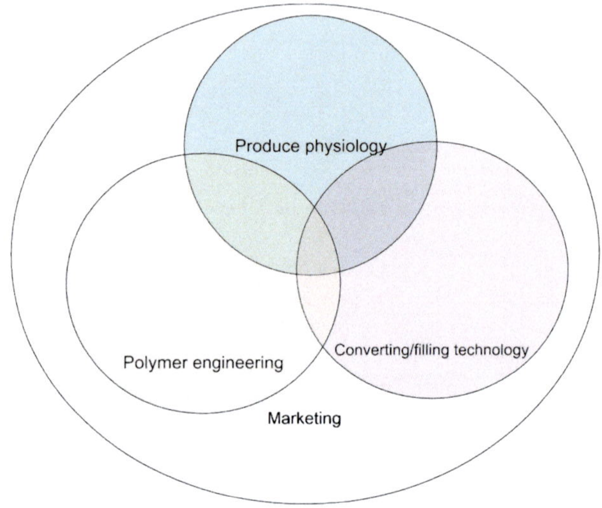

FIG. 1 Sciences impacting modified atmosphere package design.

necessary to the normal respiratory metabolism of fresh fruits and vegetables. However, normal atmospheric concentrations of O_2 encourage and facilitate senescence of many fruits and vegetables and enhance degradation of quality (Brandenburg and Zagory, 2009).

Reducing O_2 concentrations below about 8% around many fresh fruits and vegetables slows their respiration rate and therefore slows the rates at which they ripen, age, and decay. Reducing O_2 concentration can, in products that are prone, reduce oxidative browning reactions, which can be of particular concern in precut leafy vegetables. Reduced O_2 can delay compositional changes such as fruit softening, pigment development, toughening of some vegetables (such as asparagus and broccoli), and development of flavor (Kader, 1986).

Ethylene is normally produced in various levels by many kinds of ripening fruit. In addition, fruit varieties vary in their degree of sensitivity to ethylene. Elevated CO_2 (approximately 2% or more) can help reduce the damaging effects of ethylene by reducing the sensitivity of plant tissues to ethylene (Herner, 1987; Kader et al., 1988). This may be one of the primary benefits of modified atmospheres for many commodities.

Elevated CO_2 can, like reduced O_2, slow respiratory processes thereby extending shelf life. Although the effects of elevated CO_2 on respiration are not as dramatic as those of low O_2, high CO_2, and low O_2 together can, in some cases, reduce respiration more than either gas alone (Kader et al., 1988).

CO_2 at increased concentrations (>10%) has been shown to suppress the growth of a number of decay-causing fungi and bacteria. For example, 15%–20% CO_2 is routinely applied around strawberries during shipment primarily to suppress growth of the mold *Botrytis cinerea*. It is important to note these levels of CO_2 do not suppress some human pathogenic bacteria of potential concern on fresh produce. For example, *Clostridium botulinum* and *Listeria monocytogenes* are relatively resistant to the effects of CO_2 (Farber, 1991). There is some concern that elevated CO_2 could suppress spoilage microorganisms that would otherwise signal microbial growth and product spoilage while allowing potentially hazardous pathogens to continue to grow. For this reason, MAP should always work in conjunction with an excellent program of refrigeration, sanitation, and quality assurance. In addition, too much CO_2 can be damaging to plant tissues of certain fruits and vegetables as they differ dramatically in their tolerance to CO_2.

Fruit and vegetable senescence

One of the keys to understanding and optimizing MAP is that plant tissues in fresh fruits and vegetables are still alive after harvest and even after minimal processing. To remain alive, they must derive energy from their stored reserves, primarily through the process of respiration. Respiration involves the consumption of O_2 and the consequent production of energy, heat, CO_2, and moisture vapor. Different fruits and vegetables, and even different varieties of a given fruit or vegetable, will vary in their rates of respiration. Those that have high respiration rates (such as asparagus, strawberries, and broccoli) tend to be most perishable while those with low respiration rates (such as root vegetables, apples, onions, and potatoes) tend to be least perishable. In fresh-cut produce, the respiration rate is also strongly dependent on cut size. The smaller the cut size, the greater is the respiration rate. Increases in storage and handling temperature can have the greatest impact on respiration rate, with increases that can more than double for every increase of 10°C. It is critical to note that the increase in respiration rate with either temperature or cut size is not a linear progression and sometimes can reach exponential rates.

The best way to reduce respiration rate is to reduce the temperature to the minimum tolerated by the product and to maintain that temperature. All biological processes proceed more slowly at lower temperatures. Most fruits and vegetables will maintain their best quality at temperatures near 0°C. In any case, keeping fresh produce at the lowest possible temperature without causing freezing or chilling injury is the surest way to maintain quality and shelf life. As a supplement to good temperature control, MAP can further extend quality and shelf life.

One of the primary beneficial effects of MAP, in addition to the inhibition of enzymatic browning and ethylene action on produce, is related to the slowing of plant respiration in low O_2 environments. When the respiration rate is reduced, fruits and vegetables become less perishable.

Air is about 21% O_2, 0.03% CO_2, 0.9% Argon, and the remainder nitrogen (N_2) and water vapor. If the makeup of the atmosphere inside a package is

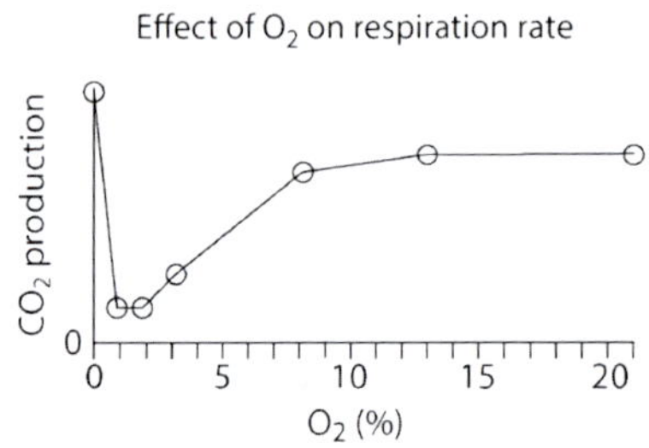

FIG. 2 Effect of oxygen concentration on fruit or vegetable respiration rate. *(Adapted from Zagory, D., 1996. Physiology and microbiology of fresh produce in modified atmosphere packages. In: Society of Manufacturing Engineers Symposium, Fundamentals of Modified Atmosphere Packaging, December 4–5, 1996. Monterey, CA.)*

something other the concentration of O_2 inside the package falls below about 8%, will the plant respiration start to slow (Fig. 2).

This suppression of respiration continues until O_2 reaches about 1%–3% for most fruits or vegetables. If the concentration of O_2 drops below 1%–3% (depending on the product and the temperature), anaerobic (fermentative) metabolism replaces normal aerobic metabolism and large amounts of CO_2, alcohols, and aldehydes, which create off-flavors and off-odors, are produced. If the CO_2 concentration gets too high, production of undesirable volatiles and physiological injury can occur. Reduced O_2 and elevated CO_2 together can reduce respiration more than either alone (Zagory and Kader, 1989) (Fig. 3). In addition, elevated CO_2 can suppress plant tissue sensitivity to the effects of the ripening hormone ethylene. Brown (1922) reported that CO_2 concentrations of 10% or more can retard the sporulation and/or growth of numerous fungal decay organisms, while O_2 typically has little effect on plant pathogen growth or survival at levels appropriate for use in MAP.

Under hypobaric (low pressure) conditions, extremely low O_2 partial pressures have been shown to be effective in decay control (Burg, 2004). Additional benefits of MAP include preservation of vitamins, particularly vitamin C (Zagory and Kader, 1989).

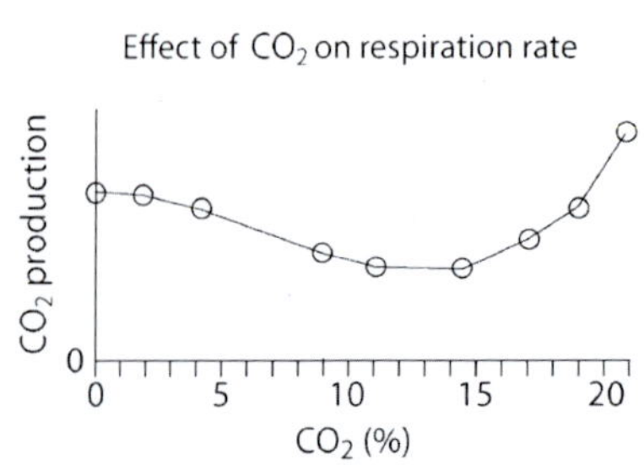

FIG. 3 Effect of carbon dioxide concentration on fruit or vegetable respiration rate. *(Adapted from Zagory, D. 1996. Physiology and microbiology of fresh produce in modified atmosphere packages. In: Society of Manufacturing Engineers Symposium, Fundamentals of Modified Atmosphere Packaging, December 4–5, 1996. Monterey, CA.)*

Due to the slowing of oxidative metabolism associated with reduced O_2 and elevated CO_2 the ripening processes of many fruits and vegetables are retarded (Burg and Burg, 1967). Ethylene, a colorless, odorless, tasteless gas is a plant hormone that has multiple physiological effects when present at concentrations below one part per million (ppm) and in certain items such as kiwifruit just a few ppb. Ethylene can induce rapid and irreversible ripening and softening of climacteric fruits such as avocados, tomatoes, kiwi, apples, stone fruit, mangoes, and others. Ethylene may also cause yellowing of many vegetables such as spinach, broccoli, cucumbers, and celery. Ethylene causes russet spot of iceberg lettuce, consisting of small, oval brown spots that develop on the lettuce midribs. Ethylene may also induce formation of bitter tasting isocoumarins in carrots, sprouting of potatoes, and toughening of asparagus (Reid, 1985). Elevated CO_2 can prevent or delay all of these effects by reducing the sensitivity of plant tissues to ethylene. In addition, low O_2 (below 2%–4%) can reduce ethylene production by plant tissues. Synthesis of some anthocyanin pigments is also retarded by reduced O_2, which slows color development of some fruits such as pomegranate.

In order to take into account all of the above, it is critical that when measuring shelf life that more than just a visual determination is made (Table 1); O_2/CO_2 measurements must be taken and analyzed and organoleptic evaluations are needed.

TABLE 1 **Physiological effects of reduced O_2 and elevated CO_2 on fruits and vegetables (Kader et al., 1989)**

	General effects of	
Cause of deterioration	*Reduced O_2*	*Elevated CO_2*
Respiration rate	(>1%) –	(<15%–20%) –
	(<1%) +	(>15%–20%) +
Ethylene action	–	–
Chlorophyll degradation	–	–
Anthocyanin development	–	–
Carotenoid biosynthesis	–	–
Enzymatic browning	– (near 0%)	–
Off-flavors	+ (<1%)	+ (>15%–20%)
Vitamin C loss	–	–
Fungal growth	– (<1%)	– (>10%–15%)
Bacterial GROWTH	– or 0	– or 0

– = decrease or inhibit, 0 = no effects, + = stimulate or increase.

It is also critically important to understand that there are conditions under which MAP will not provide a benefit. First and foremost MAP is only optimally effective if there is consistent temperature management throughout the entire life cycle of the product. This includes processing as well as the entire distribution channel. Lack of temperature control will result in produce physiological variations that will impact the effectiveness of the packaging system. In addition, MAP will never improve the quality of the incoming raw material product. Under ideal circumstances the best that can be achieved is maintenance of the existing quality level throughout the desired shelf life. In real-world applications, often MAP will maintain quality for the majority of the targeted shelf life, but due to parameter variations during distribution, quality will suffer at the very end of the desired shelf life. Since MAP will never improve incoming product quality the need for optimal postharvest handling procedures is paramount.

In addition, MAP cannot enhance the safety of fresh fruits and vegetables. Most of the human pathogens of concern are either not affected by MAP, or MAP may actually enhance their growth. MAP will never improve the quality of incoming poor quality products. At best, MAP can slow the processes of deterioration and so extend quality and shelf life, but MAP cannot reverse those processes.

Respiration rate quantification

MAP depends upon the respiratory activity, as measured by respiration rate (i.e., O_2 uptake and CO_2 emission), of the enclosed product as a driving force for the atmosphere modification within the package. This, in concert with the permeability of the packaging material, is used to maintain atmospheres within desired limits. It is the continued depletion of O_2 and/or the release of CO_2 (and water vapor) by the product that enables the modified atmosphere to become established within a sealed package. Factors that must be controlled or incorporated include film or technology permeability, film area, film thickness, temperature, type and weight of the product, and the associated respiration rate dependent upon cultivar, cut size, and temperature of the product.

The dependence of MAP on the physiological properties of the product requires a quantitative understanding of the respiration of specific fresh produce. Produce respiration is determined through testing. The condition, age, and temperature of the raw material product can significantly impact the outcome of the respiration test. There are a number of independent resources for product respiration calculations including University of California Davis (http://postharvest.ucdavis.edu) The JSB Group, LLC. (www.jsbgroup.com) and Perfotec (http://perfotec.com).

The primary method for determining the required modified atmosphere package oxygen transmission rate (OTR) is seen in the following equation. Parameters in addition to respiration rate (RR) include package surface area, product weight, structure thickness, and target final modified atmosphere. A thorough understanding of the equation and how each parameter can affect the package OTR is critical for optimal package design. It is the interplay of these parameters that dictates that there is no one package that fits all applications. MAP must be designed on an individual application basis. Once the produce respiration is quantified and the balance of parameters is determined the target OTR can be calculated. Establishment of the target OTR allows the design process to move from the physiological research into the polymer science and packaging research.

OTR for a target atmosphere is:

$$\mathrm{OTR} = \mathrm{RR}_{\mathrm{O_2}} \times t \times W/A \times \left(\mathrm{O}_{2\,\mathrm{air}} - \mathrm{O}_{2\,\mathrm{pkg}}\right),$$

where

OTR = film O_2 permeability (oxygen transmission rate) per mil,
RR = respiration rate (O_2 consumption rate in ml/kg h),
t = film thickness (mils, 1 'mil' = 1/1000 in or 25.4 μm),
W = product weight (kg),
A = film surface area (cm^2),
O_{2pkg} = desired O_2 concentration in the package (% O_2 target atmosphere).

Note: OTR in this example is calculated on a per mil basis (25 μm). Therefore, when a 2 mil thickness film is used, the OTR on a per mil basis must double to achieve the equivalent package OTR.

In order to achieve the target MA, the packaging films must be permeable to gases. Specifically they must have the required gas transmission properties to achieve the targeted OTR. The movement of gases across films depends on several physical factors that are related through Fick's Law as follows:

$$J_{\mathrm{gas}} = \frac{A \times \Delta C_{\mathrm{gas}}}{R}$$

where

J_{gas} = total flux of gas (cm^3/s),
A = surface area of the film (cm^2),
ΔC_{gas} = concentration gradient across the film,
R = resistance of the film to gas diffusion (s/cm).

The gas flow across a film increases with increasing surface area and with increasing concentration gradient across the film. The gas flow across the film decreases with increasing film resistance to gas diffusion. Gases diffuse through polymeric films at different rates. Carbon dioxide diffuses between two to five times faster than O_2. The ratio, within a polymer, film, or structure of CO_2TR to OTR is termed the beta value. Polymeric films therefore have a beta value of 2:1 to 5:1, with an average of 3:1. The beta value of an MA package will have a direct bearing on the final modified atmosphere achieved within the package.

The higher the beta value, the lower the CO_2 level will tend to be for any given O_2 partial pressure. For example, in polymer films with a low beta value like 2.5 it is possible to achieve a low O_2 level (2%, a gradient of approximately 19% across the package film) in combination with a mid CO_2 level (7%, a gradient of 7% across the package film). Notice that the gradient of O_2 to CO_2 is roughly the inverse of the beta value. Use of a film with an appropriate beta value is essential in obtaining the specific ideal modified atmosphere for each individual product.

Polymer engineering

Polymers

There are a variety of polymers used in fresh produce MAP. A portion of these polymers is used in primarily flexible packaging structures, a portion are used in primarily rigid packaging structures, and a portion are found in both applications. Each polymer has physical, chemical, and transmission rate properties that are unique to that polymer. Design of a packaging structure entails matching the specific polymer properties to the requirements of the MAP application. To address physiological requirements of fresh-cut modified atmosphere applications, polymer gas transmission rates, specifically, OTR and CO_2TR are key attributes.

This being said, the ability of a polymer to transmit, absorb, and manage ethylene is gaining in importance commercially in response to increased research into this area and the development of a more complete understanding of the behavior of the package/product system. In addition, most polymers used in fresh produce packaging are water vapor barriers and, as such, residual water, humidity, and water vapor need to be controlled through other technologies. Recently, significant work is being carried out to explore, develop, and commercialize polymers that have a significantly higher water transmission rate (WVTR) than traditional packaging films. Polymers such as polystyrene and Nylon are showing promise.

Biopolymers are also receiving a lot of research and development attention. Polymers such as polylactic acid (PLA), polyhydroxyalkanoate (PHA), polynucleotides, polyamides, polysaccharides, polyoxoesters, polythioesters, polyanhydrides, polyisoprenoids, and polyphenols are all being evaluated as potential candidates for substitution of synthetic plastics in fresh produce applications. Cellulose polymers, such as cellophane, are even being rebranded and experiencing a resurgence in use and applications. One of the challenges in the biopolymer/sustainability area is combining the transmission requirements for produce with the desire for sustainability and compostability. Understanding the difference between, and requirements for, industrial and home composting is essential in determining the functionality and applicability of bio-based polymers. Examining the entire life cycle of the polymer and package is also critical in determining not only the best polymer for the application, but also the total cost of the technology.

Films and structures

As with most food packaging, produce MAP uses not only individual polymers, but combinations of polymers, as well as other substrates such as paper and mesh in the form of films and structures. These films and structures can consist of homopolymer monolayer films, blended monolayer films, coextruded films, or combinations of films laminated and/or glued together. Determination of the gas transmission rate of the complete package (total film structure) therefore requires knowledge of not only the individual polymers but also how they are combined. A blend of polymers within one layer of film, referred to as a blend, will yield a transmission that is the weighted average of the transmission of the individual polymer components. Individual distinct layers of pure polymers, or blends, created within a single total film structure, are referred to as a coextrusion. Coextrusions can be used 'as is' or combined with additional coextrusions or films to create a lamination (Fig. 4). Since gases will move through each independent layer sequentially, the transmission of each independent layer of polymer must be determined. From this data, the following equation can be used to calculate the overall structure transmission rate.

Calculating the OTR

$$\text{OTR} = \frac{1}{\frac{t1}{\text{OTR}_1} + \frac{t2}{\text{OTR}_2} + \frac{t3}{\text{OTR}_3}}$$

where

t = thickness of the individual layer.
OTR = oxygen transmission rate of the individual layer per 1 mil of thickness.

Based upon this equation the total structure transmission can never be greater than the lowest individual layer transmission. This limiting factor establishes a practical ceiling for polymer-modified atmosphere packaging. Depending on the type of structure and all of the converting, geometric and marketing requirements this ceiling may be as low as 175 cc/100 in^2/mil/atm/day (2713 cc/m^2/mil/atm/day) or as high as 900–1000 cc/100 in^2/mil/atm/day (13,950–15,500 cc/m^2/mil/atm/day).

The specific combination of polymers, blends, coextrusions, and laminations is governed by numerous parameters in addition to the physiological requirements mandated by the product being packaged. These additional functional requirements often act in an opposing manner. Therefore to effectively

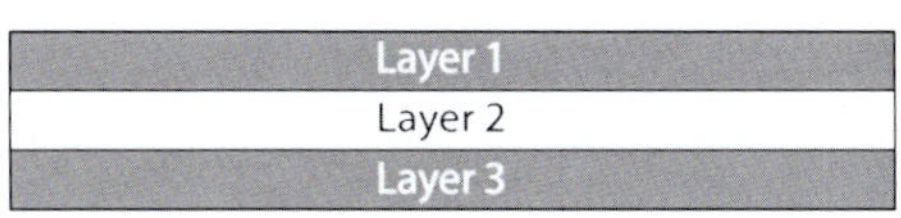

FIG. 4 Example of multilayer coextruded films and structures for OTR calculations later.

design a modified atmosphere package it is necessary to have a full understanding of all the desired features of the package. Requirements impacting polymer and film choice can include: stiffness, sealability, esthetics, clarity, graphics, dimensions, economics, sustainability, runnability, packaging format, coefficient of friction, antimicrobial additives, and thickness. Examination of just one of the parameters demonstrates the results of these opposing parameters. As is listed in Table 2, polymers that exhibit increased stiffness properties also exhibit lower gas transmission rate properties. Therefore if structure stiffness is of paramount importance then the overall gas transmission rate of the package may be low. If too low, it may not meet the requirements of the packaged produce. We can see that this parameter alone has a significant impact on overall package design. Rarely though does one have the luxury in examining one parameter in isolation therefore it is critical to have a thorough understanding of all applicable parameters and their relative importance.

Perforations

The increase in demand for higher respiring fresh produce, outside the traditional leafy green market, has necessitated a significant increase in modified atmosphere packaging gas transmission rates. A method for achieving high OTR packaging structures (not limited by the upper end of polymer gas transmission rates) while meeting stiffness constraints is microperforation technology. This technology employs the science of placing microscopic holes in the packaging structure through the use of a laser or electrostatic discharge. With microperforation technology, the gas transmission rate of the modified atmosphere package is governed by the configuration of holes and their individual geometry and size. The hole size and configuration can vary with the specific perforation method, but all microperforations are essentially invisible to the naked eye and range from 40 to 200 μm in diameter. The smaller the hole size, the greater number of holes that must be placed in a structure to maintain overall transmission rates. There is a limit however as the diameter of the perforation should not be less than the thickness of the film. This is due to the laser energy dissipates though the thickness of the film. This creates a cone shape hole where the diameter of the perforation on the inside surface of the film may be less than the diameter on the outside surface of the film. If the outside diameter is too narrow and the film too thick you may actually not have a hole all the way through the film (Fig. 5).

It is essential to have a complete understanding of the package geometry as well as produce geometry and "bulk density" so that holes are not blocked or obstructed in any way. This is critical to the success and control of gas transmissions. Microperforations also have transmission rate limitations; however for microperforation technology, the limitations are at the lower end of the range. Typical OTR for microperforated packaging are from 250 cc/100 in^2/mil/atm/day and above. Gas transmission rates of microperforated structures

TABLE 2 Lists of the common polymers used in produce-modified atmosphere packaging

Polymer	Abbreviation	Characteristic	Typical OTR (cc/100 in^2/mil/atm/day)	Application Most Commonly Used
Low-density polyethylene	LDPE	General Purpose Polymer	450–500	Both
Linear low-density polyethylene	LLDPE	Increased stiffness	480–500	Both
Linear medium-density polyethylene	MDPE	Increased stiffness, lower OTR, decreased clarity	300–350	Both
High-density polyethylene	HDPE	Relatively stiff, opaque	150	Rigid
Ultra-low-density polyethylene	ULDPE	High OTR, decreased stiffness	900	Flexible
Plastomer metallocenes	–	Very high OTR, soft	1100	Flexible
Amorphous polyethylene terapthalate	APET	Clear, rigid	5	Rigid
Polyvinyl chloride	PVC	Clear, rigid	10	Rigid
Polypropylene	PP	Decreased clarity	300	Both
Ethylene vinyl acetate	EVA	Increased sealability	600–900	Flexible
Polystyrene	PS	Increased stiffness	350	Rigid

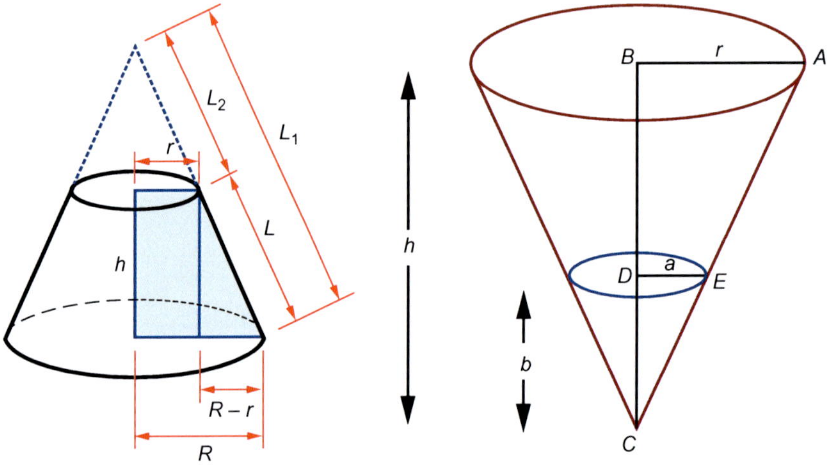

FIG. 5 Microperforation hole geometry.

are determined by the gas diffusion properties through the combined effect of the individual micro holes, their corresponding placement, and in certain cases the OTR of the perforated structure. As the targeted OTR requirements are lower the number of holes decreases. Since microperforated structures cannot have less than one hole, the gas transmission rate through a single hole dictates the lower transmission rate level. Modified atmosphere microperforation packaging with only one hole can also be problematic. With two or more holes there is less risk from hole blockage as well as there is more uniform gas flow throughout the package, mitigating the possibility of package microatmospheres. In order to avoid only one microperforated hole, alterations to the other key OTR control parameters, such as hole diameter, product weight, or package dimension, should be considered.

The diffusion rates of various gases through microperforations are very similar. In effect the diffusion rates of CO_2 and O_2 are virtually the same: a beta value very near one. Therefore for a target of 2% O_2, a 19% CO_2 level will be anticipated. The beta value difference between polymeric and microperforated structures creates significantly different final modified atmospheres. It is not possible with a microperforated film to achieve a modified atmosphere consisting of low O_2 concentrations and low-to-moderate CO_2 concentrations Vallejo and Beaudry (2010). Conversely if low O_2 concentrations in combination with high CO_2 concentrations are required, then packaging comprising of engineered polymers is not suitable. This parameter needs to be accounted for in the packaging design process.

The majority of microperforations are incorporated into the package at the packaging supplier. One company, Perfotec, has developed a unit that can be placed directly on the filling line.

Macroperforations, which are easily visible to the naked eye, should not be mistaken for microperforations. The gas movement through the larger visible holes utilized in macroperforation technology is too great to consistently modify and control the gas level within the package, unless the package is exceedingly large. Therefore attempting to create a low O_2 modified atmosphere is not feasible. This does not mean, however, there is not a function and need for macroperforated structures. Since the gas transmission through a perforation is so high, a macroperforated structure will virtually never become essentially anaerobic (i.e., O_2 levels falling to $<0.5\%$), even under temperature abuse situations. A fresh-cut produce package should not become anaerobic under any circumstances; it is the highest priority. If atmospheric levels of O_2 do not significantly impact shelf life of the fresh-cut product, then macroperforation technology may be most appropriate. Historically, this has been the technology of choice for mushroom packaging. Recent advances in microperforation technology have provided mushroom growers and processors with some alternatives to the use of macroperforations.

Breathable membranes

There have been a number of types of membrane technologies that have been developed over the years.

Membranes generally are patches, or labels, placed over a die cut hole within the package. They are high transmission rate polymers that allow for the gas transmission of the package to be independent of the package makeup. This allows them to be used in both flexible as well as rigid applications.

One of the most innovative to be commercialized is the Landec Apio BreatheWay® membrane. This technology allows the package designer not only to dictate the beta value of the membrane; allowing for specific final O_2/CO_2 design targets, but it can adjust its transmission rate with temperature increases. This is a unique feature in modified atmosphere package design that allows the package to increase its transmission rate during temperature increases, when the produce is respiring at an increased level.

As stated by Clarke (2011), BreatheWay® membranes are a flexible and enabling technology that allows the decoupling of the packaging media from providing the permeability needed by the produce, while at the same time providing the atmospheres necessary for preserving produce. The temperature switch function of side chain crystalline (SCC) polymer will maintain the atmospheres through moderate temperature fluctuations in the cold chain (Fig. 6).

Gas flushing

As previously discussed, modified atmosphere packaging for fresh-cut products relies on the relationship between produce respiration and package transmission rate to alter the atmosphere within the package. This process generally takes a

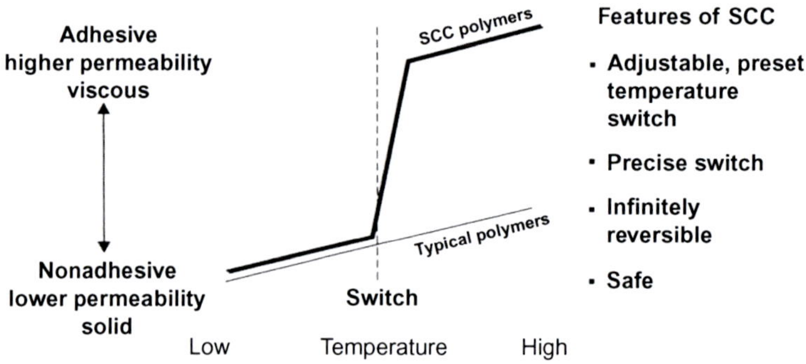

FIG. 6 Temperature sensing capabilities with side chain crystalline (SCC) polymers, which are more crystalline (i.e., solid and having low permeability) at low temperatures, compared to their more amorphous state at higher temperatures.

number of days to reach the target modified atmosphere and equilibrium. For produce items that are prone to enzymatic browning reactions, that is, "pinking," this gradual decent may be too long since pinking can be exacerbated by O_2 levels above 3%. Gas flushing in modified atmosphere packaging of fresh-cut products establishes an initial low alternative atmosphere within the package, which can be beneficial in reducing enzymatic browning/pinking reactions. Gas flushing, however, is not a substitute for proper package design and it does not overcome failure due to package leaking (i.e., "leakers"). Gas flushing is only optimized when it is employed in combination with proper package design and a leak-free package. Since the primary goal of gas flushing is to reduce the initial O_2 level within the package, N_2 is the most effective and economic gas. This being said, there are times when an initial shot of CO_2 can be beneficial in creating an increased level of CO_2 to immediately inhibit the growth of spoilage organisms. Certain pad technologies in which CO_2 is released when liquid escapes from the product (i.e., "purge") and is absorbed into the pad are useful in this application. As with gas flushing, this technology is to be used in concert with properly designed MAP and not a substitute for proper package design.

Converting technology

Packaging format

As previously mentioned, effective modified atmosphere package design hinges on the proper balance of physiological, polymer, converting, and marketing requirements. Issues such as package configuration, package stiffness, graphics, filling method, economics, environmental impact as well as ancillary requirements such as cook-in the package, antimisting, resealability, and compostability all impact the design and makeup of the MA package. Many of these requirements are similar to those found in traditional packaging applications.

The significant difference with MAP is the impact that these requirements have on the gas transmission rate properties of the package.

The majority of these requirements fall under the science of converting technology. The science of converting technology combines the raw material polymers, films, adhesives, inks, and additives in the proper sequence to create the desired package. Depending upon the format, packaging can be developed separate or in concert with automatic filling equipment. The result of these various converting requirements is the development of a wide variety of packaging formats. There is no right or wrong format. Rather it is important to choose the format with optimum suitability based on all of the outlined requirements. Packaging formats can range from the very basic monolayer preformed side-weld bag to the very complex multilayer coextruded reverse print lamination and thermoformed multilayer tray with peelable lidding structure. Table 3 lists some of the more common packaging formats. It is important to note that this is not an all-inclusive list since there are many variations to these common formats that optimize specific desired properties to the specific biology of the produce being packaged.

TABLE 3 Common packaging formats

Package format	Suitable for MAP	Natural aspiration	Prone to leakers if properly manufactured	Common products packaged
Side-weld premade bag	Somewhat	Somewhat	Yes	Carrots
Premade pouch	Yes	No	No	
Vertical form fill and seal (VFFS)	Yes	No	Structure dependent	Leafy greens
Premade, Stand up pouch (SUP)	Yes	No	No	
VFFS SUP	Yes	No	No	
Thermoformed tray with attachable lid	No	Yes	N/A	In-store fresh-cut fruit
Clamshell tray	No	Yes	N/A	Berries
Thermoformed tray with sealable lid	Yes	No	Lidding dependent	
Tray with overwrap	Package dependent	Yes	Yes	Cut squash
Macroperforations	No	Yes	N/A	Mushrooms

Flexible versus rigid packaging

As the names imply, the fundamental difference between flexible versus rigid packaging relates to the stiffness, and often thickness, of the respective packaging configuration. The relative stiffness of a given package is controlled by the choice of polymers and their respective stiffness properties as well as the thickness of the structure and geometry. In modified atmosphere packaging there is another significant difference between rigid and flexible packaging. This relates to the effective surface area available to gas diffusion. In rigid packaging the polymers required to maintain the rigid form of the package generally exhibit low to extremely low gas transmission rates. These low transmission rates in combination with the thickness required to maintain the package form effectively create a gas barrier structure. This dictates that when determining the effective "breathable" surface area of the package, the rigid portion of the package generally cannot be included in the calculations. This means that the flexible lidding material sealed to the top of the tray must take on the entire burden of gas transmission. Therefore, when comparing a flexible package with a rigid tray type package, the gas transmission level of the effective surface area of the flexible package can have a significantly lower overall transmission rate. This is an application where the use of a breathable membrane is advantageous when incorporated into the rigid tray portion of the package to increase the overall breathability of the package structure.

When a rigid tray is used in combination with a rigid lid, both components of the package are effective gas barriers. Therefore if a hermetic seal is created between lid and tray, effective gas transmission will not take place and depending upon the type and quantity of produce anaerobic conditions can rapidly develop.

Often, a rigid tray and lid combination is not designed to create a complete seal. In this situation the package is not a designed modified atmosphere package but rather "natural aspiration" package; meaning that, depending on the effectiveness of the seal, the atmosphere inside could range from anaerobic to ambient. When designing packaging for the produce market, it is important to decide if you are designing a modified atmosphere package or a "natural aspiration" package. Both can be effective packaging formats depending upon initial requirements and desired outcome. Shelf life optimization can only be achieved through MAP technology.

No matter which format, flexible or rigid, is chosen, if optimal MAP is desired, then the controlled and quantifiable transmission of gases through the package must be considered in concert with the physiological characteristics of the produce being packaged. If there is no control or quantification of gas transmission rates, then optimal modified atmospheres cannot be guaranteed and the packaging system is thereby out of control. There are a number of ways an out-of-control packaging system can be created, including improperly quantified produce physiological properties, an improperly specified package, and out of specification raw materials. However, the most common reason is due to a leaking package. If the package does not have a leak-free seal, then gases will immediately begin to pass through the leak. Depending upon the size of the leak, the impact could range from missing the optimal target modified atmosphere to allowing the package to remain at ambient conditions. In either case optimal shelf life and quality will not be achievable.

Therefore one of the most important packaging parameters that must be considered is the selection of the sealant layer polymer and configuration. Choice of the correct sealing polymer and format is dependent upon operating parameters including package machine type, filling speed, package configuration, seal configuration, as well as product type, and its weight (mass). Potential sealant layer polymers exhibit a wide range of seal characteristics. Common choices of sealant polyolefin sealant polymers include low-density polyethylene, ethyl vinyl acetate, ultra-low-density polyethylene, and plastomer metallocenes. Each polymer has its own sealing characteristics including ultimate seal strength, hot tack strength, seal initiation temperature, ability to seal through contamination, and oxygen transmission rate. Careful consideration should be given in order to optimize the specific polymer characteristics to the specific requirements of the package to include all necessary properties.

In certain circumstances, a peelable seal is desired. This can be in either a bag or tray configuration; however, it is much more common in a tray configuration. There are a number of technologies that can be employed to achieve an optimal peelable seal. As Gorny and Brandenburg (2003) point out, peelable lidding stock technologies are generally complex; therefore a dialog with a packaging expert is recommended. The three most commonly available peelable technologies are controlled contamination, dissimilar resins, and controlled delamination. Each has its own advantages and disadvantages. Gorny et al. (2003) describe the basic attributes of the three technologies.

(a) *Controlled contamination*: This technology utilizes a small amount of contaminant resin within the sealant layer to achieve a controlled weak seal or "peel seal." Common contaminant resins are various grades of polybutylene. This technology is used predominately for applications where the film is peelable to itself. Through engineered blending technology the film can be designed to be peelable within a specific temperature range. This results in peelable areas, as well as nonpeelable areas within the same finished packaging material.
(b) *Dissimilar resins*: This technology utilizes a dissimilar resin between the peelable lidstock and the bottom (thermoformed) tray stock. This technology is most commonly used in lidstock applications where a peelable lid is sealed to a rigid or flexible thermoformed bottom web or rigid container. The choice of peelable sealant layers is directly dependent upon the sealant layer of the bottom material. Whenever possible, the top and bottom sealant layers should be designed together so the optimum match is made and the desired peel seal is achieved.

(c) *Controlled delamination*: This technology involves achieving an initial fusion seal between the sealant layers of the lidstock and the bottom film (i.e., web). The "peel seal" is then achieved through the delamination of the coextruded sealant film. The delamination within the coextrusion occurs between the sealant layer and the next inner layer (often the core layer). It is possible to achieve a frosty or "tamper evident" seal, as well as a resealable (i.e., "reseal") structure. This technology is often found in medical packaging.

As Gorny et al. (2003) conclude, it is critical to get the correct match between the peelable lidding material and the bottom web. There are a wide variety of polymers and constructions available, depending on the type of bottom web sealant layer, peel seal requirements, as well as the look and feel of the total structure. Companies familiar with peelable lidding materials should be consulted so that all of the necessary variables are taken into account.

Packaging equipment

Packaging equipment used for MAP is as varied as the packaging types themselves. Packaging equipment can range from the basic handheld impulse sealer to processor-controlled automatic form fill and seal lines. Within this range of options, however, are two fundamental categories; vertical and horizontal. Figs. 7 and 8 are examples, respectively, of a vertical and horizontal filing line. Both styles of packaging equipment can be configured for either manual or automatic operation.

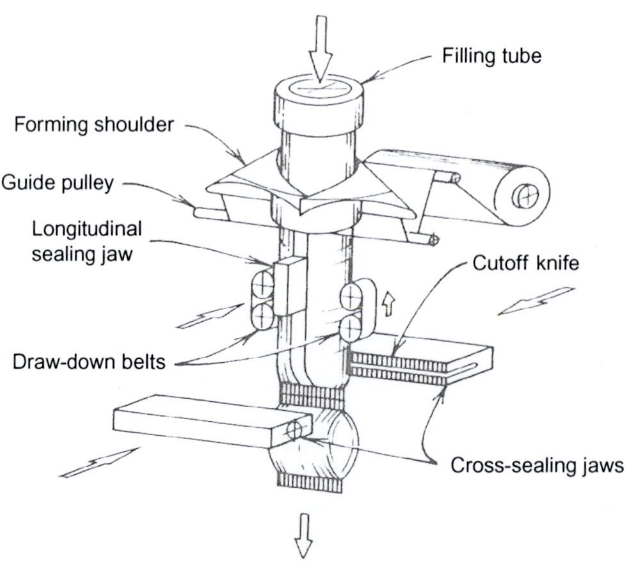

FIG. 7 A vertical form fill and seal (VFFS) operation (Gorny et al., 2003).

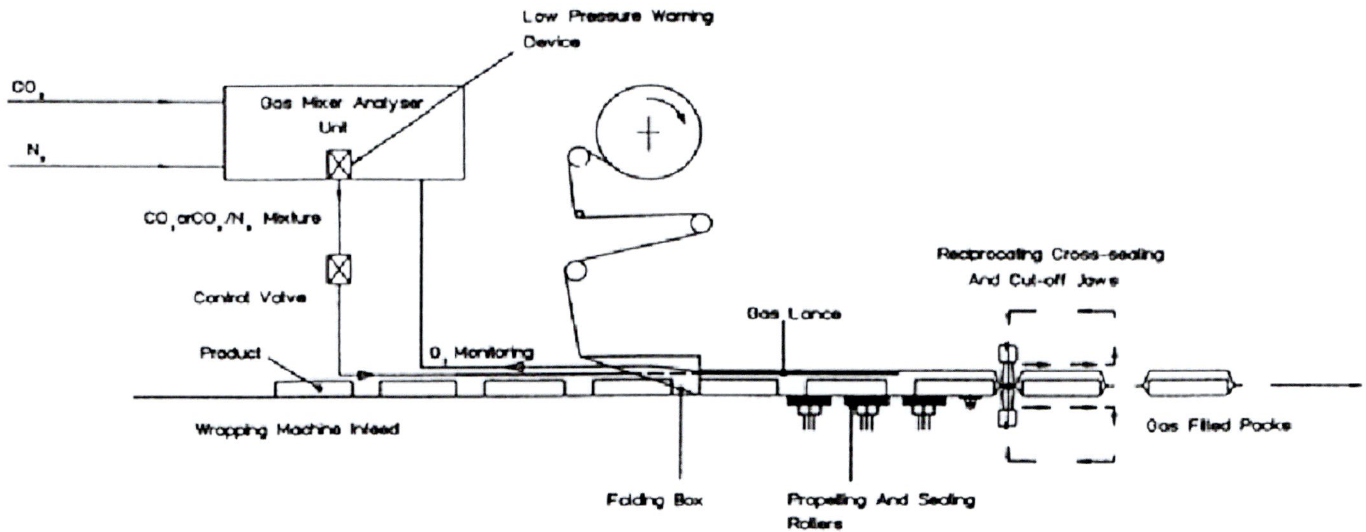

FIG. 8 A schematic diagram of horizontal form-fill-seal machine with box-motion sealing head (Soltani et al., 2015).

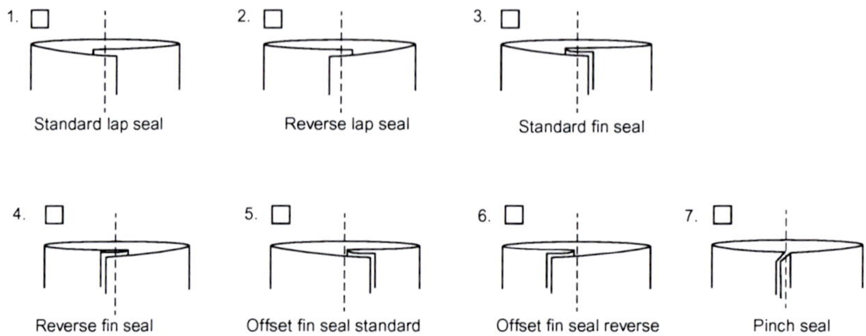

FIG. 9 Examples of vertical form fill and seal (VFFS) seal configurations (http://www.21food.com).

Although there are exceptions, generally vertical packaging machines are designed to run flexible style packaging and horizontal machines are designed to run rigid style packaging. Within the vertical configuration there are two seal configurations: fin and lap as seen in Fig. 9. Determining which seal type to be used is a critical step in the package design since the seal type can impact polymer selection, graphics, and material usage. The most significant difference between the two types is that the fin seal configuration seals only the inside of the package whereas the lap seals the inside to the outside of the package. Therefore when a lap seal configuration is used the polymers on both the inside and outside of the package must be sealable.

Traditionally, applying heat from the outside to the inside of package during sealing has been the sealing method of choice. This technology uses a combination of heat, pressure, and time (dwell) as the method to weld the sealant layer polymer together. During this welding process, the sealant layer polymer transitions from solid to liquid and back to solid, thus creating the seal. The 'hot tack' of a seal is the strength of the seal when the polymer is still in the molten or liquid phase, whereas the ultimate seal strength, as the name implies, is the final resolidified seal strength.

Recently, ultrasonic sealing has been making inroads into fresh-cut produce packaging. As stated by Herrmann Ultrasonics (2017), with ultrasonic sealing the heat required for melting is only generated inside the thermoplastic sealing layer. For localized conversion of vibrations to friction heat, the anvil profiles are mostly linear and include radii or small plateaus. These profiles ensure focusing of the energy input to the target material, therefore achieving short sealing times between 100 and 200 milliseconds. With ultrasonic sealing, heat is generated at the inside of the film, not by heat input from the outside as is the case with heat sealing,

Additives

Additive technology can have a very significant impact on modified atmosphere package design. There are a number of different additives that are frequently incorporated into MAP applications. Each has its own distinctive purpose and function. These can include materials for antifogging, slip and antiblock, antimicrobials, pigments, UV inhibitors, as well as absorbents.

Antifog

As Gorny and Brandenburg (2003) discuss, antifog technology is often incorporated into produce packaging to prevent water condensation on the inside of the package, which would obscure a potential purchaser's view of the product. Antifog technologies functionality is accomplished by coating the interior surface of the flexible packaging material with compounds that reduce water surface tension or reduce the ability of the water to adhere to the packaging material and thus cause the condensed water to run off the interior surface of the package. There are two categories of antifog technology: applied coatings and sealant layer incorporation, with each yielding distinct advantages and disadvantages. Applied coatings are applied to the sealing surface of the finished package and historically have been superior to the incorporation method in that they provide superior optics and antifog properties. Within the category of applied antifog coatings there are two predominant formats: registered applied and flood coating. In the registered applied system, the coating placement is controlled so that the antifog coating does not fall into the sealing area thereby does not impact the sealing properties of the structure. In order to utilize this system, specialized equipment is necessary to achieve registration (i.e., alignment) of the antifog coating to the package surface, as well as to achieve good adhesion without damaging the sealing surface. Not all converting companies have the equipment to properly register and apply these coatings. Flood coating does not apply the antifog 'in register,' but rather coats the entire sealing surface, thereby eliminating the need for the specialized register application equipment.

The main disadvantage with applied coatings is that the coatings are applied to the nontreated surface, and most applied coatings require a surface treatment to guarantee strong adhesion to the packaging structure. Therefore antifog coating pick off or transfer may occur. Coordination and good communication with both the antifog supplier and the converter is critical when planning to use this type of antifog coating.

Antifog sealant layer blends incorporate or blend an antifog compound into the sealant layer during the film manufacturing process. This system significantly reduces the cost of adding antifog technology to produce packages since no specialized converting equipment is needed. It thus has advantages both for the converter and end user. The major disadvantage to antifog sealant layer blends is that the antifog is, by default, in the sealing area of the bag. Therefore weaker seals and a greater likelihood of package leakers are possible. However, selecting compatible sealant layer formulations can often overcome this issue. The optics and performance of this type of antifog technology may be inferior to coated antifogs. In addition, in order to properly function, the incorporated antifog must move to the surface "bloom" and thus can potentially cause a significant reduction in the gas transmission rates of the package.

Slip and antiblock

Slip and antiblock are key components of most packaging, especially flexible packaging that are filled using automatic filling equipment. The addition of slip and antiblock compounds into the polymer film structures reduces the natural tackiness of certain polymers as well as increases the film's resistance to blocking (sticking together). This is especially important if vertical form fill and seal (VFFS) equipment is used to form and seal the package. The ability of the packaging structure to slide over the metal forming collars and tubes is essential for proper structure formation. Like antifog additives, slip and antiblock can significantly alter the gas transmission rate properties of the packaging structure. Care must be taken to properly quantify and minimize the effect upon gas transmission rates. Improper application of antiblock additives can also impact package forming and sealability, leading to an increased incidence of improperly sealed packages and leakers.

Antimicrobial films

Similar to the aforementioned additive applications, antimicrobial additives also bloom to the surface of the inside layer of the package. Therefore the issues of gas transmission rate reduction and sealing ability are equally relevant. Their method of operation is by contact of the antimicrobial component to the packaged product. Although commonplace in protein MAP, they are not widely employed in produce MAP. This is due to the relatively high amount of product surface area in relation to package surface area in fresh produce applications. This high ratio prevents the majority of the product to contact the microbial agent, thus making these systems relatively ineffective. New technologies are currently under development with potentially exciting results and improved effectiveness is possible.

Absorbing compounds

A wide variety of absorbing compounds are also being evaluated and, in some cases, have been commercialized. Moisture, ethylene, O_2, CO_2, and odors are all compounds that absorbers are targeting. There are a number of challenges with absorbing compounds including FDA approval for direct food contact, their capacity and selectivity for sorption of the target compounds, and consumer acceptance if a sachet is used. A full understanding of what these compounds can and cannot do is critical in deciding if they should be incorporated into the packaging structure.

In all additive applications, careful consideration must be given when considering the incorporation of any additive into a modified atmosphere package. Although the addition may enhance or add certain desirable features or properties they will almost surely detract or negatively impact other properties or functions. The key property in fresh produce applications that is most likely to be negatively affected is the overall gas transmission rate properties of the packaging system.

Applications

New produce items and geometries

As the fresh produce market has matured beyond the initial category of leafy green packaging, developers of MAP technologies have responded. New and novel types of vegetables, vegetable geometries, fruit, and unique combinations of different types fresh produce, ready-to-eat meals, and microwave and steam-in microwave applications are all examples of new applications and markets. Within the category of new produce items and geometries there has been tremendous growth in the past few years. New items such as cut Brussels sprouts, kale, power greens, and cauliflower rice have been introduced and are becoming commonplace. Spiralized geometries of root vegetables and summer squash are also becoming very popular in the marketplace. In each case, optimal MAP needs to be developed to insure both the optimal shelf life and quality.

As with all new developments, each new application challenges the science and technology of MAP and adds its own demands and requirements.

Microwave and steam-in

As Gorny and Brandenburg (2003) point out one of the fastest growing segments of fresh-cut produce packaging is the use of microwaveable and steam-in microwaveable packages. There are a number of venting technologies available in the marketplace, including perforations, controlled delamination slits, valves, and scoring. In considering which technology is best for your application, it is important to keep the steam venting technology from affecting the modified atmosphere perforation technology. The venting utilizes holes but their design size and functionality are significantly different from the MA perforation. To try and combine the two technologies will only result in both technologies becoming compromised.

In addition to the functional requirements generated by cooking product inside the package, there is an additional and often hidden requirement of migratory compliance. Polymers that are direct-contact food compliant and act as "functional barriers" at ambient temperatures may or may not remain compliant or serve as proper barriers under elevated temperatures. In addition, as the internal package temperature rises, fewer and fewer polymers remain both functionally and legally compliant. It is therefore critical to engage in a detailed dialog with the packaging supplier and to make sure that all applicable regulations have been identified. The following sections of the U.S. Code of

Federal Regulations (CFR's) may be applicable when dealing with microwave and steam-in microwave applications.

- 21CFR § 177.1350
- 21CFR § 177.1390
- 21CFR § 177.1395
- 21CFR § 177.1520

It is important to assure that, before developing a microwave product, you conduct in-depth discussions with your packaging supplier to assure that the above-mentioned issues have been addressed and that you are in full regulatory compliance. Not all packaging materials are suitable for all microwave applications.

Varietal blends and novel produce combinations

The growth of varietal blends and novel combinations is a direct result of processors continually searching and investigating ways to different their product line. This need to differentiate has led to an explosion of vegetable and leafy green combinations, "power blends" of vegetable stew and soup mixes, as well as varieties and combinations of tropical fruits. These new opportunities place additional challenges and requirements on the technology of MAP. Specifically, each separate and unique raw material has its own physiological properties combined with its own individual optimal target atmosphere. Therefore not all fresh-cut fruit and vegetables are compatible. In other words what may be an optimal target for one product may be very detrimental to another. Compromises in atmosphere targets will most likely be required; however, it is critical to minimize the compromises. Instead, companies are quantifying the physiological properties of their individual raw materials and then grouping like products in order to create desirable blends and combinations without having to sacrifice quality and shelf life.

Ready meals

The combination of produce, protein, and carbohydrate into a fresh ready-to-eat (i.e., "ready") meal format has increased significantly over the past few years. In addition, plant proteins are being added to kits so that the Vegan consumer has more access to proteins. This technology is still more common in Europe than in other world markets but is rapidly gaining mainstream popularity. This type of packaging requires different types of modified atmosphere technology for each type of food category. Hence, these applications generally require multicompartment trays to separate each food product with unique target atmospheres in each compartment. Since within a single package there are different and unique food groups, effective package design must cross over and combine often competing requirements. Coordination between the packaging designer and food scientists from each of the participating food groups is essential.

One requirement that has come out of all of these new applications is the need to incorporate and coordinate the package design process into and with the new product development process. Leaving the package design until the end of the product development process prevents the optimization of both product and package. Since this has not been the norm historically, reeducation and enhanced communication between the new product development group within food company and packaging designers is required.

Future research directions

It is an exciting time within the MAP industry. There are emerging technologies and opportunities that will have a far-reaching impact on the marketplace. Issues such as sustainability in packaging and the impact that packaging has on current food safety issues are already providing both tremendous challenges and opportunities. Successes will be defined by how well they incorporate all of the desired requirements into MAP without diluting its fundamental purpose. A package that tries to become all things to all applications becomes mediocre at best with respect to any one requirement.

References

Brandenburg, J., Zagory, D., 2009. Modified and controlled atmosphere packaging technology and applications. In: Yahia, E.M. (Ed.), Modified and Controlled Atmospheres for Transportation, Storage and Packaging of Horticultural Commodities. CRC Press, Taylor & Francis.

Brody, A.L., Marsh, K.S., 1997. In: Brody, A.L., Marsh, K.S. (Eds.), The Wiley Encyclopedia of Packaging Technology, second ed, John Wiley & Sons, Inc., New York.

Brown, W., 1922. On the germination and growth of fungi at various temperatures and in various concentrations of oxygen and carbon dioxide. Ann. Bot. 36, 257–283.

Burg, S.P., 2004. Postharvest Physiology and Hypobaric Storage of Fresh Produce. CAB International, Wallingford, UK.

Burg, S.P., Burg, E.A., 1967. Molecular requirements for the biological activity of ethylene. Plant Physiol. 42, 114–152.

Clarke, R., 2011. Breatheway® membrane technology and modified atmosphere packaging. In: - Brody, A.L., Zhuang, H., Han, J.H. (Eds.), Modified Atmosphere Packaging for Fresh-Cut Fruits and Vegetables. Wiley-Blackwell, pp. 185–208.

Farber, J.M., 1991. Relative effect of CO_2 on the growth of foodborne microorganisms. J. Food Prot. 54, 58–70.

Gorny, J.R., Brandenburg, J., 2003. Packaging Design for Fresh-cut Produce. International Fresh-Cut Produce Association, Alexandria VA.

Gorny, J.R., Brandenburg, J., Allen, M., 2003. Packaging design for fresh-cut produce. International Fresh-cut, Produce Association, p. 122.

Herner, R.C., 1987. High CO_2 effects on plant organs. In: Weichmann, J. (Ed.), Postharvest Physiology of Vegetables. Marcel Dekker, New York, p. 239.

Herrmann Ultrasonics, 2017. Accessed at: https://www.herrmannultrasonics.com/en-us/ultrasonic-sealing/.

Kader, A.A., 1986. Biochemical and physiological basis for effects of controlled and modified atmospheres on fruits and vegetables. Food Technol. 40, 99–100. 102–104.

Kader, A.A., Zagory, D., Kerbel, E.L., 1988. Modified atmosphere packaging of fruits and vegetables. Crit. Rev. Food Sci. Nutr. 28, 1–30.

Kader, A.A., Zagory, D., Kerbel, E.L., 1989. Modified atmosphere packaging of fruits and vegetables. CRC Crit. Rev. Food Sci. Nutr. 28, 1–30.

Reid, M.S., 1985. Ethylene in postharvest technology. In: Kader, A.A., Kasmire, R.F., Mitchell, F.G., Reid, M.S., Sommer, N.F., Thompson, J.F. (Eds.), Postharvest Technology of Horticultural Crops. University of California, Cooperative Extension Special Publication 3311, 192 pp.

Soltani, M., Alimardani, R., Mobli, H., Mohtasebi, S.S., 2015. Modified atmosphere packaging: a progressive technology for shelf-life extension of fruits and vegetables. J. Appl. Packag. Res. 7, 33–59.

Vallejo, F., Beaudry, R.M., 2010. Comparing the function of perforated and non-perforated film in MAP: impact on the quality of broccoli. In: Proceedings of 9th International Controlled Atmosphere Conference. Acta Hort.Vol. 857, pp. 503–512.

Zagory, D., Kader, A.A., 1989. Quality maintenance in fresh fruits and vegetables by controlled atmospheres. In: Jen, J.J. (Ed.), Quality Factors of Fruits and Vegetables: Chemistry and Technology. American Chemical Society, Washington, D.C, pp. 174–188.

Zagory, D., 1996. Physiology and microbiology of fresh produce in modified atmosphere packages. In: Society of Manufacturing Engineers Symposium, Fundamentals of Modified Atmosphere Packaging, December 4–5, 1996. Monterey, CA.

Chapter 11

Innovations in packaging technologies for produce

Eva Almenar
School of Packaging, Michigan State University, East Lansing, MI, United States

Introduction

Deterioration in produce is caused by enzymatic, physical, chemical, and microbiological changes since produce after harvest is still an active biological system. Endogenous factors that contribute to these changes include processes like respiration and transpiration, ethylene sensitivity, enzymatic activity [e.g., polyphenoloxidase (enzymatic browning), pectin methylesterase (cell wall degradation), amylase (starch-sugar conversion)], and initial microbial load (types and amounts). Exogenous factors also contribute to the aforementioned changes. These include temperature, relative humidity, ethylene, microorganisms, oxygen, and light. Endogenous and exogenous factors often cause a net reduction of the shelf life and of the nutritional content of produce, and in some cases affect product safety as well.

Modified atmosphere packaging (MAP) can reduce or delay the aforementioned changes, thus diminishing the loss of quality and safety, and consequently maximizing freshness and shelf life of produce. Hence, MAP has been used for several decades to extend produce marketability. MAP is a packaging technology based on the replacement of the ambient air (78.1% nitrogen (N_2), 20.9% oxygen (O_2), 0.04% carbon dioxide (CO_2), and 0.93% Argon (Ar) plus others including water vapor) inside the package with a single gas or a specific mixture of gases that can lead to produce shelf life extension (Almenar and Wilson, 2016). The most adequate gas or mixture of gases is selected based on the type of produce to be packaged, taking into account respiration rate, ethylene production and/or sensitivity, and so on. Low O_2 is desired as it can slow the respiratory rate, ethylene production, enzymatic browning, and growth of aerobic microorganisms, while high CO_2 is used due to its direct antimicrobial capacity (Almenar and Wilson, 2016). However, the depleted O_2 and/or enriched CO_2 levels required to induce some specific changes (e.g., reduced microbial growth) may not be the most beneficial for the commodity in terms

Controlled and Modified Atmospheres for Fresh and Fresh-Cut Produce
https://doi.org/10.1016/B978-0-12-804599-2.00012-0

of other physiological responses (Zagory and Kader, 1988). Additionally, the maintenance of optimal O_2 and CO_2 levels to reduce or delay undesirable changes is challenging during handling and transportation due to temperature fluctuations. Furthermore, the effectiveness of MAP on reducing or inhibiting growth of pathogens depends on the commodity-pathogen combination (González-Buesa et al., 2014; Page et al., 2016) and a general statement about its effectiveness can therefore not being made until more work is carried out. As an alternative, innovative packaging technologies like active packaging, in-package cold plasma, coatings, and novel approaches to adjust film permeability to temperature changes have been created to address the issues associated with MAP. In addition, other innovative packaging technologies like intelligent packaging have been conceived to fulfill new market requirements for produce due to changes in retail and distribution practices associated to globalization, changing consumer preferences and demands [e.g., preservative-free food, improved convenience, enhanced healthfulness, greater freshness and safety, communicating information about the product before consumption], alternative purchasing options (e.g., internet), and so on.

Active packaging

Scope

Active packaging can be defined as the packaging technology where certain additives, known as "active compounds" are incorporated into the packaging material or placed within the packaging container in order to interact directly with the perishable product and/or its environment to extend its quality and/or safety. The EU legislation (Regulation 450/2009/EC, currently no US equivalent available) defines "active materials and articles" as "materials and articles that are intended to extend shelf life or to maintain or to improve the conditions of the packaged food. They are designed to deliberately incorporate components that would release or absorb substances into or from the packaged food or environment surrounding the food. According to this legislation, "component" is an individual substance or a combination of individual substances which cause the active function of a material or article, including the products of an in situ reaction of those substances; it does not include the passive parts such as the material they are added to or incorporated into." (http://www.foodcontactmaterials.com/eu/2009%20450.pdf).

Current mechanisms that make packages active are (1) placing the active compound inside the package along with the product (e.g., sachets and labels), and (2) making the active compound part of the material (film or tray) (blended in the bulk polymer matrix or applied to the package as a coating) (Almenar et al., 2006). The interaction between the active compound and the perishable product can be direct or indirect. Active compounds can be grouped into two types depending on the type of interaction between the active compound and the commodity: (1) scavenging compounds (e.g., O_2 scavengers, ethylene scrubbers, moisture absorbers), and (2) releasing compounds (e.g., CO_2 emitters, antimicrobials). Thus an active package enhances quality and safety through either the scavenging of compounds that are involved in produce deterioration processes or the release of compounds that can mitigate the effect of factors involved in produce deterioration. The selection of the active compound depends on the specific mechanism causing the deterioration process to be counteracted. In the last decade, there has been an increase in the tendency to incorporate these compounds into the materials that constitute the package rather than within the package in the form of sachets, pads, or labels as shown by recent developments in active packaging covered later. This change has mainly been driven by processors/packers and consumers as it allows manufacturers to continue using the packing equipment commonly used for nonactive packaging while avoiding the issues associated with consumers finding foreign objects within the package that could be ingested by mistake. In principle, every conventional packaging material used by the food industry (e.g., paper, plastic, metal, etc.) could be used to develop active packaging. However, plastics are preferred due to the many advantages that they possess (Almenar et al., 2012). Among these, bio-based plastics and biodegradable plastics are the main lines of current research in active packaging (Atares and Chiralt, 2016; Kuorwel et al., 2011).

Almenar et al. (2006) thoroughly reviewed the academic work and the commercial products related to the development of active packaging for fruits and vegetable up to 2005. Since then, many studies have been done in the field and new commercial products have been launched. Thus this book chapter focuses on those recent developments. However, relevant past published research work on active packaging for fruits and vegetable is highlighted and used for discussion when needed. Furthermore, evaluation of new packaging technologies for produce must include in vivo studies instead of only in vitro studies since results obtained using a food system may differ from those obtained using a nonfood system (e.g., petri dish with agar) (Wieczynska et al., 2016). Thus packages and packaging materials that have not been tested on produce are not included in this book chapter. The following only covers selected types of active packaging for produce. The selection has been made based on amount of academic research work done and on current use by industry.

Types

Modified moisture packaging

The physiological processes of transpiration and respiration result in the movement of water molecules from the plant tissue to its surrounding environment. When fruits and vegetables are packaged in a sealed container, these water molecules can stay inside the package or leave the package depending on the water vapor permeability of the packaging material, the surrounding temperature and moisture and, to a lesser extent, on the produce's sorption properties (Powers and Calvo, 2003). If the water molecules stay, the relative humidity in the package approaches 100% in a few hours and then remains at that level (Ben-Yehoshua et al., 1983; Shirazi and Cameron, 1992; Yue Bi et al., 2014). Under these near saturation conditions, the water molecules will condense on both the produce and the packaging material and will most likely end up building up liquid water that will pool in the bottom of the package (Fig. 1A). This occurs due to the changing environmental conditions (e.g., temperature fluctuations, cool chain breaks, temperature cycling in coolers) that the packaged produce is exposed to from the packinghouse to the consumer household (Brown et al., 2016; Koutsimanis et al., 2015; Linke and Geyer, 2013). Fig. 2 shows the temperature cycling in a walk-in cooler. The buildup of liquid water is higher when antifog additives are present in the packaging material, as well as in packages of

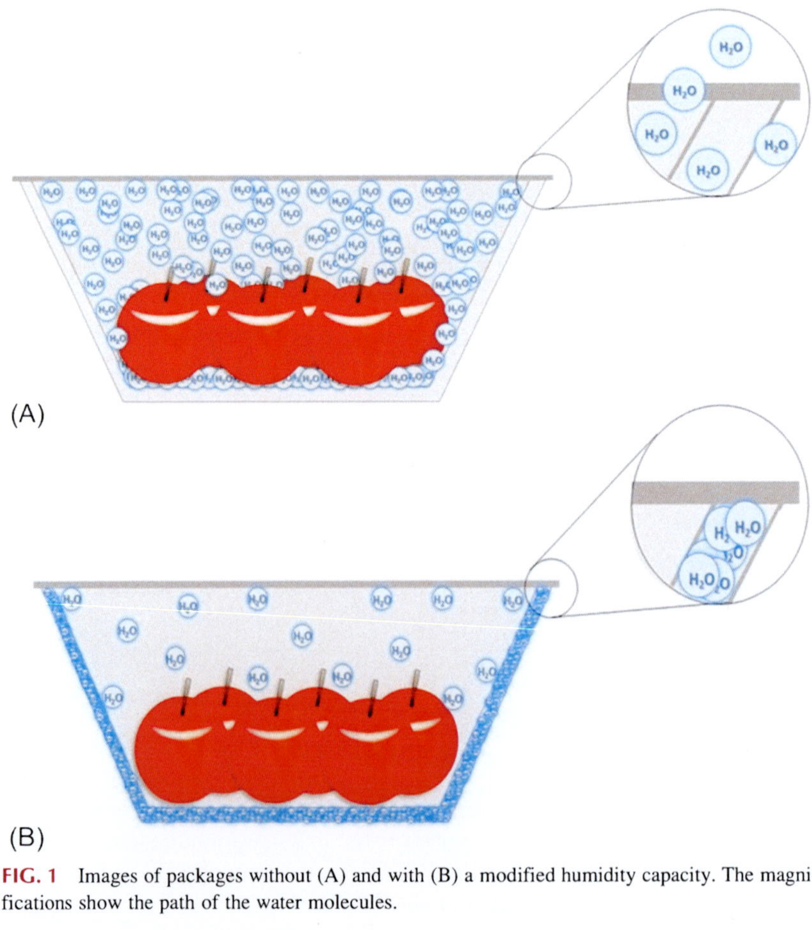

FIG. 1 Images of packages without (A) and with (B) a modified humidity capacity. The magnifications show the path of the water molecules.

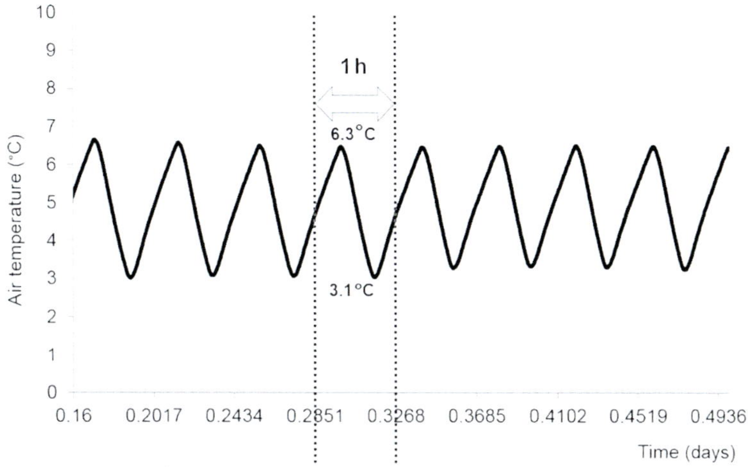

FIG. 2 Temperature cycling in a walk-in cooler that shows that temperature changes of approximately 3°C occur every hour.

fresh-cut produce compared to whole produce. The consequences of this moisture excess are loss of firmness, discoloration, leakage of solutes, inhibition of produce gas exchange, enhanced microbiological growth, accelerated senescence, and safety issues for the produce, while the package suffers from low gas permeability through the packaging material leading to an unfavorable atmosphere, fogging of films, and consumer rejection (Joyce and Patterson, 1994; Zagory and Kader, 1988; Linke and Geyer, 2013; Mahajan et al., 2008). In contrast, if the generated water molecules leave the package, the produce will lose water, which will lead to wilting, shriveling, loss of firmness, crispness, and succulence, increased respiration, weight loss, and an unmarketable product (Almenar et al., 2008; Ben-Yehoshua and Rodov, 2003; Kader, 1987). Therefore proper control of the relative humidity inside packages containing fruits and vegetables is critical.

Film perforation can be an approach to manage in-package relative humidity. However, the large perforation size or high perforation number required to reduce relative humidity will most likely prevent the development of optimal gas compositions for shelf life extension due to the misbalance in perforation effects on water vapor and gas permeabilities (Mir and Beaudry, 2016). Alternatively, films with high water vapor permeability can be used to reduce in-package relative humidity (Pesis et al., 2000). However, this approach does not allow controlling the in-package relative humidity since this also depends on the relative humidity surrounding the package, which varies notably between transportation, storage, and retail market (Koutsimanis et al., 2015). Furthermore, shrink wrapping, which can be considered as another approach for control of condensation (Ben-Yenhoshua, 1989), is not feasible for some types of whole produce or fresh-cut produce. Therefore the solution was the development of a packaging technology with its own moisture management capacity (Fig. 1B).

Modified humidity packaging is a type of active packaging that entails a change in the in-package relative humidity by using compounds that can either adsorb or absorb water molecules. These active compounds can be either incorporated into the packaging material or be placed within the packaging container. Modified humidity packages can be categorized into two types depending on how the moisture is managed:

- Humidity-regulating packages generate and maintain a desired relative humidity in the package headspace by using compounds that reach steady state when filled with water molecules. These compounds are named moisture controllers. They utilize the phenomenon of water absorption to assimilate water molecules through their bulk. The source of these water molecules is those given off by the produce in the case of produce packaging.
- Humidity-sorbing packages reduce the in-package relative humidity by using compounds that accumulate water molecules at their surface or assimilate water molecules through their bulk, but without generating any characteristic relative humidity. The former of these two compound types are named moisture adsorbers while the latter are moisture absorbers. The moisture adsorbers attract the water molecules given off by the produce to their outside surface while the moisture absorbers become filled with the water molecules given off by the produce or the buildup liquid water in the package. Therefore either phenomenon adsorption or absorption can be the basis for humidity-sorbing packaging.

The goal in developing modified humidity packaging is to reduce the relative humidity from the value near 100% created in a sealed package containing produce to levels that are below those favorable for microorganisms to grow and/or those able to induce water-related disorders. However, target humidity levels need to be high enough to minimize wilting, shriveling, and other changes resulting from the loss of water from the produce. Therefore, first, an optimal in-package relative humidity has to be selected. This is specific to the crop to be packaged since crops differ in water potential, optimal storage temperature, surface properties, major postharvest microorganisms, and water-related disorders (Rodov et al., 2010). Information on optimal storage conditions for several crops, including relative humidity values, can be found in Gross et al. (2016). Other factors that relate to the produce as well as factors related to the package, the moisture adsorber/absorber, and the environment have to be taken into consideration to develop appropriate modified humidity packaging. Fig. 3 compiles most of these factors. The full list can be found in published mathematical models that predict the relative humidity in sealed packages containing moisture absorbents along with produce (Evelo and Horst, 1996; Song et al., 2001). One should note, however, that these models have limitations.

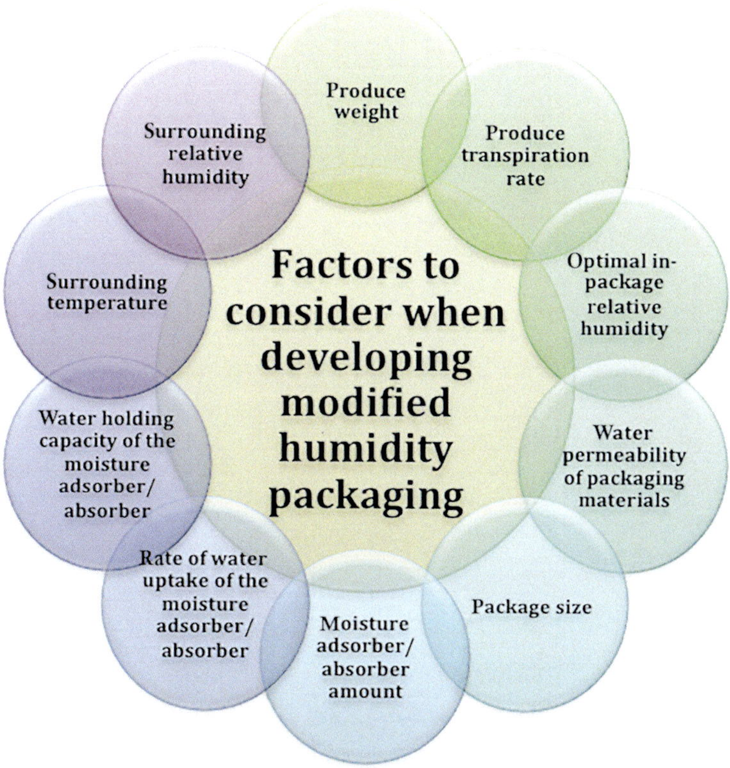

FIG. 3 Factors to be taken into consideration when developing modified humidity packaging.

For example, the model of Evelo and Horst (1996) omits heat balances, while the model of Song et al. (2001) overpredicts the effect of the absorbent on initial periods. Essentially, the ability to control relative humidity is based on the rate of water absorption by the moisture controller relative to the rate of transpiration by the produce when they are both sealed inside a package with low water vapor permeability. Thus a specific amount of moisture controller is required per gram of product to maintain a stable relative humidity (Shirazi and Cameron, 1992). The permeability to water vapor of commercial films used to package produce can be found in Almenar and Auras (2010).

Tables 1 and 2 compile and summarize relevant past and current published research work on humidity-sorbing packaging and humidity-regulating packaging for produce, respectively, in plastic packaging formats (e.g., low-density polyethylene bags). Compounds used to develop humidity-sorbing packaging include silica gel, zeolites, metal-organic frameworks, calcium chloride, and fluff pulp. Moisture controllers used to develop humidity-regulating packaging include sugar alcohols like sorbitol, humectant salts like sodium chloride, and

TABLE 1 Studies on humidity-sorbing packaging for fruits and vegetables

Compound	Compound carrier as defined by authors	Produce	Amount (g absorbent/ kg fruit)	In-package relative humidity (%)	Improved quality compared to control (no moisture controller)	Promoted deterioration compared to control (no moisture controller)	References
Silica gel	Paper sacks	Mushrooms (*Pleurotus ostreatus*)	20, 33, 46, and 100	Not specified	None	Texture deterioration and tissue leakage	Villaescusa and Gil (2003)
	Woven polyester sachet	Green bell pepper (*Capsicum annuum* L.)	9	Not specified	Reduced decay. Shelf life extended by 7 days	None	Singh et al. (2014)
Natural mineral compound	Polyethylene film	Guava (*Psidium guajava* L.)	Not specified	Not specified	Reduced decay. No water condensation	None	Combrin et al. (1990)
Modified XY zeolite	Breathable pouches	Raspberries (*Rubus idaeus* L.)	100	Not specified	Less decayed fruit. Less condensation	None	Toivonen et al. (2002)

Basolite A520	Tyvek sachet	Fresh-cut lettuce (*Lactuca sativa* L.) var. *longifolia*	4	Not specified	Reduced bacterial counts and fungal populations	None	Chopra et al. (2016)
Basolite C300	Tyvek sachet	Fresh-cut lettuce (*Lactuca sativa* L.) var. *longifolia*	4	Not specified	Reduced bacterial counts	None	Chopra et al. (2016)
$CaCl_2$	Spunbonded polyethylene packs	Red-ripe tomato (*Lycopersicon esculentum* L.)	77	25	None	16% weight loss	Shirazi and Cameron (1992)
	Not specified	Bell pepper (*Capsicum annum*)	5 g/1 fruit	80–88	None	Increased weight loss. Softer fruit. Higher water deficit. Lower water potential	Ben-Yehoshua et al. (1983)

TABLE 2 Studies on humidity-regulating packaging for fruits and vegetables

Compound	Compound carrier as defined by authors	Produce	Amount (g absorbent/ kg fruit)	In-package relative humidity (%)	Improved quality compared to control (no moisture controller)	Promoted deterioration compared to control (no moisture controller)	References
Sorbitol	Tyvek paper pouches	Mushrooms (*Agaricus bisporus*)	50, 100, 150, 200, 250	Not specified	Better color and lower microbial populations (150)	Higher weight loss and lower moisture content (50, 100, 150, 200, 250)	Roy et al. (1995)
	Paper sacks	Mushrooms (*Pleurotus ostreatus*)	67, 100, 133	Not specified	None	No improvements in quality and high weight loss	Villaescusa and Gil (2003)
	Tyvek paper pouches	Mushrooms (*Agaricus bisporus*)	50, 100, 150	80–95	Improved color	None	Roy et al. (1996)
Sorbitol + bentonite + $CaCl_2$	Bag	Mushrooms (*Agaricus bisporus*)	20, 40, 60	Not specified	Lower browning index (20) Little condensation (20, 40, 60)	Increased weight loss (20, 40, 60) Lower browning index (40 and 60)	Mahajan et al. (2008)

NaCl	Tyvek paper pouches	Mushrooms (*Agaricus bisporus)*	50, 100, 150	84–87	None	Over-drying	Roy et al. (1996)
	Mesh-cloth bags	Muskmelon (*Cucumis melo* L.)	40 g/12 fruits	Not specified	None	None	Yahia and Rivera (1992)
	Active layer of a tray	Mushrooms (*Agaricus bisporus*)	18% on a weight basis in tray active layer/250 g fruit	93	Less discolored caps and exposed gills (subjective determination as per visual observation)	Double weight loss. Higher condensation on tray walls	Rux et al. (2015)
	Spunbonded polyethylene packs	Red-ripe tomato (*Lycopersicon esculentum* L.)	44	80	Reduced mold growth. Increased shelf life from 5 to 17 days	None	Shirazi and Cameron (1992)
NaCl + hygroscopic ionomer	Active layer of a tray	Strawberries (*Fragaria x ananassa* Duch.) and tomatoes (*Lycopersicon esculentum* L.)	12% on a weight basis in tray active layer/250 g fruit	89.8	No condensation	Higher weight loss	Rux et al. (2016)

Continued

TABLE 2 Studies on humidity-regulating packaging for fruits and vegetables—cont'd

Compound	Compound carrier as defined by authors	Produce	Amount (g absorbent/kg fruit)	In-package relative humidity (%)	Improved quality compared to control (no moisture controller)	Promoted deterioration compared to control (no moisture controller)	References
Hygroscopic ionomer	Active layer of a tray	Strawberries (*Fragaria x ananassa* Duch.) and tomatoes (*Lycopersicon esculentum* L.)	15% on a weight basis in tray active layer/250g fruit	99.6	No condensation	Higher weight loss	Rux et al. (2016)
Super absorbent polymer (40:60 *w*/w) + $NaHCO_3$	Sachet	Mushrooms (*Pleurotus Ostreatus*)	40	95	Higher visual quality indexes. Better maintained firmness. No microbial growth. Extended shelf life from 4 to 14 day. Reduced moisture condensation	Higher weight loss	Yue Bi et al. (2014)

Super absorbent polymer (60:40 w/w) +ß-lactose	Sachet	Mature green tomato (*Lycopersicon esculentum* L.)	3.3	93	Higher visual quality indexes. Better maintained firmness. No microbial growth. Extended shelf life from 21 to 35 day. Reduced moisture condensation	Higher weight loss	Yue Bi et al. (2014)
Cellulose	Sheets of tissue paper	Eggplant fruit (*Solanum melongena* L.)	4 sheets/12–14 fruits	93	Reduced chilling injury and fungal growth. Higher firmness	None	Fallik et al. (1995)

super absorbent polymer. These compounds are generally evaluated individually, but in one study they have been blended to design a desired moisture controller (Mahajan et al., 2008). In general, these compounds are enclosed in pouches, sachets, or pads that are then placed inside of the package, but in a couple of studies they have been incorporated directly into the packaging material (Rux et al., 2015; Rux et al., 2016). Materials used to house the sorbants must be highly permeable to water vapor and include paper, woven polyester, spun-bonded polyolefin, mesh cloth, and breathable or perforated films. Several of the studies listed in Tables 1 and 2 show that the use of humidity-sorbing packaging and humidity-regulating packaging can improve produce shelf life by lowering microbial load, retaining color, and maintaining firmness. Thus some of the proposed approaches have been used to commercialize produce (e.g., sachet of humectant salt in the US tomato market) (Rooney, 1995). Other studies report no quality improvement or even accelerated produce deterioration compared with controls having no moisture absorber/adsorber. However, it is noteworthy that the majority of the work compiled in these tables is the result of experiments with random amounts of absorber/controller.

The incorporation of water vapor absorbers/controllers into the packaging material should be more broadly investigated since this approach maximizes the surface area for absorption/adsorption for a given amount of compound and consequently, higher and/or faster water vapor absorption/adsorption could be expected. Additionally, this would allow manufacturers to continue using the packing equipment commonly used for nonactive packaging and consumers would appreciate the absence of foreign objects within the package that could be ingested by mistake. Incorporation of sorbants into the packaging material could be done using conventional plastic processing technologies such as extrusion and coating, although only adsorbents/absorbents with a high thermal resistance could be used in the case of extrusion. Additionally, the plastic containing the adsorbent/absorbent should be a poor barrier to water and be protected by a water-barrier outer layer to limit water passage from the surrounding atmosphere (López-Rubio et al., 2004). Modified humidity packaging made of packaging materials rather than plastic has rarely been developed. Patterson and Joyce (1993) designed a package based on carton able to take up water vapor when the temperature drops (relative humidity rises) and release water vapor when the temperature rises in response to the lower relative humidity. This package was tested on carrots cooled to 3°C from 10°C or 22°C (Patterson et al., 1993).

Information on commercial products (i.e., sachets to absorb water vapor, drip-absorbent pads, lids to absorb liquid water) available to develop modified humidity packaging can be found in Table 3. The most common moisture-absorbing product, drip-absorbent pads, consists of paper fluff and/or a super-absorbent polymer located between two layers of a microporous or nonwoven polymer. More advanced moisture-absorbing materials include lids and trays containing the absorbent compound protected by mesh cloth to allow the liquid

TABLE 3 Commercially available products for development of active packaging for fruits and vegetables

Brand name	Company	Format	Purpose
Dri-loc	Novipac	Pad	Absorption of juice on the package bottom
Fresh-R-Pax	Maxwell Chase Technologies, LLC	Pad, lid, pouch	Absorption of juice on the package bottom
MiniPax	Multisorb Technologies	Sachet	Moisture absorption and odor control
StripPax	Multisorb Technologies	Sachet	Moisture absorption
AGELESS Type ZPT	Mitsubishi Gas Chemical America	Sachet	Oxygen absorber
OxyFRESH	EMCO Packaging	Sachet	Oxygen emitter/carbon dioxide absorber
AGELESS Type GL	Mitsubishi Gas Chemical America	Sachet	Oxygen absorber/carbon dioxide emitter
UltraZapXtendaPak	Paper Pak Industries	Pad	Carbon dioxide emitter/absorption of juice
UltraFreshSystem	Bunzi Distribution	Package	Carbon dioxide emitter/absorption of juice
Green bags	Evert Fresh	Bag	Ethylene absorption
It's fresh!	Food Freshness Technology Limited	Sheet	Ethylene absorption
PEAKfresh	PEAKfresh	Bag, carton liner, pallet cover	Ethylene absorption
Dri-Fresh ABV	Sirane	Pad	Microbial growth control/absorption of juice
ioFresh freshness extender	IoTech International	Pad	Microbial growth control/absorption of juice
Irgaguard	BASF Catalysts LLC	Powder	Microbial growth control

Continued

TABLE 3 Commercially available products for development of active packaging for fruits and vegetables—cont'd

Brand name	Company	Format	Purpose
OZOfresh	EMCO Packaging	Sachet	Microbial growth control
Sanocoat	Mondi	Coating	Microbial growth control
N/A	LINPAC	Coating	Microbial growth control
ANTI BACTERIAL BOPP FILM	NOW Plastics	Film	Microbial growth control

water to interact with the absorbent but not the produce. Trays made of multilayer materials with an absorbing core layer like LINstar and LINdri trays, both manufactured by LINPAC Packaging Ltd., are currently available for meat and fish products, but not for produce.

Ethylene-removing packaging

Ethylene is a gas molecule that occurs naturally in produce. It is considered a plant hormone due to its effect on ripening at low concentrations and its dose-dependent effect (Zagory, 1995). Changes caused in produce by ethylene can be either beneficial or detrimental depending on the type of produce and its ripening stage. Detrimental effects of ethylene include accelerated respiration rate and softening, production of bitter compounds, abscission of leaves and flowers, hastened toughening of vegetables, increased pathogen susceptibility, promotion of discoloration, sprouting stimulation, russet spotting, all factors that result in a net reduction of the produce shelf life (Payasi and Sanwal, 2010; Saltveit, 1999; Zagory, 1995). These deleterious effects can be enhanced in an enclosed space such as that provided by packaging through the accumulation of ethylene as it is generated by the produce. Typically, unless there is some form of scrubbing, ethylene exposure rises over time.

Although packaging technologies like MAP allow the reduction of ethylene synthesis and response by exposing the produce to depleted O_2 and/or enriched CO_2 levels, the required gaseous levels may not be the most beneficial for the crop in terms of other physiological responses (Zagory and Kader, 1988). Additionally, generation and maintenance of optimal O_2 and CO_2 levels to reduce ethylene synthesis and response is challenging during handling and transportation, limiting the effectiveness of this technology.

As an alternative to reducing ethylene production and/or response, ethylene can be removed from the package headspace while being produced. Ethylene-removing packaging is a type of active packaging that allows this due to the presence of compounds that can adsorb, absorb, counteract, or chemically alter ethylene. This packaging technology makes possible simultaneous ethylene reduction and exposure to desired O_2 and CO_2 levels during produce transport, handling, and sales. Thus for high-value, ethylene-sensitive produce, ethylene-removing packaging can be a cost-effective technology to extend shelf life.

Compounds with ethylene-removing capacity have been compiled and discussed in excellent literature reviews focused on ethylene removal (Keller et al., 2013; Matinez-Romero et al., 2007). The rate at which these compounds can remove ethylene and the total amount of ethylene that they can remove are both critical to obtain and maintain ethylene levels below deleterious ones for desired storage periods. Desired rate and total amount will be dependent on the type of commodity as crops differ in sensitivity to ethylene. Information on ethylene production and sensitivity of several crops is available in the Agriculture Handbook Number 66 (Gross et al., 2016). Additionally, factors that relate to the package including permeability to ethylene and dimensions have to be taken into consideration to develop appropriate ethylene-removing packaging. Information on the permeability to ethylene of plastics commonly used to package produce can be found in East et al. (2015).

Applications of compounds with ethylene-removing capacity used to develop ethylene-removing packaging are listed in Table 4. This table compiles and summarizes relevant past and current published research work on ethylene-removing packaging in plastic packaging formats. Experimental conditions used by several authors varied regarding the polyolefin type [polyethylene (PE) and polypropylene (PP)], film thickness, temperature, relative humidity, type and amount of ethylene scrubber, carrier type, integration into the package, type of commodity, and so on. Compounds with ethylene-removing capacity used to develop ethylene-removing packaging include potassium permanganate ($KMnO_4$), activated charcoal, zeolites, and palladium (Pd). Charcoal and zeolites have also been used as carriers for $KMnO_4$ and Pd as have aluminum oxide, clays, and glass beads. In general, the active compound, immobilized in its carrier, has been placed in a sachet, petri dish, or tray, which is then placed inside of the package. However, one study covered the incorporation of the ethylene scrubber directly into the packaging material (Esturk et al., 2014). Regardless of the experimental conditions, all of the studies listed in Table 4 concluded that polyolefin packaging, together with an ethylene scrubber and low temperature, led to a considerable increase in fruit shelf life by reducing softening and color changes, among others. Ethylene content was enormously reduced in all studies. However, in most cases, ethylene content is not below the threshold level (0.1 μL/L) for biological activity (Kader, 1985).

Few types of nonplastic ethylene-removing packaging have been developed. Bhutia et al. (2011) designed a package based on corrugated fiberboard boxes

TABLE 4 Studies on ethylene-removing packaging for fruits and vegetables

Compound	Compound carrier as defined by authors	Produce	Amount (g scavenger/ amount of fruit)	Package design	Storage conditions (time, RH, and temperature)	Ethylene concentration (μL/L) W and W/O scavenger	Improved quality compared to control (no ethylene scavenger)	Promoted deterioration compared to control (no ethylene scavenger)	References
$KMnO_4$	$KMnO_4$ on aluminum oxide placed on a shallow tray	Apples (*Malus pumila*) cv. Granny Smith	180/1	PE bag	14 day at 0.5°C followed by 13 weeks at 5°C followed by 7 day at 20°C	W/O: 90–140 W: 0.7–5	Less in-package CO_2, bitter pit, and superficial scald	None	Shorter et al. (1992)
	$KMnO_4$ on porous clay placed inside a sachet	Strawberries (*Fragaria* x *ananassa* L.)	1.5/1 basket	Basket inside PE bag	11 day at 2°C	W/O: trace W: not detected	Better maintained firmness and TSS	None	Picon et al. (1993)
	$KMnO_4$ on aluminum oxide placed on a dish	Japanese pears (*Pyrus pyrifolia*)	150/1	LDPE bag	31 weeks at 0°C followed by 7 day at 20°C	W/O: 0.10–0.14 W: 0.01	Delayed loss of green color, and flesh and core browning, Juicier, crispier, and highly acceptable fruit by sensory panel	None	Szczerbanik et al. (2005)
	Expanded sachets containing vermiculite impregnated with $KMnO_4$	Papaya (*Carica papaya)* cv. "Sunrise Golden"	0.5, 1, 1.5, or 2/3	LDPE bag	25 day at 90% and 10°C followed by 6 day at 21°C	Not specified	Extension of preclimacteric period. Less in-package CO_2, color change, fresh matter loss, softening, electrolyte leakage, and total soluble solids change	None	Silva et al. (2009)

	Glass beads coated with $KMnO_4$ placed on a petri dish	Fresh-cut iceberg lettuce (*Lactuca sativa* L.)	Not specified/ 0.1 kg	PP/EVA tray sealed with high barrier film	9 day at 7°C followed by 6 day at 21°C	W/O: 1.7 W: ~0.0	Lower russet sprouting	None	Lopez-Galvez et al. (2015)
Charcoal	Granular-activated carbon inside sachet	Tomato (*Lycopersicon esculentum* L.) cv. Beef	5/40	OPP bag	28 day at 8°C and 90%	W/O: 20 W: 11	Delayed color, softening, weight loss, CO_2 and O_2 changes. Less decay and off-flavor development. Higher scores for sweetness, juiciness, odor, and flavor by sensory panel	None	Bailen et al. (2006)
	Granular-activated carbon with palladium acetate (1%) inside sachet	Tomato (*Lycopersicon esculentum* L.) cv. Beef	5/40	OPP bag	28 day at 8°C and 90%	W/O: 20 W: 8	Delayed color, firmness, weight loss, CO_2 and O_2 changes. Less decay and off-flavor development. Higher scores for sweetness, juiciness, odor, and flavor by sensory panel	None	Bailen et al. (2006)
	AACP-$PdCl_2$-$CuSO_4$ inside sachet	Broccoli florets (*Brassica oleracea* var. *italica*) cv. Youxiu	1, 2, or 3/ 0.1 kg	Plastic basket with a dish containing 500 mL 1 M KOH covered and sealed with PE film	6 day at 20°C and 90%	W/O: 0.91 W 1: 0.21 W 2: 0.16 W 3: 0.15	Delayed yellowing, chlorophyll loss, soluble protein degradation, accumulation of free ACC and endogenous ethylene production. Shelf life extended by 2 day	None	Cao et al. (2015)

Continued

TABLE 4 Studies on ethylene-removing packaging for fruits and vegetables—cont'd

Compound	Compound carrier as defined by authors	Produce	Amount (g scavenger/ amount of fruit)	Package design	Storage conditions (time, RH, and temperature)	Ethylene concentration (μL/L) W and W/O scavenger	Improved quality compared to control (no ethylene scavenger)	Promoted deterioration compared to control (no ethylene scavenger)	References
Zeolites	Film resulting from mixing masterbatch PE/F 104010—Tazetut with PE PETILEN F2-12 resin (8%w/w)	Broccoli florets (*Brassica oleracea* var. *Italica*)	Not specified/ 0.3 kg	LDPE bag	20 day at 4°C and 75%	W/O: 61.8 W 1: 0.33	Higher headspace oxygen and sensory scores for odor and overall appearance. Shelf life extension of 15 day	Higher headspace carbon dioxide	Esturk et al. (2014)
Palladium	Palladium-impregnated zeolite (registered as e+ Ethylene Removal) placed in a petri dish	Preclimacteric avocado (*Persea americana* Mill.) cv. Hass	5/3	Hermetically sealed PP box	21 day at 5°C followed by 6 day at 20°C	W/O: 0.70 W 1: 0.02	Firmness and color retention at 5°C. Greater maintenance of an hexose and its polyol at 20°C	None	Meyer and Terry (2010)

with cushioning paper shredding and four sachets containing $KMnO_4$-impregnated material to evaluate the effect of ethylene absorption on marketability extension of sapota at different maturity stages. The presence of the ethylene scrubber delayed the ripening-related rise in ethylene and CO_2 production, weight loss, softening, and decay, with a maximum response in mature sapota compared to half-ripe and ripe sapota (Bhutia et al., 2011). This study shows the potential of extending mature sapota marketable life up to 13 days at ~25°C if the fruit is stored in corrugated fiberboard boxes containing sachets with $KMnO_4$. In contrast, tomatoes in cardboard boxes containing an ethylene absorbent based on $KMnO_4$ (e-Fresh pack, Donghu Farm Hann) and sealed with PE film for 24 h and then stored without the film for 15 days at 20°C had the same changes in quality as the control (no e-Fresh pack) over storage (Park et al., 2016). However, the use of two 1-methylcyclopropene kits instead of e-Fresh pack resulted in tomatoes with delayed softening and color changes (Park et al., 2016). The use of a palladium chloride-based scrubbing system to extend marketability of sliced kiwifruit and banana, chopped broccoli, and hand-torn pieces of spinach at 20°C was studied by Abe and Watada (1991). Palladium chloride in paper packets sorbed all of the produced ethylene until the end of storage, maintained firmness, and reduced CO_2 production in sliced banana and kiwifruit, and minimized degradation of chlorophyll in spinach pieces, but not in chopped broccoli. Palladium was also tested by Elmi et al. (2011) to extend the shelf life of preclimacteric avocado in commercial settings. Avocados were packaged in crates (4 kg) with or without sheets containing palladium-impregnated zeolite [registered as e+ ethylene removal sheets (Johnson Matthey Plc., United Kingdom)], palletized, and transported to the United Kingdom in a standard sea container with controlled atmosphere (4 kPa O_2, 6 kPa CO_2) for 5 weeks at 5°C. Upon arrival in the United Kingdom, the fruit was treated with e+ ethylene removal powder or not and stored in boxes for 4 weeks at 5°C. The treatment of avocado with e+ ethylene removal sheet and powder together resulted in greater reduction of ethylene levels, softening, and color changes compared with either material alone (Elmi et al., 2011).

The e+ ethylene removal technology is the active component of the It's Fresh! filter (Food Freshness Technology Holdings Limited, United Kingdom). This technology is unique in its method of ethylene removal as it actively removes and locks away the ethylene molecules while other technologies oxidize them. Fig. 4 shows the many possible formats of this ethylene-removing product. The same figure shows how the It's Fresh! filter is being used in combination with corrugated and plastic packaging formats to extend the quality, freshness, and flavor of fruit and provide shelf life extension, more evenly ripened fruit, and a better consumer experience for berries, cherries, tomatoes, pears, stone fruit, avocado, broccoli, and asparagus (http://www.itsfresh.com/). Currently, this filter can be found attached to a soaker pad (as used for berries) to allow simultaneous ethylene and moisture absorbency when needed. This and

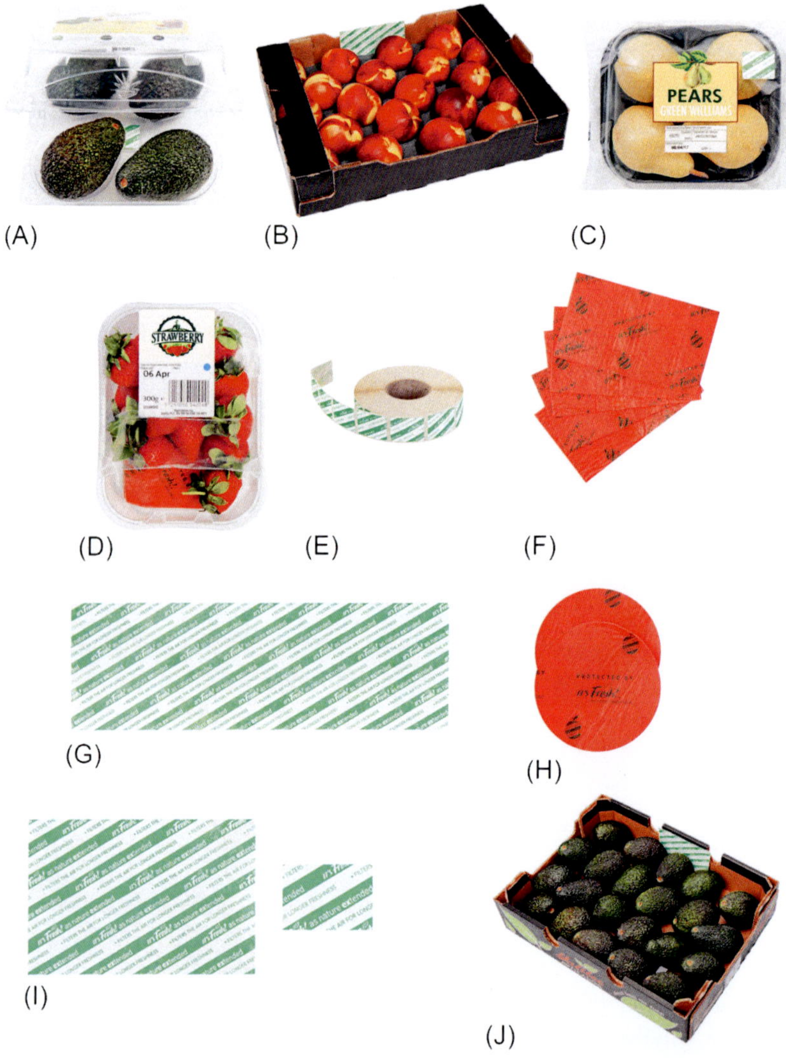

FIG. 4 (A) Filter in avocado retail pack, (B) transit filter in nectarine export carton, (C) filter in pear retail pack, (D) filter integrated into a berry pad in strawberry retail pack, (E) self-adhesive labels on roll, (F) filter integrated into a berry pad, (G) large transit filter, (H) filter integrated into a round berry pad, (I) small transit and retail filter, (J) transit filter in avocado export carton. *(Used with permission from It's Fresh, Ltd.)*

other commercial products available for ethylene-removing packaging can be found in Table 3. The most common ethylene-removing product is a sheet of a laminated ethylene remover composite material that is inserted into the package; however, bags, carton liners, and pallet covers are also available. Retailers like Morrisons, Waitrose, Tesco and Marks & Spencer (United Kingdom),

Carrefour (France), Walmart, Albertsons, and Safeway (United States), and their supply chain suppliers have used the It's Fresh! filters to extend the shelf life of packaged and bulk fruit.

Oxygen-removing packaging and carbon-dioxide-removing packaging

The physiological process of respiration results in the consumption of oxygen (O_2) and the production of carbon dioxide (CO_2). The amount of O_2 consumed and CO_2 produced depends on the type of produce and on specific conditions, such as packaging. When fruits and vegetables are packaged in a sealed container, the ambient air inside the package is modified due to the interplay between produce respiration, which replaces O_2 with CO_2, film permeability to O_2 and CO_2, and storage conditions (Almenar et al., 2006). The resulting atmospheres are usually not optimal for produce since most commercial films do not allow for the necessary CO_2 and O_2 exchange between package and environment (Zagory and Kader, 1988). Sometimes, O_2 and CO_2 levels go beyond the limits tolerated by produce, which results in physiological disorders, off-flavor development, increased susceptibility to decay, and other anaerobiosis-related phenomena (Almenar et al., 2007a; Kader et al., 1989). Furthermore, the high CO_2 levels can also cause package collapse (Lee, 2016). However, increased/decreased CO_2/O_2 levels within the limits tolerated by produce are desired in the package headspace since they can slow down respiration, decrease ethylene biosynthesis and action, reduce color changes, enhance texture, minimize vitamin loss, and delay bacteria and fungal growth (Almenar et al., 2006; Kader et al., 1989). Information on recommended CO_2/O_2 levels for shelf life extension of several crops is available in online databases like the Agriculture Handbook Number 66 (Gross et al., 2016) and the UC Davis website on "Commodity Fact Sheets" (UCDavis, 2016). An additional benefit of the increase of CO_2 in the package headspace is that this can help prevent the partial vacuum caused by high-respiring produce due to their fast consumption of O_2 (López-Rubio et al., 2004).

Commercial films can be microperforated to obtain improved gas exchange that reduces CO_2 buildup and increases O_2 inside a package containing produce (Almenar et al., 2007a; Koutsimanis et al., 2015). However, they may not provide the gas exchange required to respond to produce respiration rate changes caused by a cold chain with fluctuating temperatures and consequently may lead to anaerobiosis (Alique et al., 2003; Koutsimanis et al., 2015). Alternatively, films with a poor gas barrier can be used to mitigate too high/low levels of CO_2/O_2, respectively, developed in packages containing produce. However, they may not allow the development of the target-modified atmospheres that produce can benefit from due to a high exchange of CO_2 and O_2 between package and environment. Rather than changing the film permeability in order to modify the package atmosphere, the ambient air inside the package can be replaced with a specific mixture of CO_2/O_2 by direct flushing prior to package sealing (active MAP). However, it is important to note that initial gas levels will evolve over time if respiration rate and O_2 and CO_2 permeation are not balanced and/or temperature is not constant.

Active packaging with the capability of scavenging CO_2 or O_2 can be used to obtain optimal CO_2 and/or O_2 levels for produce over storage. This type of packaging couples CO_2 or O_2 scavenging capabilities with film permeability and product respiration rate to control in-package CO_2 and/or O_2 levels. Thus factors that relate to the produce as well as factors related to the package, the CO_2 or O_2 scavenger, and the environment have to be taken into consideration to develop appropriate carbon-dioxide-removing packaging and oxygen-removing packaging. Fig. 5 compiles most of these factors. Information on respiration rates of several crops and on permeabilities to CO_2 and O_2 of plastics commonly used to package produce can be found in Kader et al. (1989) and Almenar and Auras (2010), respectively. Information on absorption kinetics (e.g., absorption capacity, absorption rate constant, energy of activation, effect atmosphere, relative humidity, and temperature) of different CO_2 or O_2 scavengers can be found in Tewari et al. (2002), Charles et al. (2006) and Lee (2016).

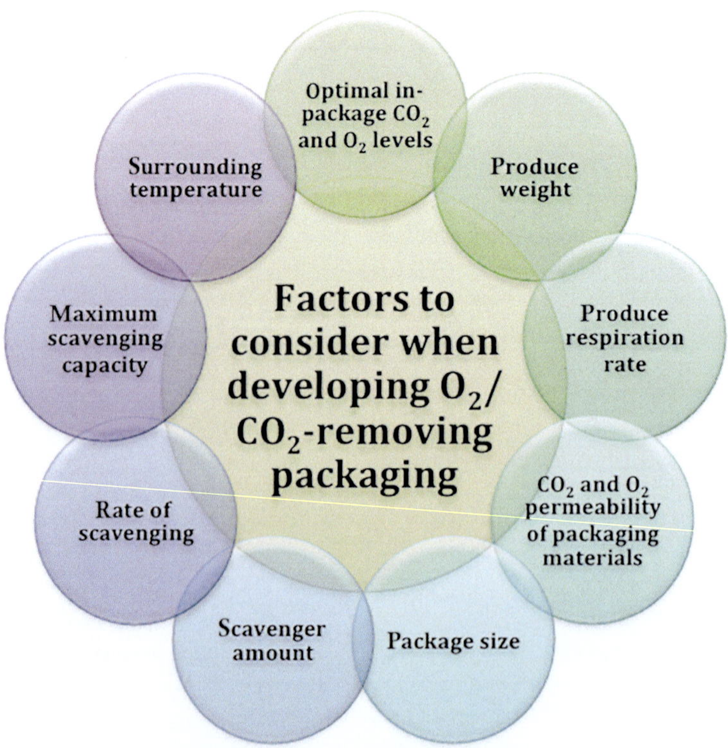

FIG. 5 Factors to be taken into consideration when developing carbon-dioxide-removing packaging and oxygen-removing packaging.

Tables 5 and 6 compile and summarize relevant past and current published research work on oxygen-removing packaging and carbon-dioxide-removing packaging for produce, respectively, in plastic packaging formats. Only iron powder has been used to develop oxygen-removing packaging while several scavenging compounds like activated charcoal, sodium carbonate, and sodium glycinate have been utilized when developing carbon-dioxide-removing packaging. In general, both types of scavengers are enclosed in sachets that are then placed inside of the package, but in one study several CO_2 scavengers are incorporated directly into an agar-based film (Wang et al., 2015). Materials used as enclosing systems for the scavenging compounds are highly permeable to O_2 and CO_2 and include woven polyester, muslin cloth, and agar. The studies listed in Table 5 report that the use of oxygen-removing packaging can improve produce shelf life by maintaining pH, color, and TSS due to the faster reduction of headspace CO_2 and O_2 levels compared to control packages. However, final gas compositions of controls and treatments do not differ much and are not optimal for produce in most cases. Additionally, these studies are limited to a couple of crops and therefore oxygen-removing packaging should be further investigated in order to determine its effectiveness on extending shelf life of other perishables. The studies listed in Table 6 report that the use of carbon-dioxide-removing packaging can significantly reduce headspace CO_2 and consequently, slow changes in firmness, color, TSS, pH, bacterial growth, flavor, and browning which result in an extended produce shelf life. In general, the developed packages maintained proper O_2 and CO_2 levels or avoided the development of CO_2 levels that can cause injury in produce.

Information on commercial sachets containing O_2 and/or CO_2 scavengers and emitters available to modify the O_2 and CO_2 levels in the package headspace can be found in Table 3. For example, OxyFRESH (EMCO Packaging) sachets emit O_2 into the package while absorbing CO_2 simultaneously. The controlled release of oxygen is tailored to the respiration rate of the packaged produce. Achieved shelf life extension of strawberries, cherries, potatoes, and broccoli when used OxyFRESH is claimed by the company (emcopackaging.com/index.php/products/oxygen-emitters). A different approach is the Ultra-ZapXtendaPak pad (Paper Pak Industries, United States), which releases CO_2 when activated by the absorption of the juices (liquid water) given off by fresh-cut produce. The company claims that the active absorbent slows down produce respiration and microbial growth as well as absorbs excess juice in packages of fresh-cut tomatoes, onions, watermelon, pineapple, cantaloupe, and honeydew (http://paperpakindustries.com/food/ultrazap_xtendapak_onions.php). This pad is commercialized individually or as part of UltraFreshSystem. The latter is a package available by Bunzl Distribution (United States) that is made of a custom-designed Sonoco tray that features an insert that quickly channels the liquid away from the product and toward the pad and allows better, faster, and lower temperature seals while being made with less plastic than alternative produce trays as claimed by the company (http://paperpakindustries.com/food/ultrafreshsystem.php).

TABLE 5 Studies on oxygen-removing packaging for fruits and vegetables

Compound	O_2 scavenger as defined by authors	Produce	Amount (possible maximum amount (ml) of scavenged O_2/kg fruit)	Package design	Storage time (weeks)/temperature (°C)	Final in-package headspace compositions (CO_2/O_2) W and W/O scavenger	Improved quality compared to control (no O_2 scavenger)	Promoted deterioration compared to control (no O_2 scavenger)	References
Iron powder	ATCO-210	Strawberries (*Fragaria x Ananassa*) cv. Camarosa	210/0.18	PVC/PE trays sealed with BOPP	5/4	W: ~40/1 W/O: ~40/1	No pH change during 3 weeks. Reduced CO_2 and O_2 contents up to week 3. Higher sensory scores for appearance	None	Kartal et al. (2012)
	ATCO-210	Strawberries (*Fragaria x ananassa*) cv. Camarosa	210/0.18	PVC/PE trays sealed with microperforated BOPP (7 and 9 holes)	5/4	W: ~15/3 W/O: ~15/3	No pH change during 3 weeks. Reduced CO_2 after week 3. Reduced O_2 up to week 3	None	Kartal et al. (2012)

	ATCO-100 and ATCO-210	Strawberries (*Fragaria x ananassa*) cv. Camarosa	100/0.2; 210/0.2;	PLA tray with nonspecified closure	4/4	W 100: ~26/6 W 210: ~26/6 W/O: ~36/0 (week 3)	Reduced CO_2 content (respiration rate), lower pH, higher TSS, less cell membrane damage and color change, higher firmness, less fungal spoilage, and better sensory scores	None	Aday and Caner (2013)
	ATCOLH100	Tomato (*Lycopersicon esculentum* L.) cv. Grace	100/0.8	LDPE pouch	1/20	W: ~5/7 W/O: ~4/7	Reduction of the transient period duration and elimination of the CO_2 peak	None	Charles et al. (2003)

TABLE 6 Studies on carbon dioxide-removing packaging for fruits and vegetables

Compound	CO_2 scavenger as defined by authors	Produce	Amount (number of sachets/number of fruits)	Package design	Storage time (weeks)/temperature (°C)	Final in-package headspace compositions (CO_2/O_2) W and W/O scavenger	Improved quality compared to control (no CO_2 scavenger)	Promoted deterioration compared to control (no CO_2 scavenger)	References
Charcoal	2g of granular charcoal wrapped with muslin cloth	Avocado (*Persea americana*) vs. Pollock	1/3	LDPE bag (0.05-mm thickness with potassium-based C_2H_4 scavenger)	4/12	W: ~3.5/3.5 W/O: ~5/3.5	Shelf life extended by 12 day based on delayed softening and better maintained TSS and "b" value	None	Illeperuma and Nikapitiya (2002)
46% sodium carbonate peroxyhydrate + 24% sodium carbonate + 14% sodium chloride + 16% bentonite clay	EMCO-A	Strawberries (*Fragaria x ananassa*) cv. Camarosa	1/200g	PLA tray sealed with a nonspecified film	4/4	W: ~16/5 W/O: ~48/5	Lower increase in pH, electro conductivity and better maintained TSS and firmness. Higher acceptability by sensory panel	None	Aday et al. (2011)
50% sodium carbonate peroxyhydrate + 20% sodium carbonate +14% sodium chloride +16% bentonite clay	EMCO-B	Strawberries (*Fragaria x ananassa*) cv. Camarosa	1/200g	PLA tray sealed with a nonspecified film	4/4	W: ~14/5 W/O: ~48/5	Lower increase in pH, electroconductivity, and better maintained TSS and firmness. Higher acceptability by sensory panel	None	Aday et al. (2011)

Limestone	Kraft paper envelops	Spinach (*Spinacia oleracea*)	Not specified	PET coated with aluminum oxide with air, 95% O_2, or 15% $CO_2+5\%O_2$	1/4 followed by 0.5/10	W: ~22.18/0.16 W/O: ~0.03/0.13	None	Increased respiration rate and growth of *Escherichia coli* O157:H7	Page et al. (2013)
Sodium carbonate	Agar-based multifunctional film with 10.6 g of sodium carbonate	Shiitake mushrooms (*Lentinula edodes*)	0.0355 m^2/200 g	HDPE tray inside a bag made of OPP with eight perforations of 98-μm diameter	1/10	W: ~2/3 W/O: ~19/3	Better maintained firmness, higher flavor score, and lower bacteria load	Higher weight loss	Wang et al. (2015)
Sodium glycinate	Agar-based multifunctional film with 9.67 g of sodium glycinate	Shiitake mushrooms (*Lentinula edodes*)	0.0355 m^2/200 g	HDPE tray inside a bag made of OPP with eight perforations of 98-μm diameter	1/10	W: ~9/3 W/O: ~19/3	None	Higher weight loss, browning, and bacteria load	Wang et al. (2015)
Sodium carbonate + Sodium glycinate	Agar-based multifunctional film with 5.3 g of sodium carbonate and 4.85 g of sodium glycinate	Shiitake mushrooms (*Lentinula edodes*)	0.0355 m^2/200 g	HDPE tray inside a bag made of OPP with 8 perforations of 98-μm diameter	1/10	W: ~14/3 W/O: ~19/3	Better maintained firmness and higher flavor score	Higher weight loss	Wang et al. (2015)
Not specified	Ageless CO_2 adsorber	Pear (*Pyrus communis* L.) vs. Silver bell	1/1	Bag made of MA-A film	28/1	W: ~8/6 W/O: ~8/6	Less internal browning during 3 months	Yellower and softer fruit	Nugraha et al. (2015)

Antimicrobial packaging

Spoilage and pathogenic microorganisms can alter produce quality and safety, respectively. Spoilage microorganisms include fungi and bacteria, the former more commonly found in fruits and the latter in vegetables (Sholberg and Conway, 2016). A list of the main postharvest diseases of fruits and vegetables is available in the Agriculture Handbook Number 66 (Sholberg and Conway, 2016). Spoilage microorganisms shorten the shelf life of produce, which may force consumption of the commodity before it has reached its optimal palatability. In addition, they spoilage change the sensory and nutritional properties of the commodity, and consequently, decrease consumer satisfaction and/or turn the commodity into food waste. In contrast, pathogens (e.g., *Escherichia coli*, *Salmonella*, *Listeria monocytogenes*) do not shorten produce shelf life but cause illnesses and death, which additionally results in significant expenses for health care and in low levels of productivity (Scharff, 2012). Heat, pressure, or other means that are commonly used to drastically reduce microbial load in other food categories cannot be used in produce because it is sold as a raw item. Thus alternative means to reduce microorganisms in produce have become a major priority for the produce industry. Sanitizing agents are commonly used to reduce the microbial loads prior to packaging (Page et al., 2016; Sholberg and Conway, 2016). However, their direct utilization on the produce surface by dipping or spraying may result in a rapid diffusion through the produce tissue that limits their effect on surface microbiota (Almenar et al., 2006). Ideally, the sanitizer at its minimum inhibitory concentration should be exposed to the produce surface for the required period in order to avoid microbe growth. Antimicrobial packaging can make this possible.

Antimicrobial packaging involves incorporating an active compound with an antimicrobial capacity into the package in order to retard, reduce, or inhibit microbial growth during a desired period of time. Compounds with antimicrobial activity have been compiled and discussed in excellent literature reviews and book chapters focused on antimicrobials for produce and on antimicrobial packaging for all types of food categories (Lucera et al., 2016; Mari et al., 2016; Otoni et al., 2016; Van Long et al., 2016). These compounds can be either volatile or nonvolatile. The majority of the antimicrobial packages developed for produce are release systems for volatile compounds. Use of volatiles enhances the interaction of the compound with the produce relative to nonvolatile compounds that rely on direct contact. In addition, a more even distribution of the antimicrobial over the commodity is achieved by the use of a volatile compound. The method of integration of the antimicrobial compound into the package has been changing over time. Initially, its incorporation was through the use of a filter paper (Wang and Buta, 2003). Subsequently, antimicrobial volatiles were incorporated in sachets (Almenar et al., 2007b) and lately, they are blended into the polymer bulk or coated onto the inner surface of the packaging material using a matrix that acts as a carrier to make the antimicrobials part of the packaging material (Muriel-Galet et al., 2013). Most recently, absorbent pads with a dual function, water absorption that activates antimicrobial release, are being evaluated by academia and used by industry (Almenar and Wilson, 2016). More information on emitting sachets and absorbent pads for antimicrobial food packaging applications can be found in Otoni et al. (2016).

The volatile is released from the paper, sachet, polymer matrix, or coating into the package headspace; diffuses throughout the air until reaching the fruit or vegetable surface; and then desorbs into it. The amount of antimicrobial in the package headspace depends on multiple factors other than fruit sorption and includes the amount released from the emitting source (e.g., diffusion through the polymer, desorption from the polymer surface), the amount absorbed into the packaging material, the amount permeated through pores in the case of microperforated packages, and temperature (Almenar et al., 2007b). Therefore release kinetics and diffusion kinetics of the volatile need to be taken into consideration. Additionally, the type of microorganism/s, initial microbial load, and volatile minimum inhibitory concentration need to be considered when developing antimicrobial packaging. A mathematical model that predicts the mass transfer phenomena that take place in an antimicrobial package for produce has been developed and validated experimentally (Cerisuelo et al., 2012). In this study, the package consists of PP film coated with a thin layer of EVOH-29 with 5% carvacrol as antimicrobial and contains ready-to-eat salad. The authors suggest that this model can lead to the mathematical simulation and prediction of similar packaging approaches as long as experimental data for all the parameters and coefficients involved are collected (Cerisuelo et al., 2012).

As mentioned previously, the ideal scenario is when the produce surface is exposed to the antimicrobial at its minimum inhibitory concentration for the period of time required to control the target organism. This is possible in the case of antimicrobial packaging with a controlled release of the antimicrobial compound. This type of antimicrobial packaging avoids the fast consumption of the antimicrobial, exposes the produce to the desired volatile minimum inhibitory concentration, allows antimicrobial release during a desired period, reduces exposure of the commodity to high levels of volatile and possible alteration of the produce sensory characteristics, and facilitates reproducible results on produce shelf life extension. Techniques used for controlled release in active packaging can be found in Nerin et al. (2016).

Table 7 compiles and summarizes relevant past and current published research work on antimicrobial packaging for fruits and vegetables in plastic and paper packaging formats. Compounds evaluated for antimicrobial packaging include alcohols (e.g., ethanol), aldehydes (e.g., trans-cinnamaldehyde), ketones (e.g., 2-nonanone), acids (e.g., cinnamic acid), chlorine dioxide, and plant essential oils (e.g., oregano oil). These studies are either about one single compound or compare a few of them, but in one study they have been blended to investigate synergism or antagonism (Ayala-Zavala et al., 2008). Alumina,

TABLE 7 Studies on antimicrobial packaging for fruits and vegetables

Compound	Compound carrier as defined by authors	Produce	Amount (μL/L antimicrobial/kg of fruit)	Package design	Storage conditions (time, RH, and temperature)	Ethylene concentration in package headspace at the end of storage	Improved quality compared to control (no antimicrobial)	Promoted deterioration compared to control (no antimicrobial)	References
Isopropyl alcohol	Filter paper strip	Sliced kiwifruit (*Actinidia deliciosa*)	300/not specified	PS tray with snap-fit lid	21 day at 15°C	Not specified	Microbial growth reduction	None	Wang and Buta (2003)
1-Propanol	Filter paper strip	Sliced kiwifruit (*Actinidia deliciosa*)	300/not specified	PS tray with snap-fit lid	21 day at 15°C	Not specified	Microbial growth reduction	None	Wang and Buta (2003)
Methyl alcohol	Filter paper strip	Sliced kiwifruit (*Actinidia deliciosa*)	300/not specified	PS tray with snap-fit lid	21 day at 15°C	Not specified	None	None	Wang and Buta (2003)
Ethyl alcohol	Filter paper strip	Sliced kiwifruit (*Actinidia deliciosa*)	300/not specified	PS tray with snap-fit lid	21 day at 15°C	Not specified	Microbial growth reduction	None	Wang and Buta (2003)
Methyl jasmonate	Filter paper strip	Sliced kiwifruit (*Actinidia deliciosa*)	2.24, 11.2 or 22.4/not specified	PS tray with snap-fit lid	15 day at 15°C	Not specified	Microbial growth inhibition with 11.2 and 22.4, and microbial growth reduction with 2.24	Higher levels of sugars and organic acids	Wang and Buta (2003)
	Filter paper strip	Sliced tomato (*Lycopersicon esculentum* L.) cv. Mill	22.4/0.1	PP tray heat-sealed with not specified film	15 day at 5°C in darkness	Not specified	Reduced microbial growth, less in-package CO_2, better maintained firmness, increased lycopene vit. C, and total phenolic contents	Less in-package O_2 level, decreased L* and h*	Ayala-Zavala et al. (2008)

Ethanol	Filter paper strip	Fresh-cut tomato (*Lycopersicon esculentum* L.) cv. Mill	300/0.1	PP tray heat-sealed with not specified film	15 day at 5°C in darkness	Not specified	Reduced microbial growth, less in-package CO_2, inhibited ethylene production, better maintained firmness and L* and h*	Decreased lycopene content	Ayala-Zavala et al. (2008)
Methyl jasmonate + ethanol	Filter paper strip	Fresh-cut tomato (*Lycopersicon esculentum* L.) cv. Mill	322.4/0.1	PP tray heat-sealed with not specified film	15 day at 5°C in darkness	Not specified	Reduced microbial growth, less in-package CO_2, inhibited ethylene production, better maintained firmness, increased vit. C, and total phenolic contents	None	Ayala-Zavala et al. (2008)
Tea tree oil	Filter paper strip	Fresh-cut tomato (*Lycopersicon esculentum* L.) cv. Mill	75/0.1	PP tray heat-sealed with not specified film	15 day at 5°C in darkness	Not specified	Reduced microbial growth, less in-package CO_2 and ethylene, better maintained firmness, increased vit. C and total phenolic contents	None	Ayala-Zavala et al. (2008)
Garlic oil	Filter paper strip	Fresh-cut tomato (*Lycopersicon esculentum* L.) cv. Mill	333/0.1	PP tray heat-sealed with not specified film	15 day at 5°C in darkness	Not specified	Reduced microbial growth, less in-package CO_2 and ethylene, increased vit. C and total phenolic contents	None	Ayala-Zavala et al. (2008)
	Tree/paper fiber cellulose tea bags containing ß-cyclodextrin-encapsulated garlic oil	Sliced tomato (*Lycopersicon esculentum* L.) cv. Mill	0.25, 0.5, or 1 g/0.1	PS container sealed with EVA film	15 day at 5°C	Not specified	Lower microbial load (0.5, or 1 g/0.1)	None	Ayala-Zavala and Gonzalez-Aguilar (2010)

Continued

TABLE 7 Studies on antimicrobial packaging for fruits and vegetables—cont'd

Compound	Compound carrier as defined by authors	Produce	Amount (μL/L antimicrobial/kg of fruit)	Package design	Storage conditions (time, RH, and temperature)	Ethylene concentration in package headspace at the end of storage	Improved quality compared to control (no antimicrobial)	Promoted deterioration compared to control (no antimicrobial)	References
Eugenol	Sterile gauze	Table grapes (*Vitis vinifera* L.) cv. Crimson Seedless	0.5 mL/0.15–0.17	OPP bag	35 day at 1°C and 90% RH in darkness	Not specified	Decreased ethylene content, weight loss, maturity index value, rachis deterioration, and microbial growth. Delayed changes in firmness, h* and L*. 3 weeks shelf life extension	Lower/higher in-package O_2 and CO_2 levels	Valverde et al. (2005)
	Sterile gauze	Table grapes (*Vitis vinifera* L.) cv. Autumn Royal	0.075 or 0.15 mL/not specified	OPP bag	56 day at 1°C and 90% RH followed by 2 day at 20°C and 90% RH	Not specified	Delayed changes in a* and L*, TSS, TA, sugars, acids, and total phenolics and anthocyanins in skin. Increased total antioxidant activity in both skin and pulp, and total phenolics in pulp Less fungal growth and higher sensory scores in fruit and rachis aspect and in firmness	Lower sensory scores in juiciness and sweetness	Valero et al. (2006)
	Nonwoven PLA sachet containing 1 g of PLA pellets impregnated with the active compound	Organic wild rocket (*Diplotaxis tenuifolia* L.)	0.1 g/0.1	PET tray wrapped with microperforated PP film	6 day at 5°C	Not detected	Better retention of green color. Lower off-odor and earthy odor scores	None	Wieczynska et al. (2016)

Menthol	Sterile gauze	Table grapes (*Vitis vinifera* L.) cv. Crimson Seedless	0.5 mL/0.15–0.17	OPP bag	35 day at 1°C and 90% RH in darkness	Not specified	Decreased ethylene content, weight loss, maturity index value, rachis deterioration, and microbial growth. Delayed changes in firmness, h* and L*. 3 weeks shelf life extension	Lower/higher in-package O_2 and CO_2 levels	Valverde et al. (2005)
Thymol	Sterile gauze	Table grapes (*Vitis vinifera* L.) cv. Crimson Seedless	0.5 mL/0.15–0.17	OPP bag	35 day at 1°C and 90% RH in darkness	Not specified	Decreased ethylene content, weight loss, maturity index value, rachis deterioration, and microbial growth. Delayed changes in firmness, h* and L*. 3 weeks shelf life extension	Lower/higher in-package O_2 and CO_2 levels	Valverde et al. (2005)
	Sterile gauze	Table grapes (*Vitis vinifera* L.) cv. Autumn Royal	0.075 or 0.15 mL/not specified	OPP bag	56 day at 1°C and 90% RH followed by 2 day at 20°C and 90% RH	Not specified	Delayed changes in L*, TSS, TA, sugars, acids, and total antioxidant activity, phenolics, and anthocyanins in skin. Increased total antioxidant activity and phenolics in pulp Less fungal growth and higher sensory scores in fruit and rachis aspect and in firmness	Lower sensory scores in juiciness and sweetness	Valero et al. (2006)
Cinnamic acid	Filter paper	Fresh-cut melon (*Cucumis melo* L.) var. Cantalupensis	0.15 or 0.30 g/L/0.25	PP tray heat-sealed with OPP film	10 day at 5°C and 90% RH	Not specified	Lower respiration rate, microbial growth, loss of total pholyphenols and vit. C	Higher ethylene production at the beginning of storage	Silveira et al. (2015)

Continued

TABLE 7 Studies on antimicrobial packaging for fruits and vegetables—cont'd

Compound	Compound carrier as defined by authors	Produce	Amount (μL/L antimicrobial/kg of fruit)	Package design	Storage conditions (time, RH, and temperature)	Ethylene concentration in package headspace at the end of storage	Improved quality compared to control (no antimicrobial)	Promoted deterioration compared to control (no antimicrobial)	References
2-nonanone	Metallocene PE sachet containing 0.1 g of alumina F-1 80/100 mesh impregnated with 2-nonanone	Wild strawberries (*Fragaria vesca*) cv. Reina de los Valles	0.0025, 0.005 or 0.01 mL/0.04	PP/EVOH/PP cup sealed with PET/PP lid with 3 microperforations	3 day at 10°C and 77% RH and 3 day at 22°C and 50% RH	0.05–0.5 mg/L depending on the amount of 2-nonanone and temperature	Delayed changes in TA, TSS, and anthocyanin content. Less weight loss, off-flavors, and *Botrytis* growth	None	Almenar et al. (2007b)
	Metallocene PE sachet containing alumina F-1 80/100 mesh impregnated with 2-nonanone	Wild strawberries (*Fragaria vesca* L.) cv. Reina de los Valles	0.0001, 0.0005, 0.001, 0.002 or 0.003 mL/0.04	PP/EVOH/PP cup sealed with PET/PP lid with 3 microperforations	4 day at 10°C and 77% RH	0.012–0.075 μL/L depending on the amount of 2-nonanone	Less weight loss and *Botrytis* growth. Delayed changes in TSS. Enhanced strawberry flavor	Higher TA depletion	Almenar et al. (2009a,b)
Carvacrol	Nonwoven PLA sachet containing 1 g of PLA pellets impregnated with the active compound	Organic wild rocket (*Diplotaxis tenuifolia* L.)	0.1 g/0.1	PET tray wrapped with microperforated PP film	6 day at 5°C	0.0021 μg/L	None	None	Wieczynska et al. (2016)
Oregano oil (80% carvacrol)	EVOH-29 as coating on PP film	Blend of leafy greens (lettuce, carrot, and red cabbage)	5% or 10%/not specified	PP/EVOH bag with a nonspecified modified atmosphere	5 day at 4°C followed by 3 day at 8°C	Not specified	Reduced spoilage and pathogenic microorganisms. Preferred visual appearance by sensory panel (5% oil)	Decreased acceptability with increased oil concentration	Muriel-Galet et al. (2012)

	EVOH-29 as coating on PP film	Blend of leafy greens (lettuce, carrot, and red cabbage)	7.5%/not specified	PP/EVOH bag flushed with 4% O_2 + 12% CO_2	5 day at 4°C followed by 3 day at 8°C	Not specified	Reduced growth of spoilage bacteria and fungi	Lower sensory scores for "odor acceptability" at the beginning storage	Muriel-Galet et al. (2013)
Trans-anethole	Nonwoven PLA sachet containing 1 g of PLA pellets impregnated with the active compound	Organic wild rocket (*Diplotaxis tenuifolia* L.)	0.1 g/0.1	PET tray wrapped with microperforated PP film	6 day at 5°C	2.35 μg/L	Lower off-odor and earthy odor scores	None	Wieczynska et al. (2016)
Cinnamal-dehyde-enriched cinnamon essential oil	Paper coated with paraffin containing the essential oil	Cultivated strawberries (*Fragaria ananassa* L.)	Not specified	Coated paper shaped into a tray	7 day at 4°C	Not specified	Reduced fungal growth for strawberries type B but not type A	None	Rodriguez et al. (2007)
Trans-cinnamaldehyde	Paper coated with paraffin containing the essential oil	Cherry tomatoes vr. Lupita	3% or 6%/0.25	Coated Kraft paper shaped into a box	7 day at 21°C	Not specified	Reduced *Alternaria* growth and pH increment	Reduced water activity	Rodriguez-Lafuente et al. (2010)
Oregano oil	Paper coated with paraffin containing the essential oil	Cherry tomatoes vr. Lupita	5.5%/0.25	Coated Kraft paper shaped into a box	7 day at 21°C	Not specified	Reduced *Alternaria* growth and pH increment	Strong flavor of oregano and bad taste	Rodriguez-Lafuente et al. (2010)
Cinnamon oil	Paper coated with paraffin containing the essential oil	Cherry tomatoes vr. Lupita	3% or 6%/0.25	Coated Kraft paper shaped into a box	7 day at 21°C	Not specified	Reduced *Alternaria* growth and pH increment	Reduced water activity	Rodriguez-Lafuente et al. (2010)
	Bottom of tray coated with active ingredient	Peach fruit (*Prunus persica* Sieb and Zucc (Jesca clone)	0.062 g/m^2 (oil/active solution)/ 0.990	Macroperforated PET tray coated with active ingredient with snap-fit lid	12 day at 21° C	Not specified	Reduced fungal growth and change in color difference, and activity of catalase	None	Montero-Prado et al. (2011)

Continued

TABLE 7 Studies on antimicrobial packaging for fruits and vegetables—cont'd

Compound	Compound carrier as defined by authors	Produce	Amount (μL/L antimicrobial/kg of fruit)	Package design	Storage conditions (time, RH, and temperature)	Ethylene concentration in package headspace at the end of storage	Improved quality compared to control (no antimicrobial)	Promoted deterioration compared to control (no antimicrobial)	References
	Self-adhesive PP label	Peach fruit (*Prunus persica* Sieb and Zucc (Jesca clone)	0.062 g/m^2 (oil/active solution)/0.990	Macroperforated PET tray with snap-fit lid containing a self-adhesive PP label	12 day at 21° C	Not specified	Reduced fungal growth, weight loss, changes in color difference and firmness, and activities of lipoxygenase, polyphenol oxidase, and catalase. Higher sensory scores	None	Montero-Prado et al. (2011)
Chlorine dioxide	Rinse	Cultivated strawberries (*Fragaria ananassa* L.) cv. Camarosa	5 or 10 ppm/0.2	PVC/PE trays sealed with PET/PP after addition of two sachets, one containing chabazite zeolite and the other containing silica gel, and flushing consisting of 8% O_2 + 15% CO_2	21 day at 3°C	Not specified	Reduced changes in pH, TSS, firmness, and in-package O_2 and CO_2	Lower "general acceptability" by sensory panel	Aday and Caner (2011)
Citral	EVOH-29 as coating on PP film	Blend of leafy greens (lettuce, carrot, and red cabbage)	5% or 10%/not specified	PP/EVOH bag with a nonspecified modified atmosphere	5 day at 4°C followed by 3 day at 8°C	Not specified	Reduced spoilage and pathogenic microorganisms	Low acceptability by consumer panel	Muriel-Galet et al. (2012)
	EVOH-29 as coating on PP film	Blend of leafy greens (lettuce, carrot, and red cabbage)	7.5%/not specified	PP/EVOH bag flushed with 4% O_2 + 12% CO_2	5 day at 4°C followed by 3 day at 8°C	Not specified	Reduced growth of spoilage bacteria and fungi	Lower sensory scores for "odor acceptability" at the beginning storage	Muriel-Galet et al. (2013)

ß-cyclodextrin, and PLA pellet have been used as carriers for some of these compounds (Almenar et al., 2007b, 2009a; Ayala-Zavala and Gonzalez-Aguilar, 2010; Wieczynska et al., 2016). These compounds are generally enclosed in pouches or sachets, or spiked onto a filter paper strip or sterile gauze that are then placed inside of the package, but in a few studies they have been incorporated directly into the packaging material, either plastic or paper, using the coating technique (Montero-Prado et al. 2011; Muriel-Galet et al., 2012, 2013; Rodriguez-Lafuente et al., 2010). All of the studies listed in Table 7 show that the use of antimicrobial packaging can improve produce shelf life by lowering microbial load, retaining color, and maintaining firmness and weight. As shown in Table 7, the more the amount of compound the more the greater the delay in deterioration for the tested antimicrobial range (Almenar et al., 2007b; Valverde et al., 2005; Wang and Buta, 2003). One should note, however, that the microbial load reduction is less than one log in some cases. Also, the effect of antimicrobial packaging for produce on the growth of pathogens should be further investigated since there are very few in vivo studies and these concluded little or no effect of antimicrobial packaging on the growth of pathogens (Muriel-Galet et al., 2012; Silveira et al., 2015). It is also noteworthy that the majority of the studies where a sensory evaluation has been carried out show lower sensory scores for the produce exposed to the antimicrobial in specific parameters, especially odor, due to the detection of the characteristic scent of the antimicrobial at some point of the study (either initially or at the end of the storage period). Therefore the choice of an antimicrobial with a pleasant odor and that is compatible with the flavor profile of the produce under study is very important (e.g., garlic oil for leafy green but not for peaches). The best solution seems to be the use of one of the compounds that is already present in the aroma profile of the studied commodity like some authors have done (Almenar et al., 2007b, 2009a). Additionally, use of food additives or substances with a GRAS status, approved specific migration limits, and dietary exposure should also be considered when developing antimicrobial packaging (Nerin et al., 2016). The increased consumption of fresh-cut produce due to the perfect combination of freshness, health, and convenience has increased the interest in developing antimicrobial packaging for fresh-cut produce as shown in Table 7. One should note, however, that this may be the most challenging type of active produce packaging to develop since microorganisms rapidly proliferate due to easy access to nutrients resulting from processing (Page et al., 2016).

Information on commercial products (i.e., powders, sachets, pads, films) available to develop antimicrobial packaging can be found in Table 3. OZO-fresh (EMCO Packaging) sachets can emit a controlled and sustained amount of ozone into a package. Ozone is a colorless gas with antimicrobial properties against spoilage and pathogenic microorganisms due to its powerful oxidizing capacity (http://www.emcopackaging.com/index.php/divisions/ozone-santisation). As for antimicrobial pads, ioFresh freshness extender pads by

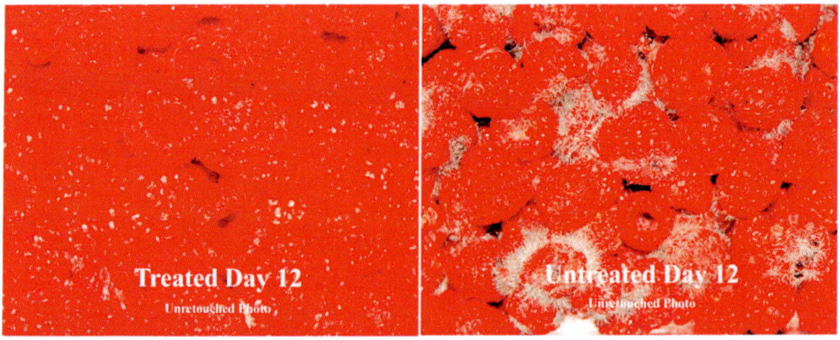

FIG. 6 Raspberry shelf life extension using ioFresh freshness extender pads by ioTech International. *(Photos courtesy of ioTech International.)*

ioTech International contain an iodine-based solution that releases iodine when activated by moisture once the pad is inserted into packages containing produce. This antimicrobial pad has shown shelf life extensions of up to 1 month for blackberries, strawberries, and raspberries (Fig. 6 (photo courtesy of ioTech International)) (https://iotech.io/iofresh/). Commercial antimicrobial coatings include Sanocoat by the Mondi Company (http://www.mondigroup.com/products/desktopdefault.aspx/tabid-1461) that can be applied to paper, plastics, and composites and antimicrobial varnished by LINPAC that can applied to trays and films after manufacture (https://www.linpacpackaging.com/en/news/201205/linpac-packaging-launches-innovative-solution-deliver-food-packaging-safety-and-drive). As for antimicrobial films for produce, NOW Plastics offers an antibacterial BOPP film as part of Everfresh specialty produce films (http://www.nowplastics.com/index.php?id=28).

Odor-control packaging

Off-flavors and off-odors are developed by produce when exposed to very low O_2 concentrations, as a result of anaerobic respiration, or to very high CO_2 levels, as a result of fermentative metabolism (Kader, 2016). The accumulation of ethanol, acetaldehyde, and ethyl acetate seems to be the primary cause for off-flavor and off-odor development (Li and Kader, 1989). Active packaging technologies may improve the overall quality and increase shelf life of produce with the selectivity of scalping these compounds. This is the motivation behind the development of odor-control packaging. There are not many studies focused on odor-control packaging for produce compared to other active packaging technologies.

DeEll et al. (2006) investigated the effect of sorbitol-containing sachets on the quality of broccoli stored in MAP in the presence of volatile absorbers containing $KMnO_4$. After 29 days of storage at 1°C, broccoli heads in packages with sorbitol had better appearance, firmness, and odor ratings compared with the controls. Although higher weight loss occurred in the presence of sorbitol, the level did not affect produce marketability. Overall, the use of sorbitol with $KMnO_4$ in MAP enhances the removal of off-flavors and maintains the quality and marketability of the produce. Pesis et al. (2005) reported that the addition of a sachet with a $KMnO_4$-based ethylene absorber in PE bags containing clusters of ethylene-pretreated bananas prior to air evacuation resulted in the absorption of acetaldehyde and ethanol.

Intelligent packaging

Scope

Intelligent packaging can be defined as the packaging technology that records, regulates, or controls the condition of the perishable product by sensing the environment inside or outside the package to ensure quality and safety during transportation, distribution, and/or retail, or that communicates another type of information related to the packaged perishable product. The EU legislation (Regulation 450/2009/EC, currently no US equivalent available) defines "intelligent materials and articles" as "materials and articles that monitor the condition of the packaged food or the environment surrounding the food." (http://www.foodcontactmaterials.com/eu/2009%20450.pdf).

Monitoring produce quality during transportation and storage gives additional information for better predicting the quality and safety of the commodity and can give important information for logistic control of the produce supply chain. In addition, intelligent packaging can ensure produce shelf life extension indirectly if it is simply used to check the effectiveness and the integrity of MAP or active packaging systems (Almenar and Wilson, 2016). Hence, the use of intelligent packaging will significantly rise in the next years. According to the Active and Intelligent Packaging Industry Association, the active and intelligent packaging industry is expected to double in size from €15 billion in 2015 to €32 billion in 2022 (Eagle, 2016). Demand for intelligent packaging, in particular, is expected to grow by double-digit percentages each year, hitting $1.3 billion by 2017 (http://www.foodproductiondaily.com/Packaging/US-intelligent-packaging-demand-to-hit-3.5bn-by-2017).

Intelligent devices or systems used to make packages intelligent include sensors, indicators, marketing communicators, and tracking devices. They are always part of the material that forms the package instead of being placed inside the package along with the product as can occur in active packaging (Almenar and Wilson, 2016). Examples of systems currently used to communicate produce quality and/or safety or to extend shelf life or to track or simply as a marketing tool are time-temperature indicators (TTIs), quality indicators, leak detectors, radiofrequency identification devices, temperature-compensation membranes, quick response codes (QR codes), and so on. One example of each of these devices that meets one of the aforementioned functions (e.g., tracking, communicate produce quality/safety, shelf life extension, marketing tool) is covered as follows.

Types

Tracking devices: radio frequency identification

Radio frequency identification (RFID) tags are a type of tracking device that allows tracing and identifying the packaged produce. RFID involves the use of radio waves for this identification and traceability. It works by having a reader emit radio waves that capture data from the RFID tag, which then can then be read by a computer (software). Unlike traditional barcodes, RFID tags do not require line of sight, meaning that the product does not have to face the reader to transmit data. This also allows multiple tags to be read simultaneously. For instance, the Spanish company Ecomovistand built a returnable packaging and transportation unit for fresh produce that incorporates embedded RFID tags to automatically track shipping units as they progress through distribution, noting locations, times, and cold chain status (Martínez-Sala et al., 2009). RFID is currently used by the US produce industry for produce identification and traceability. However, RFID tags need to be protected if their reuse is desired. Singh et al. (2010) studied the ability of RFID tags applied on reusable plastic containers commonly used for carrying produce from growers to retailers to withstand abuse in physical distribution and storage environment. The authors concluded that drops, vibration, high temperature washing, and detergent collectively degrade the RFID tag identification capability when reused, with the process of sanitation having a higher impact. The authors suggested the encapsulation of the tag as a protection method for its reuse. RFID can be coupled with other intelligent system to go beyond tracking and identification. Vergara et al. (2007) reported that an RFID reader with onboard gas sensors can be used to monitor apple quality attributes (e.g., ethylene, acetaldehyde, and ethanol) during transport and vending.

Sensors: temperature-compensating membranes

Packages capable of varying their permeability to O_2 and CO_2 in response to the changes in temperature occurring during storage, distribution, and marketing of produce are commercially available. This packaging approach is particularly important for fruits and vegetables since the temperature effect on film permeability and produce respiration rate is very different, hence, leading to undesired too high/too low concentrations of CO_2/O_2, respectively. This type of intelligent packaging is able to sense temperature changes in the surrounding environment and to react to those by changing its permeability. This change in permeability is caused by a membrane, the BreatheWay membrane (Fig. 7 (photo courtesy of Apio, Inc.)) that is attached to a cutout section with a specific diameter in a flexible bag or tray through its pressure-sensitive adhesive to act as a sole passageway for O_2 and CO_2. This membrane consists of a cellulose substrate coated with a polymer (registered under the trademark Intelimer) that is customized according to the type of produce and the package size to passively maintain

FIG. 7 The BreatheWay membrane. *(Photo courtesy of Apio, Inc.)*

the ideal O_2 and CO_2 levels independently of the surrounding temperature (Singh, 2016 (oral communication)). Intelimer belongs to a family of materials known as side chain crystallizable polymers, which exhibit abrupt thermal transitions associated with the side chain rather than the polymer backbone. Side chain crystallizable polymers exhibit melting transitions between 4°C and 20°C which matches the range of temperatures that produce is exposed to (Stewart et al., 1993). This feature allows the film O_2 transition rate to increase rapidly as temperature increases, thereby avoiding anaerobic conditions due to temperature abuse.

Indicators: time-temperature indicators

A time-temperature indicator (TTI) is a tag or label that displays a visual summary of the time–temperature history of the package to which it is attached. Thus TTIs can monitor the freshness of packaged produce. TTIs show a visible response to a temperature change in the form of a mechanical deformation, color development, or color movement. This allows producers, retailers, and consumers to check at a glance whether perishable products have been properly transported and stored. An example of a TTI that has been used in retail is Fresh Code manufactured by Varcode United States Inc. The contents of this barcode change as it is exposed to temperature abuse (http://www.varcode.com/portfolio_item/freshcode_tti/). TTIs such as these give data in addition to marked best-by dates, can be used to identify when these dates no longer correlate to freshness (Almenar and Wilson, 2016).

Marketing communicators: QR codes

QR codes printed on food packages can be used in both advertising and marketing. These codes can be scanned by a smartphone's camera and then decrypted by an app on the phone (Fig. 8). The code will send consumers to the corresponding website of the manufacturer. This virtual space has endless possibilities for the manufacturer in terms of advertising and marketing the packaged product where the QR code is located along with other products. Hence, QR codes can become one of the new "sales tools" for packaged food. In addition, this type of smart packaging can notably enhance consumer experience since it enables shoppers to be more informed about the products they are consuming. For example, consumers can learn the origins of fruits and vegetables from the QR code or retailers can use the QR codes to identify and withdraw packages of overripe produce or produce implicated in an outbreak. This would allow increasing the quality and safety of the produce purchased and hence, produce packages with QR codes have the potential to increase consumers' trust in a brand.

In-package cold plasma

Nonthermal plasma or cold plasma is an emerging tool for sterilization of food. It is based on the generation of highly reactive species like ions, free radicals, and photons to inactivate foodborne pathogens (Pankaj et al., 2014). In-package cold plasma consists of the generation of cold plasma inside a sealed package to inactivate the foodborne pathogens of the packaged product. Therefore this technology allows a nonthermal method for killing microorganism after the food is packaged (Pankaj et al., 2014). The fact that cold plasma is an effective sterilizing tool under normal atmospheric conditions opens a door for food products that are commercialized as raw items like produce. Studies focused on in-package cold plasma of strawberries, cherry tomatoes, and spinach have

FIG. 8 QR codes can be scanned by a smartphone's camera and then decrypted by an app on the phone.

shown microbial load reductions without affecting produce physicochemical quality (Klockow and Keener, 2009; Misra et al., 2014a,b). However, cold plasma can affect the permeability of the packaging material (Lee, 2010). Thus it is critical to evaluate the effect of cold plasma on the properties of the package (barrier and mechanical) besides on those of the produce since the package properties will affect produce shelf life (e.g., headspace composition).

Coatings

A coating can be defined as an edible continuous matrix applied directly on the food product. This matrix can be prepared by mixing food components such as polysaccharides, proteins, and lipids with water or ethanol, which evaporates. Plasticizers and other additives can be added to improve the matrix properties. Table 8 compiles some commercial coatings for produce including their main constituent type. The work done by academia in terms of coatings is compiled in excellent literature reviews that cover the state of the art of this field (Dhall, 2013; Quiros-Sauceda et al., 2014; Atares and Chiralt, 2016).

Coatings cannot replace the primary package since they do not provide many of the benefits of packaging, including protection from mechanical damage. However, they can be coupled with the primary packaging to enhance quality, safety, and nutritional content. Generally, coatings are used to prevent moisture migration and improve food appearance. Additionally, they can be used as a carrier of active compounds like antioxidants (Quiros-Sauceda et al., 2014), antimicrobials (Valencia-Chamorro et al., 2011), and nutraceuticals (Hernández-Muñoz et al., 2006, 2008; Almenar et al., 2009b) to be consumed with the food, thus enhancing safety or even nutritional and sensory attributes. The formulation of the coating-forming solution is critical as it can enhance or shorten product shelf life. Therefore, when formulating this

TABLE 8 Commercial coatings for produce

Name	Main component/s
NatureSeal	Modified cellulose polymers
Nutri-Save	Carboxymethyl chitosan
Pro-long	Sucrose polyesters of fatty acids and sodium salts of carboxymethyl cellulose
Sealgum and Spraygum	Gum acacia and gelatin
Shellac	Resin secreted by the female lac bug
Waxes (beeswax, paraffin, and carnauba)	Long-chain fatty acids

solution, the specifics of the type of produce (e.g., respiration and transpiration rates) and outside environmental conditions (e.g., temperature and relative humidity) need to be taken into consideration. Produce that is coated in the United States includes whole apples (always coated), fresh-cut apples (generally coated), whole oranges (generally coated), cherries (sometimes coated), grapes (stem-free; always coated), and dried fruits like raisins (always coated).

Novel approaches to adjust film permeability to temperature changes

A different approach to adjust film permeability to temperature changes in order to accommodate the respiratory needs of produce in a cold chain with alternating temperatures can be found in Koutsimanis et al. (2015). These authors developed a package that allows a rapid increase of the CO_2 concentration (while reducing the O_2 concentration) when stored at low temperature, and an immediate adjustment to temperature rises in the cold chain in order to maintain optimal gas compositions. The packaging approach was validated using fresh cherries under domestic (road transport) and international (ocean transport) supply chain conditions. The authors concluded that the new package can reduce fungal growth, firmness and color changes, and weight loss, and consequently, extend cherry shelf life.

Conclusions and future trends

Apart from the produce quality, safety, and shelf life extension benefits related to active packaging, coatings, in-package cold plasma, and so on, and the fulfillment of emerging market requirements for produce with intelligent packaging, these innovative packaging technologies offer considerable potential as a marketing tool and brand differentiation for produce companies. Thus their market presence is expected to increase. These innovative packaging technologies will continue evolving to help society overcome challenges related to safety (e.g., outbreaks), shelf life extension, temperature fluctuations in the produce supply chain, changes in distribution and retail practices associated with globalization, consumer preferences and demands, new shopping options (e.g., vending machines, internet), counterfeiting, environmental concerns, legislation, and food waste. In fact, challenges for these technologies to be overcome may result from the use of Internet as a shopping channel. The packages will have to ensure quality and safety while informing about these by receiving and storing data, checking authenticity and product condition, and delivering messages to let consumers and others know. Thus advancements in these packaging technologies are expected to grow and most likely will help to reduce hurdles related to cost and technology efficiency in produce packaging applications in years to come. Advancements in these packaging technologies are expected to result from blends of active compounds or combinations of technologies to address several of the deterioration changes that simultaneously occur in produce. Among the active packaging technologies, antimicrobial packaging is likely to advance the most in the years to come. This is driven by safety issues stemming from increased consumption of produce and persistent opportunities for contamination. Studies will be focused on natural bioactive compounds due to consumer demand for naturally occurring compounds and their controlled release to avoid alteration of the produce sensory characteristics. Materials with the capability of reducing the environmental impact of packaging waste are desired by consumers, industry, and government, and therefore, will be a focus of research and development for these innovations in packaging technologies for produce.

Acknowledgments

The author wish to express her most sincere gratitude and appreciation to Ms. Sarah Prentice, Mr. Simon Lee, and It's Fresh!, Ltd., Mr. Rodger Kolsky and ioTech International, Dr. Shehbaz Singh and APIO, Inc. for their contribution to this book chapter (photos and comments). She also would like to thank M.S. student Christopher T. Wilson for his review and feedback on the part of the book chapter dealing with active packaging.

References

Abe, K., Watada, A.E., 1991. Ethylene absorbent to maintain quality of lightly processed fruits and vegetables. J. Food Sci. 56, 1589–1592.

Aday, M.S., Caner, C., 2011. The application of 'active packaging and chlorine dioxide' for extended shelf life of strawberries. Packag. Technol. Sci. 24, 123–136.

Aday, M.S., Caner, C., 2013. The shelf life extension of fresh strawberries using an oxygen absorber in the biobased package. LWT- Food Sci. Technol. 52, 102–109.

Aday, M.S., Caner, C., Rahvali, F., 2011. Effect of oxygen and carbon dioxide absorbers on strawberry quality. Postharvest Biol. Technol. 62, 179–187.

Alique, R., Martínez, M.A., Alonso, J., 2003. Influence of the modified atmosphere packaging on shelf life and quality of Navalinda sweet cherry. Eur. Food Res. Technol. 217, 416–420.

Almenar, E., Auras, R., 2010. Permeation, sorption, and diffusion in poly(lactic acid). In: Auras, R., Lim, L.-T., Selke, S.E.M., Tsuji, H. (Eds.), Poly(Lactic Acid): Synthesis, Structure, Properties and Applications. Wiley and Sons, New York, pp. 155–179.

Almenar, E., Wilson, C.T., 2016. Advances in packaging fresh produce. Food Sci. Technol. 30 (3), 38–40.

Almenar, E., Hernández-Muñoz, P., Lagarón, J.M., Catalá, R., Gavara, R., 2006. Advances in packaging technologies for fresh fruit and vegetables. In: Noureddine, B., Norio, S. (Eds.), Advances in Postharvest Technologies of Horticultural Crops. Research Signpost Publisher, Kerala, pp. 87–112.

Almenar, E., Del-Valle, V., Hernández-Muñoz, P., Lagarón, J.M., Catalá, R., Gavara, R., 2007a. Equilibrium modified atmosphere packaging of wild strawberries. J. Sci. Food Agric. 87, 1931–1939.

Almenar, E., Del Valle, V., Catala, R., Gavara, R., 2007b. Active packaging for wild strawberry fruit (*Fragaria vesca* L.). J. Agric. Food Chem. 55, 2240–2245.

Almenar, E., Samsudin, H., Auras, R., Harte, B., Rubino, M., 2008. Postharvest shelf life extension of blueberries using a biodegradable package. Food Chem. 100, 120–127.

Almenar, E., Catala, R., Hernandez-Muñoz, P., Gavara, R., 2009a. Optimization of an active package for wild strawberries based on the release of 2-nonanone. LWT Food Sci. Technol. 42, 587–593.

Almenar, E., Hernández-Muñoz, P., Gavara, R., 2009b. Evolution of selected volatiles in chitosan-coated strawberries (*Fragaria × ananassa*) during refrigerated storage. J. Agric. Food Chem. 57, 974–980.

Almenar, E., Siddiq, M., Merkel, C., 2012. Packaging for processed food and the environment. In: Rahman, M.S., Ahmed, J. (Eds.), Handbook of Food Process Design. Wiley-Blackwell Publishing Limited, Oxford, pp. 1369–1405.

Atares, L., Chiralt, A., 2016. Essential oils as additives in biodegradable films and coatings for active food packaging. Trends Food Sci. Technol. 48, 51–62.

Ayala-Zavala, J.F., Gonzalez-Aguilar, G.A., 2010. Optimizing the use of garlic oil as antimicrobial agent of fresh-cut tomato through a controlled release system. J. Food Sci. 75, M398–M405.

Ayala-Zavala, J.F., Oms-Oliu, G., Odriozola - Serrano, I., Gonzalez-Aguilar, G.A., Alvarez-Parrilla, E., Martin-Belloso, O., 2008. Bio-preservation of fresh-cut tomatoes using natural antimicrobials. Eur. Food Res. Technol. 226, 1047–1055.

Bailen, G., Guillen, F., Castillo, S., Zapata, P.J., Serrano, M., Valero, D., Martinez-Romero, D., 2006. Use of activated carbon inside modified atmosphere packages to maintain tomato fruit quality during cold storage. J. Agric. Food Chem. 54, 2229–2235.

Ben-Yehoshua, S., Rodov, V., 2003. Transpiration and water stress. In: Bart, J.A., Bretch, J.K. (Eds.), Postharvest Physiology and Pathology of Vegetables, Second ed. Marcel Dekker, Inc., New York, pp. 111–159.

Ben-Yehoshua, S., Shapiro, B., Chen, Z.E., Lirie, S., 1983. Mode of action of plastic film in extending life of lemon and bell pepper by alleviation of water stress. Plant Physiol. 73, 87–93.

Ben-Yenhoshua, S., 1989. Individual seal-packaging of fruit and vegetables in plastic film. In: Brody, A. (Ed.), Controlled/Modified Atmosphere/Vacuum Packaging of Foods. Food and Nutrition Press, Inc., Trumbull, CT, pp. 101–118.

Bhutia, W., Pal, R.K., Sen, S., Jha, S.K., 2011. Response of different maturity stages of Sapota (*Manilkara achras* Mill.) cv. Kallipatti to in-package ethylene absorbent. J. Food Sci. Technol. 48, 763–768.

Brown, W., Ryser, E., Gorman, L., Steinmaus, S., Vorst, K., 2016. Transit temperatures experienced by fresh-cut leafy greens during cross-country shipment. Food Control 61, 146–155.

Cao, J., Li, X., Wu, K., Jiang, W., Qu, G., 2015. Preparation of a novel PdCl2-CuSo4-based ethylene scavenger supported by acidified activated charcoal power and its effects on quality and ethylene metabolism of broccoli during shelf life. Postharvest Biol. Technol. 99, 50–57.

Cerisuelo, J.P., Muriel-Galet, V., Bermudez, J.M., Aucejo, S., Catala, R., Gavara, R., Hernandez-Muñoz, P., 2012. Mathematical model to describe the release of an antimicrobial agent from an active package constituted by carvacrol in a hydrophilic EVOH coating on a PP film. J. Food Eng. 110, 26–37.

Charles, F., Sanchez, J., Gontard, N., 2003. Active modified atmosphere packaging of fresh fruits and vegetables: modeling with tomatoes and oxygen absorber. J. Food Sci. 68, 1736–1742.

Charles, F., Sanchez, J., Gontard, N., 2006. Absorption kinetics of oxygen and carbon dioxide scavengers as part of active modified atmosphere packaging. J. Food Eng. 72, 1–7.

Chopra, S., Dhumal, S., Abeli, P., Ryser, E.T., Beaudry, R., Almenar, E., 2016. Moisture Regulation to Control Microbial Growth on Packaged Produce. FreshTech Learning Center of the 2016

United Fresh Convention, Chicago, Illinois. http://www.unitedfreshshow.org/food-safety-posters (accessed 25.11.16).

Combrin, J.C., de Kock, S.L., van Eeden, C.J., 1990. Effect of postharvest treatment and packaging on the keeping quality of fresh guava fruit. Acta Hortic. (275), 639–640.

DeEll, J.R., Toivonen, P.M.A., Cornut, E., Roger, C., Vigneault, C., 2006. Addition of sorbitol with KMnO4 improves broccoli quality retention in modified atmosphere packages. J. Food Qual. 29, 65–75.

Dhall, R.K., 2013. Advances in edible coatings for fresh fruits and vegetables: a review. Crit. Rev. Food Sci. Nutr. 53, 435–450.

Eagle, J., 2016. Active, intelligent packaging industry to double to €32bn in 2022. In: Beverage Technology and Materials. William Reed Business Media Ltd. http://www.beveragedaily.com/Processing-Packaging/Active-intelligent-packaging-industry-to-double-to-32bn-by-2022?utm_source=newsletter_daily&utm_medium=email&utm_campaign=15-Nov-2016&c=MU%2FFM8HGRWPLKNrHd4Gn2Q%3D%3D&p2 (accessed 16.11.16).

East, A.R., Samarakoon, H.C., Pranamornkith, T., Bronlund, J.E., 2015. A review of ethylene permeability of films. Packag. Technol. Sci. 28, 732–740.

Elmi, F., Meyer, M.D., Terry, L.A., 2011. Extension of avocado storability using e+® ethylene remover coated sheets in sea containers. Acta Hortic. 945, 325–330.

Esturk, O., Ayhan, Z., Gokkurt, T., 2014. Production and application of active packaging film with ethylene absorber to increase the shelf life of broccoli (*Brassica oleracea* L. var. Italica). Packag. Technol. Sci. 27, 179–191.

Evelo, R.G., Horst, J., 1996. Modified atmosphere packaging of tomatoes: controlling gas and humidity. Packag. Technol. Sci. 9, 265–273.

Fallik, E., Temkin-Gorodeiski, N., Grinberg, S., Davidson, H., 1995. Postharvest Biol. Technol. 5, 83–89.

González-Buesa, J., Page, N., Kaminski, C., Ryser, E., Beaudry, R., Almenar, E., 2014. Effect of non-conventional atmospheres and bio-based packaging on the quality and safety of *Listeria monocytogenes*-inoculated fresh-cut celery (*Apium graveolens* L.) during storage. Postharvest Biol. Technol. 93, 29–37.

Gross, K.C., Wang, C.Y., Saltveit, M.E., 2016. The commercial storage of fruits, vegetables, and florist and nursery stocks. In: Agriculture Handbook Number 66. U.S. Department of Agriculture, Washington, D.C. http://ucanr.edu/datastoreFiles/234-2927.pdf (accessed 27.10.16).

Hernández-Muñoz, P., Almenar, E., Ocio, M.J., Gavara, R., 2006. Effect of calcium dips and chitosan coatings on postharvest life of strawberries (*Fragaria* x *ananassa*). Postharvest Biol. Technol. 39, 247–253.

Hernández-Muñoz, P., Almenar, E., Del Valle, V., Velez, D., Gavara, R., 2008. Effect of chitosan coating combined with postharvest calcium treatment on strawberry (*Fragaria* × *ananassa*) quality during refrigerated storage. Food Chem. 110, 428–435.

Illeperuma, C.K., Nikapitiya, C., 2002. Extension of the postharvest shelf life of 'Pollock' avocado using modified atmosphere packaging. Fruits 57, 287–295.

Joyce, D., Patterson, B., 1994. Postharvest water relations in horticultural crops: principles and problems. In: Chapman, B.R., Highley, E., Johnson, G.I. (Eds.), ACIAR Proc., Vol. 50, Postharvest Handling of Tropical Tree Fruits, pp. 228–238.

Kader, A.A., 1985. Ethylene induced senescence and physiological disorders in harvested horticultural crops. HortSci. 20, 54–57.

Kader, A.A., 1987. Respiration and gas exchange of vegetables. In: Weichman, J. (Ed.), Postharvest Physiology of Vegetables. Marcel Dekker, Inc., New York, p. 25.

Kader, A.A., 2016. Controlled atmosphere storage. In: Gross, K.C., Wang, C.Y., Saltveit, M.E. (Eds.), Agriculture Handbook Number 66: The Commercial Storage of Fruits, Vegetables, and Florist and Nursery Stocks. U.S. Department of Agriculture, Washington, DC http://ucanr.edu/datastoreFiles/234-2927.pdf (accessed 27.10.16).

Kader, A.A., Zagory, D., Kerbel, E.L., 1989. Modified atmosphere packaging of fruits and vegetables. Crit. Rev. Food Sci. Nutr. 28, 1–30.

Kartal, S., Aday, M.S., Caner, C., 2012. Use of microperforated films and oxygen scavengers to maintain storage stability of fresh strawberries. Postharvest Biol. Technol. 71, 32–40.

Keller, N., Ducamp, M.N., Robert, D., Keller, V., 2013. Ethylene removal and fresh product storage: a challenge at the frontiers of chemistry. Toward an approach by photocatalytic oxidation. Chem. Rev. 113, 5029–5070.

Klockow, P.A., Keener, K.M., 2009. Safety and quality assessment of packaged spinach treated with a novel ozone-generation system. LWT Food Sci. Technol. 42, 1047–1053.

Koutsimanis, G., Harte, J., Almenar, E., 2015. Development and evaluation of a new packaging system for fresh produce: a case study on fresh cherries under global supply chain conditions. Food Bioprocess Technol. 8, 655–669.

Kuorwel, K.K., Cran, M.J., Sonneveld, K., Miltz, J., Bigger, S.W., 2011. Antimicrobial activity of biodegradable polysaccharide and protein-based films containing active agents. J. Food Sci. 76, R90–R102.

Lee, K.T., 2010. Quality and safety aspects of fresh-cut products as affected by various physical manipulations of packaging materials. Meat Sci. 86, 138–150.

Lee, D.S., 2016. Carbon dioxide absorbers for food packaging applications. Trends Food Sci. Technol. 57 (Part A), 146–155.

Li, C., Kader, A.A., 1989. Residual effects of controlled atmospheres on postharvest physiology and quality. J. Am. Soc. Hortic. Sci. 114, 629–634.

Linke, M., Geyer, M., 2013. Condensation dynamics in plastic film packaging of fruits and vegetables. J. Food Eng. 116, 144–154.

Lopez-Galvez, F., Ragaert, P., Haque, M.A., Eriksson, M., van Labeke, M.C., Devlieghere, F., 2015. High oxygen atmospheres can induce russet spotting development in minimally processed iceberg lettuce. Postharvest Biol. Technol. 100, 168–175.

López-Rubio, A., Almenar, E., Hernández-Muñoz, P., Lagarón, J.M., Catalá, R., Gavara, R., 2004. Overview of active polymer-based packaging technologies for food applications. Food Rev. Int. 20, 357–386.

Lucera, A., Conte, A., Del Novile, M.A., 2016. Volatile compounds used in active packaging systems. In: Barros-Velazquez, J. (Ed.), Antimicrobial Food Packaging. Elsevier Inc., New York, pp. 319–327

Mahajan, P.V., Rodrigues, F.A.S., Motel, A., Leonhard, A., 2008. Development of a moisture absorber for packaging of fresh mushrooms. Postharvest Biol. Technol. 48, 408–414.

Mari, M., Bautista-Baños, S., Sivakumar, D., 2016. Decay control in the postharvest system: role of antimicrobial and plant volatile organic compounds. Postharvest Biol. Technol. 122, 70–81.

Martínez-Sala, A.S., Egea-López, E., García-Sánchez, F., García-Haro, J., 2009. Tracking of returnable packaging and transport units with active RFID in the grocery supply chain. Comput. Ind. 60, 161–171.

Matinez-Romero, D., Bailen, G., Serrano, M., Guillen, F., Valverde, J.M., Castillo, P.Z.S., Valero, D., 2007. Tools to maintain postharvest fruit and vegetable quality through the inhibition of ethylene action: a review. Crit. Rev. Food Sci. Nutr. 47, 543–560.

Meyer, M.D., Terry, L.A., 2010. Fatty acid and sugar composition of avocado, cv. Hass, in response to treatment with an ethylene scavenger or 1-methylcyclopropene to extend storage life. Food Chem. 121, 1203–1210.

Mir, N., Beaudry, B.M., 2016. Modified atmosphere packaging. In: Gross, K.C., Wang, C.Y., Saltveit, M.E. (Eds.), Agriculture Handbook Number 66: The Commercial Storage of Fruits, Vegetables, and Florist and Nursery Stocks. U.S. Department of Agriculture, Washington, D.C http://ucanr.edu/datastoreFiles/234-2927.pdf (accessed 27.10.16).

Misra, N.N., Keener, K.M., Bourke, P., Mosnier, J.-P., Cullen, P.J., 2014a. In-package atmospheric pressure cold plasma treatment of cherry tomatoes. J. Biosci. Bioeng. 118, 177–182.

Misra, N.N., Patil, S., Moiseev, T., Bourke, P., Mosnier, J.P., Keener, K.M., Cullen, P.J., 2014b. In-package atmospheric pressure cold plasma treatment of strawberries. J. Food Eng. 125, 131–138.

Montero-Prado, P., Rodriguez-Lafuente, A., Nerin, C., 2011. Active label-based packaging to extend the shelf life of "Calanda" peach fruit: changes in fruit quality and enzymatic activity. Postharvest Biol. Technol. 60, 211–219.

Muriel-Galet, V., Cerisuelo, J.P., Lopez-Carballo, G., Lara, M., Gavara, R., Hernandez-Muñoz, P., 2012. Development of antimicrobial films for microbial control of packaged salad. International. Int. J. Food Microbiol. 157, 195–201.

Muriel-Galet, V., Cerisuelo, J.P., Lopez-Carballo, G., Aucejo, S., Gavara, R., Hernandez-Muñoz, P., 2013. Evaluation of EVOH-coated PP films with oregano essential oil and citral to improve the shelf life of packaged salads. Food Control 30, 137–143.

Nerin, C., Silva, F., Manso, S., Becerril, R., 2016. The downside of antimicrobial packaging: migration of packaging elements into food. In: Barros-Velazquez, J. (Ed.), Antimicrobial Food Packaging. Elsevier Inc., New York, pp. 81–93

Nugraha, B., Bintoro, N., Murayama, H., 2015. Influence of CO_2 and C_2H_4 adsorbents to the symptoms of internal browning of the packaged "Silver Bell" pear (*Pyrus communis* L.). Agric. Agric. Sci. Procedia 3, 127–131.

Otoni, C.G., Espitia, P.J.P., Avellana-Bustillos, R.J., McHugh, T.H., 2016. Trend in antimicrobial food packaging systems: emitting sachets and absorbent pads. Food Res. Int. 83, 60–73.

Page, N., Curlew, S., Afrifah, K., Wu, C., Ryser, E., Beaudry, R., Almenar, E., 2013. Exploring novel packaging technologies on *Escherichia coli* growth in baby spinach. In: International Controlled and Modified Atmosphere Research Conference (CAMA). Trani, Italy.

Page, N., González-Buesa, J., Ryser, E., Harte, J., Almenar, E., 2016. Interactions between sanitizers and packaging gas compositions and their effects on the safety and quality of fresh-cut onions (*Allium cepa* L.). Int. J. Food Microbiol. 218, 105–113.

Pankaj, S.K., Bueno-Ferrer, C., Misra, N.N., Milosavijevic, V., O'Donnell, C.P., Bourke, P., Keener, K.M., Cullen, P.J., 2014. Applications of cold plasma technology in food packaging. Trends Food Sci. Technol. 35, 5–17.

Park, C.-Y., Kim, Y.-J., Shin, Y., 2016. Effects of an ethylene absorbent and 1-methylcyclopropene on tomato quality and antioxidant contents during storage. Hortic. Environ. Biotechnol. 57, 38–45.

Patterson, B.D., Joyce, D.C., 1993. A package allowing cooling and preservation of horticultural produce without condensation or desiccants. International Patent Application PCT/AU93/00398.

Patterson, B.D., Jobling, J., Moradi, S., 1993. Water relations after harvest—new technology helps translate theory into practice. In: Proceedings of the Australasian Postharvest Conference, University of Queensland, Gatton, Queensland, Australiapp. 99–102.

Payasi, A., Sanwal, G.G., 2010. Ripening of climacteric fruits and their control. J. Food Biochem. 34, 679–710.

Pesis, E., Aharoni, D., Aharon, Z., Ben-Arie, R., Aharoni, N., Fuchs, Y., 2000. Modified atmosphere and modified humidity packaging alleviates chilling injury symptoms in mango fruit. Postharvest Biol. Technol. 19, 93–101.

Pesis, A., Arie, R.B., Feygenberg, O., Villamizar, F., 2005. Ripening of ethylene-pretreated bananas is retarded using modified atmosphere and vacuum packaging. HortSci. 40, 726–731.

Picon, A., Martinez-Javega, J.M., Cuquerella, J., Del Rio, M.A., Navarro, P., 1993. Effects of precooling, packaging film, modified atmosphere and ethylene absorber on the quality of refrigerated Chandler and Douglas strawberries. Food Chem. 48, 189–193.

Powers, T.H., Calvo, W.J., 2003. Moisture regulation. In: Ahvenainen, R. (Ed.), Novel Food Packaging Techniques. CRC Press LLC, Boca Raton, Florida, pp. 172–185.

Quiros-Sauceda, A.E., Ayala-Zavala, J.F., Olivas, G.L., Gonzalez-Aguilar, G.A., 2014. Edible coatings as encapsulating matrices for bioactive compounds: a review. J. Food Sci. Technol. 51, 1674–1685.

Rodov, V., Ben-Yehoshua, S., Aharoni, N., Cohen, S., 2010. Modified humidity packaging of fresh produce. In: Janick, J. (Ed.), Horticultural Reviews. In: Vol. 37. Wiley-Blackwell Publishing Limited, Oxford, pp. 281–329.

Rodriguez, A., Batlle, R., Nerin, C., 2007. The use of natural essential oils as antimicrobial solutions in paper packaging. Par II. Prog. Org. Coat. 60, 33–38.

Rodriguez-Lafuente, A., Nerin, C., Batlle, R., 2010. Active paraffin-based paper packaging for extending the shelf life of cherry tomatoes. J. Agric. Food Chem. 58, 6780–6786.

Rooney, M.L., 1995. Active packaging in polymeric films. In: Rooney, M.L. (Ed.), Active Food Packaging. Springer, Boca Raton, pp. 74–110.

Roy, S., Anantheswaran, R.C., Beelman, R.B., 1995. Sorbitol increases shelf life of fresh mushrooms stored in conventional packages. J. Food Sci. 60, 1254–1259.

Roy, S., Anantheswaran, R.C., Beelman, R.B., 1996. Modified atmosphere and modified humidity packaging of fresh mushrooms. J. Food Sci. 61, 391–397.

Rux, G., Mahajan, P.V., Geyer, M., Linke, M., Pant, A., Sangerlaud, S., Caleb, O.J., 2015. Application of humidity-regulating tray for packaging of mushrooms. Postharvest Biol. Technol. 108, 102–110.

Rux, G., Mahajan, P.V., Geyer, M., Linke, M., Pant, A., Sangerlaud, S., Caleb, O.J., 2016. Humidity-regulating trays: moisture absorption kinetics and applications for fresh produce packaging. Food Bioprocess Technol. 9, 709–716.

Saltveit, M.E., 1999. Effect of ethylene on quality of fresh fruits and vegetables. Postharvest Biol. Technol. 15, 279–292.

Scharff, R.L., 2012. Economic burden from health losses due to foodborne illness in the United States. J. Food Prot. 75, 123–131.

Shirazi, A., Cameron, A., 1992. Controlling relative humidity in modified atmosphere packages of tomato fruit. HortSci. 27, 336–339.

Sholberg, P.L., Conway, W.S., 2016. Postharvest pathology. In: Gross, K.C., Wang, C.Y., Saltveit, M.E. (Eds.), Agriculture Handbook Number 66: The Commercial Storage of Fruits, Vegetables, and Florist and Nursery Stocks. U.S. Department of Agriculture, Washington, D.Chttp://ucanr.edu/datastoreFiles/234-2927.pdf. Accessed 20 November 2016.

Shorter, A.J., Scott, K.J., Ward, G., Best, D.J., 1992. Effect of ethylene absorption on the storage of Granny Smith apples held in polyethylene bags. Postharvest Biol. Technol. 1, 189–194.

Silva, D.F.P., Salomao, L.C.C., Siqueira, D.L., Cecon, P.R., Rocha, A., 2009. Potassium permanganate effects in postharvest conservation of the papaya cultivar Sunrise Golden. Pesq. Agrop. Brasileira 44, 669–675.

Silveira, A.C., Moreira, G.C., Artes, F., Aguallo, E., 2015. Vanillin and cinnamic acid in aqueous solutions or in active modified packaging preserve the quality of fresh-cut cantaloupe melon. Sci. Hortic. 192, 271–278.

Singh, S., 2016. BreatheWay: Intelligent Packaging Technology for shelf life extension of fresh produce. Seminar held at the School of Packaging at Michigan State University (04.08.16).

Singh, J., Singh, S.P., Desautels, K., Saha, K., Olsen, E., 2010. An evaluation of the ability of RFID tags to withstand distribution of fresh produce in the RPC pooling system. Packag. Technol. Sci. 23, 217–226.

Singh, R., Giri, S.K., Kotwaliwale, N., 2014. Shelf-life enhancement of green bell pepper (*Capsicum annuum* L.) under active modified atmosphere storage. Food Packag. Shelf life 1, 101–112.

Song, Y., Lee, D.S., Yam, K.L., 2001. Predicting relative humidity in modified atmosphere packaging systems containing blueberry and moisture absorbent. J. Food Process. Preserv. 25, 49–70.

Stewart, R.F., Mohr, J.M., Budd, E.A., Phan, L.X., Arul, J., 1993. Temperature-compensating films for modified atmosphere packaging of fresh produce. In: El-Nokaly, M. et al., (Ed.), Polymeric Delivery Systems. In: ACS Symposium SeriesAmerican Chemical Society, Washington, DC, pp. 232–243.

Szczerbanik, M.J., Scott, K.J., Paton, J.E., Best, D.J., 2005. Effects of polyethylene bags, ethylene absorbent, and 1-methylcyclopropene on the storage of Japanese pears. J. Hortic. Sci. Biotechnol. 80, 162–166.

Tewari, G., Jayas, D.S., Jeremiah, L.E., Holley, R.A., 2002. Absorption kinetics of oxygen scavengers. Int. J. Food Sci. Technol. 37, 209–217.

Toivonen, P.M.A., Kempler, C., Stan, S., 2002. The use of a natural clay adsorbent improves quality retention in three cultivars of raspberries stored in modified atmosphere packaging. J. Food Qual. 25, 385–393.

UCDavis, 2016. UCDavis Website on "Produce Fact Sheets" http://postharvest.ucdavis.edu/Commodity_Resources/Fact_Sheets/ retrieved on 10/27/16.

Valencia-Chamorro, S.A., Palou, L., Del Rio, M.A., Perez-Gago, M.B., 2011. Antimicrobial edible films and coatings for fresh and minimally processed fruits and vegetables: a review. Crit. Rev. Food Sci. Nutr. 51, 872–900.

Valero, D., Valverde, J.M., Martinez-Romero, D., Guillen, F., Castillo, S., Serrano, M., 2006. The combination of modified atmosphere packaging with eugenol and thymol to maintain quality, safety and functional properties of table grapes. Postharvest Biol. Technol. 41, 317–327.

Valverde, J.M., Guillen, F., Martinez-Romero, D., Castillo, S., Serrano, M., Valero, D., 2005. Improvement of table grapes quality and safety by the combination of modified atmosphere packaging (MAP) and eugenol, menthol, or thymol. J. Agric. Food Chem. 53, 7458–7464.

Van Long, N.N., Joly, C., Dantigny, P., 2016. Active packaging with antifungal activities. Int. J. Food Microbiol. 220, 73–90.

Vergara, A., Llobet, E., Ramirez, J.L., Ivanov, P., Fonseca, L., Zampolli, S., Scorzoni, A., Becker, T., Marco, S., Wollenstein, J., 2007. An RFID reader with onboard sensing capacity for monitoring fruit quality. Sensors Actuators B 127, 143–149.

Villaescusa, R., Gil, M.I., 2003. Quality improvement of *Pleurotus* mushrooms by modified atmosphere packaging and moisture absorbers. Postharvest Biol. Technol. 28, 169–179.

Wang, C.Y., Buta, J.G., 2003. Maintaining quality of fresh-cut kiwifruit with volatile compounds. Postharvest Biol. Technol. 28, 181–186.

Wang, H.J., An, D.S., Rhim, J.-W., Lee, D.-S., 2015. A multi-functional biofilm used as an active insert in modified atmosphere packaging for fresh produce. Packag. Technol. Sci. 28, 999–1010.

Wieczynska, J., Luca, A., Kidmose, U., Cavoski, I., Edelenbos, M., 2016. The use of antimicrobial sachets in the packaging of organic rocket: impact on microorganism and sensory quality. Postharvest Biol. Technol. 121, 126–134.

Yahia, E.M., Rivera, M., 1992. Modified atmosphere packaging of muskmelon. LWT- Food Sci. Technol. 25, 38–42.

Yue Bi, C.Y., Yao, B.N., Konan, H.K., Tano, K., 2014. Modified atmosphere and humidity packages of mushrooms (*Pleurotus ostreatus*) and tomatoes (*Lycopersicon esculenttum* var. petomech) to avoid water condensation. J. Chem., Biol. Phys. Sci. 4, 3247–3260.

Zagory, D., 1995. Ethylene-removing packaging. In: Rooney, M.L. (Ed.), Active Food Packaging. Springer, Boca Raton, pp. 38–54.

Zagory, D., Kader, A.A., 1988. Modified atmosphere packaging of fresh produce. Food Technol. 42, 70-74 & 76-77.

Chapter 12

Industrial advances of CA/MA technologies: innovative storage systems

David Bishop*, James Schaefer* and Randolph Beaudry†

*Storage Control Systems, Inc., Sparta, MI, United States, †Department of Horticulture, Michigan State University, East Lansing, MI, United States

Modified atmospheres have been described as having three primary effects on plant physiology of significant commercial relevance (Beaudry, 1999, 2000):

1. Inhibition of ethylene action (via low O_2 and high CO_2)
2. Inhibition of cut surface browning (via low O_2)
3. Inhibition of decay (via high CO_2)

Additionally, modified atmospheres can reduce chlorophyll breakdown via high CO_2 (and as a result of ethylene inhibition by both low O_2 and high CO_2) and suppress respiration by low O_2. While low O_2 can inhibit respiration (Kidd and West, 1927; Jurin and Karel, 1963), this inhibition has never been unambiguously demonstrated to prolong storability. There are many reports of elevated CO_2 suppressing respiration, but this appears to be solely an effect of ethylene inhibition, rather than a direct effect of CO_2 on the respiratory machinery (Kubo et al., 1990).

Innovation in the application of controlled atmosphere (CA) and modified atmosphere packaging (MAP) primarily revolves around more effective and precise delivery of the respiratory gases O_2 and CO_2. Nevertheless, important contributions to storability in these environments are also being made through the judicial use of bioactive molecules, the most important of which is the ethylene action inhibitor, 1-methylcyclopropene (1-MCP).

Areas of innovation in CA

1-MCP

The discovery of 1-MCP (Sisler and Blankenship, 1996) promised improved control over ethylene responses in many crops. However, due to limitations

Controlled and Modified Atmospheres for Fresh and Fresh-Cut Produce
https://doi.org/10.1016/B978-0-12-804599-2.00013-2

imposed by the biology of the crop or the demands of the market place, the scope of benefit of 1-MCP was eventually seen as being rather limited after several years of experience with this compound (Sozzi and Beaudry, 2007). In fact, after 20 years, apple fruit constitute 95% of the market for 1-MCP (Nazir Mir, personal communication). There are niche applications for avocado, pear, persimmon, and banana among others, but the fit for the physiology and consumer demand really is best for apple. Of all the fruits that ripen in response to ethylene, apple is somewhat unique in that consumer expectations are for an eating experience provided by an immature fruit (i.e., crunchy and tart) rather than a ripe fruit, which is mellow and soft. Thus inhibition of ripening is beneficial to apple, but much less so for other fruit. Nevertheless, innovation in the application of 1-MCP continues with the development of preharvest sprays for the suppression of preharvest drop in apple and the inhibition of ripening postharvest (De Ell and Ehsani-Moghaddam, 2010) and the use of multiple applications during and/or after harvest (De Ell and Ehsani-Moghaddam, 2013; Mir et al., 2001).

Dynamic controlled atmosphere

Work by Tucker and Laties (1985) suggested that extremely low levels of O_2 (e.g., <0.2 kPa) within bulky fruit tissues would support normal respiratory metabolism. Gran and Beaudry (1993) demonstrated that apple fruit could sustain partial pressures of O_2 in the range of 1.0–0.3 kPa without inducing fermentation, but would maximally inhibit ethylene action (Burg and Burg, 1967). These levels were significantly lower than commercially applied atmospheres at the time, but suggested that lower O_2 levels were potentially useful. Technological advances would be needed, however, to maintain O_2 partial pressures in this range. Subsequently, innovations in the construction of CA rooms such as sealed, foam-filled wall panels, improved door sealing systems, and pressure control features provided by breather bags and modern CO_2 scrubbing systems have improved room tightness dramatically since the early days of CA. The impact has been an improvement in the capacity for regulation of O_2 and CO_2. In addition, the development of efficient and inexpensive equipment for O_2 regulation (e.g., pressure swing adsorption and hollow fiber membrane technologies) in the 1980s and 1990s and the advent of computer-based systems for room monitoring and control have also contributed to improved atmosphere control (Bishop, 1990; Dilley, 1990, 2010; Mditshwa et al., 2018). Improved regulation of CA room atmospheres opened the door for the development of technologies that would take advantage of new developments in sensors for the creation of dynamically controlled atmospheres. Wolfe et al. (1993) first described a dynamic controlled atmosphere (DCA) system in which computer-controlled sensors for O_2 and CO_2 would be used to assess respiration and thereby regulate that process through automatic atmosphere manipulation.

DCA for apple storage has increased over the last 5 years, leading to reductions in O_2 concentrations from 1.2–2.0 to 0.4–0.8 kPa O_2. Storage at <1.0 kPa O_2 extends storage, but requires monitoring of fruit health to avoid damage. Competing DCA systems exist. Currently, three DCA technologies are in development or are commercially available. Each of the three DCA technologies is implemented in roughly the same fashion: the O_2 in the CA room is initially established at a safe level and the O_2 is slowly decreased until the sensor system indicates that the atmosphere is stressing the fruit. At this point, the O_2 level is raised to the previous, nonstressful level. These systems have the advantage of being able to continually probe the lower oxygen limit, which apparently can decline after some period in low oxygen storage. The O_2 levels in storages controlled by these systems have reached unprecedented low levels, often below 0.5%. At these low O_2 levels, ethylene action is essentially completely inhibited (Burg and Burg, 1967; Solomos, 2017) and some postharvest disorders, such as CO_2 injury, may be suppressed to some extent (De Ell and Lum, 2017; Mditshwa et al., 2018). DCA has been demonstrated to slow ripening very effectively, being better at retarding firmness loss in apple than standard CA or even ultralow oxygen (ULO) storage (Thewes et al., 2015). It also effectively suppresses superficial scald without the use of the antioxidant, diphenylamine (DPA) (Zanella, 2003).

The first system to be described uses the emission of the fermentation product ethanol to control O_2 (Schouten et al., 1997). Developed as the Storex (The Netherlands) system, this technology operates by sensing ethanol as fruit turn anaerobic, but has not yet reached widespread commercialization. The Storex system uses a specialized fruit sample chamber. At this time, this chamber cannot be placed inside the CA room in North America due to patent restrictions of a competing technology (i.e., SafePod). Testing fruit in a chamber outside the CA room raises the possibility of atmosphere and temperatures differences between test fruit and fruit in storage.

The second DCA technology uses chlorophyll fluorescence of the fruit to detect the lower oxygen limit (Prange et al., 2002). Harvest Watch (Isolcell, Italy) is a commercial system using 3 chlorophyll fluorescence sensors to monitor stress indirectly in 15–20 representative fruit. The physiological connection between chlorophyll fluorescence and the tolerance of the fruit to low O_2 is not well understood. The Isolcell technology is perhaps the most widespread DCA technology in use at this time (Mditshwa et al., 2018).

The third DCA technology uses a chamber connected to or within the room to determine the respiratory quotient (RQ, the ratio of CO_2 production to O_2 uptake), which is taken as a measure of fermentation (Bessemans et al., 2016; Schaefer and Bishop, 2014). One of the two current RQ technologies is The Van Amerongen or My Dynamic (The Netherlands) system. Introduced in 2014, this RQ-based DCA technology measures the RQ of whole stores, but requires expensive ancillary equipment, new store construction, stoppage of cooling and CA control for a period of time, and can conceivably put the entire store at risk when testing lower O_2 limits (LOL). The second RQ-based DCA system is known as the SafePod (United States) (Schaefer and Bishop, 2014). The SafePod (US patent 10143210, Canada CA 2746152) allows fruit to be stored safely at lower O_2 concentrations <1 kPa, lowering metabolic processes and maintaining the quality of fruit for longer. The SafePod contains new generation of O_2 and CO_2 sensors allowing more accurate and stable measurement of O_2/CO_2. The SafePod monitors the respiration of a representative sample of the stored fruit in real time, alerting operators to changes in crop health during storage under low O_2 conditions. It can be used to determine how to best optimize the O_2 and CO_2 regimes required to extend the storage life of fruit. The SafePod's advantage over the other technologies is that it provides the ability to respond to changing oxygen requirements during fruit storage. In part, this reflects the ability to use the SafePod to measure the rate of respiration of the fruit sample at the beginning of the store to help the operator determine the long- or short-term storability of the entire store. Studying respiration characteristics can help determine weaker stores (i.e., more mature and/or stressed fruit) from strong stores (less mature, healthier fruit), which helps operators rank their market timing.

Innovations of SafePod overcome many practical issues of existing technology. SafePod is a 0.4 m^3 chamber, fitted with internal gas sensors, filled with ~750 fruit. It is placed in the CA store that is being monitored with free exchange of gases with the store environment through ventilation valves. Electronic closure of valves allows accurate periodic measurement of fruit respiration and RQ within the chamber by measuring gas content at regular intervals, providing accurate estimates of the respiratory behavior of the fruit in the sample, which are taken to represent the fruit in the store. RQ of apples stored above the LOL range from 0.9 to 1.1 and increase to 2–5 under low O_2 stress. The RQ is measured in an airtight chamber, when the safepod isolation valve closes. Any presence of lime in the CA room to scrub CO_2 will not interfere with measurements. A further innovation of SafePod is that the atmosphere can be allowed to become more extreme than the rest of the store to explore the storage LOL limits without exposing the complete consignment to potentially damaging conditions, as is the case with the My Dynamic and Isolcell systems. In addition, SafePod can readily be integrated into new or old CA rooms and the controls can be integrated with installations using the post-2012 SCS 6000 CA monitoring system. The system can also be "stand alone." Additionally, a modified stand-alone lab-based "LabPod" has been developed using the same O_2 and CO_2 sensing and control technologies. This system is suitable for CA research in commercial or laboratory facilities.

Accuracy of gas composition measurement is essential for RQ-based technologies to work. Success depends on the sensors, chamber design including airflow both within the chamber and exchange with the surrounding atmosphere. High sensor sensitivity (i.e., to three decimal places) is needed to increase reading stability by amplifying and smoothing data output. This provides an additional benefit of reducing time required to make stable readings of

CO_2 and O_2 and calculation of RQ. Barometric pressure compensation in the master PLC controller is a needed feature to handle analyzer accuracy during weather changes. The effect of small sensor deviation, when the barometric pressure in the outside atmosphere shifts, is compensated for to calculate and accurate RQ.

While all three technologies have the potential to adequately control O_2 and provide low O_2 benefits, some distinctions may be important. The use of ethanol as a marker, while physiologically relevant, has the potential to be misleading. If only a portion of the fruit is undergoing fermentation, the entire room is subject to regulation based on the response of those few. Further, once ethanol accumulates in a fruit, it takes many days for it to dissipate, slowing the response time of the system. Finally, ethanol is synthesized as a product of ripening, not just in response to fermentation, leading to the possibility of "false positive" signals. The use of chlorophyll fluorescence is somewhat indirect, being in response to changes in the energy status of the chloroplast, rather than the respiratory machinery of the fruit cells. Both the chlorophyll fluorescence and the SafePod RQ measurement systems rely on samples held in the CA room or in a chamber attached to the room and so introduce the possibility that the samples may not be representative of the remainder of fruit in the room.

Sensor technology for ethylene and other biomarkers

Portable ethylene detectors that do not rely on gas chromatography emerged in the 2010s that are able to reliably and specifically detect ethylene levels as low as 0.1 ppm. Technologies evaluated include ring-down spectroscopy (Wahl et al., 2006) and nanoporous gold (Shekarriz and Allen, 2008) detectors that permit real-time monitoring of ethylene levels in gas streams pulled from rooms. The commercial advantage for CA storage has not been effectively realized for this technology, although it may prove of some benefit in making decisions on the length of storage expectations for fruit just after harvest and perhaps for decisions regarding opening CA rooms for moving fruit into the marketing stream.

The use of volatile biomarkers is also being investigated for their suitability for predicting storability and for market decisions (Rudell et al., 2009; Rudell and Watkins, 2011), but this concept has not yet been successfully commercialized.

Areas of innovation in MAP

In 2005, fresh-cut salads were estimated to be valued at approximately $3 billion in the United States (Calderon, 2006). The 2014 estimate of the value of packaged fresh-cut produce in the United States by the Produce Marketing Association is $27 billion. Ironically, Platenius (1946) identified a need for investigating the keeping quality of chopped vegetables in packages 60 years before this widespread adoption. In what is referred to as "active" or "intelligent" packaging techniques, packages may be flushed with specific gas mixtures designed to obtain a desired initial atmospheric composition, gases may be actively released or scavenged in the package, a partial vacuum can be imposed, biologically active materials can be incorporated in the package, and so on (Ben-Yehoshua et al., 2005).

Modeling

The development of meaningful models for the design of packages to yield target atmospheres is essentially a mature aspect of MAP technology, but one that continues to remain relatively inaccessible to some produce packaging firms. The earliest work in this area dates back to the late 1950s and early 1960s (Jurin and Karel, 1963; Workman, 1959) and further developed in the 1980s and 1990s (Cameron and Yang, 1982; Cameron et al., 1989; Emond et al., 1991). Cameron et al. (1994) added the feature of temperature accommodation into predictive models, permitting the prediction of O_2 and CO_2 across a range of temperatures once appropriate biological data regarding the temperature sensitivity of respiration was created/incorporated. Models advanced to include estimating the risk of fermentation (Talasila and Cameron, 1995), dynamic changes in the package headspace as a function of time and temperature (Hertog et al., 1998), and the presence and abundance of perforations (Hirata et al., 1996; Fishman et al., 1996). Lacking is a meaningful, publicly accessible database that permits the package designer to incorporate biological data into the design process. Nevertheless, packaging for many commodities is extremely successful and target O_2 and CO_2 levels are reliably obtained using models for initial design and trials under real-world conditions.

Perforation technology

The development of technologies for creating precise holes in packaging films is relatively recent. The perforations are typically between 50 and 250 μm in diameter. Two popular methods of making perforations include electrostatic discharge and laser ablation. Electrostatic discharge is rapid, but does not allow for precise perforation location or shape, although the process is less expensive than lasers. Laser perforations are usually uniformly circular but may not be cylindrical. In fact, they tend to be wider at the upper surface facing the laser and narrower at the bottom. Another method of perforations using hot needles is less commonly used for MAP. Another alternative for establishing perforations is through the incorporation of ceramic materials, zeolites, volcanic rocks, $CaCO_3$, or SiO_2, all of which change packaging permeability.

Perforations can be placed into essentially any film, permitting the use of polymers having little permeability to either O_2 or CO_2 and, if used, would cause anaerobiosis and fermentation. Consequently, nylon and other films with high water permeability in MAP application can be used to enhance water vapor management (Ben-Yehoshua et al., 2005). Some perishables, notably tomatoes and peppers, benefit from slightly lower humidity in the package atmosphere, being less prone to decay development (Shirazi and Cameron, 1987, 1992). One of the potentially useful characteristics of perforated packages is that perforated packages tend to deliver higher CO_2 concentrations for a given level of O_2 (Vallejo and Beaudry, 2010). This may be highly beneficial if the package is to be used to suppress fungal decay or chlorophyll breakdown but may be inappropriate for those products damaged by CO_2 levels above 5%.

Bioactive compounds

Addition of bioactive materials to the package is not limited to MAP, but has potential to contribute to the atmosphere modification or to interact with the product in such a way as to minimize deterioration or decay. For atmosphere modification, technologies that have been evaluated include absorption of oxygen (e.g., iron powder) or CO_2 (e.g., calcium hydroxide and activated carbon), release of CO_2 (e.g., sodium bicarbonate), absorption of water vapor (e.g., silica gel, starch, salts, CaO), absorption of ethylene ($KMnO_4$, activated carbon, zeolites, metal organic frameworks), and its release (Ethrel). Further, the benefits of 1-MCP release into packages have also been evaluated (Nazir Mir, personal communication). A banana-packaging concept termed "RipeLock" employs packaging that creates an elevated CO_2 atmosphere and, with an appropriately timed application of 1-MCP, extends the yellow life of ripening fruit. Commercial rollout is underway with this technology (Edagi et al., 2014; Mir, 2017).

The inclusion of antimicrobial compounds in MAP has been a goal for many years, The influence of vapors of hydrogen peroxide, acetic acid, ethanol, chlorine dioxide, sulfur dioxide, and of aroma volatiles such as 2-nonanone and various aldehydes has been investigated with relation to stored produce (Aharoni and Stadelbacher, 1973; Hamilton-Kemp et al., 1992; Leepipattanawit et al., 1997; Sapers et al., 2003; Song et al., 1996; Vaughn et al., 1993). Despite extensive research, widespread commercial adoption of antimicrobials in packages of produce has yet to occur. There are several hurdles to the implementation of this sort of technology. In some cases, the volatiles alter the aroma of the product to which they are applied. Another potential pitfall is toxicity to the tissues. The inclusion of antimicrobial compounds such as silver and immobilized bacteriocides (e.g., penicillin) in the packaging film itself has also been investigated. However, the limited contact between the film and the surface of the product prevents effective decontamination of either human or plant pathogens.

Sensors

The incorporation of sensors in the packaging films has significant merit. In theory, the sensor system can alert store managers or consumers to unsafe food or can be used to develop a system to respond automatically to or to the package environment or regulate package atmosphere. One interesting example of a sensor, RipeSense, was developed to indicate the ripeness of fruit in a package. This technology was successful technically, but failed to meet market demands for low cost. Smyth et al. (1999) introduced the idea of using an ethanol sensor as an indicator of fermentation in packages. However, no commercial product emerged from this innovation. Another interesting concept involved the idea of creating a temperature-sensitive "valve" for packages that would open when the packages were exposed to temperatures above their design threshold (Patterson and Cameron, 1991). However, the concept had limited utility in that there was no simple way of protecting the film from high temperatures prior to being formed into a package.

Biodegradable plastics

Disposal of plastics used in packages has presented a significant problem to consumers and to communities required to manage the waste. In the United States, plastics (of which, fresh produce packaging forms a small portion) were reported to comprise approximately 10%–12% of the waste stream on a weight basis in 1998 by the US Environmental Protection Agency. Additionally, the half-life of most polymers used in packages is measured in decades, if not longer. There are several biodegradable materials that have been formed into functional polymers, the most successful of which is likely poly(lactic)acid (PLA) (Almenar et al., 2008), which has the potential to add value to the packaged product (Koutsimanis et al., 2012). Additional opportunities exist for the development of films developed from whey protein or polysaccharides, but little to no commercial success is in evidence to date.

Prospects for the future

If the last 20 years serves as an example of the advances possible in CA and MAP, it is very likely that new developments will continue to emerge. Major breakthroughs like the discovery of the benefits of atmosphere modification to perishables, gas-selective properties of polymers, and ethylene action inhibitors are behind us, but refinements in the application of the principles of CA and MCP will continue. In CA, it seems likely that additional sensor technologies will emerge and that they will be integrated in the control systems managing stored fruits and vegetables. Similarly, sensors and monitoring systems to improve control and decision making in MAP systems are likely. The major hurdle in this application is to reduce implementation costs sufficiently to deliver enough value to the consumer that he will recognize the contribution of the technology and pay more for it. Alternatively, if the innovation does not cost more than a small fraction of the total packaging cost, it may prove to be sufficiently competitive to displace existing packaging systems. It is also likely that some potential for new, somewhat unanticipated technologies relating to food safety, traceability, and value-added activities contribute to both CA and MAP.

References

Aharoni, Y., Stadelbacher, G.J., 1973. The toxicity of acetaldehyde vapor to postharvest pathogens of fruits and vegetables. Phytopathology 63, 544–545.

Almenar, E., Samsudin, H., Auras, R., Harte, B., Rubino, M., 2008. Postharvest shelf life extension of blueberries using a biodegradable package. Food Chem. 100, 120–127.

Beaudry, R.M., 1999. Effect of O_2 and CO_2 partial pressure on selected phenomena affecting fruit and vegetable quality. Postharvest Biol. Technol. 15, 293–303.

Beaudry, R.M., 2000. Responses of horticultural commodities to low oxygen: limits to the expanded use of modified atmosphere packaging. HortTechnology 10, 491–500.

Ben-Yehoshua, S., Beaudry, R.M., Fishman, S., Jayanty, S., Mir, N., 2005. Modified atmosphere packaging and controlled atmosphere storage. In: Ben-Yehoshua, S. (Ed.), Environmentally Friendly Technologies for Produce Quality. CRC Press, Boca Raton, FL, pp. 61–112.

Bessemans, N., Verboven, P., Verlinden, B.E., Nicolaï, B.M., 2016. A novel type of dynamic controlled atmosphere storage based on the respiratory quotient (RQ-DCA). Postharvest Biol. Technol. 115, 91–102. https://doi.org/10.1016/j.postharvbio.2015.12.019.

Bishop, D., 1990. Controlled atmosphere storage. In: Dellino, C.V.J. (Ed.), Cold and Chilled Storage Technology. Van Nostrand Reinhold, New York, pp. 68–98.

Burg, S.P., Burg, E.A., 1967. Molecular requirements for the biological activity of ethylene. Plant Physiol. 42, 114–152.

Calderone, V., 2006. Valley's Agriculture Is Ever Evolving. Ag Outlook, Salinas, CA. June 12, 2006.

Cameron, A.C., Yang, S.F., 1982. A simple method for the determination of resistance to gaseous diffusion in plant organs. Plant Physiol. 70, 21–29.

Cameron, A.C., Boylan-Pett, W., Lee, J., 1989. Design of modified atmosphere packaging systems: modeling oxygen concentrations within sealed packages of tomato fruits. J. Food Sci. 54, 1413–1416. 1421.

Cameron, A.C., Beaudry, R.M., Banks, N.H., Yelanich, M.V., 1994. Modified-atmosphere packaging of blueberry fruit: modeling respiration and package oxygen partial pressures as a function of temperature. J. Amer. Soc. Hort. Sci. 119, 534–539.

De Ell, J.R., Ehsani-Moghaddam, B., 2010. Preharvest 1-methylcyclopropene treatment reduces soft scald in "Honeycrisp" apples during storage. Hortscience 45, 414–417.

De Ell, J.R., Ehsani-Moghaddam, B., 2013. Effects of rapid consecutive postharvest 1-methylcyclopropene treatments on fruit quality and storage disorders in apples. Hortscience 48, 227–232.

De Ell, J.R., Lum, G.B., 2017. Effects of low oxygen and 1-methylcyclopropene on storage disorders of "Empire" apples. Hortscience 52, 1265–1270.

Dilley, D.R., 1990. Historical aspects and perspectives of controlled atmosphere storage. In: Calderon, M., Barkai-Golan, R. (Eds.), Food Preservation by Modified Atmospheres. CRC Press, Boca Raton, FL, pp. 187–196.

Dilley, D.R., 2010. Controlled atmosphere storage—chronology and technology. Proc. 9th Int'l Controlled Atmosphere Conference. Acta Hort. 857, 495–502.

Edagi, F.K., D.M. Becerra, N. Mir, E. McCaskey, F.M. Terra and R.L. McGee. 2014. Methods of handling papaya. U.S. Patent Application Publication US2014/0242235A1.

Emond, J.P., Castaigne, F., Toupin, C.J., Desilets, D., 1991. Mathematical modeling of gas exchange in modified atmosphere packaging. Trans. ASAE 34, 239–245.

Fishman, S., Rodov, V., Ben-Yehoshua, S., 1996. Mathematical model for perforation effect on oxygen and water vapor dynamics in modified-atmosphere packages. J. Food Sci. 61, 956–961.

Gran, C.D., Beaudry, R.M., 1993. Determination of the lower oxygen limit for several commercial apple cultivars. Postharvest Biol. Tech. 3, 259–267.

Hamilton-Kemp, T.R., MacCracken, C.T., Loughrin, J.H., Anderson, R.A., Hildebrand, D.F., 1992. Effect of some natural volatile compounds on the natural pathogenic fungi *Alternaria alternata* and *Botrytis cinerea*. J. Chem. Ecol. 18, 1083–1091.

Hertog, M.L.A.T.M., Peppelenbos, H.W., Evelo, R.G., Tijskens, L.M.M., 1998. A dynamic and generic model of gas exchange of respiring produce: the effects of oxygen, carbon dioxide and temperature. Postharvest Biol. Technol. 14, 335–349.

Hirata, T., Makino, Y., Ishikawa, Y., Katusara, S., Hasegawa, Y., 1996. A theoretical model for designing a modified atmosphere packaging with a perforation. Trans. ASAE 39, 1499–1504.

Jurin, V., Karel, M., 1963. Studies on control of respiration of McIntosh apples by packaging methods. Food Technol. 17, 104–108.

Kidd, F., West, C., 1927. A relation between the concentration of oxygen and carbon dioxide in the atmosphere, rate of respiration, and length of storage of apples. Food Investigation Bd. Rpt. London 1925, 41–42.

Koutsimanis, G., Getter, K., Behe, B., Harte, J., Almenar, E., 2012. Influences of packaging attributes on consumer purchase decisions for fresh produce. Appetite 59, 270–280.

Kubo, Y., Inaba, A., Nakamura, R., 1990. Respiration and C_2H_4 production in various harvested crops held in CO_2-enriched atmospheres. J. Amer. Soc. Hort. Sci. 115, 975–978.

Leepipattanawit, R., Beaudry, R.M., Hernandez, R.J., 1997. Control of decay in modified-atmosphere packages of sliced apples using 2-nonanone vapor. J. Food Sci. 62, 1043–1047.

Mditshwa, A., Fawole, O.A., Opara, U.L., 2018. Recent developments on dynamic controlled atmosphere storage of apples—a review. Food Packaging Shelf Life 16, 59–68.

Mir, N. 2017. Active compound formulation package and its subsequent release for use on plant and plant parts. U.S. Patent 9561894.

Mir, N.A., Curell, E., Khan, N., Whitaker, M., Beaudry, R.M., 2001. Harvest maturity, storage temperature, and 1-MCP application frequency alter firmness retention and chlorophyll fluorescence of "Redchief Delicious" apple fruit. J. Amer. Soc. Hort. Sci. 125, 618–624.

Patterson, B.D. and A.C. Cameron. 1991. Modified-atmosphere packaging. Australian Provisional Patent Application No. PK6567, filed (6 June 1991). Commonwealth Scientific and Industrial Research Organization (CSIRO), Patterson, B.; and Michigan State University, Cameron, A.C. International Patent Application No: PCT/AU92/00267 filed (6 October 1992).

Platenius, H., 1946. Films for produce: their physical characteristics and requirements. Mod. Packaging 20 (2), 139–143. 170.

Prange, R.K., DeLong, J.M., Leyte, J.C., Harrison, P.A., 2002. Oxygen concentration affects chlorophyll fluorescence in chlorophyll-containing fruit. Postharvest Biol. Technol. 24, 201–205.

Rudell, D.R., Watkins, C.B., 2011. Predicting storage disorders by developing diagnostic toolboxes. New York Fruit Quarterly 19 (4), 21–24.

Rudell, D.R., Mattheis, J.P., Hertog, M.L.A.T.M., 2009. Metabolic change precedes apple superficial scald symptoms. J. Agric. Food Chem. 57, 8459–8466.

Sapers, G.M., Walker, P.N., Sites, J.E., Annous, B.A., Eblen, D.R., 2003. Vapor-phase decontamination of apples inoculated with *Escherichia coli*. J. Food Sci. 68, 1003–1007.

Schaefer, J.C. and D.J. Bishop. 2014. Dynamic controlled system and method for controlled atmosphere room. US Patent 10143210.

Schouten, S.O.P., Prange, R.K., Verschoor, J., Lammers, T.R., Oosterhaven, J., 1997. Improvement of quality of "Elstar" apples by dynamic control of ULO conditions. Proc. 7th CA Res. Conf. Postharvest Hortic. Ser. 16, 71–78.

Shekarriz, R., Allen, W.L., 2008. Nanoporous electrocatalysis for ethylene monitoring and control. European J. Hort. Science 73, 171–176.

Shirazi, A., Cameron, A.C., 1987. Modified humidity packaging: a new concept for improving the success of modified atmosphere packaging of fresh produce. Hortscience 22, 1055 (Abst).

Shirazi, A., Cameron, A.C., 1992. Controlling relative humidity in modified-atmosphere packages of tomato fruit. Hortscience 27, 336–339.

Sisler, E. and S. Blankenship. 1996. Method of counteracting an ethylene response in plants. U.S. Patent No. 5518988 (May 21, 1996). Washington, DC: U.S. Patent and Trademark Office.

Smyth, A.B., Talasila, P.C., Cameron, A.C., 1999. An ethanol biosensor can detect low-oxygen injury in modified atmosphere packages of fresh-cut produce. Postharvest Biol. Technol. 15, 127–134.

Solomos, T., 2017. Aspects of the biology and physics underlying modified atmosphere packaging. In: Yildiz, F., Wiley, R.C. (Eds.), Minimally Processed Refrigerated Fruits and Vegetables, Food Engineering Series. Springer Science+Business Media LLC, pp. 17–51. https://doi.org/10.1007/978-1-4939-7018-6_2.

Song, J., Leepipattanawit, R., Deng, W., Beaudry, R.M., 1996. Hexanal vapor is a natural, metabolizable fungicide: inhibition of fungal activity and enhancement of aroma in apple slices. J. Amer. Soc. Hort. Sci. 121, 937–942.

Sozzi, G., Beaudry, R.M., 2007. Current perspectives on the use of 1-methylcyclopropene in tree fruit crops: an international survey. Stewart Postharvest Reviews 3 (2), 1–16. https://doi.org/10.2212/spr.2007.2.8.

Talasila, P.C., Cameron, A.C., 1995. Modeling distribution of steady-state oxygen levels in modified-atmosphere packages: an approach for designing safe packages. J. Food Process Eng 18, 199–217.

Thewes, F.R., Both, V., Brackmann, A., Weber, A., de Oliveira Anese, R., 2015. Dynamic controlled atmosphere and ultralow oxygen storage on "Gala" mutants quality maintenance. Food Chem. 188, 62–70.

Tucker, M., Laties, G., 1985. The dual role of oxygen in avocado respiration: kinetic analysis and computer modeling of diffusion-affected respiratory oxygen isotherm. Plant Cell Environ. 9, 117–127.

Vallejo, F., Beaudry, R.M., 2010. Comparing the function of perforated and non-perforated film in MAP: impact on the quality of broccoli. Proc. 9th Int'l Controlled Atmosphere Conference. Acta Hort. 857, 503–512.

Vaughn, S.F., Spence, G.F., Sasha, B.S., 1993. Volatile compounds from raspberry and strawberry fruit inhibit postharvest decay fungi. J. Food Sci. 58, 793–796.

Wahl, E.W., Tan, S.M., Konlikov, S., Kharlamov, B., Rella, R.R., Crosson, E.R., Biswell, D., Paldus, B.A., 2006. Ultrasensitive ethylene post-harvest monitor based on cavity ring-down spectroscopy. Opt. Express 14 (4), 1673–1684.

Wolfe, G.C., Black, J.L., Jordan, R.A., 1993. The dynamic control of storage atmospheres. In: Proceedings of the 6th Int'l CA Conference, Ithaca, NY, pp. 323–332.

Workman, M., 1959. The status of polyethylene film liners to provide modified atmosphere for the storage of apples. Eastern Fruit Grower 23 (6), 10–14.

Zanella, A., 2003. Control of apple superficial scald and ripening—A comparison between 1-methylcyclopropene and diphenylamine postharvest treatments, initial low oxygen stress and ultralow oxygen storage. Postharvest Biol. Technol. 27, 69–78.

Further reading

Cameron, A.C., Talasila, P.C., Joles, D.J., 1995. Predicting the film permeability needs for modified-atmosphere packaging of lightly processed fruits and vegetables. Hortscience 30, 25–34.

Chapter 13

Success stories for CA/MA

Xi Chen, Chenyi Xu and Nazir Mir
MirTech, Inc., Somerset, NJ, United States

Introduction

Fresh produce industry in US alone has more than $60 billion in annual sales and is one of the fastest growing industries due to consumer trend toward healthy and fresh food (Anonymous, 2015a, 2016a). Postharvest loss in the fresh produce industry, including fruit and vegetables, is a worldwide problem, and it is estimated to be up to 25% in the developed countries and up to 50% in the developing countries (Hodges et al., 2011; Muth and Karns, 2011). Some of the causes for these losses are predominantly related to ripening, senescence due to aging, microbial spoilage, and moisture/high humidity-related issues (Watada et al., 1990; Schlimme and Rooney, 1994). There is a substantial loss that could be addressed with modified atmosphere packaging (MAP)-like technologies (Kader and Watkins, 2000; Gorny, 2001; Farber et al., 2003; Irtwange, 2006).

The use of packaging for produce items is common in fresh produce industry for reasons varying from esthetics, moisture transmission, and barrier properties, to freshness maintenance-focused functional needs (Lee et al., 2008). MAP combines the esthetic value of packaging with functional properties, which thereby helps maintain quality and extend shelf life. There are growing interests to use flexible polymeric films with improved gas and moisture transmission properties for a number of produce applications. While high barrier films (those with low transmission rates) are usually used to protect shelf-stable foods against oxygen and water vapor, those films are not suitable for packaging fresh produce. Oxygen below tolerable levels or CO_2 above tolerable levels can both lead to injury on both fruits and vegetables, resulting in decreased shelf life (Kader, 1997; Beaudry, 2000). High oxygen transmission films are particularly necessary for fresh produce where the produce is still respiring, taking in O_2 and giving out CO_2, and thus its packaging requires polymeric films with selective O_2 and CO_2 permeability to match the requirements of different products with specific respiring rates.

Controlled and Modified Atmospheres for Fresh and Fresh-Cut Produce
https://doi.org/10.1016/B978-0-12-804599-2.00014-4

The selection of film and optimization of the transmission property of the film is thus critical and is dovetailed based on the requirements of the respiring product at different storage conditions. It is important to recognize that while modifying gas transmission through packaging atmosphere can improve storability of some fruits and vegetables, it has the potential to induce undesirable effects as well (Mir and Beaudry, 2016).

Currently, a variety of MAP technologies are successfully used for fresh fruits and vegetables, both whole and cut. The focus of this chapter is to provide a glimpse on some of the successful MA/CA technologies for three types of products: (1) MAP for fresh salads and fresh-cut fruits, (2) MAP for banana, and (3) MAP with product specific shelf life attributes in conjunction with microwavable steam cooking function.

MAP for fresh salads and fresh-cut fruits

MAP utility for fresh-cut salads and fruits

Market assessment

The US fresh-cut fruits and vegetables market is one of the fastest growing segments and the retail dollar and volume sales are continuously increasing (Anonymous, 2014a). Sales of fresh-cut fruits and vegetables in key US retailers in 2015 were \$2.2 and \$1.7 billion, respectively, and sales of bag salad were \$3.7 billion. If the sales volume in food service sector is included, the total sales would be \$12.6 billion (Cook, 2016).

Examples of commonly available fresh-cut vegetables include cleaned, chopped, and shredded lettuce; washed and trimmed spinach and leafy greens; broccoli and cauliflower florets; shredded cabbage; shredded carrots and carrot sticks; celery sticks; whole peeled, sliced, and diced onions; peeled, sliced, diced potatoes and other roots; sliced mushrooms; sliced, diced jicama, squash, cucumber; fresh peeled and sliced garlic. Examples of commonly available fresh-cut fruits include apples, watermelon, other melons, honeydews, citrus, tropicals, pineapple, cantaloupe, and mixed fruits. These products are packaged in various forms of packaging including resealable bags, single-serve snack packs, rigid containers and bowls, and so on (Cantwell, 2006). A significant proportion of these products employ MAP.

Physiological concerns for fresh-cut products

Due to the nature of these products, they have different physiological concerns that limit their shelf life. Some concerns include overripening, water loss, and enzymatic discoloration of cut surfaces, microbial decay, senescence and mechanical damage during processing, handling, and transportation (Watada et al., 1990; Schlimme and Rooney, 1994).

MAP technologies for fresh salads and fresh-cut fruits and vegetables

For fresh salad and fresh-cut products, specific gas and moisture transmission properties of packaging materials are achieved by one or combination of: (1) modification of polymeric nature of the film, (2) addition of fillers such as $CaCO_3$ and SiO_2 during film extrusion, and (3) perforations (Mir, 2009).

Modifying polymeric nature of film: Based on the novel polymer characteristics, new films are being constantly introduced to the market place. For instance, high (6%–18%) ethylene vinyl acetate content, low-density polyethylene (Elvax) films (DuPont, Wilmington, DE), oriented polypropylene (OPP) laminates (BP Amoco, Lisle, IL), styrene butadiene block copolymer (K-Resin) films (Phillips Chemical Company, Houston, TX, Perftech), ultra low-density ethylene octene copolymer (Attane series) films (Dow Chemical Company, Midland, MI), and polyolefin plastomer octene copolymer (Affinity series) films (Dow Chemical Company, Midland, MI) have become available in recent years. A new trend in flexible film business is to offer materials as blends of two or three different polymers, in which each polymer performs a specific function, such as contributing strength, transparency, or improved gas transmission, to meet specific product requirements. Similarly, films can be laminated to achieve specified properties.

Addition of fillers: The plastic polymers can be mixed with inert inorganic materials such as $CaCO_3$ and SiO_2, or organic minerals such as nanoclays (Othman, 2014). Such films possess adjusted transmission properties by creating perforations to meet the specific needs of the application; for example, biaxially oriented PP film with fillers can create small perforations generally in the range of 0.14–1.4 μm (Mizutani, 1989). The gas and vapor transmission properties of these films can be manipulated by adjusting the particle size of the filler and the degree of stretching.

Perforation of flexible film: Perforation of flexible films implies holes drilled in the film to enhance gas transmission, especially improved O_2 transmission. If the size of the perforation is between 50 and 300 μm, the films are considered microperforated, and if the perforation exceeds 300 μm, it is called macroperforated films. Perforation can be accomplished in any film, as long as the proper selection of polymer is carried out. There are several technologies available for perforating plastic films. Macroperforation is usually done by mechanical puncturing through cold needle or melting at directed spots by hot needle and flame technologies. Microperforation of flexible films is carried out through high-energy electrical discharge and laser technologies.

Microperforation of plastic films is commonly achieved by CO_2 laser systems. The laser energy generated through such a system is absorbed by a polymeric film in directed spots; the film in those spots gets heated, melted, and eventually evaporated leaving perforations in the polymeric film. Although the principle of laser-drilled perforations is common in all the plastic films, there is a strong interaction between the nature of the polymeric film and its laser absorption characteristics. Some films may need a very small dose of energy to melt while for others it may almost be impossible to drill a hole under commercial requirements (Mir, 2009).

Following is an example of microperforated MAP for fresh-cut produce (adopted from Preco Laser System LLC) to demonstrate the effectiveness of microperforation on modifying gas transmission (Chow, 2015). This example shows an application where fresh-cut melon and citrus are contained in plastic cup with PE lidding films. Table 1 clearly shows that with microperforation, product-specific oxygen transmission rate (OTR) is achievable by packaging film with limited surface area.

MAP for banana

MAP utility for bananas—market assessment

Bananas, as the most globally consumed fruit crop benefit from MAP all along its extensive value chain, reducing food waste, reducing shipping costs, and enhancing the consumer eating experience. The value chain begins in developing countries such as Ecuador, Guatemala, Costa Rica, the Philippines, and other banana producing countries where bananas are harvested, packed in LLDPE-perforated liners in 40-lbs boxes, and shipped in controlled atmosphere (CA) refrigerated containers to developed countries. Along the journey, 7 days to the United States and 21 days to Europe, endogenous ethylene from stressed, bruised, or advanced maturity fruit triggers premature fruit ripening. As fruit ripens, fruit metabolism increases O_2 consumption and generation of CO_2,

TABLE 1 Microperforations on lid for fresh fruit containers

Fresh fruit lid perforation		
Product	*6 oz. melon*	*6 oz. citrus*
Container ID (in)	4	4
Lid area (in^2)	12.6	12.6
OTR required	1000	320
OTR for 2 mil permeable PE	254	254
OTR on lid without perf	29	29
# of perfs on lid	4	2
OTR for 2 mil lid with perf	958	314

OTR is measured in cc/100 sq. in./24 h at 23°C 0% R.H.
Adapted from Chow, C., 2015. Microperforations for Fresh Cut Produce Packaging. Preco Inc. https://www.yumpu.com/en/document/view/36872677/micro-perforation-for-fresh-cut-produce-packaging-preco-inc (accessed 01.08.17).

causing fruit to enter an anaerobic phase producing "gray peel" which in turn results in rejection of fruit at the receiving port. During shipment in refrigerated containers, exogenous ethylene produced by stressed fruits also triggers premature ripening in nonstressed fruits, resulting in entire refrigerated 40-feet container (18,480-kgs) of fruits which are too mature upon arrival in port to be sold. Consumers typically purchase bananas at a stage of 5–6, just prior to optimal eating; however, since fruit are quickly moving toward stage 7 (over ripe), fruits have a narrow range of optimal freshness. According to the USDA/ERS, US consumers experience 35%–53% wastage of fresh bananas (Muth and Karns, 2011).

In 2015 the top 15 banana exporting countries produced an estimated 869 MM boxes (42-lbs/box) of fruit with top Central/South American exporters producing 764 MM boxes. These Latin American producers in this group use a macroperforated LLDPE bag inside the box to reduce moisture loss during shipment, which is estimated at USD 0.10/bag or USD 76 MM. Other major banana markets not included in the top 15 exporting countries include India (1447 MM boxes), China (634 MM boxes), Brazil (362 MM boxes), and Indonesia (281 MM boxes) as per FAOSTAT 2013 data (FAO, 2014, 2015; Workman, 2016). These nonexport banana countries could certainly benefit from MAP considering, for example, the estimated 60% losses from field to consumer in India. In summary, nearly 90% of the 764 MM boxes from top Central/South American producers and perhaps 25%–50% of the top nonexporters or between 681 MM to 1362 MM boxes could benefit from MAP. Assuming both top exporters and top nonexport producers comprise the available market, the current addressable market value for bags would be USD 144MM to USD 212MM. An additional premium would be added to the current USD 0.10/bag based upon the value added by the MA packaging delivered benefits.

MAP for controlling the O_2/CO_2 levels within banana boxes could fit into the supply chain in the shipment of fruit in the tropics, in refrigerated containers during transit to long-distance markets such as Europe, to distribution centers, and finally to retail. Replacing CA refrigerated storage during ocean transit to delay fruit maturation with MA packaging would eliminate an important cost element associated with shipment. MA packaging would also extend the freshness of fruit at retail by reducing shrinkage.

We can consider the following value added MA packaging (Table 2):

$$\text{MA packaging} = \text{total cost savings} + \text{value add}$$

The banana market has shifted dramatically in the last 5 years from major banana brand owners controlling the market to regional/national owners exporting directly to retailers, thus capturing a greater portion of the value. At the same time, retailers have captured a greater portion of the value of the market as the retail segment consolidates and negotiates directly with national banana producers, bypassing the major banana brand owners. With more power consolidated at retail, driving cost down, for example, less shrinkage due to

TABLE 2 Value-added MA packaging

	Value propositions	Cost savings $/box	Value add $/box
Control of fruit respiration and delayed maturation	Replacement of CA refrigeration	1.00	
	Reduces shrink at retail (10%)	2.10	–
	Adds 5 days of freshness (5% sales lift)	–	1.00

bruising, fungal contamination, and fresher fruit, MAP will be carefully evaluated to confirm savings and value add. Exporters, whether national producers or brand owners will also look for cost savings, especially by eliminating CA in refrigerated containers, and ripen turns during transit.

Participants all along the value chain potentially will realize cost savings and/or capture value with MA packaging. Producers will effectively differentiate their product, exporters will reduce the container cost by moving away from CA, retailers will reduce shrink (in store loss due to bruising), and customer who will experience less waste and enjoy fresher tasting fruit.

MA/CA technologies for bananas

This section lists selected MA/CA technologies for bananas. It is important to note that to achieve the desired transmission properties, some methods mentioned in the previous section for fresh salads and fresh-cut fruits are also employed for bananas.

RipeLock™ technology

RipeLock™ is a patented technology from AgroFresh Inc., for banana. It combines specially engineered, laser microperforated MAP bag for banana with the ethylene managing benefits of 1-Methylcyclopropene (1-MCP) (Anonymous, 2016b). Respiration control by MAP bags during ripening and distribution plus the addition of ethylene-sensitivity control by AgroFresh's 1-MCP technology during ripening deliver enhanced fruit quality up and down the value chain. The RipeLock™ system maximizes the marketable fruit life for brand owners and retailers, and enhances the quality of the eating experience for consumers (Anonymous, 2015b).

Bananas are packed in RipeLock™ bags at the farms in producing countries and kept tightly sealed. The fruit is shipped in refrigerated vessels to the importing country where fruit will be consumed. The MAP bags are designed to transmit ethylene and 1-MCP through the bag. The banana is ripened with ethylene and subsequently treated with 1-MCP inside ripening rooms in the destination market near the end of the ripening cycle. The specially designed banana bags not only develop the optimum atmosphere for bananas but also allow both ethylene and 1-MCP to interact with the fruit evenly without opening the bags, making the technology highly conducive to banana handling and logistics.

The RipeLock™ MAP bag is designed with high gas (O_2/CO_2) transmission properties to keep the respiring banana at an optimal gas range during a wider storage/distribution temperature of 58–70 °F. The addition of ethylene management with 1-MCP helps slow the color progression and hold fruit firmer, thereby providing an extended yellow-life while the banana is inside the bag and extends the benefit even after removal of the bananas from MAP bags.

The combination of RipeLock™ 1-MCP with MAP works synergistically to extend yellow-life by 4–6 days over other technologies, and help consistently deliver higher color fruit with extended shelf life. Also, in European markets, RipeLock™ MAP bags have shown extended banana green life, thereby effectively in replacing CA and Banavac on intermediate shipping length routes (Anonymous, 2016c).

BreatheWay® packaging technology

BreatheWay® technology was developed by Landec and licensed to Chiquita by Landec's subsidiary, Apio Inc. (Anonymous, 2009, 2014). The BreatheWay® membrane technology is commercially used for premium single finger bananas sold in convenience and coffee shops (Monk, 2006; Anonymous, 2014).

It utilizes specialty patented polymer patch applied over a small cutout section in a flexible film bag of any polymer (Anonymous, 2008). The area of the patch is tailored based on produce respiration requirement. The BreatheWay® membrane patch is made of patented crystallizable polymer that changes morphology from crystalline to amorphous as the temperature shifts higher, thereby increasing the permeability of both O_2 and CO_2 (Anonymous, 2008). This approach works well with respiring produce, such as banana, as the respiration increases at higher temperature and would help maintain the optimum atmosphere inside the banana bag, throughout the fluctuating distribution temperatures.

The CO_2 permeability of the membrane is around four times higher than the O_2 permeability and increases further as temperatures increase from 0°C to 10° C, resulting in a 1.8 × increase in permeability, while still maintaining a constant CO_2/O_2 permeability ratio (Anonymous, 2008).

The single finger offering has about 22–24 fingers packed in the bag containing BreatheWay® membrane in the tropics. The bags are left unsealed for ethylene transmission during ripening and once the ripening is triggered in the ripening rooms in the importing country, the bags are sealed. The membrane technology helps maintain the yellow-life of the fruit for close to10 days inside the bag.

Other MA technologies for bananas

KeepRipe paks™ technology:

The KeepRipe Paks™ is a MA technology developed by Dole for their 22 count single finger offering. The KeepRipe bags are polyethylene based, and are precision laser perforated to create a higher CO_2 and lower O_2 that would help maintain lower color and prevent sugar spotting. Shelf life is extended as much as 7 days inside the bag (Anonymous, 2017).

Controlled ripening technology (CRT):

CRT is developed by DelMonte for single-serving bananas. The CRT bags are polypropylene based and aimed for single-serving bananas in vending machines and convenience stores. The CRT bags are extremely clear and delays ripening to extend both green and yellow-life. CRT packaging is designed to create a brighter color and sweet banana with shelf life extended up to 3 days longer than normal (Riemenschneider, 2011).

MAP for microwavable steam cooking product

There is a growing consumer demand for microwavable fresh produce cooking meals because of their convenience, versatility, and high nutritional value. These types of products have expanded rapidly in US market as high value-added products (Cook, 2016).

One such technology invented by the authors of this chapter, Supreme Steam® packaging system utilizes packages with built-in customized designed microperforations pattern, which can provide desirable internal atmosphere with low O_2 and high CO_2 to extend shelf life of perishable food products including fresh produce. With selection of microwavable safe polymer materials, it also allows the food products to be cooked within the package in microwave oven. During microwave cooking, the package is pressurized by steam to ensure uniform heat distribution, and the excessive steam can be vented through the microperforation to reduce super heating and ensure safety. Thus, microperforation in these packages can control the package atmosphere for quality maintenance during storage, and regulate pressure during microwave cooking. This technology is used on a wide range of products including vegetables; poultry; fish; home meal replacement products; and products with sauces, seasonings, butter, and cheese (Mir, 2016). In addition to microperforation, other venting systems can also be used such as slits.

Limitations of MAP

To date, the major aim of MAP technology has been to maintain the visual and textural characteristics that define quality (Kader et al., 1989). Recent outbreaks of food-borne illness due to consumption of contaminated fresh packaged fruit and vegetable products in the United States have raised concerns that MAP effects on produce microbiology need to be evaluated (Oliveira et al., 2015; Callejón et al., 2015). In the United States, major microbial pathogen outbreaks in fresh-cut fruits and vegetables include *Shigella sonnei*, *Listeria monocytogenes*, *Escherichia coli*, *Clostridium botulinum*, and *Salmonella* spp. (Hurst, 1995; Scallan et al., 2011). The growth of these organisms on fresh produce has been linked to the overmodified package environments and abusive storage conditions (Hintlian and Hotchkiss, 1986; Oliveira, M., et al., 2015).

The type of microorganisms and growth rate will depend on the product, O_2, CO_2, water vapor, and temperature levels (Brackett, 1989). Low O_2 and/or high CO_2 partial pressures used in MA packages can retard the growth of gram-negative bacteria and favor the development of gram-positive organisms such as lactic acid bacteria (Brackett, 1989; Carlin et al., 1990). In some cases, atmospheric modification creates environments where pathogenic microorganisms can proliferate while inhibiting growth of spoilage microorganisms (Farber, 1991). Thus contaminated food that lacks off-odors or other signs of decomposition may be unwittingly consumed. Temperature abuse during distribution may lead to the development of certain microorganisms (Carlin et al., 1990). It is likely that fresh-cut vegetables are more susceptible to bacterial growth than fresh vegetables due to mechanical processing damage and associated plant cell leakage; for example, King et al. (1991) reported higher bacterial counts on processed lettuce when compared to lettuce heads after storage for 15 days at 2.8°C.

It is important to recognize that while atmosphere modification can improve storability of some fruits and vegetables, it has the potential to induce undesirable effects as well. If O_2 levels decline below concentrations required to sustain aerobic respiration, fermentation and off-flavors may occur (Kays, 1991; Richardson and Kosittrakun, 1995; Martínez-Sánchez et al., 2011). Risks include not only the loss of product quality through fermentative metabolism, but also the growth of potential human pathogens that thrive under anaerobic conditions (Hintlian and Hotchkiss, 1986). Similarly, if CO_2 concentrations exceed a level tolerated by the plant tissues, injury is likely to occur. Ranges of nondamaging O_2 and CO_2 levels have been published for a numbers of fruits and vegetables (Kader, 1997, Beaudry, 1999, 2000), minimally processed products (Gorny, 1997), and flowers and ornamentals (Reid, 1997).

Future trends of MAP

There are still extensive research and development involved in MAP to overcome the abovementioned limitations. New developments focus on two areas: (1) improvement in manufacturing and processing of packaging materials, and (2) addition of functional attributes such as antimicrobials into packaging materials.

(1) Packaging materials: One good example is microperforation of laminated films in which each layer offers different strength, transparency, or gas transmission to meet specific product requirements. But the adhesive of the laminate can interfere with the laser energy absorption process and may dramatically reduce the speed of the laser perforation process. In some instances, the adhesive may serve as a strong sink for the laser energy, thus making it impossible to drill a hole through all layers of the laminate. Thus it is conceivable to think that the development of operating perforation procedures for films which vary in chemical structure due to variations in polymer blending or adhesion of various layers in a laminate can often be extremely complex, time consuming, or impractical for commercialization. The implementation of a perforation protocol for these relatively new film structures depends as expected on their ability to solve the packaging problems and at a cost that would suit the utilizing industry.
(2) Functionality: There are growing interests in adding active and intelligent functionalities into MAP. Attempts include embed gas absorbers/scavengers (e.g., CO_2, ethylene absorber), or volatile releasing active compounds (e.g., antimicrobials and antioxidants) into packaging film; add time–temperature indicator, RFID label, or moisture indicator in the package.

Conclusions

CA/MA technologies and packaging systems are great tools to preserve postharvest storage quality of fresh produce so that to benefit the whole supply chain, and in some cases add value to the commodity during marketing. MAP systems are product specific and may benefit quality retention through protection from handling abuse and moisture loss and restriction of metabolism. MAP also has limitations and not all products can benefit from MAP, and for those that do may differ in their responses to atmospheres generated. To overcome the limitations of MAP, efforts are still needed to improve packaging manufacturing methods and add active functionality to packaging.

References

Anonymous, 2008. BreatheWay® Membrane Technology. http://www.breatheway.com/pdf/BreatheWay_SS_overview6_08.pdf. (Accessed 18 January 2017).

Anonymous, 2009. BreatheWay® Affiliates & Partners. http://www.breatheway.com/affiliates/. (Accessed 18 January 2017).

Anonymous, 2014. Commercial Applications Expanding for Landec BreatheWay Produce Packaging. http://ir.landec.com/releasedetail.cfm?ReleaseID=828097. (Accessed 1 August 2017).

Anonymous, 2014a. U.S. Fresh-Cut Fruit and Vegetable Market. http://www.pma.com/content/articles/2014/08/us-fresh-cut-fruit-and-vegetable-market. (Accessed 1 November 2017).

Anonymous, 2015a. Fresh Produce Industry Overview–Success Factors of Produce Report for Produce Marketing Association. Produce Marketing Association, pp. 6–10.

Anonymous, 2015b. RipeLock Technology. https://www.agrofresh.com/ripelock-technology. (Accessed 29 November 2016).

Anonymous, 2016a. FMI Report: Fresh Produce Growth Driven by Mega Trends. http://www.fmi.org/news-room/latest-news/view/2016/06/22/fmi-report-fresh-produce-growth-driven-by-mega-trends. (Accessed 1 August 2017).

Anonymous, 2016b. RipeLock™ Quality System Controls Ripening for Better Bananas from Harvest to Home. https://www.agrofresh.com/documents/40978/41162/RipeLock+Quality+System+Controls+Ripening+for+Better+Bananas+from+Harvest+to+Home/95bb8d82-6243-420e-8c6a-bb6f5c669854. (Accessed 1 August 2017).

Anonymous, 2016c. AgroFresh aims to revolutionize banana practices. https://www.agrofresh.com/documents/40978/41162/AgroFresh+aims+to+revolutionize+banana+practices/e6b6d8e5-a3ed-4c01-9ac9-c726422f06ff. (Accessed 1 May 2017).

Anonymous, 2017. Going Bananas Over Bananas? Dole KeepRipe Bananas enable you to control the ripening. http://www.jkings.com/pdf/keep-ripe-bananas-revised.pdf. (Accessed 1 November 2017).

Beaudry, R.M., 1999. Effect of O_2 and CO_2 partial pressure on selected phenomena affecting fruit and vegetable quality. Postharvest Biol. Technol. 15, 293–303.

Beaudry, R.M., 2000. Responses of horticultural commodities to low oxygen: limits to the expanded use of modified atmosphere packaging. HortTechnology 10, 491–500.

Brackett, R.E., 1989. Changes in the microflora of packaged fresh broccoli. J. Food Qual. 12, 169–181.

Callejón, R.M., et al., 2015. Reported foodborne outbreaks due to fresh produce in the United States and European Union: trends and causes. Foodborne Pathogens Dis. 12 (1), 32–38.

Cantwell, M., 2006. Fresh-Cut Products: Overview and Challenges. http://ucce.ucdavis.edu/files/datastore/234-1368.pdf. (Accessed 1 November 2017).

Carlin, F., Nguyen-the, C., Hilbert, G., Chambroy, Y., 1990. Modified atmosphere packaging of fresh, "ready-to-use" grated carrots in polymeric films. J. Food Sci. 55, 1033–1038.

Chow, C., 2015. Microperforations for Fresh Cut Produce Packaging. Preco Inc. https://www.yumpu.com/en/document/view/36872677/micro-perforation-for-fresh-cut-produce-packaging-preco-inc. (Accessed 1 August 2017).

Cook, R.A., 2016. Fresh-Cut/Value-Added Produce Marketing Trends. https://arefiles.ucdavis.edu/uploads/filer_public/fb/7b/fb7b6380-cdf9-4db5-b5d2-993640bcc1e6/freshcut2016cook20160926final.pdf. (Accessed 1 November 2017).

FAO, 2014. Banana Market Review and Banana Statistics 2012–2013. http://www.fao.org/docrep/019/i3627e/i3627e.pdf. (Accessed 12 May 2016).

FAO, 2015. Banana Market Review 2013–2014. http://www.fao.org/fileadmin/templates/est/COMM_MARKETS_MONITORING/Bananas/Documents/Banana_Information_Note_2014-_rev.pdf. (Accessed 1 August 2017).

Farber, J.M., 1991. Microbiological aspects of modified atmosphere packaging technology. J. Food Prot. 54, 58–70.

Farber, J.N., Harris, L.J., Parish, M.E., Beuchat, L.R., Suslow, T.V., Gorney, J.R., Garrett, E.H., 2003. Microbiological safety of controlled and modified atmosphere packaging of fresh and fresh-cut produce. Compr. Rev. Food Sci. Food Saf. 2 (s1), 142–160.

Gorny, J.R., 1997. A summary of CA and MA requirements and recommendations for fresh-cut (minimally-processed) fruits and vegetables. In: Gorny, J. (Ed.), Proceeding of the Seventh International Controlled Atmosphere Research Conference. In: Vol. 5. University of Davis, pp. 30–66.

Gorny, J.R., 2001. A summary of CA and MA requirements and recommendations for fresh-cut (minimally processed) fruits and vegetables. In: VIII International Controlled Atmosphere Research Conference 600.

Hintlian, C.B., Hotchkiss, J.H., 1986. The safety of modified atmosphere packaging: a review. Food Technol. 40, 70–76.

Hodges, R.J., Buzby, J.C., Bennett, B., 2011. Postharvest losses and waste in developed and less developed countries: opportunities to improve resource use. J. Agric. Sci. 149, 37–45.

Hurst, W.C., 1995. Sanitation of lightly processed fruits and vegetables. Hort. Science 30, 22–24.

Irtwange, S., 2006. Application of modified atmosphere packaging and related technology in post-harvest handling of fresh fruits and vegetables. Agric. Eng. Int. 4, 1–13.

Kader, A.A., 1997. Biological bases of O_2 and CO_2 effects on postharvest life of horticultural perishables. In: Vegetables and Ornamentals, CA'97 Proceedings. University of California, Davis, CA.

Kader, A.A., Watkins, C.B., 2000. Modified atmosphere packaging—toward 2000 and beyond. HortTechnology 10, 483–486.

Kader, A.A., Zagory, D., Kerbel, E.L., 1989. Modified atmosphere packaging of fruits and vegetables. Crit. Rev. Food Sci. 28, 1–30.

Kays, S.J., 1991. Postharvest physiology of perishable plant products. Van Nostrand Reinhold, New York.

King Jr., A.D., Magnusen, J.A., Torok, T., Goodman, N., 1991. Microbial flora and storage quality of partially processed lettuce. J. Food Sci. 56, 459–461.

Lee, D.S., Yam, K., et al., 2008. Overview of food packaging systems. In: Lee, D.S., Yam, K., Piergiovanni, L. (Eds.), Food Packaging Science and Technology. Taylor and Francis Group, LLC, New York, pp. 1–11.

Martínez-Sánchez, A., et al., 2011. Low oxygen levels and light exposure affect quality of fresh-cut Romaine lettuce. Postharvest Biol. Technol 59 (1), 34–42.

Mir, N.A., 2009. Perforated films. In: Yam, K.L. (Ed.), The Wiley Encyclopedia of Packaging Technology. John Wiley & Sons Inc, Hoboken, NJ, pp. 486–488.

Mir, N.A., 2016. How FreshTech "Supreme Steam®" Cook System Works. http://www.freshtechinc.com/science-how-supreme-steam-works.php. (Accessed 18 January 2017).

Mir, N.A., Beaudry, R.M., 2016. Handbook 66: Modified atmosphere packaging, United States Department of Agriculture Publication. In: Gross, K.C., Wang, C.Y., Saltveit, M.E. (Eds.), The Commercial Storage of Fruits, Vegetables, and Florist and Nursery Stocks, Agriculture Handbook Number 66. U.S. Department of Agriculture, Washington, DC. http://ucanr.edu/datastoreFiles/234-2927.pdf. (Accessed 27 October 2016).

Mizutani, Y., 1989. Microporous polypropylene sheets. Ind. Eng. Chem. Res. 32, 221–227.

Monk, D., 2006. Shaking the Tree. http://www.breatheway.com/news/2006.06.12.aspx. (Accessed 11 November 2016).

Muth, M.K., Karns, S.A., 2011. Consumer-Level Food Loss Estimates and Their Use in the ERS Loss-Adjusted Food Availability Data. USDA. http://www.ers.usda.gov/media/134674/tb1927.pdf. (Accessed 1 August 2017).

Oliveira, M., et al., 2015. Application of modified atmosphere packaging as a safety approach to fresh-cut fruits and vegetables—a review. Trends Food Sci Technol 46 (1), 13–26.

Othman, S.H., 2014. Bio-nanocomposite materials for food packaging applications: types of biopolymer and nano-sized filler. Agric. Agric. Sci. Proc. 2, 296–303.

Reid, M.S., 1997. A summary of CA and MA requirements and recommendations for ornamentals and cut flowers. CA. 97Vol. 4 Vegetables and Ornamentals, pp. 129–136.

Richardson, D.G., Kosittrakun, M., 1995. Off-flavor development of apples, pears, berries and plums under anaerobiosis and partial reversal in air. In: Rouseff, R.L., Leahy, M.L. (Eds.), Fruit Flavors, Biogenesis, Characterization and Authentication. American Chemical Society, Washington, DC, pp. 211–275.

Riemenschneider, P., 2011. Del Monte Unveils Controlled Ripening Technology for Bananas. http://www.thepacker.com/fruit-vegetable-news/del_monte_unveils_controlled_ripening_technology_for_bananas_122124264.html. (Accessed 1 November 2017).

Scallan, E., et al., 2011. Foodborne illness acquired in the United States—major pathogens. Emerg. Infect. Dis. 17 (1), 7.

Schlimme, D.V., Rooney, M.L., 1994. Packaging of minimally processed fruits and vegetables. In: - Wiley, R.C. (Ed.), Minimally Processed Refrigerated Fruits and Vegetables. Chapman and Hall, New York, NY, pp. 135–182.

Watada, A., Abe, K., Yamauchi, N., 1990. Physiological activities of partially processed fruits and vegetables. Food Technol. 44, 116–122.

Workman, D., 2016. Bananas Exports by Country. http://www.worldstopexports.com/bananas-exports-country/. (Accessed 18 December 2016).

Part II

CA/MA requirements and recommendations for fresh and fresh-cut fruits

Chapter 14.1

Pome fruits: Apple quality and storage

Jennifer R. DeEll
Ontario Ministry of Agriculture, Food and Rural Affairs, Simcoe, ON, Canada

Quality characteristics

There are three main quality characteristics of apple: appearance, texture, and flavor.

Visual appearance is a large component of apple quality. There are specific commercial requirements for skin color based on cultivar and these can range from fully green to 100% red blush for cultivars such as 'Granny Smith' or 'Delicious,' respectively. There are also bicolor requirements for cultivars like 'Gala' and 'Braeburn,' while consumers prefer 'Golden Delicious' with a pale yellow skin.

High quality apples are typically free from blemish and defects. Bruising, stem punctures, physiological disorders, and disease are not acceptable (although the organic market may be a little more tolerant of these).

Most consumers demand firm crisp apples, but optimum firmness values vary with cultivar, such as 81–99 N (18–22 lb-force) for 'Granny Smith' and 54–63 N (12–14 lb-force) for 'Golden Delicious.' Flavor is another important component of apple quality, and sweetness and acidity also vary with cultivar.

Controlled and Modified Atmospheres for Fresh and Fresh-Cut Produce
https://doi.org/10.1016/B978-0-12-804599-2.00015-6

'Granny Smith' is known for high acidity (0.8%–1.2% malate), while 'Fuji' is very sweet (≥20% soluble solids concentration).

Main causes of quality loss

Respiration rate is proportional to deterioration, so factors resulting in high fruit respiration will result in rapid deterioration. Respiration increases exponentially with increasing temperature, leading to rapid fruit quality loss. Therefore low temperature is the primary way to retain apple quality.

Loss in quality also occurs due to disease development. The main postharvest diseases of apple are gray mold (*Botrytis cinerea*) and blue mold (*Penicillium* spp.), which enter the fruit primarily through damaged areas. Several other pathogens can appear in stored apples, coming from infections in the orchards but remaining latent until the fruit weakens in storage.

Physiological disorders can result in significant apple quality loss. Some disorders develop in apples on the tree, such as water core, in which sorbitol fills the intercellular air spaces. Other disorders develop on the tree as well as during storage, such as bitter pit and lenticel breakdown or blotch. There are also disorders that develop during storage, related to senescence, chilling, or atmospheric stress. Storage disorders of apples will be discussed in "Problems under suboptimal conditions" section.

Compositional changes that alter appearance, texture, or fruit flavor can also result in quality loss. Fruit softening to below minimum levels for consumer acceptance, water loss leading to shrivel, or bruising due to poor handling all result in reduced apple quality.

Optimum storage conditions

There are many benefits to using controlled atmosphere (CA) storage for apples. These include delayed ripening, reduced fruit respiration and ethylene production, improved firmness and acidity retention, control of senescent-related disorders and superficial scald, limited pathogen development, and overall extension of apple storage life. Only apples harvested at proper maturity and of good quality will result in successful CA storage. Common CA storage conditions for major apple cultivars are presented in Table 1.

Oxygen levels below 1% are becoming more popular due to recent advancements in dynamic CA and improved accuracy of the required gas analyzers. Lower O_2 concentrations increase apple sensitivity to CO_2, so reduced CO_2 levels become more important. In general, the CO_2 concentration should not be much above the O_2 level.

Postharvest treatment of apples with 1-methylcyclopropene (1-MCP) to inhibit ethylene and delay ripening is common practice in many countries. Unfortunately, 1-MCP has also been shown to increase susceptibility to CO_2 injury in sensitive apple cultivars, so CA storage regimes must be adjusted accordingly.

TABLE 1 Common controlled atmosphere storage conditions for major apple cultivars[a]

Cultivar	O_2 (%)	CO_2 (%)	Temperature (°C)	Comments
Ambrosia	1.2–2.5	1–2%	0–1	
Braeburn	1.2–2	<1	1–2	Sensitive to CO_2; CA delayed in some regions
Bramley's Seedling	1	5	4–4.5	Sensitive to CO_2; CA delayed in some regions
Cortland	2–3	2–3	0–2	
Cox's Orange Pippin	1–3	0.5–2	3–4	Lower CO_2 required with lower O_2
Cripps Pink	1–2	1–2	0–1	
Delicious (Red)	1–2	1–2	0	<1% O_2 with dynamic CA
Elstar	1–2	1–3	1–2	
Empire	1.5–2.5	1–2	2	Sensitive to CO_2; if not treated with diphenylamine keep CO_2 <0.5% for first month of storage
Envy	3	0–1	0.5	Sensitive to CO_2; slow stepwise cooling is beneficial
Fuji	1.5–2	0.5–1	0–1	Sensitive to CO_2
Gala	1–2	0.5–3	0–1	<1% O_2 with dynamic CA
Gloster	1.5–3	2–3	0–1	
Golden Delicious	1–2	1–3	0–1	<1% O_2 with dynamic CA; slow stepwise cooling required if treated with 1-methylcyclopropene
Granny Smith	1.5–2	0.5–2	0–1	<1% O_2 with dynamic CA
Honeycrisp	2–3	1–3	3	Sensitive to CO_2; CA not recommended in some regions; if not treated with diphenylamine keep CO_2 <0.5% for first month of storage or delay CA

Continued

TABLE 1 Common controlled atmosphere storage conditions for major apple cultivars—cont'd

Cultivar	O_2 (%)	CO_2 (%)	Temperature (°C)	Comments
Idared	2	2–3	1	
Jazz	1.5–2	1–2	0.5–1	Slow stepwise cooling is beneficial
Jonagold	1–3	2–4	0–2	
Law Rome	1.5–2	2–3	0	
McIntosh	2–3	2–3 for 1 month, then 4–5	2–3	Some strains more sensitive to low O_2, i.e., Marshall
Mutsu	1.5–2	2–3	0	
Northern Spy	1.5–2	2–3	0	
Spartan	2–3	0.5–3	0–2	6% CO_2 used in some regions for long-term storage
Stayman	2–3	2–5	0–1	

[a] *The author thanks Randy Beaudry, Richard Colgan, Ian Crouch, Jim Mattheis, and Chris Watkins for reviewing and/or contributing to this table.*

Problems under suboptimal conditions

Storage disorders can develop in apples due to atmospheric stress and/or chilling. Low oxygen injury results in loss of flavor, fermentation-related odors, purpling and splitting of the skin, and ultimately severe flesh browning (Fig. 1).

CO_2 injury can be expressed internally or externally. Flesh cavities and/or brown heart are characteristic of internal CO_2 injury. Alternatively, external CO_2 injury typically consists of well-defined, partly sunken, rough bronze lesions on the apple skin (Fig. 2).

Chilling-related disorders include low-temperature breakdown, core browning, and internal or flesh browning (Fig. 3). These disorders generally increase in incidence and severity as time in storage increases, but they are also influenced by apple cultivar, weather, and growing conditions. Soft scald is related to chilling sensitivity as well as climate. This disorder is characterized by sharply defined, irregular shaped, smooth areas of soft brown lesions of the skin (Fig. 4).

Superficial scald is a major storage disorder, which can be substantially reduced by CA storage and low O_2. Diffuse browning of the skin develops after several months of storage and becomes more intensive at room temperature.

FIG. 1 Skin purpling and splitting due to low oxygen injury (right), as compared to a similar healthy apple (left).

FIG. 2 External CO_2 injury, as characterized by well-defined, partly sunken, rough bronze lesions on the apple skin.

FIG. 3 Core and flesh browning in an apple from CA storage.

FIG. 4 Soft scald, which can be induced by low temperatures in susceptible apples.

Senescent breakdown is related to harvesting overmature fruit and exacerbated by higher than optimal storage temperatures. Deep browning of the flesh begins directly underneath the skin and the whole fruit may become dry and mealy.

References

DeEll, J., 2012. Controlled atmosphere storage guidelines and recommendations for apples. Factsheet Order No. 12-045, Agdex 211/66, Ontario Ministry of Agriculture, Food and Rural Affairs (Canada).

DeEll, J., 2017. Storage Disorders of Apples, 3rd ed. Poster published by Storage Control Systems Inc., Michigan (USA). http://www.storagecontrol.com/products/disorders-poster/.

Hardenburg, R.E., Watada, A.E., Wang, C.Y., 1986. The Commercial Storage of Fruits, Vegetables, and Florist and Nursery Stocks. Agriculture Handbook No. 66 (revised). United States Department of Agriculture.

Kupferman, E., 2001. Controlled atmosphere storage of apples and pears. Washington State University—Tree Fruit Research and Extension Center, Postharvest Information Network. http://postharvest.tfrec.wsu.edu/emk2001d.pdf.

Thompson, A.K., 2010. Controlled Atmosphere Storage of Fruits & Vegetables, second ed.

Watkins, C.B., Kupferman, E., Rosenberger, D.A., 2016. Apple, In: Gross, K.C., Wang, C.Y., Saltveit, M.E. (Eds.), The Commercial Storage of Fruits, Vegetables, and Florist and Nursery Stocks. Agriculture Handbook No. 66, U.S. Department of Agriculture, Washington, D.C. http://ucanr.edu/datastoreFiles/234-2927.pdf (accessed 27.10.16), pp. 176-194.

Webster, T., Cross, J., Berrie, A., Johnson, D., Biddlecombe, T., Pennell, D., Luton, M., Guest, G., Ashdown, L., 2016. Optimal storage conditions, in: Apple Best Practice Guide, AHDB Horticulture (a division of the Agriculture and Horticulture Development Board). United Kingdom, http://apples.ahdb.org.uk/post-harvest-section9.asp.

World Food Logistics Organization, 2008. Apples, controlled atmosphere storage, in: WFLO Commodity Storage Manual. Global Cold Chain Alliance, Virginia (USA). http://www.gcca.org/commodity-storage-manual/.

Chapter 14.2

Pome fruits: Pears

Yan Wang[a]
Mid-Columbia Agricultural Research and Extension Center, Oregon State University, Hood River, OR, United States

Quality characteristics

European pears are harvested when mature and must be kept in an unripened (preclimacteric) state during storage/shipping, yet be capable of ripening properly when placed at room temperature by consumers. Consumers enjoy pears when fruit have ripened to a buttery-juicy texture with characteristic pear flavor. Pear fruit are generally resistant to ripening at harvest and require exposure to a period of chilling or ethylene conditioning in order to develop the capacity to produce ethylene internally at a sufficient rate to activate and complete the ripening process when subsequently maintained at room temperatures.

a. ✠ Deceased.

Controlled and Modified Atmospheres for Fresh and Fresh-Cut Produce
https://doi.org/10.1016/B978-0-12-804599-2.00016-8

The postharvest chilling or ethylene conditioning also promote uniform ripening of individual fruit as well as synchronize ripening among fruit in a lot. The amount of postharvest chilling requirement differs among cultivars and is influenced by fruit harvest maturity, orchard location, and production year. For example, it may need 30–90, 0–30, 30, 15–25 days of postharvest chilling for proper ripening for 'd'Anjou,' 'Bosc,' 'Comice,' and 'Bartlett' pears, respectively. Ethylene conditioning pears with 100-ppm ethylene gas for 2–3 days at room temperature (fruit core temperature $> 18°C$) can reduce or replace the chilling requirement for ripening capacity and uniform ripening. Harvest maturity is an extremely important factor affecting storage potential and eating quality of European pears after storage. While immature pears are lack of flavor and more susceptible to shrivel, overmature fruit undergo a faster quality deterioration with developing a dry-coarse instead of buttery-juicy texture and have a greater susceptibility to physiological disorders and fungal decay during storage. Recommended range of fruit firmness as a reliable harvest maturity index differs among pear cultivars (Table 1).

Main causes of quality loss

The main causes of quality loss during postharvest storage and transportation of European pears are superficial scald, deterioration in texture and flavor, senescence disorders, and friction discoloration. Other physiological disorders caused by high CO_2 and/or low O_2 injuries may occur in CA storage.

Superficial scald

Superficial scald, characterized as brown or black patches on fruit skin during or after storage, is a major factor influencing postharvest management of certain European pear cultivars. The current commercial control for scald of 'd'Anjou' pear in the US Pacific Northwest (PNW) relies on the prestorage application of an antioxidant 1,2-dihydro-6-ethoxy-2, 2, 4-trimethyl-quinoline (ethoxyquin) while 'Packham's Triumph' pear in the Southern Hemisphere by diphenylamine (DPA), either as a drench, line spray, thermofogging, or in treated paper wraps. However, the commercial control for scald by antioxidants is not adequate since significant storage loss due to scald occurs in certain years. Many factors such as harvest maturity and excess heat in growing seasons affect fruit susceptibility and the efficacy of the antioxidants. While these antioxidant materials are not registered for use on pears in some pear production areas (like California), the European Union withdrew authorization for plant protection products containing ethoxyquin and DPA in 2009. CA storage at low O_2 (i.e., $\leq 1\%$ O_2) controls scald; however, other disorders are increased by the same low-O_2 conditions. 1-Methylcyclopropene (1-MCP) inhibits scald development in European pears; however, it interferes with pear's ability to ripen normally after storage. Alternative antioxidants, low-oxygen CA storage, and 1-MCP treatment are currently investigated for superficial scald control.

TABLE 1 **Harvest maturity, storage life, and optimal CA conditions for major cultivars of European pears kept at -1.1 to 0°C and 90%–95% relative humidity**

Cultivar	Harvest flesh firmness (N)	Air storage life (months)	CA O_2%	CA CO_2%	CA storage life (months)
Abate Fetel	58–67	3–4	1–2	0–1	4–6
d'Anjou	58–67	5	1–2.5 (0.8–1.0)	0–1 (0–0.1)	6–8 (8–10)
Bartlett	67–84	2–3	1–2	0–0.5	3–6
Bosc	62–71	4	1–2.5	0.5–1.5	5–7
Comice	49–58	4	1.5–4	0.5–4	5–6
Concord	49–58	4–5	0.5–2	0–1	6–7
Conference	49–58	5–6	1–2.5	0.5–1.5	7–8
Forelle	58–67	4–5	1–2	0–1.5	6–7
Packham's Triumph	62–71	6	1.5–2	1.5–2.5	7–8
Rocha	60–70	5	3	0.7	10
Winter Nelis	58–67	7–8	1–3	0–1	8–10

Source: Reproduced and modified from Bai, J., Prange, R.K., Toivonen, M.A., 2009. Pome fruits, in: Yahia, E.M. (Ed.), Modified and Controlled Atmospheres for the Storage, Transportation, and Packing of Horticultural Commodities. CRC Press, Taylor & Francis Group, Boca Raton, FL, pp. 267–286; Chen P.M., 2004. Pear, in: Gross, K.C., Wang, C.Y., Saltveit, M.E. (Eds.), The Commercial Storage of Fruits, Vegetables, and Florist and Nursery Crops. USDA, ARS, Agriculture Handbook 66, Washington D.C. pp. 471–480; Hansen, E., Mellenthin, W.M., 1979. Commercial handling and storage practices for winter pears. Oregon Agr. Expt. Sta. Spec. Rpt. 550; Kader, A.A., 2007. Controlled atmospheres, in: Mitcham, E.J., Elkins, R.B. (Eds), Pear Production and Handling Manual. University of California, Division of Agriculture and Natural Resources, Publication #3483, pp. 175–177; Kupferman, E., 2003. Controlled atmosphere storage of apples and pears. Acta Hort., 600, 729–735; Sugar, D., 2007. Postharvest handlings of winter pears, in: Mitcham, E.J., Elkins, R.B. (Eds), Pear Production and Handling Manual. University of California, Division of Agriculture and Natural Resources, Publication #3483, pp. 175–177.

Deterioration in texture and flavor

Upon gaining the capacity to soften after a period of chilling, pears can develop into either a buttery-juicy (melting) texture or a dry-coarse (mealy) texture within 5–7 days at room temperatures depending on storage condition and duration. For example, at appropriate storage temperature of −1.1°C, 'd'Anjou' pears develop a buttery-juicy texture after ripening following 3–5 months storage in regular air (RA), 5–8 months in normal CA (i.e., 1%–2.5% O_2), and 8–10 months in low-O_2 CA (i.e., 0.8%–1% O_2). In contrast, fruit may develop a dry-coarse texture after ripening following >5 months storage in RA, >8 months in normal CA, and >10 months in low-O_2 CA. Fruit texture and flavor after storage are also affected by harvest maturity, fruit nutrition, and culture factors. For example, late-harvested fruit or fruit low in calcium content generally have a short storage life with a good buttery-juicy texture.

Senescence disorders

Significant losses may occur after long-term storage or long-distance shipping due to senescence disorders including senescent scald and core breakdown of pears. Senescent fruit are more susceptible to decay. CA/MA and ethylene inhibitors such as 1-MCP (pre- or postharvest treatments) and AVG (aminoethoxyvinylglycine) (preharvest treatment) reduce senescence disorders of pears.

Friction discoloration

The pear cultivars with tender skin such as 'd'Anjou' and 'Comice' are subject to friction discoloration (belt burn) during packing. The susceptibility of pears to friction discoloration increases as storage duration extends. Cold fruit are less prone to discoloration than those that have warmed during packing. A sucrose ester edible coating (Semperfresh™) applied to fruit as a line spray gives protection of fruit from friction discoloration occurring further down the packing line. Slowing down the packing speed is also used by industry to minimize friction discoloration.

Optimum storage conditions

Pear fruit begin aging and quality loss once removed from the trees and this physiological process is greatly slowed by storing fruit at low temperatures. CA is considered as an important and efficient supplemental treatment to low temperatures for further extending storage quality of pears. Both of low temperatures and CA reduce the rates of fruit respiration and ethylene production. Accordingly, the storage life of pears is maximized by cooling fruit as soon as possible after harvest, maintaining a stable low temperature during packing and storage, and establishing a CA environment promptly after cooling.

Optimum temperature

Storage quality of pears is very sensitive to temperature during precooling and long-term storage. Rapid removal of field heat during precooling is essential for a successful long-term storage of pears. It is recommended to cool down to fruit core temperature below 5°C within 1–2 and 2–3 days of harvest for 'Bartlett' and 'd'Anjou' pears, respectively. During the temperature pull-down period, storage room temperatures at −3.5°C to −2.0°C can be used but should be raised to −1°C as the fruit pulp temperature approach the target temperature. Pear fruit can be safely stored at temperatures slightly above their freezing point. The minimum temperature that the fruit can tolerate should be maintained consistently during the storage period. The actual freezing temperature is influenced by the soluble solids content (SSC) of the fruit, such as the freezing point at SSC of 10%–12% would be approximately −1.7°C to −2.2°C. The storage life of 'd'Anjou' and 'Bartlett' pears is 35%–40% longer at −1°C than at 0°C. Typically, pears are stored at temperature of −1.1°C.

Optimum relative humidity (RH)

Compared to apples, pear fruit are very susceptible to moisture loss and consequently shrivel during long-term storage. Hence, it is advisable to precool fruit fast after harvest and maintain RH levels of 90%–95% in cold storage rooms. Fruit packed in perforated polyethylene box liners are generally free from shrivel. However, fruit stored in bulk bins are at risk of shrivel even at the target RH after long-term storage. Lining the sides and covering the tops of bins after cooling with polyethylene film prevent shrivel without adversely delaying cooling. It is not recommended to introduce water vapor to storage rooms at −1.1°C for reducing moisture loss of pears because of ice formation on fruit and floors. Edible coatings are used during presizing to reduce moisture loss of pears stored in bulk bins in the PNW.

CA/MA considerations

European pears are mainly stored under CA conditions in long-term commercial storage. The appropriate O_2 and CO_2 levels vary among different pear cultivars and may vary with growing district. Generally, the optimal CA range for most of the major cultivars is 1%–2.5% O_2 and 0–1.5% CO_2 at −1.1 to 0°C and 90%–95% RH (Table 1). A low-O_2 CA at 0.8%–1% O_2 + < 0.1% CO_2 has also been used for 'd'Anjou' pears for up to 8–10 months storage in the PNW. A stepwise low-O_2 CA procedure is used as an alternative to ethoxyquin for controlling superficial scald of 'd'Anjou' pears: storing in 0.7%–0.8% O_2 + < 0.1% CO_2 for 100–120 days before transfer to a normal CA for another 100–120 days. A short-term (i.e., 2 weeks) high-CO_2 (i.e., 12%) treatment improves storage life of 'd'Anjou' pears in subsequent RA storage but not in subsequent CA storage; this method has not been adopted by the pear industry.

Exogenous ethylene can increase ethylene production and respiration rate and advance green color loss and flesh softening of pears stored at elevated temperatures. However, these effects are minimum at storage temperature of −1.1°C under CA conditions. It is not a practical procedure to remove ethylene from commercial CA storage of pears.

Modified atmosphere packaging (MAP) is used by the pear industry in the PNW, although limited amount, to store pears for 1–4 months plus transportation for 1–3 weeks. MAP with an equilibrium atmosphere at ~10%–15% O_2 + ~4%–6% CO_2 extends storage quality of 'Bartlett,' 'Bosc,' and 'Comice' at −1.1°C and maintains fruit quality during transportation at 0–2°C for 1–3 weeks.

Benefits

The benefits of CA on pears result from the reduced O_2 atmosphere; CO_2 at the safe concentration (<1.5%) has little effect. (1) Pear cultivars differ widely in the length of storage life with maintaining the ability to ripen to a buttery-juicy texture with good flavor. The main advantage of CA storage for pears is to maintain fruit ripening capacity over a longer period thus extending the market season. It is generally to extend the postharvest life of pear cultivars in CA by ~50%–100% relative to RA storage at the same storage temperature and RH (Fig. 1). For example, the storage life of 'Bartlett,' 'Abate Fetel,' 'd'Anjou' pears with excellent ripening capacity is about 3, 4, and 5 months in RA and up to about 6, 6, and 8 months in CA, respectively, at −1.1°C and 90%–95% RH (Table 1). (2) CA at 1%–2.5% O_2 reduces the rates of respiration and ethylene production, color change from green to yellow, and losses of firmness and acidity and therefore maintains better flavor of pears. (3) CA storage tends to protect pears from superficial scald, although it is not a reliable method by itself as a commercial tool. Nevertheless, CA can improve the efficacy of antioxidants on controlling scald. (4) CA inhibits/reduces senescence disorders of pears in long-term storage. (5) The CA conditions in which pears can tolerate during long-term storage may suppress some pathogens of the side rot type, but may not affect the most postharvest diseases significantly.

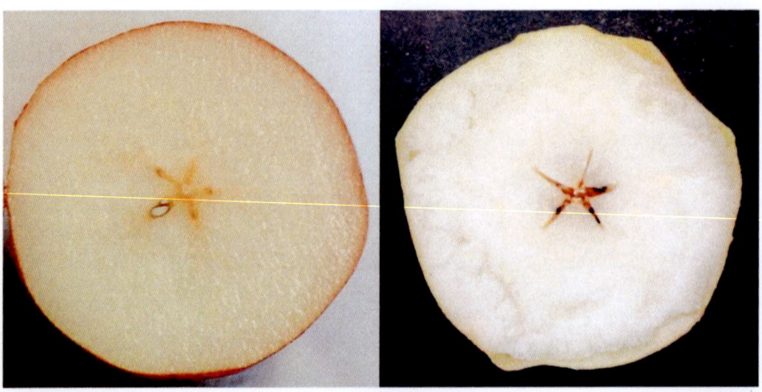

FIG. 1 A 'd'Anjou' pear develops a butter-juicy texture after ripening at room temperature following 8 months in CA (1.5% O_2 + 0.1% CO_2) at −1.1°C (left). A 'd'Anjou' pear develops a dry-coarse texture after ripening at room temperature following 8 months in RA at −1.1°C (right). *(Photo: Y. Wang.)*

To obtain the maximum benefits from CA for pears, three essential procedures should be executed. First, fruit should be harvested early in the maturity range; CA cannot repair the condition of late-harvested fruit. Second, O_2 concentration should be pulled down to the recommended level as fast as possible once fruit core temperature is reduced to below ~2°C. The current nitrogen generator enables O_2 to be reduced to below 5% within a day or two. Commercially, a week from the harvest until approaching the target CA conditions is considered as "rapid" CA. After the target atmosphere has been established, it should be maintained throughout the storage period. Delaying CA after harvest or terminating it prematurely will greatly reduce the benefits on preserving fruit quality from CA. At last, for packed fruit with polyethylene liners in CA storage, the film should be perforated with adequate ventilation (i.e., 12 3.2-mm holes on a standard liner).

Tolerances

To maximize the benefits of CA, it is desirable to have the lowest O_2 level and highest CO_2 level that the fruit will tolerate without adverse effects. The low O_2 limit (LOL) for mature-green 'Bartlett' pears at which aerobic respiration shifts to anaerobic respiration is about 0.3% O_2 at 0°C, which increases with higher temperatures. The LOL measured by chlorophyll fluorescence is about 0.2%–0.3% and 0.75%–0.85% for 'd'Anjou' and 'Abate Fetel' pears, respectively, in dynamic CA storage at 0.5°C. Pears are highly susceptible to CO_2 injury especially at O_2 level < 5%. The safe CO_2 concentration for pears is normally < 1.5% and reduces with O_2 level declining (Table 1). Therefore scrubbing of CO_2 in pear CA storage rooms is necessary. Tolerance to CO_2 may vary with harvest maturity, growing district, and production year. For example, early- and mid-season California-grown 'Bartlett' pears may tolerate CO_2 concentration up to 5% while at 1.5%–2% O_2 for 6 months storage. In contrast, late-season fruit should be kept below 1% CO_2 due to their greater susceptibility to CO_2 injury, especially after 3 months storage. Late-harvested 'd'Anjou' pears especially in cool growing seasons are more susceptible to CO_2 injury.

O_2 below 1% or CO_2 at 30-60% can be tolerated by 'Bartlett' pears for up to 10 days at 0–5°C or up to 4 days at 10°C. d'Anjou' pears tolerate a CA condition with 1% O_2 + 15% CO_2 plus heat for 6h. Such conditions are insecticidal for some insects and may have potential as quarantine treatments.

At O_2 10-15%, 'Bartlett,' 'Bosc,' and 'Comice' in the PNW tolerate CO_2 4%–6% in MAP during storage at −1.1°C for several months plus transportation at 0–2°C for 1–3 weeks.

Injuries

(1) CO_2 injury. Internal browning in pears is a typical CO_2 injury under low O_2 atmosphere which involves flesh and/or core. It has different nomenclatures in the literature such as cortical browning, flesh browning, brown heart, brown

FIG. 2 CO_2 injury in 'Bartlett' (upper), 'Starkrimson' (bottom-left), and 'd'Anjou' (bottom-right) pears. *(Photo: Y. Wang.)*

core, pithy-browning core, and so on (Fig. 2). In contrast to core breakdown caused by senescence, the texture of internal browning tissue caused by high CO_2/low O_2 normally appears as dry and with randomly localized cavities in severe cases. (2) Low-O_2 injuries. 'd'Anjou' pears may develop a peel disorder, black speck (Fig. 3), at O_2 below ~1.0%. 'd'Anjou' pears may loss ripening capacity following Dynamic CA (DCA) storage at $O_2 \approx 0.5\%–0.6\%$. Pears may develop off-flavor due to the accumulation of acetaldehyde, ethanol, and ethyl acetate at O_2 below their LOL.

Problems under suboptimal conditions

Handling and storing European pears under suboptimal conditions may accelerate the declines in fruit firmness and peel color, as a consequence, increase susceptibility to mechanical injuries such as friction discoloration; shorten storage life with ripening capacity; increase superficial scald, senescence disorders, and decay.

FIG. 3 Low-O_2 injury, black speck, on 'd'Anjou' pears. *(Photo: Y. Wang.)*

Comments

Although CA is routinely employed in long-term storage of European pears, significant storage losses due to physiological disorders including CO_2 injury occur every year. The response of fruit to CA conditions varies with harvest maturity, orchard, and season. The preharvest factors influencing pear CA storage efficacy warrant further research. DCA has a potential to be an alternative to antioxidants for scald control, but it could generate other peel and internal physiological disorders and the loss of ripening capacity in pears. The physiological mechanisms underlying low-O_2 damages are not clear and multiple factors may contribute to the low-O_2 injury susceptibility. As an alternative to antioxidants, the combination of 1-MCP prestorage treatment and low-O_2 CA at 0.8%–1.0% O_2 + < 0.1% CO_2 has been proved to be efficient for controlling scald of 'd'Anjou' pears in commercial trails. However, initiating ripening capacity of the 1-MCP treated fruit following CA storage is still a challenge at the present.

Asian pears including cultivars of both climacteric (such as 'Yali' and 'Tsuli') and nonclimacteric (such as 'Nijisseiki' and 'Kosui') respiration and ethylene production patterns appear to be less responsive to CA conditions regarding to extending storage quality. Nevertheless, O_2 level of 1%–3% for 'Nijisseiki' and 3%–5% for 'Yali' pears may help retain firmness and delay changes in skin color in long-term storage. Most of Asian pear cultivars are sensitive to CO_2 injury, i.e., >2%, when stored for more than 1 month at 0°C.

Further reading

Bai, J., Prange, R.K., Toivonen, M.A., 2009. Pome fruits. In: Yahia, E.M. (Ed.), Modified and Controlled Atmospheres for the Storage, Transportation, and Packing of Horticultural Commodities. CRC Press. Taylor & Francis Group, Boca Raton, FL, pp. 267–286.

Chen, P.M., 2004. Pear. In: Gross, K.C., Wang, C.Y., Saltveit, M.E. (Eds.), The Commercial Storage of Fruits, Vegetables, and Florist and Nursery Crops. USDA, ARS, Agriculture Handbook 66, Washington D.C, pp. 471–480.

Chen, P.M., Mellenthin, W.M., 1981. Effects of harvest date on ripening capacity and postharvest life of 'd'Anjou' pears. J. Am. Soc. Hortic. Sci. 106, 38–42.

Chen, P.M., Varga, D.M., 1989. 'Black Speck', a superficial disorder of 'd'Anjou' pears after prolonged CA storage. In: Fellman, J.K. (Ed.), Controlled Atmospheres for Storage and Transport of Perishable Agricultural Commodities. Washington State Univ, Pullman, WA, pp. 146–156.

Chen, P.M., Varga, D.M., 1997. Determination of optimum controlled atmosphere regimes for the control of physiological disorders of 'd'Anjou' pears after short-term, mid-term or long-term storage. In: Mitcham, E.J. (Ed.), Apples and Pears. Postharvest Hort. Series No. 16 (2). Univ. California, Davis, CA, pp. 85–90.

Crisosto, C.H., 2004. Asian pear. In: Gross, K.C., Wang, C.Y., Saltveit, M.E. (Eds.), The Commercial Storage of Fruits, Vegetables, and Florist and Nursery Crops. USDA, ARS, Agriculture Handbook 66, Washington, DC, pp. 205–209.

Hansen, E., Mellenthin, W.M., 1979. Commercial handling and storage practices for winter pears. Oregon Agr. Expt. Sta. Spec. Rpt. 550.

Kader, A.A., 1989. Mode of action of oxygen and carbon dioxide on postharvest physiology of 'Bartlett' pears. Acta Hort. (258), 161–167.

Kader, A.A., 2007. Controlled atmospheres. In: Mitcham, E.J., Elkins, R.B. (Eds.), Pear Production and Handling Manual. University of California, Division of Agriculture and Natural Resources, Publication #3483, pp. 175–177.

Ke, D., Vangorsel, H., Kader, A.A., 1990. Physiological and quality responses of Bartlett pears to reduced O_2 and enhanced CO_2 levels and storage-temperature. J. Am. Soc. Hortic. Sci. 115, 435–439.

Kupferman, E., 2003. Controlled atmosphere storage of apples and pears. Acta Hort. (600), 729–735.

Mattheis, J.P., Felicetti, D., Rudell, D.R., 2013. Pithy brown core in 'd'Anjou' pear (*Pyrus communis* L.) fruit developing during controlled atmosphere storage at pO_2 determined by monitoring chlorophyll fluorescence. Postharvest Biol. Technol. 86, 259–264.

Sugar, D., 2007. Postharvest handlings of winter pears. In: Mitcham, E.J., Elkins, R.B. (Eds.), Pear Production and Handling Manual. University of California, Division of Agriculture and Natural Resources, Publication #3483, pp. 175–177.

Vanoli, M., Grassi, M., Rizzolo, A., 2016. Ripening behavior and physiological disorders of 'Abate Fetel' pears treated at harvest with 1-MCP and stored at different temperatures and atmospheres. Postharvest Biol. Technol. 111, 274–285.

Villalobos-Acuña, M.G., Mitcham, E.J., 2008. Ripening of European pears: the chilling dilemma. Postharvest Biol. Technol. 60, 17–23.

Wang, C.Y., Mellenthin, W.M., 1976. Effect of different handling methods on cooling rate and moisture loss of 'd'Anjou' pears stored in bins. HortScience 11, 397–398.

Wang, Y., 2016. Storage temperature, controlled atmosphere, and 1-methylcyclopropene effects on α-farnesene, conjugated trienols, and peroxidation in relation with superficial scald, pithy brown core, and fruit quality of 'd'Anjou' pears during long-term storage. J. Am. Soc. Hortic. Sci. 141, 177–185.

Wang, Y., Sugar, D., 2013. Internal browning disorder and fruit quality in modified atmosphere packaged 'Bartlett' pears during storage and transit. Postharvest Biol. Technol. 83, 72–82.

Wang, Y., Sugar, D., 2015. 1-MCP efficacy in extending storage life of 'Bartlett' pears is affected by harvest maturity, production elevation, and holding temperature during treatment delay. Postharvest Biol. Technol. 103(1–8).

Xie, X., Song, J., Wang, Y., Sugar, D., 2014. Ethylene synthesis and ripening capacity of 1-MCP treated 'd'Anjou' pears affected by storage temperatures. Postharvest Biol. Technol. 97, 1–10.

Chapter 15.1

Stone fruits: Peaches, nectarines, plums, apricots

George A. Manganaris* and Carlos H. Crisosto†

*Department of Agricultural Sciences, Biotechnology and Food Science, Cyprus University of Technology, Lemesos, Cyprus, †Department of Plant Sciences, University of California, Davis, Davis, CA, United States

Quality characteristics

Controlled and Modified Atmospheres for Fresh and Fresh-Cut Produce
https://doi.org/10.1016/B978-0-12-804599-2.00017-X

Peach, nectarine, plum, and apricot have great economic impact worldwide, but are characterized by high perishability and short market life. Such commodities display a climacteric increase in ethylene production and an equally rapid rate of ripening. Interestingly, differences in the pattern of softening rate and ethylene production may exist within cultivars of the same species, i.e., stony-hard peach fruit (Begheldo et al., 2008) or plum fruit with a suppressed climacteric pattern (El-Sharkawy et al., 2007; Minas et al., 2015). Breeding programs have released a significant number of new cultivars with appreciably better appearance (i.e., lack of corky spot in peaches) and high soluble solids content. Postharvest losses are mainly due to decay and the incidence of chilling-related disorders, such as internal breakdown (IB), woolliness (mealiness), and translucency (Lurie and Crisosto, 2005; Manganaris et al., 2008a).

Main causes of quality loss

Peach and nectarine are highly susceptible to chilling injury (CI) symptoms, evident as dry fruit with a mealy or woolly texture (mealiness or woolliness), hard-textured fruit with no juice (leatheriness), fruit with flesh or pit cavity browning (internal browning), or flesh bleeding (internal reddening) and development of "off flavor," usually well before visual symptoms are evident (Crisosto and Labavitch, 2002; Lurie and Crisosto, 2005) (Fig. 1). Such disorders are genetically influenced and triggered by a combination of storage temperature and duration. Fruit size, storage atmosphere, and temperature have

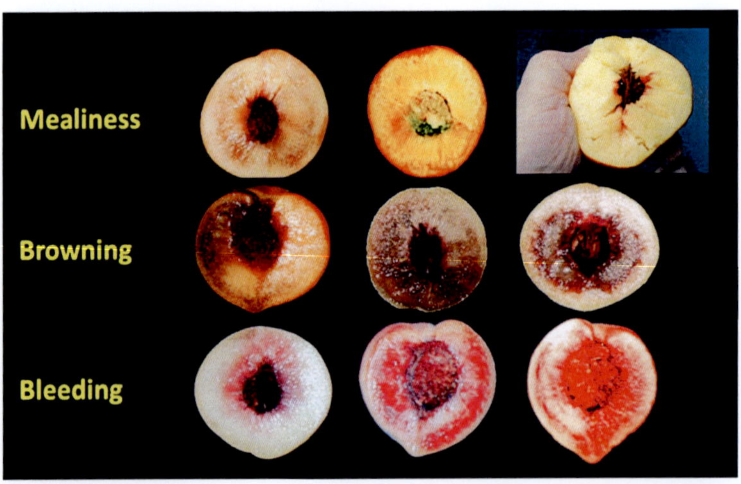

FIG. 1 Chilling-related disorders in peaches, evident as mealiness (woolliness), flesh browning (internal breakdown), and flesh reddening (bleeding). *(Source: Crisosto and Peace)*

significant effects on development of CI symptoms in peach and nectarine (reviewed in Crisosto et al., 2009). Overall, physiological disorders due to extended cold storage are key issues that limit the storage potential of peach fruit to a few weeks. In all susceptible cultivars, flavor is lost before visual CI symptoms are evident.

Plums include the European species (*Prunus domestica* L.), which is consumed fresh or dried, and the Japanese species (*Prunus salicina* Lindell), which is mainly consumed fresh. Plums are climacteric fruits that undergo rapid deterioration after ripening, including softening, dehydration, and decay. Commercial storage conditions (0–5°C and 80%–95% RH) can delay softening, but may also promote development of storage disorders, manifested as translucency, bleeding, flesh browning, and/or failure to ripen (reviewed in Manganaris et al., 2008b) (Fig. 2). Unlike to peach and nectarine, plums respond to exogenous ethylene, a key ripening regulator, and treatments with 1-MCP, an ethylene action inhibitor, delay fruit ripening (Manganaris et al., 2007, 2008a). Besides fast ripening, postharvest diseases are a main concern throughout the supply chain. Postharvest life varies among cultivars and is strongly affected by temperature.

Apricots are highly susceptible to fruit disorders, evident as gel breakdown (translucent, gelatinous mass) or internal breakdown (flesh browning). In addition, apricots are susceptible to decay by several fungal species, including *Monilinia fructicola*, *Botrytis cinerea*, *Penicillium expansum*, and *Rhizopus stolonifer*.

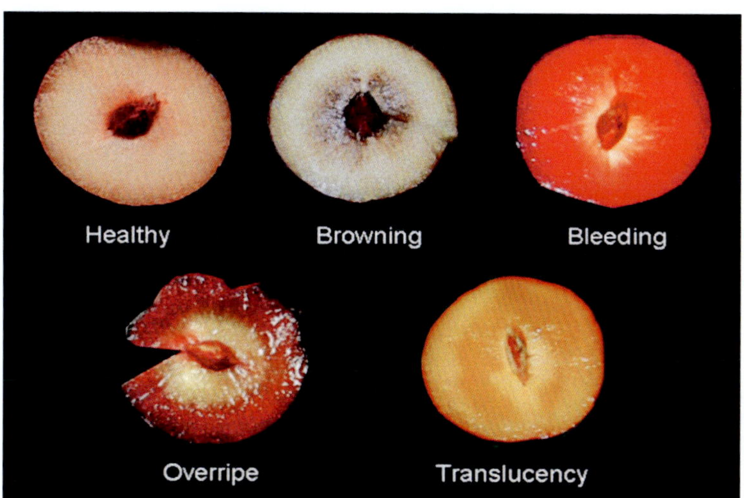

FIG. 2 Chilling-related disorders in plums, evident as flesh browning (internal breakdown), flesh reddening (bleeding), and translucency. Healthy and overripe fruit are also depicted. *(Source: Crisosto)*

Optimum storage conditions

Temperature

Prompt cooling and low-temperature storage (0°C) at high relative humidity are recommended to delay ripening and maintain fruit quality, while all stone fruits are susceptible to cold storage disorders when stored at 5°C. Rapid cooling to target temperature is essential to minimize quality loss. The suppression of ethylene synthesis using postharvest applications of 1-metylcyclopropene (1-MCP), an ethylene inhibitor, to plums will suppress ethylene production and delay ripening without affecting normal ripening at a later time (Manganaris et al., 2007, 2008c). 1-MCP has not been proven beneficial for extending the market life of peaches and nectarines (Dal Cin et al., 2006), making it important to dissect potential benefits of CA/MAP technologies.

CA/MAP considerations for peach

CA as a supplement to managing temperature and relative humidity is used widely as a postharvest technology to extend the postharvest life of horticultural products. There have been significant advances for some fleshy fruit commodities with the development of new, interactive Dynamic Controlled Atmosphere (DCA) storage systems (Zanella et al., 2008; Prange et al., 2013; Bessemans et al., 2016). However, application of CA/MAP technologies to peach and nectarine has not been proven beneficial and the limited number of studies performed does not reflect the economic importance of these commodities. Furthermore, after prolonged storage under CA regimes, fruit may have an acceptable appearance, but being inferior in sensory characteristics such as aroma and taste.

CA is particularly critical to extend the shelf life of stone fruits destined for a distant market. South Africa and Chile are important suppliers of stone fruit to the northern hemisphere from November to March. Such long-distance shipping requires maintaining fruit quality during sea freight for at least four weeks after harvest (Maré et al., 2005). Furthermore, premium quality fruit sent to distant and premium markets, such as from California to Australia, will gain appreciably higher prices that justify special treatment. Despite commercial use of CA conditions in peach, there is inconsistency in the potential benefits of reducing CI symptoms. In addition, there is great variability in the gas concentrations applied to peach and nectarine fruits and contradictory results have been reported about the optimal gas concentration (Fig. 3).

An early study documented that the combination of 1–2 kPa O_2 and 3–5 kPa CO_2 delays ripening without a pronounced effect on alleviating chilling disorders, while O_2 concentrations below 1 kPa and CO_2 above 10 kPa may induce failure to ripen, skin and flesh browning, and "off flavors" (Kader and Mitchell, 1989). Elevated concentrations of CO_2 (higher than 10 kPa) suppress fungal growth, mainly *B. cinerea* and *M. fructicola*, while short-term exposure

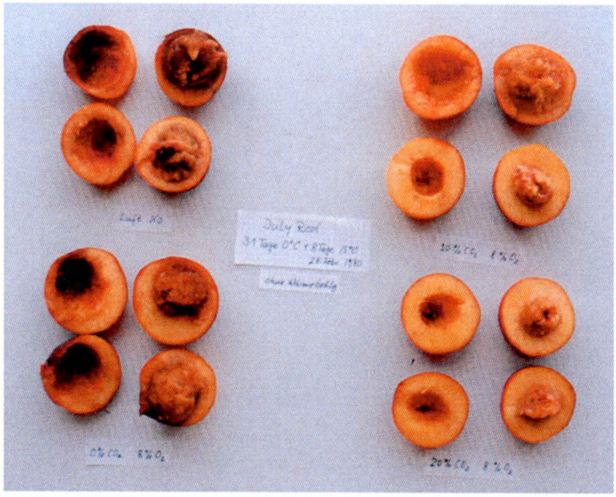

FIG. 3 Postharvest performance of 'July Red' peaches after cold storage under different controlled atmosphere regimes. *Upper panels:* Luft Ko (*left*) and 10% CO_2, 8% O_2 (*right*). *Lower panels:* 0% CO_2, 8% O_2 (*left*) and 20% CO_2, 8% O_2 (*right*). *(Source: Crisosto)*

to 15 kPa CO_2 was beneficial to midseason 'Summer Red' nectarine inoculated with brown rot (Ahmadi et al., 1999). However, when high CO_2 concentrations (10 kPa CO_2 + 10 kPa O_2) were applied for 6 weeks to three nectarine cultivars, this strategy prevented CI disorders (internal breakdown and reddening) but failed to develop increased extractable juice, rendering the fruit of unacceptable quality (Lurie, 1992).

CA technology has been used jointly with other protocols, giving promising results for extended cold storage of peach fruit. In particular, delayed storage (48 h at 20°C) of 'Flavortop' nectarines followed by CA storage (10 kPa CO_2, 3 kPa O_2) alleviated or prevented woolliness symptoms during six weeks cold storage (Zhou et al., 2000). The beneficial effect of this combined strategy was attributed to distinct mechanisms: delayed storage initiated ripening so that, after removal from storage, polygalacturonase (PG) activity was higher and pectin esterase (PE) activity was lower than in control fruits, leading to normal softening. Therefore CA storage suppressed PG activity during storage, but allowed its recovery after removal from cold storage and maintenance at room temperature.

CA storage at 5 kPa O_2 + 5 kPa CO_2 reduced CI in 'Okubao' peach fruits and delayed the reduction in key antioxidant enzymes like peroxidase (POD), superoxide dismutase (SOD), and catalase (CAT) found in the control. Such results suggest that potentially, the decreased activity of these enzymes may contribute to CI development in peach fruits (Wang et al., 2005). CA better maintained quality of 'Chiripá' peaches than air storage, through preservation of tissue integrity and decreased incidence of woolliness and decay due to pathogens (Girardi et al., 2005). CA storage was better than conventional cold storage in maintaining aroma and flavor quality of 'Rich Lady' peach fruit (Ortiz et al., 2009). In contrast, there were no significant differences in flesh firmness between air- and CA-stored (3 kPa O_2, 10 kPa CO_2) 'Rich Lady' peaches (Ortiz et al., 2012).

Peach fruit grown under deficit irrigation had better postharvest performance after two weeks storage in CA (3-4 kPa O_2 and 12–14 kPa CO_2) and additional ripening for four days at 15°C than conventionally irrigated controls; this preharvest/postharvest combination is particularly beneficial for water savings and quality maintenance (Falagán et al., 2015). Controlled Atmosphere storage (5 kPa CO_2 and 10 kPa O_2) allowed up to 45 days storage of late-season, non-melting 'Jesca' and 'Evaisa' peach fruits, but when CA storage was combined with intermittent conditioning (24 h at 20°C every six days), detrimental effects occurred (Ferrer-Mairal et al., 2012).

A comprehensive study on the effect of three CA storage regimes (2 kPa O_2/5 kPa CO_2, 3 kPa O_2/10 kPa CO_2, and 6 kPa O_2/17 kPa CO_2) on four commercially important peach and nectarine cultivars ('Big Top', 'Venus,' 'Early Rich,' and 'Sweet Dream') highlighted the significance of gas composition on flavor perception: fruit placed in 2/5 O_2/CO_2 were less flavorful than those at 3/10 or 6/17. These results can be qualitatively extended to juiciness and sweetness, since these sensory properties are strongly correlated (Cano-Salazar et al., 2013). Extended cold storage (34 days) of 'Miraflores' peaches at two different regimes (2°C and 2 kPa O_2 + 15 kPa CO_2 or 2 kPa O_2 + 10 kPa CO_2) reduced disease incidence and intensity and maintained better sensory properties when CO_2 did not exceed 10 kPa (Truque et al., 2012). However, excessive concentrations of CO_2 (17 kPa) combined with 6 kPa O_2 proved beneficial for some fresh peaches and nectarines shipped from Chile (Retamales et al., 1992).

Another tool to reduce CI in peach is modified atmosphere packaging (MAP). Promising effects of MAP on cold storage potential of 'Chaoyang' honey peach were observed, using three different thicknesses (15, 25, and 40 μm) of low-density polyethylene (LDPE) films (An et al., 2007). MAP inhibited the climacteric peak, reducing color change and softening. The LDPE25 bags had a steady-state atmosphere of 4 kPa O_2+5 kPa CO_2 and provided the best quality after 20 days cold storage at 2°C. MAP technology has been tested intensively on several California-bred yellow flesh peach cultivars grown in Chile, without consistent success (Zoffoli et al., 2002). Despite high CO_2 concentrations reached during simulated shipping, flesh mealiness and flesh browning development limited the potential benefits of this technology. In some commercial situations, box MAP increased the decay incidence due to lack of proper cooling and condensation during transportation. Fruit damage was also observed when MAP-packed fruit (O_2 <3 kPa and CO_2 >13 kPa) were exposed to warm temperatures during postharvest distribution (Malakou and Nanos, 2005).

CA/MAP considerations for plum

Although softening can be delayed by CA and modified atmospheres, this technology is not widely used commercially, since the benefits are not as pronounced as in other fruit species and contradictory results exist about the cost/benefit of CA storage in plums. For some cultivars, O_2 concentrations of 1–2 kPa delay ripening and CO_2 concentrations of 0–5 kPa suppress fruit softening. Oxygen concentrations under 1 kPa may induce failure to ripen and off flavors, while CO_2 >15 kPa is associated with flesh browning (Kader, 1997). Another early study found plums amenable to CA/MAP and the storage time of Japanese plums, particularly late-season cultivars, could be increased two to three months beyond the normal storage time of four to six weeks (Lurie et al., 1999). More recent studies found that 'President' and 'Valjevka' (*P. domestica* L.) fruits benefited from ULO storage (0.9 kPa O_2, 1.0 kPa CO_2) after 30 days at 3.0–3.3°C (Goliáš, 2007), while optimum CA conditions delayed fruit ripening and CI through augmentation of antioxidant metabolism and suppression of oxidative processes (Singh and Singh, 2013).

When 1-MCP treatment was followed by CA (1.8 kPa O_2+2.5 kPa CO_2), storage potential of 'Angeleno' plum was extended from 40–60 days to 80 days (Menniti et al., 2006). The effects of CA (1 kPa O_2 + 3 kPa CO_2 or 2.5 kPa O_2 + 3 kPa CO_2) on 1-MCP-treated 'Blackamber' plums were beneficial in delaying fruit ripening, retarding lipid peroxidation, and reducing the incidence and severity of CI during 5 or 8 weeks of storage compared to normal air and MAP (~10 kPa O_2 and 3.8 kPa CO_2) (Singh and Singh, 2012). This study concluded that the development of CI symptoms in plums, a manifestation of oxidative damage to the fruit tissue, was significantly reduced after 1-MCP treatment followed by CA storage. However, 1-MCP-treated 'Laetitia' plums developed physiological disorders, evident as flesh browning after 60 days CA storage (1 kPa O_2 + 3 kPa CO_2) (de Alves et al., 2009).

Maintaining plum fruit quality for 4 weeks or even longer is needed for orderly overseas marketing. New cultivars have extended the harvest season from late spring through the summer. The storage life of four Japanese plum cultivars was extended by 2–3 weeks during CA shipping compared to conventional refrigerated transportation (Maré et al., 2005). When 'Laetitia' plums were subjected to 60 days cold storage, although CA delayed fruit ripening, extended cold storage did not reduce the incidence of flesh browning (de Alves et al., 2010). In another study on the same cultivar, the recommended CA conditions were 2 kPa O_2 + 2 kPa CO_2 (Steffens et al., 2014).

The effects of different CA regimes have been proven beneficial for extended cold storage of the global cultivar 'Friar' (Crisosto and Garner, 2008) (Fig. 4). In addition, the efficacy of MAP (LifeSpan L316, FF-602, FF-504, 2.0 kPa vented area perforated, and Hefty liner) on quality and shelf life of the same cultivar stored at 0°C for 60 days was studied (Cantín et al., 2008). MAP box liners were recommended to improve market life of 'Friar' plums up to 45 days in cold storage. However, using box liners without CO_2 control may lead to cold storage disorders in fruit cold-stored for longer periods. After 60 days, fruit from the MAP box liners with higher CO_2 and lower O_2 concentrations had a greater incidence of cold storage symptoms, evident as flesh translucency, gel breakdown, and "off flavor," than

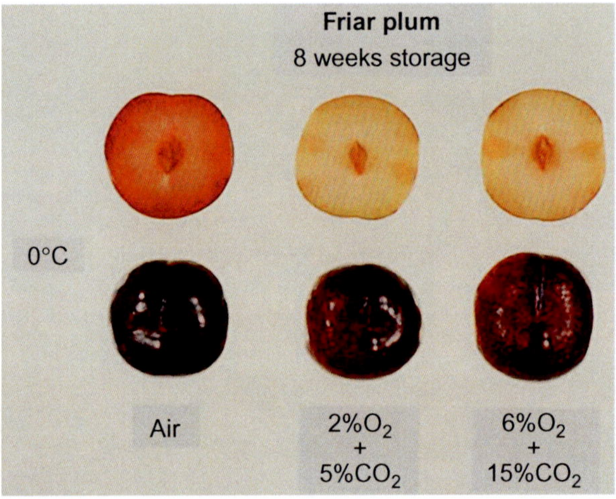

FIG. 4 The beneficial effect of controlled atmosphere in 'Friar' plums after extended cold storage. *(Source: Crisosto)*

fruit from other treatments. MAP such as Xtend and others must be hermetically sealed manually and its success depends on cultivar respiration rate, elapsed time during storage and transportation, and temperature management during transportation and retail handling.

Most studies indicated that when fruits were cold-stored at temperatures above 0°C, CA-treated fruit was firmer than air-stored fruit. However, evaluation of specific cultivars can determine the potential commercial benefits of CA at low cold temperatures, a demanding procedure due to the large number of commercial plums cultivars being distributed into the market. In addition, CA has a rather limited use for storage of plums for periods longer than 4 weeks.

CA/MAP considerations for apricot

CA and MA can delay onset of apricot flesh disorders, but have an accelerating effect if conditions are unfavorable. We have suggested that CA conditions (2-3 kPa O_2 + 2-3 kPa CO_2) can be employed, although these offer only moderate commercial benefits.

Early harvested apricot fruit did not benefit from low (1 or 2 kPa) oxygen because the fruit did not reach the optimal SSC, but in late-harvested fruits, 1 kPa O_2 at a higher temperature can be an alternative to low temperature for shipping or short-term storage (Botondi et al., 2000). 'Búlida' apricot fruits stored under three different CA regimes (20 kPa O_2 + 20 kPa CO_2, 1 kPa O_2 + 20 kPa CO_2, or 1 kPa O_2 + 0.03 kPa CO_2) were characterized by complete ethylene inhibition; however, in apricots stored in 1 kPa O_2, the ethanol concentration increased after one week storage, leading to off-flavors and rendering the fruits unsuitable for consumption (Pretel et al., 1999).

CA or MA conditions for 'Aprikoz' apricots provided better external appearance and taste than other storage conditions (Koyuncu et al., 2010). Active MA replacing the initial atmosphere of the bags with an atmosphere enriched with 20 kPa CO_2 did not modify the gas composition at equilibrium, although it shortened the time necessary for equilibrium to be reached. In 'Beliana' apricots, a slight decrease in ethylene concentration was observed inside bags with an active MA below that of bags with a passive MA (Pretel et al., 2000). A study on the effect of several packaging materials on postharvest performance of 'Tom Cot' apricot fruits indicated that the biodegradable film used can successfully store apricot fruits up to 21 days under passive MAP conditions (Peano et al., 2015). Although storing 'Goldrich' fruit under MA and CA at 1°C was better at preventing fruit softening, 1-MCP was more effective at limiting the incidence of decay, while the accumulation of CO_2 inside the Xtend packaging induced internal browning disorders (Gabioud Rebeaud et al., 2015).

Despite the fact that apricot is highly perishable, the number of studies on the efficacy of CA/MAP technologies in maintaining apricot fruit quality is limited and further studies are warranted. Such studies should additionally consider the genotype effect (slow-softening versus fast-softening cultivars); particular attention should be provided to novel apricot cultivars being launched into the market.

Comments

It is of prime importance to understand the specific conditions and limitations of using CA/MAP when marketing different stone fruit types. Furthermore, there are few or no cost/benefit analyses on whether these technologies will allow growers access to long-distance markets with high-quality fruit and low transportation costs. Postharvest technologies like CA/MAP have been tested to increase stone fruit market life by reducing cold storage disorders and senescence changes, including softening, with sometimes contrasting results. Optimal atmospheric modifications may vary greatly not only among species, but also among commercially important cultivars of the same species.

CA efficacy is greatly affected by cultivar, preharvest factors, temperature, fruit size, marketing period, and shipping time, as measured by the incidence and intensity of chilling-related disorders. As responses to atmospheric modification vary greatly among cultivars and market conditions, it is important to develop very specific information about the benefits and limitations of these techniques for marketing commercially important cultivars, particularly those destined for distant markets.

References

Ahmadi, H., Biasi, W.V., Mitcham, E.J., 1999. Control of brown rot decay of nectarines with 15% carbon dioxide atmospheres. J. Am. Soc. Hortic. Sci. 124, 708–712.

An, J.M., Zhang, M., Zhan, Z., 2007. Effect of packaging film on the quality of 'chaoyang' honey peach fruit in modified atmosphere packages. Packaging Technol. Sci. 20, 71–76.

Begheldo, M., Manganaris, G.A., Bonghi, C., Tonutti, P., 2008. Different postharvest conditions modulate ripening and ethylene biosynthetic and signal transduction pathways in stony hard peaches. Postharvest Biol. Technol. 48, 84–91.

Bessemans, N., Verboven, P., Verlinden, B.E., Nicolaï, B.M., 2016. A novel type of dynamic controlled atmosphere storage based on the respiratory quotient (RQ-DCA). Postharvest Biol. Technol. 115, 91–102.

Botondi, R., Crisà, A., Massantini, R., Mencarelli, F., 2000. Effects of low oxygen short-term exposure at 15°C on postharvest physiology and quality of apricots harvested at two ripening stages. J. Hortic. Sci. Biotechnol. 75, 202–208.

Cano-Salazar, J., López, M.L., Echeverría, G., 2013. Relationships between the instrumental and sensory characteristics of four peach and nectarine cultivars stored under air and CA atmospheres. Postharvest Biol. Technol. 75, 58–67.

Cantín, C., Crisosto, C.H., Day, K.R., 2008. Evaluation of the effect of different modified atmosphere packaging (MAP) box liners on the quality and shelf life of 'Friar' plums (*Prunus salicina* Lindell). Hort. Technol. 18, 261–265.

Crisosto, C.H., Garner, D., 2008. Effects of controlled atmosphere on plums. Central Valley Postharvest Newsletter, May, 4–6.

Crisosto, C.H., Labavitch, J.M., 2002. Developing a quantitative method to evaluate peach (*Prunus persica*) flesh mealiness. Postharvest Biol. Technol. 25, 151–158.

Crisosto, C.H., Lurie, S., Retamales, J., 2009. Stone fruits. In: Yahia, E.M. (Ed.), Modified and Controlled Atmospheres for the Storage, Transportation, and Packaging for Horticulture Commodities, Edition. CRC Press, Taylor & Francis Group, pp. 287–316.

Dal Cin, V., Rizzini, F.M., Botton, A., Tonutti, P., 2006. The ethylene biosynthetic and signal transduction pathways are differently affected by 1-MCP in apple and peach fruit. Postharvest Biol. Technol. 42, 125–133.

de Alves, E.O., Steffens, C.A., do Amarante, C.V.T., Pavanello, E.P., Brackmann, A., 2009. Ethylene management during controlled atmosphere storage of 'Laetitia' plums. Ciencia Rural 39, 2445–2451.

de Alves, E.O., Steffens, C.A., do Amarante, C.V.T., Brackmann, A., 2010. Quality of 'Laetitia' plums when affected by temperature and storage atmosphere. Revista Brasileira de Fruticultura 32, 1018–1027.

El-Sharkawy, I., Kim, W.S., El-Kereamy, A., Jayasankar, S., Svircev, A.M., Brown, D.C.W., 2007. Isolation and characterization of four ethylene signal transduction elements in plums (*Prunus salicina* L.). J. Exp. Bot. 58, 3631–3643.

Falagán, N., Artés, F., Gómez, P.A., Artés-Hernández, F., Conejero, W., Aguayo, E., 2015. Deficit irrigation strategies combined with controlled atmosphere preserve quality in early peaches. Food Sci. Technol. Int. 21, 547–556.

Ferrer-Mairal, A., Remón, S., Peiró, J.M., Oria, R., 2012. Effects of intermittent conditioning on the color and enzymatic activity of peaches during controlled atmosphere storage. J. Food Biochem. 36, 129–138.

Gabioud Rebeaud, S., Maurer, A., Cotter, P.Y., Baumgartner, D., Christen, D., 2015. Influence of temperature, 1-MCP, MA and CA on quality and aroma profiles of 'Goldrich' apricots. Acta Hortic. (1071), 297–302.

Girardi, C.L., Corrent, A.R., Lucchetta, L., Zanuzo, M.R., Costa, T.S.D., Brackmann, A., Twyman, R.M., Nora, F.R., Nora, L., Silva, J.A., Rombaldi, C.V., 2005. Effect of ethylene, intermittent warming and controlled atmosphere on postharvest quality and the occurrence of woolliness in peach (*Prunus persica* cv. Chiripá) during cold storage. Postharvest Biol. Technol. 38, 25–33.

Goliáš, J., 2007. Postharvest response of plum fruits (*Prunus domestica* L.) to low oxygen atmosphere storage. Acta Hortic. (734), 431–440.

Kader, A.A. (Ed.), 1997. Fruits other than apples and pears.Proceedings of the Seventh International Controlled Atmosphere Research Conference, July 13–18, University of California, Davis, CA, pp. 1–261.

Kader, A.A., Mitchell, F.G., 1989. Postharvest physiology peaches. In: LaRue, J.H., Johnson, R.S. (Eds.), Peaches, Plums and Nectarines. Growing and Handling for Fresh Market.pp. 158–164 University of California, Division of Agriculture and Natural Resources, Publication 3331.

Koyuncu, M.A., Dilmacünal, T., Özdemir, O., 2010. Modified and controlled atmosphere storage of apricots. Acta Hortic. (876), 55–66.

Lurie, S., 1992. Controlled atmosphere storage to decrease physiological disorders in nectarines. Int. J. Food Sci. Technol. 27, 507–514.

Lurie, S., Crisosto, C.H., 2005. Chilling injury in peach and nectarine. Postharvest Biol. Technol. 37, 195–208.

Lurie, S., Weksler, A., Lapsker, Z., Greenblat, Y., 1999. Storage properties of new plum cultivars. Alon HaNotea 53, 238–241.

Malakou, A., Nanos, G.D., 2005. A combination of hot water treatment and modified atmosphere packaging maintains quality of advanced maturity 'Caldesi 2000' nectarines and 'Royal Glory' peaches. Postharvest Biol. Technol. 38, 106–114.

Manganaris, G.A., Vicente, A.R., Crisosto, C.H., Labavitch, J.M., 2007. Effect of dips in a 1-methylcyclopropene-generating liquid formulation on 'Harrow Sun' plums stored under different temperature regimes. J. Agric. Food Chem. 55, 7015–7020.

Manganaris, G.A., Vicente, A.R., Crisosto, C.H., Labavitch, J.M., 2008a. Cell wall modifications in chilling-injured plum fruit (*Prunus salicina*). Postharvest Biol. Technol. 48, 77–83.

Manganaris, G.A., Vicente, A.R., Crisosto, C.H., 2008b. Effect of pre-harvest and post-harvest conditions and treatments on plum fruit quality. CAB Reviews: Perspectives in Agriculture, Veterinary Science. Nutrition & Natural Resources 3 (009), 10.

Manganaris, G.A., Crisosto, C.H., Bremer, V., Holcroft, D., 2008c. Novel 1-methylcyclopropene immersion formulation extends shelf life of advanced maturity 'Joanna Red' plums (*Prunus salicina* Lindell). Postharvest Biol. Technol. 47, 429–433.

Maré, L., Huysamer, M., Dodd, M.C., Truter, A.B., Kemp, A.T., Holcroft, D.M., 2005. Extension of the storage life of plums (*Prunus salicina*) using controlled atmosphere shipping. Acta Hortic. (682), 1689–1696.

Menniti, A.M., Donati, I., Gregori, R., 2006. Responses of 1-MCP application in plums stored under air and controlled atmospheres. Postharvest Biol. Technol. 39, 243–246.

Minas, I.S., Forcada, C.F., Dangl, G.S., Gradziel, T.M., Dandekar, A.M., Crisosto, C.H., 2015. Discovery of non-climacteric and suppressed climacteric bud sport mutations originating from a climacteric Japanese plum cultivar (*Prunus salicina* Lindl.). Front. Plant Sci. 6. https://doi.org/10.3389/fpls.2015.00316.

Ortiz, A., Echeverría, G., Graell, J., Lara, I., 2009. Overall quality of 'Rich Lady' peach fruit after air- or CA storage. The importance of volatile emission. LWT—Food Sci. Technol. 42, 1520–1529.

Ortiz, A., Sopeña, P., Comabella, E., Lara, I., 2012. Firmness loss and cell wall degradation after air- or CA-storage of 'Rich Lady' peaches. Acta Hortic. (940), 475–482.

Peano, C., Giuggioli, N.R., Girgenti, V., 2015. Effects of innovative packaging materials on apricot fruits (cv Tom Cot). Fruits 69, 247–258.

Prange, R.K., Wright, A.H., DeLong, J.M., Zanella, A., 2013. History, current situation and future prospects for dynamic controlled atmosphere (DCA) storage of fruits and vegetables, using chlorophyll fluorescence. Acta Hortic. (1012), 905–916.

Pretel, M.T., Serrano, M., Amorós, A., Romojaro, F., 1999. Ripening and ethylene biosynthesis in controlled atmosphere stored apricots. Eur. Food Res. Technol. 209, 130–134.

Pretel, M.T., Souty, M., Romojaro, F., 2000. Use of passive and active modified atmosphere packaging to prolong the postharvest life of three varieties of apricot (*Prunus armeniaca* L.). Eur. Food Res. Technol. 211, 191–198.

Retamales, J., Cooper, T., Streif, J., Kania, J.C., 1992. Preventing cold-storage disorders in nectarines. J. Hortic. Sci. 67, 619–626.

Singh, S.P., Singh, Z., 2012. Postharvest oxidative behaviour of 1-methylcyclopropene treated Japanese plums (*Prunus salicina* Lindell) during storage under controlled and modified atmospheres. Postharvest Biol. Technol. 74, 26–35.

Singh, S.P., Singh, Z., 2013. Controlled and modified atmospheres influence chilling injury, fruit quality and antioxidative system of Japanese plums (*Prunus salicina* Lindell). Int. J. Food Sci. Technol. 48, 363–374.

Steffens, C.A., Do Amarante, C.V., Alves, E.O., Brackmann, A., 2014. Fruit quality preservation of 'Laetitia' plums under controlled atmosphere storage. Anais da Academia Brasileira de Ciencias 86, 485–494.

Truque, E., Aguayo, E., Artés-Hernández, F., Gómez, P., Artés, F., 2012. Controlled atmosphere for the export of 'Miraflores' peaches. Acta Hortic. (962), 585–590.

Wang, Y.S., Tian, S.P., Xu, Y., 2005. Effects of high oxygen concentration on pro- and anti-oxidant enzymes in peach fruits during postharvest periods. Food Chem. 91, 99–104.

Zanella, A., Cazzanelli, P., Rossi, O., 2008. Dynamic controlled atmosphere (DCA) storage by the means of chlorophyll fluorescence response for firmness retention in apple. Acta Hortic. (796), 77–82.

Zhou, H.W., Lurie, S., Lers, A., Khatchitski, A., Sonego, L., Ben Arie, R., 2000. Delayed storage and controlled atmosphere storage of nectarines: two strategies to prevent woolliness. Postharvest Biol. Technol. 18, 133–141.

Zoffoli, J.P., Balbontin, S., Rodríguez, J., 2002. Effect of modified atmosphere packaging and maturity on susceptibility to mealiness and flesh browning of peach cultivars. Acta Hortic. (592), 573–579.

Chapter 15.2

Stone fruits: Sweet cherries (*Prunus avium* L.)

Peter M.A. Toivonen* and George A. Manganaris[†]
[*]Agriculture and Agri-Food Canada, Summerland Research and Development Centre, Summerland, BC, Canada, [†]Department of Agricultural Sciences, Biotechnology and Food Science, Cyprus University of Technology, Lemesos, Cyprus

Quality characteristics

Premium sweet cherries have large fruit size (diameter > 30 mm) and a bright and shiny appearance. Color can range from dark red ("Bing") to red ("Sweetheart") to red blush on a yellow background ("Rainier"). The stems should have a fresh green appearance with no browning. Good flavor is associated with high soluble solids and titratable acidity contents and the fruit should be firm and juicy.

Main causes of quality loss

Water loss (often visualized by increased severity of surface pitting and/or "pebbling" (Fig. 1) of the skin or as fruit shriveling in extreme cases), softening, decay and stem browning.

Controlled and Modified Atmospheres for Fresh and Fresh-Cut Produce
https://doi.org/10.1016/B978-0-12-804599-2.00018-1

Pebbling scale: 1 = none
2 = slight
3 = severe

FIG. 1 Three severity rating levels for "pebbling" on sweet cherries. This defect usually shows up after several weeks in cold storage under MAP conditions; however, it can be observed when fruit is still on the tree. "Pebbling" is a cosmetic defect, not affecting flavor and it is often mistaken for pitting when it is severe in nature.

Optimum storage conditions

Temperature

−1 to 0°C at a relative humidity of 95% or higher. Rapid cooling to target temperature using hydrocooling technology is essential to minimize quality loss and changes in sweet cherries in long-term storage/shipping (Manganaris et al., 2007).

CA/MAP considerations

Cherries must be properly cooled after packing even if they were hydrocooled before packaging (Toivonen, 2014). Modified atmosphere packages are designed for specific shipping temperatures (Kupferman and Sanderson, 2001). Fruit must be sound and free of injury and decay and insect infestations. Generally, CA is not applied to sweet cherry storage but MAP is routinely used for shipping by refrigerated truck or container or by air (Toivonen, 2014; Chockchaisawasdee et al., 2016). In addition, it has been shown that "Sweetheart" cherries held in CA (8 kPa O_2 + 10 kPa CO_2) was judged by a sensory panel to be similar in quality to cherries held in MAP for up to 27 days (Santos et al., 2014), suggesting that the more cost-effective approach of using modified atmosphere packaging is the best option in the commercial industry.

Benefits: Fruit color and intensity, green stem color and turgor, and fruit firmness are better maintained (Kupferman and Sanderson, 2001; Toivonen, 2014). Water loss and shriveling is reduced. However, surface defects of pitting and pebbling and titratable acidity loss are not reduced.

FIG. 2 Effects of low oxygen and high carbon dioxide levels on sweet cherry quality. *(Photos extracted from Lidster, P.D., Blanpied, D.G., Prange, R.K. (Eds.), 1990. Controlled-atmosphere disorders of commercial fruits and vegetables. Agriculture Canada Publication 1847/E.)*

Tolerances: For MAP, 3–10 kPa O_2 with 5–15 kPa CO_2 is effective when fruit temperature is maintained at 0–5°C (Mattheis and Reed, 1994; Meheriuk et al., 1995; Jiang et al., 2002; Tian et al., 2004).

Injuries: Both low O_2 and high CO_2 levels will cause darkening of the fruit, browning of fruit flesh, stem browning, and enhanced levels of decay (likely as a direct consequence of cellular injury). Typically the two injuries are very similar in appearance (Fig. 2).

Intersection of temperature and MA and temperatures and periods of exposure

If either O_2 or CO_2 is at atmospheric concentrations (i.e., 21 kPa or 0.03 kPa, respectively) then cherries will tolerate up to 10 kPa of CO_2 or 3 kPa of O_2, respectively, when fruit are stored at recommended temperatures (Kader and Morris, 1977). It was shown that for "Bing" cherries stored in air enriched with 0–100 kPa CO_2 at 0°C, off-flavor intensity was proportionate to a combination of time of exposure and the CO_2 concentration to which the cherries are exposed (Patterson and Melstad, 1977). Off-flavors developed in a week when fruit were stored in $\geq$80 kPa CO_2 and after a week when stored at 60 kPa CO_2, but no off-flavors developed at $\leq$40 kPa CO_2. In general, warmer storage temperatures and greater durations of exposure increase the risk of CO_2 injury. It is generally known that tolerance to either O_2 or CO_2 changes with the concentration of the other gas component (either CO_2 or O_2, respectively) and with changes in temperature and time, but relatively few data exist for sweet cherries. Wang et al. (2015) indicated that 0.5–1.5 kPa O_2 and ~10 kPa CO_2 can lead to development of anaerobic off-flavors in "Lapins" and "Skeena" sweet cherries after four and six weeks of storage at 0°C and this was associated with the accumulation of ethanol in the fruit tissue. Short exposures to as high as 95 kPa CO_2 can be tolerated up to 3 days, with no adverse effects on quality, when fruit are held at 3°C (Golding et al., 2012).

Specific atmospheres used for insect control

Nonchemical quarantine treatments, using a combination of short duration high temperatures under low O_2, elevated CO_2 atmospheric environment were developed to control western cherry fruit fly, *Rhagoletis indifferens* Curran, in sweet cherries. A chamber temperature of 45°C for 45 min or 47°C for 25 min, both under a 1% O_2, 15% CO_2, and a −2°C dew point environment was applied (Neven and Rehfield-Ray, 2006). Both treatments have been shown to provide control of all life stages of western cherry fruit fly, while preserving commodity market quality.

A specific quarantine treatment was developed to control Queensland fruit fly, *Bactrocera tryoni* Froggart, utilizing a combination of low temperature (3°C) and exposure to 95 kPa CO_2 for up to three days (Golding et al., 2012). While holding at 3°C led to 92% mortality of the insect larvae after four days, a 1–3 day exposure to the 95 kPa CO_2 quarantine treatment enhanced mortality to 97% or more. No adverse effects on quality attributes was noted in "Bing," "Lapins," "Rons," or "Stella" cultivars of sweet cherry.

Problems under suboptimal conditions

Controlled and modified atmosphere have minor effect on controlling loss of titratable acidity; however, holding or shipping cherries at temperatures even a few degrees above the recommended target of 0°C will result in accelerated loss in titratable acidity and firmness (Toivonen, 2014). Depending on the duration of time, elevated handling/shipping temperatures will also lead to increased decay and stem browning (Toivonen, 2014).

The optimal range for modified or controlled atmosphere storage of sweet cherries ranges between 3 and 10 kPa O_2 with 5–15 kPa CO_2, based on the cultivar considered. If O_2 is above 10 kPa, there will be greater losses of titratable acidity and ascorbic acid due to the higher respiration in such atmospheres (Wang et al., 2015). If O_2 is below 3 kPa, then anaerobic metabolism can be triggered and ethanol accumulation ensues, leading of off-flavors (Wang et al., 2015). These problems become more severe with the length of time the sweet cherries are exposed to the suboptimal O_2 atmospheres and so the problems for suboptimal (high or low) conditions are generally only relevant to fruit shipped long distances in MAP in containers.

The main concern for suboptimal conditions occurs when fruit is either in transit at a transfer point (e.g., from truck to airplane) or it is received at destination where good temperature control is not available. At such points in handling/transit, it is possible that packages will warm and atmospheres will become overmodified and hence out of optimal range.

Comments

There are significant differences between cultivars in regards to quality retention over long-term MAP storage (Kappel et al., 2002; Toivonen and Kappel, 2012).

There are many MAP systems for use in sweet cherry and the evidence to date suggests that most provide similar benefits to maintaining quality and controlling decay during long-term shipping simulations (Zoffoli and Rodriguez, 2014). However, some modified atmosphere packages have been found to generate steady-state O_2 levels above 10 kPa or below 3 kPa, resulting in either poor maintenance of quality or development of off-flavors, respectively (Wang et al., 2015). Therefore it is important to verify MAP performance before adopting a particular packaging system.

While high in-package CO_2 is known to have fungistatic or bacteriostatic effects (Serradilla et al., 2013), application of postharvest or in-package treatments provide added benefit for control of decay of sweet cherries under scenarios simulating long-distance shipping (Serrano et al., 2005; Wang et al., 2014; Colgecen and Aday, 2015). While one synthetic fungicide is currently registered for postharvest use in cherries (fludioxonil), a significant amount of work is being conducted recently looking at alternatives to the use of synthetic postharvest fungicides (Feliziani et al., 2013; Chockchaisawasdee et al., 2016).

References

Chockchaisawasdee, S., Golding, J.B., Vuong, Q.V., Papoutsis, K., Stathopoulos, C.E., 2016. Sweet cherry: Composition, postharvest preservation, processing and trends for its future use. Trends Food Sci. Technol. 55, 72–83.

Colgecen, I., Aday, M.S., 2015. The efficacy of the combined use of chlorine dioxide and passive modified atmosphere packaging on sweet cherry quality. Postharvest Biol. Technol. 109, 10–19.

Feliziani, E., Santini, M., Landi, L., Romanazzi, G., 2013. Pre- and postharvest treatment with alternatives to synthetic fungicides to control postharvest decay of sweet cherry. Postharvest Biol. Technol. 78, 133–138.

Golding, J.B., Jessup, A., Spohr, L., Daniels, D., Satyan, S., Pristijono, P., McGlasson, W.B., Jacas, J., 2012. Efficacy of a combination quarantine treatment at 3°C as a potential disinfestation treatment for Queensland fruit fly (*Bactrocera tyroni* (Froggat)) in cherry fruit. Acta Hortic. (934), 343–346.

Jiang, A.L., Tian, S.P., Xu, Y., 2002. Effects of controlled atmospheres with high-O_2 or high-CO_2 concentrations on postharvest physiology and storability of "Napoleon" sweet cherry. Acta Bot. Sin. 44, 925–930.

Kader, A.A., Morris, L.L., 1977. Relative tolerance of fruits and vegetables to elevated CO2 and reduced O2 levels. In: Dewey, D.H. (Ed.), Controlled atmospheres for the storage and transport of perishable agricultural commodities. Proceedings 2nd National Controlled Atmosphere Research Conference. Mich. State Univ. Hortic. Rep. 28. pp. 260–265.

Kappel, F., Toivonen, P., McKenzie, D.L., Stan, S., 2002. Storage characteristics of new sweet cherry cultivars. Hort. Sci. 37, 139–143.

Kupferman, E., Sanderson, P., 2001. Temperature management and modified atmosphere packaging to preserve cherry quality. In: Postharvest Information Network 9 pp. Viewed at: http://postharvest.tfrec.wsu.edu/EMK2001B.pdf on October 13, 2016.

Manganaris, G.A., Ilias, I.F., Vasilakasis, M., Mignani, I., 2007. The effect of hydrocooling on ripening related quality attributes and cell wall physicochemical properties of sweet cherry fruit (*Prunus avium* L.). Int. J. Refrig. 30, 1386–1392.

Mattheis, J.P., Reed, N., 1994. Response of 'Bing' sweet cherries to modified atmosphere packaging. In: Proceedings of the Oregon Horticultural Society, January 1994, pp. 22–28.

Meheriuk, M., Girard, B., Moyls, L., et al., 1995. Modified atmosphere packaging of 'Lapins' sweet cherry. Food Res. Int. 28, 239–244.

Neven, L.G., Rehfield-Ray, L., 2006. Combined heat and controlled atmosphere quarantine treatments for control of western cherry fruit fly in sweet cherries. J. Econ. Entomol. 99 (3), 658–663.

Patterson, M.E., Melstad, J.L., 1977. Sweet cherry handling and storage alternatives. In: Dewey, D.H. (Ed.), Controlled atmospheres for the storage and transport of perishable agricultural commodities. Proceedings 2nd National Controlled Atmosphere Research Conference. Mich. State Univ. Hortic. Rept. 28pp. 55–59.

Santos, A.C.A., Palma, V., Rato, A.E., Machado, G., Lozano, M., González-Gómez, D., 2014. Quality of 'Sweetheart' cherry under different storage conditions. Acta Hort. (1020), 101–110.

Serradilla, M.J., Villalobos, M.D.C., Hernández, A., Martín, A., Lozano, M., Córdoba, M.D.G., 2013. Study of microbiological quality of controlled atmosphere packaged 'Ambrunés' sweet cherries and subsequent shelf-life. Int. J. Food Microbiol. 166, 85–92.

Serrano, M., Martínez-Romero, D., Castillo, S., Guillén, F., Valero, D., 2005. The use of natural antifungal compounds improves the beneficial effect of MAP in sweet cherry storage. Innov. Food Sci. Emerg. Technol. 6, 115–123.

Tian, S.P., Jiang, A.L., Xu, Y., Wang, Y.S., 2004. Responses of physiology and quality of sweet cherry fruit to different atmospheres in storage. Food Chem. 87, 43–49.

Toivonen, P.M.A., 2014. Relationship of typical core temperatures after hydrocooling on retention of different quality components in sweet cherry. Hort. Technol. 24, 457–462.

Toivonen, P.M.A., Kappel, F., 2012. Effect of cold storage duration on shelf quality attributes of 'Bing' and five sweet cherry cultivars developed in the Summerland breeding program. Acta Hortic. (934), 1011–1016.

Wang, Y., Xi, X., Long, L.E., 2014. The effect of postharvest calcium application in hydro-cooling water on tissue calcium content, biochemical changes, and quality attributes of sweet cherry fruit. Food Chem. 160, 22–30.

Wang, Y., Bai, J., Long, L.E., 2015. Quality and physiological responses of two late-season sweet cherry cultivars 'Lapins' and 'Skeena' to modified atmosphere packaging (MAP) during simulated long distance ocean shipping. Postharvest Biol. Technol. 110(1-8).

Zoffoli, J.P., Rodriguez, J., 2014. Effect of active and passive modified atmosphere packaging of sweet cherry. Acta Hortic. (1020), 115–120.

Further reading

Lidster, P.D., Blanpied, D.G., Prange, R.K., 1990. Controlled-atmosphere disorders of commercial fruits and vegetables. In: Agriculture Canada Publication 1847/E.

Chapter 16.1

Small fruits: Strawberries

Manuel Madrid
Fruit Profits, Valencia, Spain

Quality characteristics

Main quality characteristics of strawberries are as follows:

- Size: A weight between 50 and 80 g is ideal. Larger sizes or diameters higher than 25 mm are not preferred by consumers.
- Shape: Ideally a "heart" shape, no misshapen, no deformed.
- Color: Full red color (no wider light or white color zone than three rows of seeds); dark red color can be a sign of overripe or old fruit.

Additional desirable quality characteristics:

- Absence of finger marks
- Calyx condition: Whole green calyx with no signs of desiccation
- Sugars: Brix degrees above 7%
- Acidity: no greater than 0.7% titratable acidity

Controlled and Modified Atmospheres for Fresh and Fresh-Cut Produce
https://doi.org/10.1016/B978-0-12-804599-2.00019-3

- Shiny skin
- No missing seeds (achenes). Sometimes birds eat strawberry seeds in the field. In that case, nets need to be used
- Absence of any disease symptoms
- Taste: Strawberries should be picked ripe. Sugars will not increase after harvest and early harvest will result in dull tasting, overly tart fruit. The choice of variety is important when considering taste

Main causes of quality loss

There are several causes of quality loss in strawberries:

- Careless picking and handling in the field: Commercial strawberries are handpicked in the field and directly packed in punnets. Handling should be very careful. Too much finger pressure while picking will result in damaged skin. These damaged skin areas later evolve into soft or dark tissue, reducing the value of the fruit. If possible, strawberries should be packed directly in the field. They should not be repacked.
- Delays in cooling: Strawberries should be cooled down to their optimum temperature within 1 hour after harvest (Maxie et al., 1959). Failure to do so will result in reduced shelf life and higher presence of mold. Cooling delay also results in dark and soft fruit after a few days. Strawberries should never be exposed to direct sunlight after picking. Keep boxes in a shaded place while waiting for transport to the cold storage facility.
- Wet fruit from the field (dew or rain): Strawberries should always be harvested dry. Sometimes berries are wet due to dew or recent rain. In that case, harvest is postponed until berries are dry. In many countries, strawberries are grown under plastic tunnels also called hoops (Fig. 1). In this case, dryness of the fruit and sanitation are improved as well as earliness.
- Poor field sanitation: The level of inoculum in the field should be kept low by removing damaged and decayed fruit and spraying with authorized fungicides only, respecting reentry periods and manufacturer labels. Failure to do so will result in high incidence of postharvest diseases.
- Birds: In some countries, birds like to eat strawberry seeds (achenes). In those cases, nets are used.
- Dirt: Strawberries are often grown in mounds covered with black plastic mulch to prevent the berries from touching the soil.

Optimum storage conditions

Optimum temperature

Temperature of 0°C and relative humidity of 90%–95% RH are recommended as optimum storage conditions. Avoid temperatures below 0°C. Strawberries should be cooled down to the optimum temperature 1 h after harvest. Forced air-cooling

FIG. 1 Strawberries are often grown under plastic tunnels also called hoops or hoophouses to manage the production season and reduce losses due to predation and decay.

is the most common cooling method (Fig. 2). Temperature pull down should be done in 20–30 min of entering the cooling tunnel. To avoid risk of freezing the fruit due to room temperature oscillation, the cooling tunnel temperature set point is set often at 1°C. Under optimal cooling and sanitation conditions, strawberries remain marketable for approximately 7 days, depending on variety.

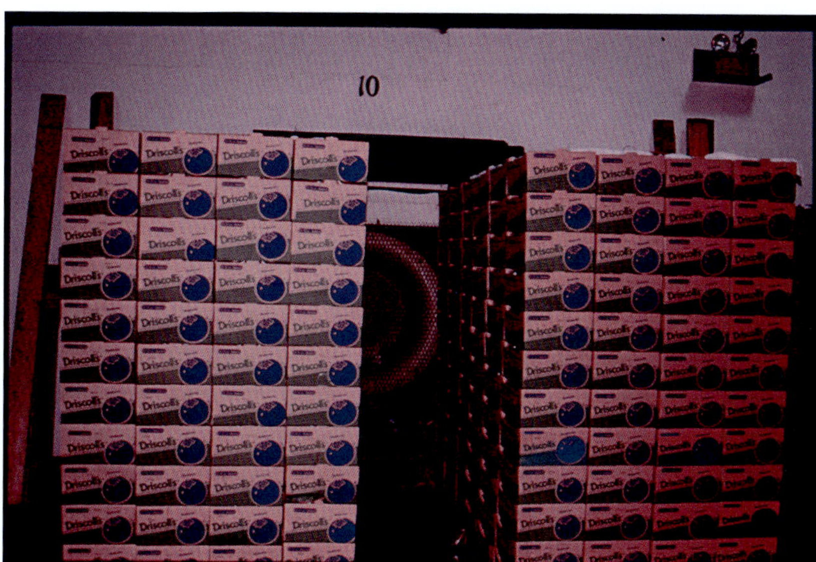

FIG. 2 Forced-air cooling is highly recommended within 1 h of harvest to avoid loss of quality and extend the shelf-life.

CA/MA considerations

MAP is used as a standard procedure in California and Mexico for distribution to the United States. A pallet-sized package is used, by which the stacked pallet is covered with a large plastic bag and sealed to a plastic film base. After sealing, the pallet cover is injected with a gas mixture to achieve 10%–15% CO_2. High CO_2 reduces the incidence and severity of decay and extend the postharvest life of strawberries.

This MAP system is often not considered to be necessary in Europe or other continents due to the close proximity of the production areas to markets.

Benefits and injuries of CA/MA control

Due to their high perishability, strawberries rarely show CA injury. Disease or advanced maturity almost always shows before CA injury can be observed (Fig. 3).

Problems under suboptimal conditions

Mold: *Botrytis cinerea* Pers. (gray mold) is the most common postharvest disease of berries (Snowdon, 1990). It is one of the most destructive diseases of strawberries. It is often latent in the field and shows up during transport. The fungus grows readily at temperatures above 4°C. It is very common during the rainy periods. It can result in downgraded fruit and, if serious enough, total loss of the crop.

FIG. 3 Harvested strawberries ready for shipment. Berries should be cooled immediately after harvest and kept at 0–1°C at all times during distribution.

Other diseases that appear sometimes are *Rhizopus* sp., *Colletotrichum* spp., and *Phytophthora*.

Darkening of strawberry skin can result from senescence due to delays in cooling, elevated storage temperatures, or excessive storage durations. The extent of darkening varies among varieties and is partly dependent on initial field color at ripeness, which is also variety dependent.

Appearance of dark marks: Delays in cooling result in more visible marks and sunken areas marking bruising when fruit was picked.

Soft tip: This condition is usually associated to moist and cold growing conditions, and to a calcium deficiency.

Juice leakage: Rough handling of punnets and improper cooling will result in "bleeding" of strawberries.

Comments

The strawberry is highly perishable and a very valuable commodity. Cooling, gentle handling, transportation at temperatures close to 0°C, and sanitation in the field are the keys to successful production and marketing (Fig. 4).

FIG. 4 Packing station operations are weighing, placing protective films or lids, labeling, palletizing, and cooling.

References

Maxie, E.C., Mitchell, F.G., Greathead, A., 1959. Quality studies of strawberries. Calif. Agric. 13 (6), 15.

Snowdon, A., 1990. A color atlas of post-harvest diseases. In: General Introduction and Fruits. Vol. 1. CRC Press, Boca Raton, FL.

Chapter 16.2

Small fruits: Raspberries, blackberries, blueberries

Manuel Madrid[a] and Randolph Beaudry[b]
[a]*Fruit Profits, Valencia, Spain,* [b]*Department of Horticulture, Michigan State University, East Lansing, MI, United States*

Quality characteristics of raspberries

- Varieties and color: There are red, yellow, orange, and black raspberries, but the most commonly traded are red raspberries. This chapter deals mostly with red raspberry postharvest handling.
- Premium fruit should weigh between 3 and 5 g each, larger sizes are even preferred. Some varieties yield larger fruit, up to 8 g per fruit. Fruit below 3 g is considered lower quality standard.
- Minimum caliber: 12 mm.

- Shape: ideally with heart shape, no misshapen, no deformed.
- Light red color. Avoid dark red fruit (this can be a sign of overripe or old fruit).
- Sugars: Brix degrees above 9%.

Controlled and Modified Atmospheres for Fresh and Fresh-Cut Produce
https://doi.org/10.1016/B978-0-12-804599-2.00020-X

- Acidity: no more than 0.7% titratable acidity.
- Shiny skin, light red color.
- Whole fruit. No broken drupelets (this may happen when fruit is not fully ripe and pickers pull the fruit roughly).
- Absence of any disease, especially Botrytis is frequent in raspberry fruit.
- Absence of disease spores (sometimes fruit is stained by yellow rust fungal spores coming from infected leaves).
- Taste: raspberries should be picked ripe (red). White or pink fruit will not develop full color and good taste.
- The choice of variety is important when considering taste and shelf life.
- Tolerance of defects:
  - Premium fruit should have less than 5% of small defects in coloration and shape.
  - Broken fruit should be less than 2%.

Main causes of quality loss

There are several causes of quality loss in raspberries.

- Not careful picking and handling in the field: Commercial raspberries are hand-picked in the field and directly packed in punnets. Handling should be very careful. Rough pulling while picking will result in broken berries, reducing the value of the crop. Only one or two layers of fruit should be put in the punnets, fruit should never be piled.
- Delays in cooling: Raspberries should be brought under cooling within 1 h of picking. Failure to do so will result in reduced shelf life and higher presence of mold. They also result in dark and soft fruit after a few days. Raspberries should never be exposed to direct sunlight after picking. Keep boxes in a shaded place while waiting for transport to the cold storage facility.
- Pulldown: raspberries should be forced-air cooled immediately after arrival at the packing station. Refrigeration equipment should be powerful enough to cool down raspberries from field temperature to 0°C within 20–30 min of entering the cooler.
- Wet fruit from the field (dew or rain). Raspberries should always be harvested dry. Sometimes berries are wet due to dew or recent rain. In that case, harvest is postponed until berries are dry. In many countries, raspberries are grown under plastic tunnels, also called hoops. In this case, dryness of the fruit and sanitation are improved as well as earliness in harvest.
- Poor field sanitation. Level of inoculum in the field should be kept low, removing decayed and dry leaves and dry vines, and opening plastic greenhouses on the sides to allow air circulation. Raspberry plants can be sprayed with authorized fungicides only, respecting reentry periods and manufacturer labels. Failure to follow the earlier discussed measures will result in high incidence of postharvest diseases.

- White drupelets. This is a physiological disorder of raspberries, showing as some white drupelets among the red drupelets of the raspberries. White drupelets can be caused by two reasons
  - Direct sunlight. Direct exposure of raspberries to sunlight will result in sunburn and white drupelets
  - Sharp changes in temperature between morning and midday, usually associated to low relative humidity in the air. This phenomenon is more frequent in winter and early spring Some varieties are more prone to this than others
- Mold growth on fruit due to excess flower nectar. During cool humid mornings, flowers may exude excessive nectar, which drips on growing and ripe fruit and decays, forming gray or black stains. This is weather and irrigation dependent. Proper greenhouse or plastic-house ventilation, and a light shaking of the plants, taking advantage of the trellis structure, will avoid nectar dripping from upper flowers on fruit below.

Optimum storage conditions

Harvest should be done early in the morning when possible to avoid high field temperatures. Make sure the punnets are clean and dust free; dust accumulates in the materials' warehouse.

Optimum storage temperature of raspberries

0°C, 90%–95% RH. Avoid temperatures below 0°C. Raspberries should be cooled down to the optimum temperature 1 h after harvest.

Forced air-cooling is the most common cooling method. Temperature pull-down should be done in 20–30 min of entering the cooling tunnel. To avoid risk of freezing the fruit, cooling tunnel temperature thermostat set-point should be set at 1°C, because thermostats usually present a temperature oscillation. Under optimal cooling and sanitation conditions, raspberries can last for 7 days, depending on variety.

Preserving the cold chain. Packaged and cooled raspberries should never be exposed to higher temperatures. Cold chain interruptions will result in condensation of the berries and increased mold and softening. Cold chain interruptions are frequent in air shipments, since some airports do not have cold room facilities or do not handle the cold chain carefully.

CA/MA considerations

Modified atmosphere packaging (MAP) is used as a standard procedure in California and Mexico for distribution to the United States. Pallet MAP is used, by which the pallet is wrapped with a plastic bag and injected with a gas concentration of 10%–15% CO_2. High CO_2 deters the growth of fungi.

This MAP system is not used in Europe or other continents on a regular basis due to the proximity of the production areas to the market.

Benefits and injuries of CA/MA control: due to their high perishability, raspberries rarely show CA injury. Disease or overmaturity in the form of dark berries shows up first, before CA injury can be observed.

Problems under suboptimal conditions

Mold: *Botrytis cinerea* Pers. is the most common postharvest disease of raspberries (Snowdon, 1990). It is often latent in the field and shows up during transport. The fungus grows readily at temperatures above 4°C and during the rainy season. It results in total loss of the crop or downgraded fruit.

Other diseases that appear less frequently are *Mucor* sp., *Rhizopus* sp., and *Alternaria* spp.

Darkening of raspberry. Delays in cooling will cause fast darkening of the fruit. This effect varies among varieties. Raspberry color evolution is also variety dependent.

Juice leakage. Rough handling of punnets and improper cooling will result in "bleeding" of raspberries. This is a very severe defect that will result in total loss of value.

In order to absorb possible humidity droplets, or juice from broken drupelets, an absorbent pad is often attached to the bottom of the punnet.

Comments

Raspberries are highly perishable and a very valuable commodity. Cooling, gentle handling, transportation at temperatures close to 0°C, integrity of the cold chain, and sanitation in the field are the keys to successful transportation and delivery to the customer.

Quality characteristics of blackberries

- Minimum caliber: 20 mm.
- Shape: long cylindrical shape, no misshapen, no deformed fruit.
- Deep black color. Avoid dark red fruit (this can be a sign of unripe fruit or mite infested fruit).
- Avoid half black half red fruit.
- Sugars: Brix degrees above 8%.
- Acidity: no more than 0.7% titratable acidity.
- Shiny skin, no dehydrated drupelets, no soft fruit.
- Whole fruit. No broken or missing drupelets.
- Absence of any disease, especially Botrytis is frequent in blackberry fruit.
- Free of attached petals, these may develop later disease.
- Taste: blackberries should be picked ripe (black). Red will not develop full color and good taste.
- The choice of variety is important when considering taste and shelf life.
- Tolerance of defects:
  - Premium fruit should have less than 5% of small defects in coloration and shape.
  - Mold, wounds, and smashed drupelets are not acceptable.

Main causes of quality loss

There are several causes of quality loss in blackberries

- Not careful picking and handling in the field: Commercial blackberries are hand-picked in the field and directly packed in punnets. Handling should be very careful. Rough pulling while picking will result in broken berries, reducing the value of the crop. Only one or two layers of fruit should be put in the punnets, fruit should never be piled up.
- Delays in cooling: Blackberries should be brought under cooling within 1 h of picking. Failure to do so will result in reduced shelf life and higher presence of mold. They also result in dark and soft fruit after a few days. Blackberries should never be exposed to direct sunlight after picking. Keep boxes in a shaded place while waiting for transport to the cold storage facility.
- Temperature pulldown: blackberries should be forced-air cooled immediately after arrival at the packing station. Refrigeration equipment should be powerful enough to cool down blackberries from field temperature to 0°C within 20–30 min of entering the cooler.
- Wet fruit from the field (dew or rain). Blackberries should always be harvested dry. Sometimes berries are wet due to dew or recent rain. In that case, harvest is postponed until berries are dry. In many countries, blackberries are grown under plastic tunnels, also called hoops. In this case, dryness of the fruit and sanitation are improved as well as earliness in harvest.

- Poor field sanitation. Level of inoculum in the field should be kept low, removing decayed and dry leaves and dry branches and opening plastic greenhouses on the sides to allow air circulation. Sometimes flower petals do not detach and cause mold growth. This is variety dependent. Blackberry plants can be sprayed with authorized fungicides only, respecting reentry periods and manufacturer labels. Failure to follow the earlier discussed measures will result in high incidence of postharvest diseases.
- Red drupelets (also called color reversion). This is the most serious physiological disorder of blackberries, showing as some red drupelets among the black drupelets of the blackberries. Red drupelets can be caused by several reasons:
  - Transport vibration. When roads are rough and trucks are not equipped with good shock absorbers, upper pallet boxes tend to vibrate, causing fruit bruising. This bruising shows up later and red drupelets among the black ones.
  - Delays in cooling. If blackberries remain warm after picking, black pigments degrade faster. It is of the utmost importance to cool down fruits within one hour after picking.
  - Genetics: some varieties are more prone to red drupelets than others.
- Partially unripe fruit—Mites. Sometimes blackberry plant material is infected with spider mites. These spider mites are located in the root or trunk structures of the plant and crawl up to the fruit, causing unripe or half ripe fruit. These spider mites are very difficult to eradicate, because treatment do not reach all crevices where the pest lives. It is essential to plant blackberry fields with mite, pest, and disease free material, and to ensure that the material remains clean.

Optimum storage conditions

Harvest should be done early in the morning when possible to avoid high field temperatures. Make sure the punnets are clean and dust free. Dust often accumulates in the materials' warehouse.

Optimum storage temperature of blackberries

0°C, 90%–95% RH. Avoid temperatures below 0°C. Blackberries should be cooled down to the optimum temperature 1 h after harvest.

Forced air-cooling is the most common cooling method. Temperature pull-down should be done in 20–30 min of entering the cooling tunnel. To avoid risk of freezing the fruit, cooling tunnel temperature set-point is set at 1°C, because thermostats usually present a temperature oscillation. Under optimal cooling and sanitation conditions, blackberries can last for 7 days, depending on variety.

Preserving the cold chain. Packaged and cooled blackberries should never be exposed to higher temperatures. Cold chain interruptions will result in condensation of the berries and increased mold and softening. Cold chain interruptions will also cause higher incidence of red drupelet disorder. Breaks in the cold chain are frequent in air shipments, since some airports do not have cold room facilities or do not handle the cold chain carefully.

CA/MA considerations

Modified atmosphere packaging (MAP) is used as a standard procedure in California and Mexico for distribution to the United States. Pallet MAP is used, by which the pallet is wrapped with a plastic bag and injected with a gas concentration of 10%–15% CO_2. High CO_2 deters the growth of fungi.

This MAP system is not used in Europe or other continents on a regular basis, due to the proximity of the production areas to the market.

Benefits and injuries of CA/MA control: due to their high perishability, blackberries rarely show CA injury. Disease or overmaturity in the form of soft berries shows up first, before CA injury can be observed.

Problems under suboptimal conditions

Mold: *Botrytis cinerea* Pers. is the most common postharvest disease of blackberries (Snowdon, 1990). It is often latent in the field and shows up during transport. The fungus grows readily at temperatures above 4°C and during the rainy season. It results in total loss of the crop or downgraded fruit. Occasionally *Rhizopus* spp. can be found.

Color reversion of blackberry. Delays in cooling will cause some drupelets to revert from black to red. The incidence is dependent upon varieties.

Juice leakage. Rough handling of punnets and improper cooling will result in "bleeding" of blackberries. This is a very severe defect that will result in total loss of value.

In order to absorb possible humidity droplets, an absorbent pad is often attached to the bottom of the punnet, usually of black color.

Comments

Blackberries are highly perishable and a very valuable commodity. Cooling, gentle handling, transportation at temperatures close to 0°C, integrity of the cold chain, and sanitation in the field are the keys to successful transportation and delivery to the customer.

Quality characteristics of blueberries

- Varieties and color: Blueberries should be deep dark blue in color when harvested. Pruine, a whitish waxy layer on the skin of the fruit, is appreciated as quality characteristic. Amount of pruine depends on the variety. Also excessive handling removes pruine and visual quality of the blueberries is reduced.
- Minimum caliber: 14 mm in diameter for premium quality blueberries.
- Shape: round shape, no misshapen, no deformed, no broken or squeezed berries.
- Epidermis should be intact after harvest. There should be no broken skin in the area of the pedicel. This defect occurs in some thin-skin varieties when harvested due to pulling.
- Deep blue color, shiny skin. Avoid red or whitish fruit (this is a sign of unripe fruit).
- Sugars: Brix degrees above 10%.
- Acidity: less than 0.7% titratable acidity.
- Crunchy texture when biting, no soft fruit.
- Absence of any disease, especially *Botrytis* spp., which is frequent in blueberry fruit.
- Taste: blueberries should be picked ripe (deep blue). White or light red fruit will not develop full color and good taste.
- The choice of variety is important when considering taste and shelf life.
- Tolerance of defects for premium produce:
  - Premium fruit should have less than 5% of small defects in coloration and shape,
  - Absence of waxy bloom (pruine) in less than 5% of the berries.

Main causes of quality loss of blueberries

There are several causes of quality loss in blueberries

- Not careful picking and handling in the field: Commercial blueberries are hand-picked in the field and packed in punnets. Handling should be very careful. Rough pulling while picking will result in broken berries, reducing the value of the crop. Only 5–6 layers of fruit should be put in the trays, fruit should never be piled.
- Careful sorting: Blueberries are sorted in the packing station, where defect fruit is removed. Avoid bumps while unloading trays into sorting machine. Avoid sharp edges or falls in sorting machinery.
- Delays in cooling: Blueberries should be brought under cooling within 2 h of picking. Failure to do so will result in reduced shelf life and higher presence of mold. They also result in dark and soft fruit after a few days. Blueberries should never be exposed to direct sunlight after picking. Keep boxes in a shaded place while waiting for transport to the cold storage facility.
- Pulldown: blueberries should be forced-air cooled immediately after sorting at the packing station. Refrigeration equipment should be powerful enough to cool down blueberries from field temperature to 0°C within 20–30 min of entering the cooler.
- Wet fruit from the field (dew or rain). Blueberries should always be harvested dry. Sometimes berries are wet due to dew or recent rain. In that case, harvest is postponed until berries are dry. In some countries, like Spain or Morocco, blueberries are grown under plastic tunnels, also called hoops. In this case, dryness of the fruit and sanitation are improved as well as earliness in harvest. In other countries like Chile, blueberries are harvested in open field.
- Poor field sanitation. The level of inoculum in the field should be kept low, removing decayed and dry leaves. Blueberry plants can be sprayed with authorized fungicides only, respecting reentry periods and manufacturer labels. Failure to follow the earlier discussed measures will result in high incidence of postharvest diseases.
- Handling. Blueberries should be handled as little as possible to avoid loss of waxy bloom, bruising and other damages to fruit skin.
- Excessive storage duration can result in severe losses. Internal darkening increases as storage duration increases. An index of degree of darkening can be a useful indicator of fruit quality (Fig. 1).

Optimum storage conditions

Make sure the trays are clean and dust free. Dust accumulates in the materials' warehouse.

Optimum storage temperature of blueberries

0°C, 90%–95% RH. Avoid temperatures below 0°C. Blueberries should be cooled down to the optimum temperature 2 h after harvest.

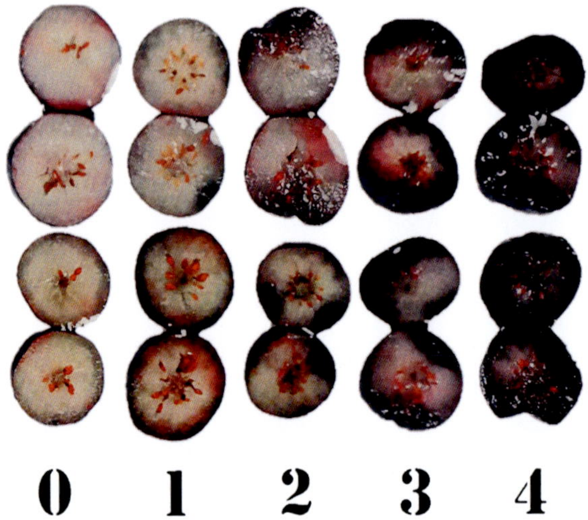

FIG. 1 Index of senescence-related internal darkening of blueberry fruit (R. Beaudry and P. Abeli).

Forced air-cooling is the most common cooling method. Temperature pull-down should be done in 20–30 min of entering the cooling tunnel. To avoid risk of freezing the fruit, cooling tunnel temperature set-point should be set at 1°C, because thermostats usually present a temperature oscillation. Under optimal cooling and sanitation conditions, blueberries can last for 21 days. Shelf life is highly dependent on variety.

Preserving the cold chain. Packaged and cooled blueberries should never be exposed to higher temperatures. Cold chain interruptions will result in condensation of the berries and increased mold and softening. Cold chain interruptions are frequent in air shipments, since some airports do not have cold room facilities or do not handle the cold chain carefully.

CA/MA considerations

Modified atmosphere packaging (MAP) is used as a standard procedure in Chile for distribution to the United States, Europe, and China. Individual MAP bag for each box is the recent standard of the industry for 5-kg boxes, but MAP pallet covers are also used frequently (Fig. 2). In other cases, punnets ready to be sold are placed in a MAP bag on a tray. Ideal blueberry CA conditions are high CO_2 (15%) and low O_2 (3%–5%), although the benefit of low O_2 is very minor. Most MAP bags do not reach optimum O_2/CO_2 levels due to the nature of the gas permeation process. A CO_2 level of 15% would be accompanied by an O_2 level of about 6% for packages in which permeation is through perforations (the sum of O_2 and CO_2 should add up to about 21%). Further, small leaks in the packages

FIG. 2 MAP pallet covers for berry fruit.

and a mismatch between permeation and respiration rates also lead to poor atmosphere control.

Under optimum O_2 and CO_2 levels, shelf life of CA or MAP-stored blueberries can reach 6 weeks, depending on the variety.

The punnet-sized MAP bag system is not used in Europe or other continents on a regular basis, due to the proximity of the production areas to the market. Use of pallet MAP is common in all major production areas.

Benefits and injuries of CA/MA control: due to their high perishability, blueberries rarely show CA injury. Disease in the form of mold or overmaturity (soft fruit with internal discoloration) shows up first, before CA injury can be observed.

Problems under suboptimal conditions

Mold: *Alternaria* spp. is the most common postharvest disease of blueberries (Snowdon, 1990). It is often latent in the field and shows up during transport. The fungus grows readily at temperatures above 4°C and during the rainy season. It can result in total loss of the crop or downgraded fruit.

Most Alternaria rots occur at the stem scar after harvest. Other diseases that appear less frequently are Anthracnose (*Colletotrichum acutatum*) and Botrytis (*Botrytis cinerea*) (Fig. 3).

Dehydration: Rough handling of punnets and improper cooling will result in dehydration of blueberries. This is a very severe defect that will result in loss of value.

In order to absorb possible humidity droplets, an absorbent pad is often attached to the bottom of the punnet.

FIG. 3 Anthracnose symptoms on blueberry fruit (R. Beaudry and P. Abeli).

Comments

Blueberries are highly perishable and a very valuable commodity. Fast cooling, minimum handling, transportation at temperatures close to 0°C, integrity of the cold chain, and sanitation in the field are the keys to successful transportation and delivery to the customer.

Reference

Snowdon, A., 1990. A color atlas of post-harvest diseases. In: General Introduction and Fruits. vol. 1. CRC Press, Boca Raton, FL.

Further reading

Gross, K.C., Wang, C.Y., Saltveit, M.E., 2016. The Commercial Storage of Fruits, Vegetables, and Florist and Nursery Stocks. Agriculture Handbook Number 66, U.S. Department of Agriculture, Washington, DC. http://ucanr.edu/datastoreFiles/234-2927.pdf.

Retamales, J., Hancock, J., 2012. Blueberries. CABI, Cambridge, MA.

Chapter 16.3

Small fruits: Table grapes

Luis Luchsinger
Postharvest Study Center (CEPOC), University of Chile, Santiago, Chile

Quality characteristics

Controlled and Modified Atmospheres for Fresh and Fresh-Cut Produce
https://doi.org/10.1016/B978-0-12-804599-2.00021-1

Table grape fruit develop as clusters (bunches) with each berry attached to the bunch stem (rachis and branches) via a pedicel, which contains vascular bundles (also known as the cap stem). There is much variation between cultivars in stem structure (i.e., length of parts, toughness, and adherence to berries). Impact or shaking of bunches may cause the loss of berries, leaving an unsightly spray of vascular strands (brush).

The cluster of table grapes is formed by the berries and the rachis. The stem unites the berry with the rachis. This union is very important to avoid loss of berries (shattering).

The most important fruit quality attributes are sugar content, expressed as °Brix or soluble solids content (SSC), titratable acidity (TA), SSC/TA ratio, freshness and green color of the rachis (dehydration), fruit color and firmness, free from decay and water berries.

Minimum soluble solids contents are specific for each cultivar, with values starting from 15.5%, and 16.5% for most varieties. The list with the minimum and umbral SSC for each cultivar is worldwide known. The umbral SSC is when the SSC/TA is greater than 20. Therefore a proper ratio of % of SSC and TA expressed as tartaric acid must be fulfilled.

Main causes of quality loss

In general, cumulative water loss during pre- and postharvest handling results in weight loss, fruit stem (rachis or peduncle and pedicels) browning (Fig. 1), berry shatter, and even shriveling of berries.

Stems are particularly susceptible to water loss due to their high surface-to-volume ratio. The high rate of respiration of stems may also be a contributor to

FIG. 1 Rachis browning and dehydration.

stem browning, as the respiration rate of the stems may be 15 times or more that of berries. Although stem browning does not affect the eating quality of the berries, it is a serious quality defect, as it reduces the overall attractiveness of the bunch. Varieties differ greatly in the rate at which postharvest stem browning occurs. The dehydration potential, measured as the percentage weight loss to affect commercially the appearance of the cluster ranged from 4% to 5%, based on the initial weight measured just after harvest.

Berries may suffer from skin and pulp browning if they become bruised during handling. Rachis and berry browning is inhibited by SO_2 treatment. Berries that have not been treated with SO_2 are also more susceptible to gradual browning of the pulp over time. This may be exacerbated using high levels of CO_2 during storage for fungicidal or quarantine purposes.

Berry loss or shatter (Fig. 2) can be a significant problem with certain cultivars of table grapes, especially with Thompson Seedless and Superior Seedless.

The primary cause of postharvest loss in table grapes is gray mold disease or Botrytis rot caused by *Botrytis cinerea*. This disease occurs wherever the crop is grown. The fungus can grow at temperatures as low as $-0.5°C$ and so may

FIG. 2 Berry shatter.

FIG. 3 Berry decay.

spread from one berry to another during storage and transportation even if adequate precooling is carried out and suitable temperatures are maintained. Botrytis rot can be identified by the characteristic 'slipskin' condition that develops, and later, by 'nests' of decayed berries encased in white mycelium (Fig. 3). Postharvest berry infection is primarily caused by conidial infection at or after veraison, although some authors suggest it may happen at the flower stage. The fungus remains quiescent in the developing fruit, with symptoms only appearing on the mature fruit. Other less important fungal postharvest diseases of table grapes include: Aspergillus rot (*Aspergillus niger*) which does not grow below 5°C, blue mold rot (*Penicillium* spp), Rhizopus rot (*Rhizopus stolonifer*), and Alternaria rot (*Alternaria alternata*).

Optimum storage conditions

Optimum temperature

The higher the SSC, the lower the berry freezing temperature. Most cultivars can be precooled and stored from −1°C to 0°C, with a RH of 85%–95%. When SSC allowed, use −1.5°C.

CA/MA considerations

- *Benefits*: high CO_2 levels (above 15%) have long been known to inhibit the growth of fungi. In the last few decades, controlled or modified atmospheres have been shown to have promise for the control of *Botrytis* in table grapes. A combination of 10% CO_2 with 3%, 6%, or 12% O_2 is suggested for up to 12 weeks storage for late harvested 'Redglobe' grapes. An atmosphere of

$10\%\ CO_2 + 6\%\ O_2$ is suggested for early commercially harvested 'Redglobe' grapes, but not to exceed 4 weeks.

- *Tolerances*: levels of CO_2 at 15% or above can completely suppress the growth of *Botrytis*; levels above 10% CO_2 for too long a period, however, were found to generate off-flavors and accelerate stem browning. Sensitivity to CO_2 was, however, dependent on cultivar and maturity. In some cases, the return to air atmosphere at the end of storage can be followed by strong rot development.
- *Injuries caused by CA/MA*: stem and berry browning, and off-flavors.
- *In case of specific atmospheres used for insect control*: several combination treatments are being developed to benefit from additive and synergistic effects. For example, an official approved protocol based on CO_2 and SO_2 fumigation is being used in California to export grapes to Australia, England, and New Zealand. This quarantine treatment kills black widow spiders in packaged table grapes.

Comments

The SO_2 is still the best method to control decay (*Botrytis* and *Penicillium*). Also helps to avoid or delay stem and rachis browning, since it is an antioxidant. The MAP or CA will not be able to compete with this in the near future.

Also, the industry does not have the policy of precooling the harvest trays, and do not have the systems to decrease the field heat right after harvest (precoolers), nor acclimatized packings. Therefore the use of MAP is almost impossible to use since cooling down the fruit after packaging would be too much time consuming, practically an impossible task, and the condensation would be too much to handle it.

Further reading

Ahumada, M.H., Mitcham, E.J., Moore, D.G., 1996. Postharvest quality of 'Thompson Seedless' grapes after insecticidal controlled-atmosphere treatments. HortScience 31, 833–836.

Artes-Hernandez, F., Aguayo, E., Artes, F., 2004. Alternative atmosphere treatments for keeping quality of 'Autumn seedless' table grapes during long-term cold storage. Postharvest Biol. Technol. 31, 59–67.

Artes, F., Gomez, P.A., Artes-Hernandez, F., 2006. Modified atmosphere packaging of fruits and vegetables. Stewart Postharvest Review. Stewart Postharvest Solutions (UK) Ltd pp. 13.

Chervin, C., Aked, J., Crisosto, C.H., 2012. Grapes. In: Rees, D., Farrell, G., Orchard, J. (Eds.), Crop Post-Harvest: Science and Technology. first ed Blackwell Publishing Ltd, pp. 187–211.

Crisosto, C.H., Mitchell, F.G., 2000. Postharvest handling systems: small fruits. I. Table grapes, Chapter 29. In: Kader, A.A. (Ed.), Postharvest Technology of Horticultural Crops. third ed. University of California, Davis DANR Publication #3311.

Crisosto, C.H., Garner, D., Crisosto, G., 2002. High carbon dioxide atmospheres affect stored 'Thompson Seedless' table grapes. HortScience 37, 1074–1078.

Crisosto, C.H., Garner, D., Crisosto, G., 2002. Carbon dioxide-enriched atmospheres during cold storage limit losses from Botrytis but accelerate rachis browning of 'Redglobe' table grapes. Postharvest Biol. Technol. 26, 181–189.

Kock, P.J., Holz, G., 1991. Colonization of table grapes by *Botrytis cinerea* in the western Cape Province. Phytophylactica 23, 73–80.

Luchsinger, L., Vargas, M., Mardones, C., Marin, J., 2000. Storage of Red Globe grapes under controlled atmosphere. 4th International symposium on table grape (Abstracts). INIA Serie Actas 6, 92 ISSN 0717-4810.

Luvisi, D.A., Shorey, H., Smilanick, J., Thompson, J., Gump, B., Knutson, J., 1992. Sulfur dioxide fumigation of table grapes. In: University of California, DANR, Bulletin 1932. University of California, Davis.

Mitcham, E., 2005. Postharvest CO_2/SO_2 Treatment for Black Widow Spider Control. Report to California Table Grape Commission.

Morris, J.R., Oswald, O.L., Main, G.L., Moore, J.N., Clark, J.R., 1992. Storage of new seedless grape cultivar with sulfur dioxide generators. American Journal of Enology & Viticulture 43, 230–232.

Nair, N.G., Allen, R.N., 1993. Infection of grape flowers and berries by *Botrytis cinerea* as a function of time and temperature. Mycological Research 97, 1012–1014.

Retamales, J., Defilippi, B., Arias, M., Manriquez, D., Aviles, R., 2000. Controlled atmosphere as a mean of preserving quality in postharvest of 'Thompson Seedless' and 'Red Globe' table grapes. 4th International symposium on table grape (Abstracts). INIA Serie Actas 6 (ISSN 0717-4810), 21.

Retamales, J., Defilippi, B.G., Arias, M., Castillo, P., Manriquez, D., 2003. High-CO_2 controlled atmospheres reduce decay incidence in Thompson Seedless and Red Globe table grapes. Postharvest Biol. Technol. 29, 177–182.

Thompson, J., Mitchell, F.G., Rumsey, T.R., Kasmire, R.F., Crisosto, C.H., 1998. Commercial cooling of fruits, vegetables, and flowers. Univ. of Calif. Div. of Agric. Nat. Res. Pub. 21567, 62.

Snowdon, A.L., 1990. Soft fruits and berry fruits (Grapes). In: A Colour Atlas of Post-harvest Diseases and Disorders of Fruits and Vegetables, vol. 1: General Introduction and Fruits. Wolfe Scientific, London. pp. 255-266.

UC Davis Postharvest Technology, 2001. Optimal controlled atmospheres for horticultural perishables: Grapes. Postharvest Horticulture Series 22A. p. 45.

USDA, 1999. United States Standards for Grades of Table Grapes (European or Vinifera Type). pp. 14.

Winkler, A.J., Cook, J.A., Kliewer, W.M., Lider, L.A., 1974. General Viticulture. University of California Press, Berkeley, CA.

Yahia, E.M., Nelson, K.E., Kader, A.A., 1983. Post-harvest quality and storage life of grapes as influenced by adding carbon monoxide to air or controlled atmospheres. J. Am. Soc. Hortic. Sci. 108, 1067–1071.

Chapter 17.1

Tropical fruits: Bananas

Eduardo Kerbel* and Manuel Madrid†
*Postharvest Technology Business Manager, Carrier Transicold, Syracuse, NY, United States,
†Fruit Profits, Valencia, Spain

Quality characteristics

Initial fruit quality is determinant for maximum transit and shelf life potential even under best handling practices following optimum temperature and controlled atmosphere regimes.

Controlled and Modified Atmospheres for Fresh and Fresh-Cut Produce
https://doi.org/10.1016/B978-0-12-804599-2.00022-3

Fruit maximum quality potential is determined at harvest by following correct fruit age and calibration parameter decisions. Any deviations from this may shorten transit and shelf life potentials.

Banana shelf life potential is not constant during the year because of differences in fruit physiology, which is affected by different growing regions, climates, and seasons.

As a result it is recommended that pulp condition and state be checked in packing plants prior to processing.

Bananas have one of the most stringent definitions of quality at several different points during their cold chain.

Top initial quality characteristics include: (1) Correct fruit age and calibration according to trip length and destination market; (2) noncreamy or soft pulp condition at harvest; (3) intact peel free of blemishes, cuts, abrasions, or insect marks; (4) absence of any signs of anthracnose; (5) crowns with a fresh appearance, that are moisture saturated, and free of any presence of rots or molds (Fig. 1); (6) flower tip well healed and free of any flower remains; and (7) fingers not too curved, deformed, or twined.

Top quality characteristics at arrival to destination markets include: (1) No sign of premature ripening or color turning; (2) deep green color; (3) firm and hard texture; (4) cucumber-like smell and latex flow from vascular tissues when cut transversally; (5) no signs of chilling on surface and particularly under the peel; (6) clean peel free of blemishes, cuts, abrasions, insect marks; (7) intact peel with no signs of splitting; and (8) clean crown free of rot or mold.

Top quality characteristics at arrival to retail and wholesale outlets include (1) Good ripening quality and uniformity; (2) adequate firmness;

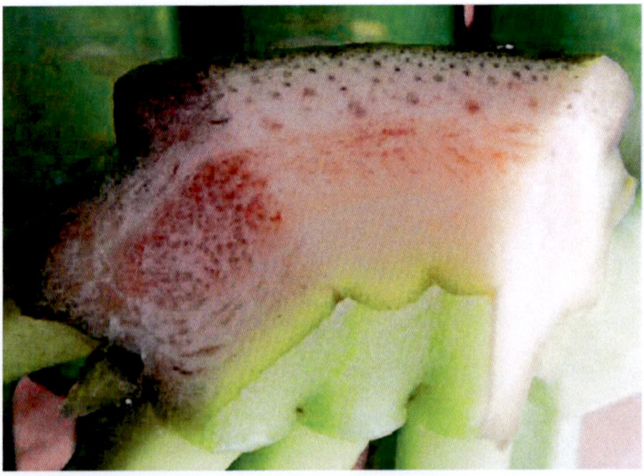

FIG. 1 Shaping the crown of the cluster properly and keeping it saturated with moisture may significantly reduce crown mold development.

(3) correspondent peel color and pulp condition; (4) clean peel free of blemishes, cuts, abrasions, insect marks; (5) intact peel with no signs of splitting; (6) clean crown free of rot or mold; and (7) fruit should not develop its characteristic sugar spots before 2.5–3 days on display.

Main causes of quality loss

- Improper bag protection and removal practices in the field.
- Too advanced age at harvest (as per trip length and destination market needs).
- Higher than maximum calibration when taken on lateral or central finger of 2nd hand.
- Overly fingers in clusters are allowed to be packed.
- Presence of flowers on tips during processing and packing.
- Incorrect delatexing procedure.
- Using too much chlorine or other sanitizers in processing water.
- Allowing creamy pulped fruit to be processed and packed.
- Fruit harvested from area with heavy black sigatoka presence.
- Fruit harvested from areas with noticeable anthracnose on the plant.
- Fruit harvested from stressed plant (drought, flooding, heavy wind).
- Incorrect handling practices at packing plant.
- Poor cluster crown shaping, application of fungicide, and dehydration (Fig. 1).
- Presence/development of crown mold and rot.
- Poor handling from plant to box leading to mechanical injuries.
- Delayed processing and cooling.
- Exposing fruit to temperature abuse and fluctuations.
- Exposing fruit to temperatures below 13.3°C (56°F), causing chilling injury.
- Exposing fruit to relative humidity under 85%, causing peel discoloration.
- Use of wrong packaging materials that do not protect properly.
- Shipping bananas with improperly handled baby bananas in same compartment/container.
- Exposing banana to any ethylene source between farm and destinations markets.
- Not using a complementary technology such as controlled or modified atmospheres during transport to distant markets to allow longer transit life extension and protection beyond what adequate temperature management can achieve.
- Exposing fruit to lower than minimum safe oxygen levels and/or higher than maximum safe CO_2 levels for bananas.
- Improper ripening conditions and protocols.
- Improper handling and temperature abuses at destination markets.

Optimum storage and transport conditions

Optimum temperature

Minimum transit, storage, or holding safe temperature should be 13.3°C (56°F).

The minimum safe temperature is also the optimum temperature for handling bananas, however, depending on the route port-of-origin-to-port-of-destination, and on the time of the year (i.e., outside ambient temperatures), the minimum safe temperature may need to be raised to 13.3°C (57°F) or even 14.5°C (58°F) to avoid chilling injury.

Any exposure of fruit outside these temperature ranges, especially after it has been cooled to set point, may shorten transit and shelf life potential.

It is far more detrimental to fruit quality to expose the fruit to temperature fluctuations, than it is to delay initial cooling.

Optimum relative humidity

Bananas should always be held, stored, or transported at a RH of 85%–95% to avoid dehydration, which when severe can negatively influence fruit quality and may trigger premature ripening or affect ripening quality and uniformity at destination.

CA/MA considerations

Bananas can benefit greatly from the use of controlled atmospheres (CA) or modified atmospheres (MA). Although temperature management is of primary importance, effective cooling and holding at the optimum temperature through the cold chain avoiding fluctuations can only limitedly protect the fruit, especially on longer trips. CA and MA can very effectively complement proper temperature management to maximize the postharvest life potential of bananas (Fig. 2).

CA or MA can be supplied and managed through special chambers or rooms, reefer containers or compartments on vessels, or through modified atmosphere packaging (MAP) using specialized equipment for atmosphere generation and monitoring. In this chapter, MA is referred to a lesser degree of control of the CA because it is based on fruit respiration and leak rate of the container. The main difference between CA and MA is the degree of control by CA whereby N_2 is separated from air to reach a low level of O2 between 1% and 5%, and the N_2/O_2 mix is flushed into the chamber or container to displace the ambient O2 and establish the desired new equilibrium for O_2. Carbon dioxide (CO_2) is sourced from the respiring produce and maintained by removing its excess via activated carbon or other means, or by venting the chamber with air or N_2. Alternatively O_2 and CO_2 can be sourced from single or mixed cylinders to constantly supply the desired and target concentrations of O_2 and CO_2. MA refers to allowing desired and target O_2 and CO_2 to be established by fruit respiration, and maintaining those concentrations by allowing differential gas

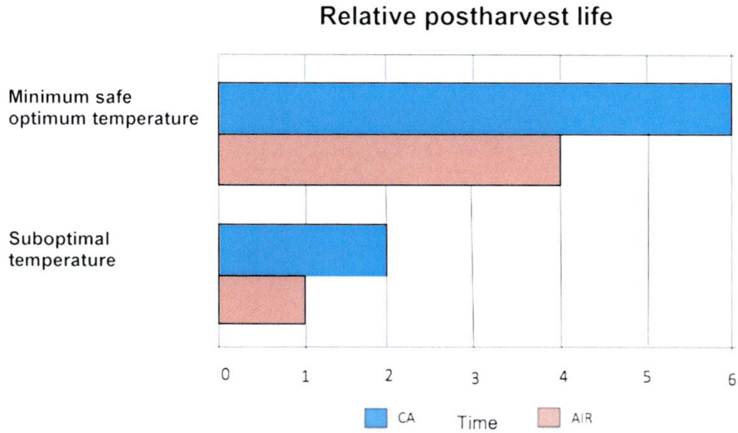

FIG. 2 Relative postharvest life is a function of handling/transport temperature when the commodity is under air or under CA/MA conditions. X-axis in weeks. Note that temperature management has a more dramatic effect on bananas than atmospheric modification.

exchange through a semipermeable plastic bag, or by manipulating fresh air intake or excess CO_2 exit through the ventilation system in reefer containers or vessel compartments.

Injection of N_2 or CO_2 can occasionally be used as an aid in both CA and MA systems to speed up pull-down of O_2 or building up of CO_2 to target levels of certain low-respiring commodities. In the case of bananas, being high-respiring fruit, added CO_2 is seldom necessary and may not provide tangible benefit.

Benefits of CA/MA:

- Prevents premature ripening and color turning in transit, especially on longer hauls (Fig. 3).
- Maintains adequate firmness and texture until the fruit is ripened.
- Protects when inadvertently exposing fruit to ethylene.
- Can reduce chilling injury.
- Maintains condition of freshness.
- Delays decay and spoilage (not very effective against Anthracnose which is secondary to mechanical/physical injury and bruising).
- Retards compositional changes and loss of nutrients.
- Maintains freshness and better fruit condition at arrival to destination ports.
- Reduces dehydration and weight loss.
- Allows successful shipments to new and more distant markets.

Tolerances

The tolerance of bananas to low O_2 and high CO_2 depends on the levels of O_2 and CO_2, on the exposure time at such levels, the temperature and fruit condition when fruit is exposed to MA or CA. In general, green bananas will tolerate

FIG. 3 The main commercial benefit of MA or CA on bananas is to prevent premature ripening and color turning during transit to destination ports. In order to avoid major quality claims, bananas need to arrive firm and in a deep solid green color condition so they can be ripened uniformly.

O_2 levels down to 2% and CO_2 levels up to 8% when being shipped in the solid green stage. O_2 and CO_2 act synergistically on bananas, the lower the O_2, the lower the CO_2, and the higher the CO_2 the higher the O_2 needed to induce injury.

CA/MA limitations and risks

Exposing green bananas to lower than safe O_2 and higher than safe CO_2 levels during transit and holding is time/temperature dependent and can result in: (1) off flavor and aroma, fermentations, fluid leakage, poor ripening quality, bland taste and flavor, (2) brown to black discolorations (Figs. 4 and 5), (3) softening while fruit is still green, (4) accelerated decay and crown mold/rot, and (5) inability to ripen adequately (Fig. 6).

O_2 and CO_2 injury on fresh produce is dependent on time of exposure, temperature, and O_2 and CO_2 concentrations. O_2 and CO_2 act sinergistically which means that low O_2 injury can be aggravated when CO_2 levels are higher than safe >7%–8%, and CO_2 injury can be aggravated when O_2 levels are lower than the minimum safe <1.5%–2.0%.

Recommendations

Oxygen levels below 8% and CO_2 levels above 3% will show tangible benefits on green bananas depending on fruit condition and trip length but optimum levels for maximum transit life potential are 2%–3% O_2 and 5%–6% CO_2 at 13.3°C to 14.5°C.

FIG. 4 O_2 injury on green bananas. O_2 levels below 1.5%–2% can cause peel discoloration and blackening depending on exposure time.

FIG. 5 CO_2 injury on green bananas. CO_2 levels beyond 8% can cause peel discoloration and blackening depending on exposure time.

Ripe bananas can benefit best from 2% to 3% O_2 and 12% to 14% CO_2 at 13.3°C to 14.5°C.

It is recommended that when CA is used for green bananas, fruit is packed in perforated liners or bags to allow free gas exchange and prevent asphyxiation due to additional atmosphere modification imposed by a liner with inadequate gas exchange, resulting in a double MA-CA effect.

FIG. 6 Damage caused by exposure to high CO_2.

CA and MA are a complement not a substitute for adequate temperature management. Banana shippers should use correct handling practices and quality considerations and should not rely on CA or MA to compensate for poor practices (Fig. 2).

Levels of CO_2 can rise up to 7%–8% CO_2 but O_2 should never go below 2%. Levels of O_2 of 4%–5% are less effective than levels of 2%–3% in terms of green life preservation and fungus control. Therefore it is important to bring the levels of O_2 as close to 2%–3% as possible.

Ethylene considerations during shipping

Bananas are extremely sensitive to ethylene and concentrations in the parts per billion range may cause premature ripening, which is highly undesirable.

Exposure of bananas to ethylene should be avoided during transport as premature ripening and fruit turning are the top causes for quality claims in destination ports worldwide.

Best postharvest handling practices for effective ethylene management and control require the following: (1) Harvest fruit at the optimum youngest age and lowest possible calibration to reduce risk of fruit reaching its threshold for ethylene sensitivity sooner than expected, (2) do not run refrigeration units on containers or trucks during loading at packing plant as this practice can inadvertently introduce ethylene from nearby sources such as combustion engines or smoke, (3) avoid shipping fruit together with other ethylene-producing commodities, (4) avoid cross contamination from other commodities being shipped in separate containers or hatches, and (5) remove ethylene during transit and holding as needed using effective scrubbing technology.

Considerations during banana ripening

Concentration of CO_2 should not exceed 1% during ripening. CO_2 meters should be installed in ripening rooms and be connected to an interface to ensure adequate and automated ventilation when CO_2 exceeds 1%. Accumulation of CO_2 during ripening should be avoided, since CO_2 interferes with ethylene in the ripening process.

It is important to have an adequate concentration of ethylene in the room. Many times ripening rooms are not completely gas tight. Levels of ethylene between 500 and 1000 ppm will ensure homogeneous ripening. Levels of ethylene below 200 ppm could result in uneven ripening, with different ripening stages in the same box or among boxes.

Bananas should be ventilated for 24 hours under normal air atmosphere between storage under CA and application of ethylene for ripening. Failure to do so will result in uneven ripening or green and yellow patches in the same banana fingers.

Level of 2%–3% O_2 and 5%–6% CO_2 are optimal for maintaining banana quality and condition during transport and were traditional specifications used when the big banana companies had their own ships for European destinations. When that changed and bananas were now shipped with third-party vessels, CA practically disappeared from banana shipments as third-party shipping lines did not provide CA in transit.

Prior to ripening, it is recommended to expose bananas for 24 h to fresh air so they can get oxygenated for adequate ripening quality and uniformity. This instruction is easily accomplished prior to ripening because Port Authorities at destination ports demand that the CA needs to be broken 24 h prior to arrival because of workers safety.

Bananas are the perfect candidate for MA or CA benefits and protection and therefore banana companies are always looking for cheaper alternatives such as the vacuum MAP bags to protect the fruit. Some protection is offered by these bags, but it is not as effective or as consistent as that provided by true CA. In the real world shipping lines offer mainly container CA which is more like an MA because it is based on fruit respiration with less control as with true CA. However, cost issues create resistance to use of CA or MA. Banana companies are still incurring millions of dollars in claims yearly from premature ripening in transit by not using MA or CA. The cost/benefit ratio of spending on MA (CA more difficult because of the established shipping lines and contracts with 3rd parties) has to be reviewed carefully.

Further reading

Ahmad, S., Perviez, M.A., Thompson, A.K., Ullah, H., 2006. Effects of storage of banana in controlled atmosphere before ethylene treatments on it ripening and quality. J. Agric. Res. 44, 219–229.

Kerbel, E., 2016. Banana and plantain. In: Gross, K.C., Wang, C.Y., Saltveit, M.E. (Eds.), The Commercial Storage of Fruits, Vegetables, and Florist and Nursery Stocks. Agriculture Handbook

Number 66 (AH66). http://www.ars.usda.gov/is/np/CommercialStorage/CommercialStorage.pdf. (Accessed 8 May 2016).

Modified Atmosphere Packaging, 2016. In: The Commercial Storage of Fruits, Vegetables, and Florist and Nursery Stocks. Agriculture Handbook Number 66 (AH66). Gross, K.C., Wang, C.Y., Saltveit, M.E. (Eds.) http://www.ars.usda.gov/is/np/CommercialStorage/CommercialStorage.pdf (Accessed 12.05.16).

International Cold Chain Technology, 2014. Controlled atmosphere and modified atmosphere guidelines, refrigerated cargo ships, and refrigerated containers. Cambridge, UK.

Postharvest Technology UC Davis. 2016. Produce fact Sheets. http://www.postharvest.ucdavis.edu/PFfruits/Banana/. (Accessed 04.16).

Sen, C., Mishra, H.N., Srivastav, P.P., 2012. Modified atmosphere packaging and active packaging of banana (Musa spp.): a review on control of ripening and extension of shelf life. J. Stor. Prod. & Posth. Res. 3, 122–132.

Turner, D.W., 2001. Bananas and Plantains. In: Mitra, S.K. (Ed.), Postharvest Physiology and Storage of Tropical and Subtropical Fruit. CAB International, London, pp. 47–84.

Chapter 17.2

Tropical fruits: Mangoes

Jeffrey K. Brecht
Horticultural Sciences Department, University of Florida, Gainesville, FL, United States

Quality characteristics

Mango fruit quality is related to harvest maturity, which determines the quality of the ripened fruit. Mangoes intended for storage or long-distance transport must be harvested when physiologically mature, but prior to ripening initiation ("mature green"). Mature mangoes have full cheeks and shoulders (Fig. 1) with peel ground color that has changed from dark green to light green or greenish yellow, depending on the cultivar. Red on the peel of some cultivars is an indicator of UV light exposure, not maturity. The flesh color also changes from greenish to cream color (Fig. 2). The soluble solids content of mature mangoes is around 8%–10% and the dry matter is almost entirely starch rather than soluble sugars. Otherwise, mango quality is related to uniformity of shape and size, and freedom from defects and decay. Defects occurring at or as a result of harvest can include various types of internal breakdown, sunburn (Fig. 3), lenticel discoloration (Fig. 4), sapburn (Fig. 5), abrasion (Fig. 6), and insect damage (Fig. 7).

Controlled and Modified Atmospheres for Fresh and Fresh-Cut Produce
https://doi.org/10.1016/B978-0-12-804599-2.00023-5

FIG. 1 Mango fruit with full cheeks and shoulders, indicating full maturity.

FIG. 2 Flesh color progression of mango fruit (left to right) from cream color in a mature but unripe fruit, advancing from yellow to orange as the fruit ripens.

FIG. 3 Sunburn on a mango fruit at harvest due to direct exposure to sunlight.

FIG. 4 Lenticel discoloration on a mango fruit, which can appear as reddish, brown or black discoloration of individual lenticels.

FIG. 5 Sapburn on a mango fruit due to the caustic nature of mango latex, which exudes from the cut stem when fruit are removed from the tree.

FIG. 6 Abrasion damage on the surface of mango fruit may occur during development on the tree due to rubbing of the fruit against branches, as well as after harvest from vibration or rubbing against rough surfaces.

FIG. 7 Mango fruit being sectioned in a packinghouse to inspect for the presence of fruit fly larvae.

Main causes of quality loss

After harvest, mangoes can develop lenticel discoloration, scald, and internal cavitation from hot water treatment. The hot water insect quarantine treatment applied to most mangoes exported to the United States (46.1°C for 60–110 min) can damage immature mangoes. Lenticel discoloration, pebbled surface appearance, peel pitting and grayish discoloration, and inability to ripen—beginning with loss of aroma development—may develop as a result of chilling injury from low temperature exposure. The chilling threshold temperature for mature green mangoes varies from 10°C to 15°C for different cultivars.

Postharvest decay is due to primarily anthracnose (*Colletotrichum gloeosporioides*; Fig. 8), which arises from latent preharvest infections, and Diplodia stem end rot (*Lasiodiplodia theobromae*; Fig. 9), which develops on physical

FIG. 8 Anthracnose decay on mango fruit caused by the fungus *Colletotrichum gloesporioides*, which arises from latent preharvest infections.

FIG. 9 Diplodia stem end rot on mango fruit caused by the fungus *Lasiodiplodia theobromae*, which develops on physical injuries.

injuries. Hot water treatment of 50–55°C for 2–5 min with or without imazalil or thiabendazole fungicide (if allowed) can reduce these decays.

Premature ripening during storage or transport is considered to be a quality loss as it results in fruit softening and is often accompanied by decay development. Its causes include advanced harvest maturity (i.e., ripening initiation before harvest), delayed cooling to the optimum temperature for storage and transport, and exposure to higher than optimum temperatures at any point during the postharvest period, including from farm to packinghouse, at the packinghouse, during storage and transport, and during destination handling. Failure to maintain at least 85% relative humidity throughout the postharvest period can result in shriveling of the outermost tissues. Exposure to any source of ethylene, including exhaust from internal combustion engines and ethylene produced by ripening or decaying fruit, will trigger mango ripening.

Any kind of physical injury can lead to discoloration, shriveling, and decay, as well as being a potential source of ethylene, which can cause premature ripening of other fruit. Potential causes of physical injuries include rough harvesting and handling of fruit after harvest, incorrect harvest containers that can either damage the fruit from abrasion or fail during handling and allow the fruit to be compressed or crushed. Injuries can also occur on a poorly designed packing line with excessive drop heights or abrupt changes in direction. Shipping containers that are inadequate to protect the fruit and incorrect palletization that allows shifting or leaning can also damage mangoes.

Optimum storage conditions

Optimum temperature

The optimum temperature for mango storage and transport is the same as the chilling threshold temperature for the particular cultivar being handled, which varies from 10°C to 15°C for mature green mangoes. As mango fruit ripen, the chilling threshold temperature decreases, so that partially ripe mangoes can be held at 5–10°C, but the potential storage duration is less than for mature green fruit.

Optimum relative humidity

Mangoes should be maintained in 90%–95% relative humidity throughout holding, storage, transport, and ripening.

CA/MA considerations

Mangoes do not exhibit great benefit from CA/MA, with the postharvest life for mature green fruit being potentially extended from 2 to 4 weeks in air to 3 to 6 weeks in CA/MA at 12.5°C. Utilizing CA/MA at 5–8°C allows tree-ripe mangoes to be transported for up to 3 weeks. Ethylene scrubbing is commonly used during long-distance transport of mangoes in either air or CA/MA.

Benefits

Ripening of mangoes is delayed or slowed by CA/MA, including color change, softening, aroma development, decrease in acidity, starch to sugar conversion, and other ripening-related compositional changes. Decay development is also slowed under CA/MA conditions. The elevated CO_2 in CA/MA also reduces chilling injury.

Tolerances

Mango tolerance of CA/MA conditions varies among cultivars, maturity and ripeness stages, and for different temperatures and durations of storage or transport. A good starting point for application of CA/MA to mature green mangoes is 3%–5% O_2 plus 5%–10% CO_2 at 12.5°C for 3–6 weeks. Partially ripe mangoes tolerate higher CO_2 and lower temperature but are more sensitive to initiation of fermentative metabolism in reduced O_2 atmosphere. 'Tommy Atkins' mangoes have been successfully stored for 3 weeks in 5% O_2 plus 10% or 25% CO_2 at 8°C. To illustrate the effect of exposure time on mango CA/MA tolerance, mature green mangoes have been shown to tolerate extreme CA/MA conditions such as are applicable to insect control—as low as 0.2% O_2 plus up to 80% CO_2—for 5 days at 20°C.

Injuries

Holding mangoes for too long in CA/MA usually simply results in ripening initiation, but extreme CA/MA conditions cause injuries. The first indication of incipient CA/MA injury is usually ethanolic off odor and off flavor related to fermentative metabolism. If visible injury has not occurred, this off odor dissipates upon transfer from CA/MA to air. Color development may be inhibited following exposure to injurious levels of CO_2. More extreme CA/MA conditions result in failure of mango fruit to recover ripening capacity. Visible injury symptoms from CA/MA include peel discoloration from high CO_2 injury. The peel of mango fruit injured by high CO_2 develop what has been described as a dull grayish green or olive-khaki appearance that appears to be brown discoloration of injured epidermal tissue overlaying normal green color from retarded ripening (Fig. 10). Injurious CO_2 levels may also cause abnormal flesh softening. Low O_2 injury causes grayish flesh color.

Problems under suboptimal conditions

The most common cause of atmosphere injury of mango fruit arises from failure to properly set the fresh air exchange vent on marine containers during long-distance transport. When this occurs, it can result in injury that is apparently a combination of low O_2 and high CO_2 damage (Fig. 11).

FIG. 10 High CO_2 on mango fruit results in development of a dull grayish green or olive-khaki color that may appear brown where it overlays the green color of unripe fruit.

FIG. 11 A combination of low O_2 and high CO_2 damage on mango fruit arising from failure to open the fresh-air exchange of a marine container during transport.

Comments

There is little economic incentive to store mangoes for extended periods of time. However, mangoes may be transported quite long distances in international trade—3–4 weeks is not uncommon. Therefore the best use of CA/MA for mangoes may be during extended marine transport. Since the supply of CA/MA-equipped marine containers is inadequate to handle the volume of mangoes in international trade, use of modified atmosphere packaging (MAP) may currently have more potential.

Further reading

Brecht, J.K., Sargent, S.A., Kader, A.A., Mitcham, E.J., Maul, F., Brecht, P.E., Menocal, O., 2017. Mango postharvest best management practices manual. In: Extension Publication HS1185. University of Florida, IFAS, Gainesville. http://edis.ifas.ufl.edu/hs1185.

Kader, A.A. 2017. Mango: Recommendations for maintaining postharvest quality. Produce Facts. Produce Fact Sheets. http://postharvest.ucdavis.edu/Commodity_Resources/Fact_Sheets/Datastores/Fruit_English/?uid=37&ds=798 (accessed 22.11.17).

Paull, R.E., Chen, C.C. 2014. Mango, in: Gross, K.C., Wang, C.Y., Saltveit, M.E. (Eds.), The Commercial Storage of Fruits, Vegetables, and Florist and Nursery Stocks. Agriculture Handbook Number Vol. 66. USDA ARS, Washington, D.C., pp. 397–461. http://www.ars.usda.gov/is/np/CommercialStorage/CommercialStorage.pdf (accessed 22.11.17)

Thompson, A.K., 2010. Recommended CA Storage Conditions for Selected Crops. Controlled Atmosphere Storage of Fruits and Vegetables. CABI. U.K., Wallingford, pp. 116–191

Yahia, E.M. 2011. Mango (*Mangifera indica* L.), in: Yahia, E.M. (Ed.), Postharvest Biology and Technology of Tropical and Subtropical Fruits. Volume Vol. 3: Cocona to Mango. Woodhead, Cambridge, U.K., pp. 492–565

Yahia, E.M., Singh, S.P., 2009. Tropical fruits. In: Yahia, E.M. (Ed.), Modified and Controlled Atmospheres for the Storage, Transportation, and Packaging of Horticultural Commodities. CRC Press, Boca Raton, Fla, U.S.A., pp. 397–461.

Chapter 17.3

Tropical fruits: Papayas

Robert E. Paull* and Jurandi G. Oliveira†
*Department of Tropical Plant and Soil Sciences, University of Hawaii at Manoa, Honolulu, HI, United States, †Universidade Estadual do Norte Fluminense, Rio de Janeiro, Brazil

Quality characteristics

The main visual quality characteristics of papaya (*Carica papaya*) are size, shape, smooth skin, and the absence of blemishes. The pear-shaped 'Solo' group weigh from 300 to 700 g and other varieties that are spherical or cylindrical in shaped range from 200 g to 10 kg (Oliveira and Vitoria, 2011). During ripening, the external yellow skin color develops from the blossom end, whereas the internal flesh coloring and softening develops from the endocarp outwards.

The sweetness of the pulp is the most important sensory quality appreciated by consumer besides color (Sivakumar and Wall, 2013). Sweetness is used as a ripening index with Hawaii requiring a minimum of 11.5% TSS (Anon, 1990).

Controlled and Modified Atmospheres for Fresh and Fresh-Cut Produce
https://doi.org/10.1016/B978-0-12-804599-2.00024-7

To meet this standard, fruit of the Solo group should have started to ripen before harvest (Akamine and Goo, 1971; Gómez et al., 2002; Sivakumar and Wall, 2013). Mature green fruit are lower in TSS and ripen poorly after harvest with little flavor. Harvest time is limited by the high sensitivity of riper fruit to bruising (Quintana and Paull, 1993) and by fruit fly egg lay.

Fruit flesh color is another important sensorial quality in papaya fruit. Immature flesh is greenish-white and changes when ripe to pale orange-yellow, salmon pink, or red depending upon the cultivar (Wall, 2006). Consumers in Western countries also apparently prefer a fruit without the heavy, musky, sweaty odor found in some Southeast Asian cultivars.

Main causes of quality loss

The three major causes of postharvest losses are mechanical injury, poor storage environment (chilling injury), and disease (Bautista-Baños et al., 2013; Paull et al., 1997; Sivakumar and Wall, 2013). Color-break fruit are very susceptible to abrasion damage, which leads to greenish sunken areas on ripe fruit (Li and Thomas, 2014; Quintana and Paull, 1993). Chilling injury symptoms include skin scald, hard lumps in the pulp around the vascular bundles, uneven ripening, inferior fruit flavor, water soaking of flesh, and increased susceptibility to postharvest pathogens (Chen and Paull, 1986; El-Tomi et al., 1974; Thompson and Lee, 1971; Sivakumar and Wall, 2013). The extent of symptom development is dependent upon cultivar and duration of exposure to temperatures below 10°C. The main pathological disorders in papaya are anthracnose (*Colletotrichum gloeosporioides*), black stem-end rot (*Phoma caricae-papaya*), phomopsis rot (*Phomopsis caricae-papayae*), phytophthora stem end rot (*Phytophthora nicotianae* var. *parasitica*), and Alternaria rot (*Alternaria alternat*) (Alvarez and Nishijima, 1987). Alternaria rot is most frequent after storing the fruit at temperatures below 12°C. Most of the diseases develop rapidly when the fruit is more than 25% ripe (Bautista-Baños et al., 2013; Nishijima et al., 1990; Sivakumar and Wall, 2013).

Unsightly skin freckles (small, brown, slightly raised lesions that may coalesce) are seasonal in occurrence and more common on the side of the fruit exposed to the sun (Oliveira et al., 2005; Oliveira and Vitoria, 2011; Reyes and Paull, 1994). The condition is associated with high rainfall and low temperatures about two months before harvest (Kaiser et al., 1996; Oliveira and Vitoria, 2011; Reyes and Paull, 1994;).

Optimum storage conditions

Optimum temperature

Storage temperature is recommended to be in the range from 7°C to 13°C and the RH should be between 90% and 95%. Commercially, fruit are most often shipped at 7–8°C as long as the storage duration does not exceed 21 days. However, research results using different criteria as to marketable life recommend 12–13°C. At 7–10°C, storage life is limited by chilling injury, while at 10–13°C the fruit ripens slowly and ripeness limits storage life (Chen and Paull, 1986; Sivakumar and Wall, 2013). Ripe, full colored fruit can be held for more than a week at 1–3°C, though the fruit will lose gloss and may shrivel due to water loss. The optimum temperature for fruit ripening is between 22.5°C and 27.5°C, with fruit taking 10–16 days to reach full skin yellowing from the color-break stage in this temperature range (An and Paull, 1990).

CA/MA considerations

A postharvest life extension of 1–1.5 days (~12%) is obtained when papaya are stored at 12°C in 1% to 1.5% O_2 for 6 days (Akamine and Goo, 1969). Low O_2 (1%–5%) with or without high CO_2 (2% to 10%) reduces decay (Hatton and Reeder, 1969) and delays ripening (Akamine, 1959; Chen and Paull, 1986). Quite high CO_2 (30%) adversely affects internal color, aroma, and flavor. While there is no residual effect of 10% CO_2 on decay control, skin degreening is delayed. At 10°C, fruit can be stored for 36 days in 8% CO_2 and 3% O_2 and still have ~5 days at 25°C for retailing (Cenci et al., 1997). Ethylene removal prior to storage has shown variable results (Nazeeb and Broughton, 1978). Fruit stored at 10°C, 98% RH, and low pressure of 20 mm Hg (~2.7 kPa) ripen more slowly than fruit from normal pressure (Alvarez, 1980). Low pressure storage also suppresses disease development (Chau and Alvarez, 1983).

CA recommendations are in the range 2%–5% O_2 in combination with 3%–8% CO_2 at 10–13°C (Yahia, 1998). Subsequent research has not led to changes in these recommendations. No large-scale commercial use has so far been reported, possibly because fruit that are harvested throughout the year and the application of CA during marketing with its added cost may not be warranted. New technologies for CA such as atmosphere controllers and more robust sensors to quantify O_2 and CO_2 concentrations (Chaix et al., 2014; Jochum et al., 2016) integrated into to modern refrigerated shipping containers require continued reevaluation of the costs and benefits of CA for papaya (Rodrigue and Notteboom, 2013). The optimum internal O_2/CO_2 requirement maybe different at different stages of ripening (Pereira et al., 2009) and implementation might have positive cost–benefit advantages.

The use of film wraps and wax coatings has the advantage in limiting weight loss, adding gloss in the case of coatings, and avoiding skin shrivel, which becomes visible upon water loss of about 7%–10% of the original fruit weight (Paull and Chen, 1989; Ali et al., 2000). Coatings, when carefully selected, do generate a modified atmosphere in the fruit and give a few days extension in postharvest life. Extension of storage life to four weeks has been reported for fruit shipped wrapped in low-density polyethylene (0.04 mm) where the gas concentration in the bag changed in 24 h to 2% to 3% O_2 and 4%–5% CO_2 at 10–12°C (Rohani and Zaipun, 2007).

Fresh-cut products

The fruit is prepared as deseeded halved fruit, slices, and chunks. Minimally processed products made from 60% to 80% yellow-skinned fruit on a tray overwrapped with plastic film can be held for up to 3 weeks at 0–4°C (Paull and Chen, 1997). It is possible to extend the shelf life of fresh-cut papaya by combining a chemical treatment with modified atmosphere packaging (MAP). Waghmare and Annapure (2013) showed that fresh-cut papaya could be held for up to 25 days at 5°C using MAP (5% O_2, 10% CO_2 and 85% N_2) after treatment with a preservative solution of calcium chloride (1% w/v) and citric acid (2% w/v) solution for 5 min. The postharvest life of fresh-cut unripe papaya pieces is about 12 days using low-density polyethylene (LDPE, thickness 37 μm) vacuum sealed pouches at 4°C, with minimum change in the product's quality and with higher sensory scores (Tirkey et al., 2014).

Shredded papaya used for green papaya salad when dipped in 1% ascorbic acid or citric acid and stored in 5% O_2 and 5% CO_2 remained in acceptable condition in a plastic bag for up to 32 h at ambient temperatures of 20°C and 87% RH (Lichanporn and Kanlayanarat, 2006). At a lower temperature, dipped shredded papaya can potentially be held for a longer period.

Insect disinfestation

Fruit fly infestation becomes a problem in papaya after the fruit has achieved 25% or greater skin yellowing (Seo et al., 1982). The damage caused by fruit flies includes small surface blemishes, destruction of the edible flesh, and spoilage from decay (Singh and Rao, 2011). Papaya can tolerate insecticidal atmospheres (0.17%–0.35% O_2, balance N_2) for up to 3 days at 20°C without the risk of fruit injury, after 3 days in low O_2 and a further 3 days in air, the fruit developed 'off-flavors' (Yahia et al., 1989). The disinfestation potential of this atmosphere has not been shown (Fig. 1).

Problems under suboptimal conditions

Off odors do occur when fruit are subject to very low oxygen levels of less than 2% or CO_2 levels higher than 8% (Sandhya, 2010; Yahia et al., 1992). This off odor dissipates when the fruit is returned to ambient conditions.

Increased Alternaria rot is observed in mature green papayas kept at chilling temperatures. Susceptibility to chilling injury varies among cultivars and is greater in mature green than ripe papayas.

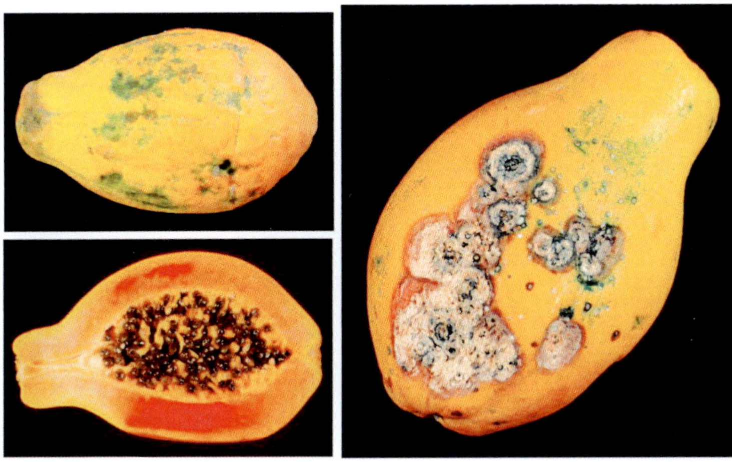

FIG. 1 Top left abrasion injury, bottom left impact injury, right Anthracnose.

Comments

Ethylene-treated papayas ripened faster (An and Paull, 1990). Ethylene treatment accelerates the rate of ripening and is not recommended commercially as the rapid softening severely limits available marketing time (An and Paull, 1990). The application of 1-MCP has not been recommended for fruit less than 25% yellow as it often leads to rubbery flesh texture when fully ripe (Ahmad et al., 2013; Manenoi et al., 2007).

Acknowledgments

This work was support by the National Institute of Food and Agriculture, U.S. Department of Agriculture, under an agreement 58-2040-5-010 through the Agriculture Research Service and Hatch Project H862.

References

Ahmad, A., Ali, Z.M., Zainal, Z., 2013. Delayed softening of papaya (*Carica papaya* L. cv. Sekaki) fruit by 1-methylcyclopropene (1-MCP) during ripening at ambient and low temperature storage conditions. Australian Journal of Crop Science 7, 750–757.

Akamine, E.K., 1959. Effects of carbon dioxide on quality and shelf life of papaya. Hawaii Agricultural Experiment Station Technical Progress Report No. 120.

Akamine, E.K., Goo, T., 1969. Effects of controlled atmosphere storage of fresh papaya (*Carica papaya*, L. Var. Solo) with special reference to shelf life extension of fumigated fruit. Tech. Bull. 144, Hawaii Agricultural Experiment Station, University of Hawaii, Hawaii.

Akamine, E.K., Goo, T., 1971. Relationship between surface color development and total soluble solids in papaya. HortScience 6, 567–568.

Ali, Z.M., Ng, C.H., Chua, S.T., Othman, R., Lazan, H., 2000. Heat treatment alleviates chilling injury symptoms, retards texture changes and maintains the capacity to produce ethylene in papaya fruits. In: Artés, F., Gil, M.I., Conesa, M.A. (Eds.), Improving Postharvest Technologies of Fruits, Vegetables and Ornamentals. International Institute of Refrigeration, Paris, pp. 519–524.

Alvarez, A.M., 1980. Improved marketability of fresh papaya by shipment in hypobaric containers. HortScience 15, 517–518.

Alvarez, A.M., Nishijima, W.T., 1987. Postharvest diseases of papaya. Plant Dis. 71, 681–686.

An, J.F., Paull, R.E., 1990. Storage temperature and ethylene influence on ripening of papaya fruit. J. Amer. Soc. Hort. Sci. 115, 949–953.

Anonymous, 1990. Standards for Hawaii-grown papaya. Hawaii Department of Agriculture, Marketing Division.

Bautista-Baños, S., Sivakumarb, D., Bello-Péreza, A., Villanueva-Arcec, R., Hernández-Lópeza, M., 2013. A review of the management alternatives for controlling fungi on papaya fruit during the postharvest supply chain. Crop Prot. 49, 8–20.

Cenci, S.A., Soares, A.G., Balbino, J.M.S., Soiya, M.L.M., 1997. Study of the storage of Sunrise 'Solo' papaya fruits under control atmospheres. In: Seventh International Controlled Atmosphere Research Conference, July 1997. University of California, Davis CA 95616. Abstract #112.

Chaix, E., Guillaume, C., Guillard, V., 2014. Oxygen and carbon dioxide solubility and diffusivity in solid food matrices: a review of past and current knowledge. Compr. Rev. Food Sci. Food Safety 13, 261–286.

Chau, K.F., Alvarez, A.M., 1983. Effects of low-pressure storage on *Colletotrichum gloeosporioides* and postharvest infection of papaya. HortScience 18, 953–955.

Chen, N.M., Paull, R.E., 1986. Development and prevention of chilling injury in papaya fruit. J. Amer. Soc. Hort. Sci. 111, 639–643.

El-Tomi, A.L., Abon-Aziz, A.B., Abdel-Kader, A.S., Abdel-Wahab, F.K., 1974. The effect of chilling and non-chilling temperatures on the quality of papaya fruit. Egyptian J. Hort. 1, 179–185.

Gómez, M., Lajolo, F., Cordenunsi, B., 2002. Evolution of soluble sugars during ripening of papaya fruit and its relation to sweet taste. J. Food Sci. 67, 442–447.

Hatton, T.T., Reeder, W.F., 1969. Controlled atmosphere storage of papaya. Proc. Trop. Reg. Amer. Soc. Hort. Sci. 13, 251–256.

Jochum, T., Rahal, L., Suckert, R.J., Poppa, J., Frosch, T., 2016. All-in-one: a versatile gas sensor based on fiber enhanced Raman spectroscopy for monitoring postharvest fruit conservation and ripening. Analyst 141, 2023–2029.

Kaiser, C., Allan, P., White, B.J., Dehrmann, F.M., 1996. Some morphological and physiological aspects of freckle on papaya (*Carica papaya* L.) fruit. J. South Afr. Soc. Hort. Sci. 6, 37–40.

Li, Z., Thomas, C., 2014. Quantitative evaluation of mechanical damage to fresh fruits. Trends Food Sci. Technol. 35, 138–150.

Lichanporn, I., Kanlayanarat, S., 2006. Effect of organic acid and modified atmosphere conditions on quality of shredded green papaya (*Carica papaya* L.). Acta Hort. (712), 729–733.

Manenoi, A., Bayogan, E.R.V., Thumdee, S., Paull, R.E., 2007. Utility of 1-methylcyclopropene as a papaya postharvest treatment. Postharvest Biol. Technol. 44, 55–62.

Nazeeb, M., Broughton, W.J., 1978. Storage conditions and ripening of papaya 'Bentong' and 'Taiping'. Sci. Hort. 9, 265–277.

Nishijima, W.T., Ebersole, S., Fernandez, J.A., 1990. Factors influencing development of postharvest incidence of *Rhizopus* soft rot of papaya. Acta Hort. (269), 495–502.

Oliveira, J.G., Vitoria, A.P., 2011. Papaya: nutritional and pharmacological characterization, and quality loss due to physiological disorders. An overview. Food Res. Int. 44, 1306–1313.

Oliveira, J.G., Pereira, M.G., Martelleto, L.A.R., Ide, C.D., 2005. Skin freckle on papaya fruit. A perspective of obtaining tolerant genotypes. Rev. Bras. Frutic. 27, 458–461.

Paull, R.E., Chen, N.J., 1989. Waxing and plastic wraps influence water loss from papaya fruit during storage and ripening. J. Amer. Soc. Hort. Sci. 114, 937–942.

Paull, R.E., Chen, W.J., 1997. Minimal processing of papaya (*Carica papaya* L.) and the physiology of halved fruit. Postharvest Biol. Technol. 12, 93–99.

Paull, R.E., Nishijima, W., Reyes, M., Cavaletto, C., 1997. A review of postharvest handling and losses during marketing of papaya (*Carica papaya* L). Postharvest Biol. Technol. 11, 165–179.

Pereira, T., Almeida, P.S.G., Azevedo, I.G., Cunha, M., Oliveira, J.G., Silva, M.G., Vargas, H., 2009. Gas diffusion in 'Golden' papaya fruit at different maturity stages. Postharvest Biol. Technol. 54, 123–130.

Quintana, M.E.G., Paull, R.E., 1993. Mechanical injury during postharvest handling of Solo papaya fruit. J. Amer. Soc. Hort. Sci. 118, 618–622.

Reyes, M.E.Q., Paull, R.E., 1994. Skin Freckles on Solo papaya fruit. Sci. Hort. 58, 31–39.

Rodrigue, J.-P., Notteboom, T., 2013. The cold chain and its logistics. In: Rodrigue, J.-P., Comtois, C., Slack, B. (Eds.), The Geography of Transport Systems. Routledge, Taylor & Francis Group LLC, Kentucky, pp. 158–187.

Rohani, M.Y., Zaipun, M.Z., 2007. MA storage and transportation of 'Eksotika' papaya. Acta Hort. (740), 303–311.

Sandhya, 2010. Modified atmosphere packaging of fresh produce: current status and future needs. LWT—Food Sci. Technol. 43, 381–392.

Seo, S.T., Farias, G.B., Harris, E.J., 1982. Oriental fruit fly: ripening of fruit and its effect on index of infestation of Hawaiian papayas. J. Econ. Ent. 75, 173–178.

Singh, S.P., Rao, D.V.S., 2011. Papaya (*Carica papaya* L.). In: Yahia, E.M. (Ed.), Postharvest Biology and Technology of Tropical and Subtropical Fruits. WP, Woodhead Publishing, Oxford, pp. 86–124.

Sivakumar, D., Wall, M.M., 2013. Papaya fruit quality management during the postharvest supply chain. Food Rev. Int. 29, 24–48.

Thompson, A.K., Lee, G.R., 1971. Factors affecting the storage behaviour of papaya fruit. J. Hort. Sci. 46, 511–516.

Tirkey, B., Pal, U.S., Bal, L.M., Sahoo, N.R., Bakhara, C.K., Panda, M.K., 2014. Evaluation of physico-chemical changes of fresh-cut unripe papaya during storage. Food Pack. Shelf Life 1, 190–197.

Waghmare, R.B., Annapure, U.S., 2013. Combined effect of chemical treatment and/or modified atmosphere packaging (MAP) on quality of fresh-cut papaya. Postharvest Biol. Technol. 85, 147–153.

Wall, M.M., 2006. Ascorbic acid, vitamin A, and mineral composition of banana (*Musa* sp.) and papaya (*Carica papaya*) cultivars grown in Hawaii. J. Food Comp. Anal. 19, 434–445.

Yahia, E.M., 1998. Modified and controlled atmospheres for tropical fruits. Hort. Rev. 22, 123–183.

Yahia, E.M., Medina, F., Rivera, M., 1989. Tolerance of mango and papaya to atmospheres containing very high levels of CO_2 and/or very low levels of O_2 as a possible insect control treatment. Proc. 5th International CA Research Conference (June 14-16, 1989) Wenatchee. Washington. 2, 77–89.

Yahia, E.M., Rivera, M., Hernandez, O., 1992. Responses of papaya to short-term insecticidal oxygen atmosphere. J. Amer. Soc. Hortic. Sci. 117, 96–99.

Chapter 17.4

Tropical fruits: Pineapples

Robert E. Paull and Nancy Jung Chen
Department of Tropical Plant and Soil Sciences, University of Hawaii at Manoa, Honolulu, HI, United States

Quality characteristics

Controlled and Modified Atmospheres for Fresh and Fresh-Cut Produce
https://doi.org/10.1016/B978-0-12-804599-2.00025-9

Pineapple fruit maturity is commercially evaluated in the field on the extent of fruit "eye" flatness, sugar content, acidity, and skin yellowing (Anon, 1968; Dull, 1971; Soler, 1992). Preharvest application of ethephon is used to increase the degree of skin yellowing and has little impact of flesh quality (Crochon et al., 1981).

Pineapples are graded by degree of skin coloration, size (weight), absence of defects and disease, and uniformity of these characteristics before packing (Soler, 1992; Newett and Rigden, 2015). Other characteristics include the needs to be mature, firm, well formed, with flat eyes, well cured broken stem (peduncle), and a minimum TSS (Anon, 1968). Immature fruit are not shipped as they do not develop full flavor, have low total soluble solids, and are more prone to chilling injury (Cancel, 1974; Paull and Lobo, 2012).

Crown size is a crucial grade component, with a ratio of crown to fruit length (0.33–1.5) being required for the higher grades. Fruit with pronounced eyes or fruitlets (prickly eyes) are normally not acceptable in Fancy grades (Lim, 1985).

Consumers judge fruit quality by skin color and aroma. A minimum reading of 12% TSS is required. A sugar-to-acid ratio of 0.9–1.3 is recommended (Soler, 1992).

Main causes of quality loss

Mechanical injury, translucency, chilling injury, and postharvest disease are the main causes of postharvest losses. Both the fruit and crown appearance need to be evaluated. Dry, brown crown leaves; dull and yellow skin appearance; mold on the fruit and the broken or cut stem; and fruit having a soft, not a firm feel are negative characteristics. Fruit bruising due to abrasion and impact damage is a major problem during harvesting, packing, and shipping of pineapple (Keetch, 1978; Paull and Lobo, 2012). Flesh translucency is associated with greater fruit sensitivity to mechanical injury indicated by leakage and oozing of cellular fluids. Translucency begins before harvest and continues after harvest (Paull and Reyes, 1996). Damaged, bruised, and overripe fruit can show 'yeasty' fermentation from already present yeasts and bacteria that grow and cause bubbles of gas and juice from cracks in the skin. Saprophytic fungi (e. x. *Penicillium* sp.) growing on the broken end of peduncle are unsightly and a marketing problem (Rohrbach, 1989). This peduncle or "butt" mold is more common on highly translucent fruit. Sunburnt fruit are more susceptible to disease (Keetch and Balldorf, 1979).

Black Rot, a common pineapple postharvest disease, causes a soft watery rot and is due to the fungus *Chalara paradoxa* (De Seynes) Sacc (Rohrbach, 1983). Infection occurs within 8–12 h and enters through the broken or cut peduncle or wounds (Rohrbach and Schmitt, 1994). The rot is controlled by minimizing bruising during handling and with postharvest fungicides (Rohrbach and Phillips, 1990).

Fruitlet core rot describes the brown to black color of the central part of an individual fruitlet and is rare in the major commercial pineapple areas. The low acid cultivars now being grown commercially are more susceptible (Rohrbach and Schmitt, 1994). This disease is caused by a complex of fungi.

Optimum storage conditions

Color-break fruit can be stored at 7–12°C for 14–21 days (Paull and Lobo, 2012). Relative humidity of 85%–95% is recommended and the higher humidity reduces water loss. Ripe fruit can be held at 7°C for about 10–14 days. Quarter-yellow fruit at harvest gain about one additional week of storage for every 6°C decrease in storage temperature (Dull, 1971). The maximum storage life at 7°C is 3–4 weeks (Paull and Lobo, 2012); however, when removed to ambient conditions, chilling-induced internal browning develops within 2–3 days of extended cold storage.

CA/MA considerations

Considerable proprietary and some public research have been undertaken on the benefits of controlled and modified atmosphere storage. Controlled atmospheres via modified O_2 levels have shown only minimal effectiveness at extending postharvest life in this nonclimacteric fruit (Akamine and Goo, 1971). Low O_2 has no effect on crown condition or decay, but delays shell color development and has reduced superficial mold growth (Akamine and Goo, 1971). Subjecting fruit to 1%–2% O_2 plus 0–10% CO_2 after shipment from Mexico to England at 8°C does delay chilling injury-induced browning (Haruenkit and Thompson, 1994). The fruit waxes currently used generate higher internal concentrations of CO_2 (up to 5%) and reduced O_2, and less chilling injury symptoms (Paull and Rohrbach, 1982; Abdullah et al., 1985). Polyethylene bagging, though difficult to preform commercially on individual fruit, results in atmospheres of 8%–10% O_2 plus 7% CO_2, and delays the appearance of chilling symptoms (Rohrbach and Paull, 1982). The tentative recommendation is 2%–5% O_2 plus 5 to 10% CO_2 (Yahia, 1998).

Fresh-cut products

The major causes of fresh-cut losses are product leakage that can be very high in translucent fruit, browning often due to chilling injury, and microbial growth (Montero-Calderon and Martin-Belloso, 2016). Varieties show significant differences in their susceptibility to these disorders (Budu and Joyce, 2003, 2005; González-Aguilar et al., 2004; Rocculi et al., 2009; Bierhals et al., 2011). Fresh-cut pineapple packs can contain a cylinder with the core removed, slices, spears, chunks, or wedges. The product has a shelf life of 7–21 days with proper temperature management in plastic trays or pouches (Marrero and Kader, 2006;

Montero-Calderón et al., 2008; Puthmee et al., 2012; Finnegan and O'Beirne, 2015b; Montero-Calderon and Martin-Belloso, 2016). Controlled atmospheres of 2%–8% O_2 and 5%–5% CO_2 have been evaluated though no significant advantages are reported for this approach in fresh-cut product (Antoniolli et al., 2007).

Mature pineapple fruit are not sterile inside due to the way that the fruit develops, yeast and bacteria can enter through the nectary gland (Okimoto, 1948; Rohrbach and Apt, 1986). Growth of the bacteria and yeast can become an acute problem following bruising and preparing fresh-cut products. Seasonal and varietal differences are seen, along with stage of fruit ripeness affecting the suitability for preparing fresh-cut product (Montero-Calderón et al., 2008; Montero-Calderon and Martin-Belloso, 2016). Fruit with low shell color (0–25%) to ripe fruit (25%–50% yellow) give better fresh-cut product quality (Finnegan and O'Beirne, 2015a,b) (Fig. 1).

Atmosphere modification for fresh-cut pineapple products is achieved by use of edible coatings, modified atmosphere, and package atmosphere adjustment. Most recommendations suggest storage at 0–5°C (Marrero and Kader, 2006; Montero-Calderón et al., 2008; Puthmee et al., 2013; Benítez et al., 2014; Jiang et al., 2016; Montero-Calderon and Martin-Belloso, 2016). Depending upon the tray plastic film, O_2 levels declined from 21% to about 10% in 20 days with the CO_2 increasing to about 10% over the same period (Montero-Calderón et al., 2008; Benítez et al., 2012). High O_2 (50%) and CO_2 (50%) does inhibit spoilage yeast growth (Zhang et al., 2013). Low O_2 is more easily attained in MA with 4%–5% O_2 and 5%–15% CO_2 being best for slices held at 1–4.5°C (Budu et al., 2007; Pan et al., 2015).

MA approaches can be coupled with chemicals to delay browning and quality loss such as ascorbic acid, acetyl-cysteine, glutathione, and volatile oils (González-Aguilar et al., 2004; Jiang et al., 2016). Numerous edible coatings have been applied to fresh-cut pineapples (Montero-Calderon and Martin-Belloso, 2016). Examples of coating include chitosan, sorbate, methyl cellulose, wheat gluten, cassava starch, and alginates (Sangsuwan et al., 2008; Viana et al., 2009; Pilon et al., 2016a). Calcium may be included in these coatings to improve firmness (Benítez et al., 2014; Pilon et al., 2016b).

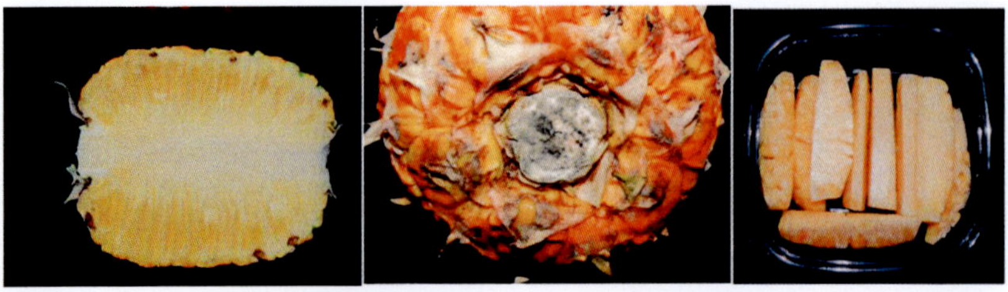

FIG. 1 Left to right showing highly translucent fruit, mold on the broken peduncle, and fresh-cut wedges showing darkening and leakage.

Problems under suboptimal conditions

The symptoms of chilling injury (CI) include wilting, drying, and discoloration of crown leaves, failure of green-shelled fruit to yellow, browning and dulling of yellow fruit, and internal flesh browning (Lim, 1985; Paull and Rohrbach, 1985). Preharvest shading and pre- and postharvest low temperature are the major factors increasing CI symptom intensity (Akamine et al., 1975; Akamine, 1976; Keetch and Balldorf, 1979; Smith, 1983). Susceptible fruit are generally lower in ascorbic acid and sugars and are opaque (Paull and Rohrbach, 1985; Swete Kelly and Bragshaw, 1993).

Comments

Pineapple fruit that are more than 50% Smooth Cayenne are not regarded as a host for tephritid fruit flies; hence, insect disinfestation is not required for importation into fly-free countries (Soler, 1992; Armstrong, 1994). Other insects such as the pineapple caterpillars, pineapple scale, pineapple fruit mite, and mealy bugs are often of quarantine interest with countries requiring commodities to be insect free.

Acknowledgments

This work was supported in part by the National Institute of Food and Agriculture, U.S. Department of Agriculture, under an agreement 58-2040-5-010 through the Agriculture Research Service and Hatch Project H862 to Robert E. Paull.

References

Abdullah, H., Rohaya, M.A., Zaipun, M., 1985. Effect of modified atmosphere on black heart development and ascorbic acid contents in 'Mauritius' pineapple (*Ananas comosus* c. 'Mauritius') during storage at low temperature. ASEAN Food J. 1, 15–18.

Akamine, E.K., 1976. Postharvest control of endogenous brown spot in fresh Australian pineapples with heat. Hort. Sci. 11, 586–588.

Akamine, E.K., Goo, T., 1971. Controlled atmosphere storage of fresh pineapples (*Ananas comosus* (L.) Merr. 'Smooth Cayenne'). Hawaii Agric. Exp. Stat. Res. Bull. 152, 8.

Akamine, E.K., Goo, T., Steepy, T., Greidanus, T., Iwaoka, N., 1975. Control of endogenous brown spot of pineapple in postharvest handling. J. Am. Soc. Horticult. Sci. 100, 60–65.

Anon, 1968. Wholesale Standards for Hawaiian-Grown Pineapple. Hawaii Department of Agriculture, Marketing Division.

Antoniolli, L.R., Benedetti, B.C., Sigrist, J.M.M., Silveira, N.F.d.A., 2007. Quality evaluation of fresh-cut 'Pérola' pineapple stored in controlled atmosphere. Food Sci. Technol. (Campinas) 27, 530–534.

Armstrong, J.W., 1994. Tropical and subtropical fruits. In: Paull, R.E., Armstrong, J.W. (Eds.), Insect pests and fresh horticultural products: Treatments and responses. CAB International, Wallingford, England, pp. 275–290.

Benítez, S., Chiumenti, M., Sepulcre, F., Achaerandio, I., Pujolá, M., 2012. Modeling the effect of storage temperature on the respiration rate and texture of fresh cut pineapple. J. Food Eng. 113, 527–533.

Benítez, S., Soro, L., Achaerandio, I., Sepulcre, F., Pujolá, M., 2014. Combined effect of a low permeable film and edible coatings or calcium dips on the quality of fresh-cut pineapple. J. Food Process Eng. 37, 91–99.

Bierhals, V.S., Chiumarelli, M., Hubinger, M.D., 2011. Effect of cassava starch coating on quality and shelf life of fresh-cut pineapple (*Ananas comosus* L. Merril cv "Pérola"). J. Food Sci. 76, E62–E72.

Budu, A.S., Joyce, D.C., 2003. Effect of 1-methylcyclopropene on the quality of minimally processed pineapple fruit. Aust. J. Exp. Agric. 43, 177–184.

Budu, A.S., Joyce, D.C., 2005. Effect of modified atmosphere packaging on the quality of minimally processed pineapple cv. 'Smooth Cayenne' fruit. J. Hortic Sci. Biotech. 80, 193–198.

Budu, A.S., Joyce, D.C., Terry, L.A., 2007. Quality changes in sliced pineapple under controlled atmosphere storage. J. Hortic Sci. Biotech. 82, 934–940.

Cancel, H.L., 1974. Harvesting and storage conditions for pineapples of the red Spanish variety. J. Agric. Puerto Rico 58, 162–169.

Crochon, M., Tisseau, R., Teisson, C., Huet, R., 1981. Effet d'une application d'Ethrel avant la recolte sur la qualite gustative de ananas de Cote d'Ivoire. Fruits 36, 409–415.

Dull, G.G., 1971. The pineapple: General. In: Hulme, A.C. (Ed.), The Biochemistry of Fruits and their Products. In: Vol. 2. Academic Press, London, pp. 303–331.

Finnegan, E., O'Beirne, D., 2015a. Characterising and tracking deterioration patterns of fresh-cut fruit using principal component analysis, Part I. Postharvest Biol. Technol. 100, 73–80.

Finnegan, E., O'Beirne, D., 2015b. Characterising deterioration patterns in fresh-cut fruit using principal component analysis: II: Effects of ripeness stage, seasonality, processing and packaging. Postharvest Biol. Technol. 100, 91–98.

González-Aguilar, G.A., Ruiz-Cruz, S., Cruz-Valenzuela, R., Rodriguez-Félix, A., Wang, C.Y., 2004. Physiological and quality changes of fresh-cut pineapple treated with antibrowning agents. LWT-Food Sci. Technol. 37, 369–376.

Haruenkit, R., Thompson, A.K., 1994. Storage of fresh pineapple. In: Champ, B.R., Highley, E., Johnson, G.I. (Eds.), Postharvest Handling of Tropical Fruits. Proceeding of International Conference, Chiang Mai, Thailand, 19–23 July 1993. ACIAR Proc #50, pp. 422–426.

Jiang, Z.D., Zheng, H., Mantri, N., Qi, Z.C., Zhang, X.D., Hou, Z.N., Chang, J.D., Lu, H.F., Liang, Z.S., 2016. Prediction of relationship between surface area, temperature, storage time and ascorbic acid retention of fresh-cut pineapple using adaptive neuro-fuzzy inference system (ANFIS). Postharvest Biol. Technol. 113(1–7).

Keetch, D.P., 1978. Bruising of pineapples. Farm. South Africa Bull. H11 (/1978), 2.

Keetch, D.P., Balldorf, D.B., 1979. The incidence of certain pineapple fruit blemishes in the eastern cape and border. Citrus Subtropic. Fruit J. 551, 12–15.

Lim, W.H., 1985. Diseases and Disorders of Pineapples in Peninsular Malaysia. MARDI Report #97, Kuala Lumpur, Malaysia. pp. 53.

Marrero, A., Kader, A.A., 2006. Optimal temperature and modified atmosphere for keeping quality of fresh-cut pineapples. Postharvest Biol. Technol. 39, 163–168.

Montero-Calderon, M., Martin-Belloso, O., 2016. Fresh-cut pineapple. In: Lobo, G., Paull, R.E. (Eds.), Handbook of Pineapple Technology: Production, Postharvest Science, Processing and Nutrition. In press.

Montero-Calderón, M., Rojas-Graü, M.A., Martín-Belloso, O., 2008. Effect of packaging conditions on quality and shelf-life of fresh-cut pineapple (*Ananas comosus*). Postharvest Biol. Technol. 50, 182–189.

Newett, S., Rigden, P., 2015. The pineapple problem solver field guide. Department of Agriculture and Fisheries, Queensland Government.

Okimoto, M.C., 1948. Anatomy and histology of the pineapple inflorescence and fruit. Bot. Gaz. 110, 217–230.

Pan, Y.G., Zhu, J.J., Li, S.Y., 2015. Effects of pure oxygen and reduced oxygen modified atmosphere packaging on the quality and microbial characteristics of fresh-cut pineapple. Fruits 70, 101–108.

Paull, R.E., Lobo, M.G., 2012. Pineapple. In: Siddiq, M. (Ed.), Tropical and Subtropical fruits. John Wiley & Sons, pp. 333–357.

Paull, R.E., Reyes, M.E.Q., 1996. Preharvest weather conditions and pineapple fruit translucency. Sci. Hortic. 66, 59–67.

Paull, R.E., Rohrbach, K.G., 1982. Juice characteristics and internal atmosphere of waxed 'Smooth Cayenne,' pineapple fruit. J. Am. Soc. Horticult. Sci. 107, 448–452.

Paull, R.E., Rohrbach, K.G., 1985. Symptom development of chilling injury in pineapple fruit. J. Am. Soc. Horticult. Sci. 110, 100–105.

Pilon, L., Tetelboim, M.C., Gallo, C.R., Sarmento, S.B.S., Spoto, M.H.F., 2016a. Effect of wheat gluten-based edible coating and irradiation on quality of fresh-cut pineapple. Acta Hortic. (1111), 212–220.

Pilon, L., Tetelboim, M.C., Gallo, C.R., Sarmento, S.B.S., Spoto, M.H.F., 2016b. Physicochemical and microbiological changes in fresh-cut pineapple coated with wheat gluten and alginate. Acta Hortic. (1111), 221–226.

Puthmee, T., Matulaprungsan, B., Kanlayanarat, S., 2012. Effect of temperature on quality of fresh-cut pineapple cultivar 'Tradseethong'. Acta Hortic. (943), 187–190.

Puthmee, T., Matulaprungsan, B., Kanlayanarat, S., 2013. Effects of trimming and storage temperature on quality of fresh-cut pineapple cultivar 'Tradseethong'. Acta Hortic. (989), 193–197.

Rocculi, P., Cocci, E., Romani, S., Sacchetti, G., Rosa, M., 2009. Effect of 1-MCP treatment and N_2O MAP on physiological and quality changes of fresh-cut pineapple. Postharvest Biol. Technol. 51, 371–377.

Rohrbach, K.G., 1983. Exotic pineapple diseases and pests and their potential for spread. In: Singh, K.G. (Ed.), Exotic Plant Quarantine Pests & Procedures for Introduction of Plant Materials. ASEAN Plant Quarantine Centre and Training Institute, Sergang, Selangor, Malaysia, pp. 145–171.

Rohrbach, K.G., 1989. Unusual tropical fruit diseases with extended latent periods. Plant Dis. 73, 607–609.

Rohrbach, K.G., Apt, W.J., 1986. Nematode and disease problems of pineapple. Plant Dis. 70, 81–87.

Rohrbach, K.G., Paull, R.E., 1982. Incidence and severity of chilling induced browning of waxed 'Smooth Cayenne' pineapple. J. Am. Soc. Horticult. Sci. 107, 453–457.

Rohrbach, K.G., Phillips, D.J., 1990. Postharvest diseases of pineapple. Acta Hortic. 26, 503–508.

Rohrbach, K.G., Schmitt, D.P., 1994. Pineapple. In: Ploetz, R.C., Zentmyer, G.A., Nishijima, W.T., Rohrbach, K.G., Ohr, H.D. (Eds.), Compendium of Tropical Fruit Diseases. American Phytopathology Society, St. Paul, Minnesota, pp. 45–55.

Sangsuwan, J., Rattanapanone, N., Rachtanapun, P., 2008. Effect of chitosan/methyl cellulose films on microbial and quality characteristics of fresh-cut cantaloupe and pineapple. Postharvest Biol. Technol. 49, 403–410.

Smith, L.G., 1983. Causes and development of black heart in pineapples. Trop. Agricult. (Trinidad) 60, 31–35.

Soler, A., 1992. Pineapple. CIRAD-IRFA, Paris, France 48pp. ISBN (2-87614-078-0).

Swete Kelly, D.E., Bragshaw, J., 1993. Effect of fruit handling and fruit coatings on black heart (internal brown spot) and other aspects of fresh pineapple quality. Acta Hortic. (334), 305–316.

Viana, E.S., Amorim, T.S., Oliveira, L.A., Reinhardt, D.H., Silveira, F.G.F., Santana, F.A., Santana, P.N., 2009. Physicochemical characteristics of minimally processed 'Pérola' pineapples treated with edible coatings. Acta Hortic. (822), 291–298.

Yahia, E.M., 1998. Modified and controlled atmospheres for tropical fruits. Horticult. Rev. 22, 123–183.

Zhang, B.Y., Samapundo, S., Pothakos, V., de Baenst, I., Sürengil, G., Noseda, B., Devlieghere, F., 2013. Effect of atmospheres combining high oxygen and carbon dioxide levels on microbial spoilage and sensory quality of fresh-cut pineapple. Postharvest Biol. Technol. 86, 73–84.

Further reading

Collins, J.L., 1968. The Pineapple. Leonard Hill, London, p. 295.

Paull, R.E., Duarte, O., 2011. Tropical fruits (Vol. 1). CAB International, Wallingford, United Kingdom pp. 400.

Chapter 18.1

Subtropical fruits: Avocados

Allan B. Woolf*, Mary Lu Arpaia[†], Bruno G. Defilippi[‡] and John P. Bower[§]
*The New Zealand Institute for Plant & Food Research Limited, Auckland, New Zealand, [†]Extension Subtropical Horticulturist, Department of Botany and Plant Sciences, University of California, Riverside, CA, United States, [‡]Depto. de Mejoramiento Genético y Biotecnología, Instituto de Investigaciones Agropecuarias—CRI La Platina, Unidad de Postcosecha, Santiago, Chile, [§]Consultant, Agassiz, BC, Canada

Quality characteristics

- Harvesting fruit at an acceptable maturity (as measured by dry matter—DM) is important both to taste and to achieving a high quality ripe product.
- Key quality aspects of unripe product are freedom from external damage, cuts, and blemishes.

Controlled and Modified Atmospheres for Fresh and Fresh-Cut Produce
https://doi.org/10.1016/B978-0-12-804599-2.00026-0

- Avocado must be ripened to be consumed but if allowed to become overripe, quality is rapidly lost, often because of decay.
- Ripe fruit should be free of decay (visible externally and in the flesh), bruising, and other causes of flesh browning such as chilling injury.

Main causes of quality loss

- Fruit maturity: Immature fruit (low DM) take longer to ripen, have more stem end rots, as well as being of generally poor flavor. Overmature fruit (late season, higher dry matter) ripen more quickly often during storage, which can lead to more decay. Minimum DM for harvest varies depending on fruit origin but ranges from 20.6% to 25% for the cultivar 'Hass' to ensure ripening and acceptable flavor and ripe fruit quality.
- Physical damage during harvest to the skin (epidermis) leading to damage to raised nodules (sometimes called lenticels), or potentially internal damage (bruising) which is typically not evident until fruit are ripe. Lenticel damage also increases postharvest water loss, resulting in increased chilling sensitivity and damage.
- Good temperature management is a critical factor for avocado fruit quality. The optimum temperature is 4–6°C, and temperatures below this can lead to external chilling injury (skin damage), while higher temperatures or long storage times lead to internal chilling injury (flesh discoloration) and premature softening. This is influenced by maturity stage (DM content) at harvest.
- Ripening temperature can influence fruit quality, with more decay at both higher temperatures and significantly lower temperatures. Temperatures >22°C can result in higher decay and inhibited ripening, while temperatures below 15°C can also result in delayed ripening and enhanced internal discoloration.
- Excessive water loss and potential for chilling damage as a result of high air flow or low ambient humidity can result in visual fruit shriveling. This is more prevalent in less mature fruit.
- Ripe fruit decay can be a significant problem, particularly where grown under conditions of higher rainfall or mist/fog. These environmental conditions will probably result in significantly higher decay pressure. The main fruit decay causing organism worldwide is the *Colletotrichum gloeosporioides* species complex, which typically infects fruit during development in the field and remains quiescent until the fruit are harvested and begin to ripen. Generally fruit decay in avocado is divided into two broad categories:
- Stem end rots can be a more significant problem, since one infection tends to result in a greater portion of the flesh being brown, and vascular browning is often associated with this type of decay, which further discolors large portions of the flesh. Members of the Botryosphaeriaceae genera are the

primary cause of stem end rots in avocado, although other fungi can also cause similar symptoms.

- Body rot is a major problem in higher rainfall conditions and is often associated with slower ripening (longer shelf life) and long storage durations. *Colletotrichum gloeosporioides* and related species are the primary causal organisms.

Optimum storage conditions

Optimum temperature

Excellent temperature management is a key factor in relation to avocado fruit quality. Fruit should be cooled as soon as possible after harvest. Some packing houses use hydrocooling on arrival to decrease fruit temperature to approximately 12°C, followed by forced air cooling after packing to achieve the desired final temperature. For 'Hass' avocado (the main traded cultivar whose skin darkens as the fruit ripen), the optimum temperature is ≈4–6°C, although early season fruit should be stored at slightly higher temperatures, and later season fruit at slightly lower temperatures. Low storage temperatures (typically <3°C) can lead to external symptoms of chilling injury (discrete patches on the skin), while higher temperatures (>8°C for ≈3–4 weeks) and/or longer times at optimum temperatures (>5–6 weeks) can lead to internal chilling symptoms (diffuse flesh browning or graying) and premature softening. The high respiration rate of avocado means that high respiratory-driven heat production results in more challenging temperature management and can thus exacerbate fruit ripening and further quality reduction if temperature management is inadequate.

For long-distance shipping some exporters have used a step-down temperature regime where fruit temperature is slowly lowered over the course of transit, to allow the fruit to acclimate to lower temperatures without inducing chilling injury.

Optimum relative humidity

>85% during storage. Cooling tunnels should be humidified during cooling to prevent water loss and excessive shriveling. However, some water loss is important to cause ripening, and possibly this is why modified atmosphere (MA) storage can lead to increased rots in some situations, since high relative humidity may lead to longer ripening times.

CA/MA considerations

Because of the limited storage life of avocados (≈4–5 weeks in air), controlled atmosphere (CA) storage (Fig. 1) is generally used to assist with reliable *transport* of fruit, i.e., in refrigerated seafreight containers rather than onshore

FIG. 1 Slowed avocado ripening by controlled atmosphere storage (CA): Air (left) and CA (4% O_2 and 6% CO_2) (right).

coolstore storage (as carried out for crops with long-term storage potential such as apples).

Although there has been research reporting significant benefits of modified atmosphere systems, results are variable, generally more challenging where decay pressure is higher, and require very careful coolchain management, particularly in market (this being a significant commercial challenge). Fruit coatings have been suggested to have potential, but there is a significant commercial challenge in that even product application must be achieved, formulations vary considerably in effectiveness, and differences in fruit physiology result in greater fruit-to-fruit variability. Additionally, fruit-ripening companies prefer fruit uncoated, as the coating can delay ripening, enhance fruit-to-fruit variability within the package, and can even result in uneven ripening within an individual fruit (Fig. 2). Thus neither MA storage nor fruit coatings have achieved significant commercial adoption.

Maximum storage duration for 'Hass' is ≈4–4.5 weeks in air and 5.5–6 weeks in CA. The growing conditions, especially tree nutrition, play a role in fruit

FIG. 2 Example of variable avocado fruit ripening ("checker boarding").

quality. In particular, fruit calcium and nitrogen are important. High calcium increases quality, while high nitrogen decreases it. Further, decay pressure as a result of wet environments is a key consideration, as is optimum temperature management. Under experimental conditions, storage times of as long as 9–10 weeks have been reported, but such durations are not routinely commercially achieved. The CA conditions used vary depending on the shipping line, CA technology used, and in some cases exporters. A widely used regime is 4% O_2+6% CO_2.

CA technology varies from limited control using bags of lime to control increasing CO_2 concentrations, to computer-managed venting of containers coupled with polymer films to remove excess CO_2 differentially to maintain stable conditions. Some containers include ethylene scrubbing.

Benefits of CA/MA

Generally the main benefit of CA is to improve out-turn quality and it is not used as a means of reducing the storage temperature. The main benefit is extending storage life by reducing the expression of internal chilling symptoms—flesh graying/diffuse flesh discoloration and associated vascular browning.

- Prevents premature ripening with its associated skin color change (green to purple/black for 'Hass')
- Slows softening during storage and transport
- Reduces internal chilling injury—flesh graying/diffuse flesh discoloration, vascular browning (distal end)
- Reduces ripe fruit decay
- Reduces water loss
- Results in more reliable out-turn on arrival in importing countries.

Tolerances

Tolerance to lower O_2 and higher CO_2 concentrations is influenced by a wide range of factors such as cultivar, duration, fruit temperature, and preharvest factors such as fruit maturity that influence the physiological state of the fruit. Oxygen tolerance is $\approx$2%, but may be as low as 1%. Carbon dioxide tolerance is $\approx$10–15%, but fruit become less tolerant to lower O_2 concentrations.

Injuries

- Low O_2 damage (Fig. 3) is expressed as skin damage, with irregular brown to dark brown patches, and may cause diffuse browning of flesh beneath affected skin. Low O_2 results in production of ethanol and acetaldehyde and therefore off-flavors.
- High CO_2 (Fig. 4) causes discoloration of the skin similar to that of low O_2 and the development of an 'off' flavor.
- Poor ripening leading to variability and increased decay.

FIG. 3 Low O_2 damage to avocado fruit.

Recommendations

General recommendations for CA atmospheres are 2–5% O_2 and 3–10% CO_2. However, there are various reports ranging from no benefit of CO_2, to others reporting higher CO_2 benefits. Generally reduced O_2 is more important than CO_2 for slowing softening, which reduces skin coloration at out-turn and thus reduced internal chilling disorders (such as diffuse flesh discoloration). While some reports suggest delays of up to 14 days to establishing CA do not influence the efficacy of the CA treatment, in general application of CA as near to harvest as possible is a best practice.

Commercial use of CA for long-distance transport is now common, although competition with 1-MCP at similar cost may reduce its use. Conversely, since 1-MCP lengthens the time taken for fruit to ripen/soften, and many markets now ethylene-ripen fruit at destination, this may discourage its widespread use.

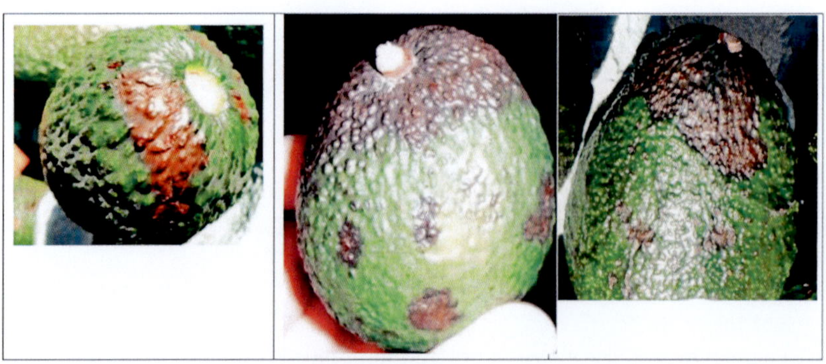

FIG. 4 Examples of high CO_2 damage to avocado fruit. Images from left to right show progression after removal from cool store. Damage is often located at the stem end of the fruit.

Controlled atmosphere shipping containers control temperature, O_2, and CO_2, with ethylene scrubbing as a further option. Previous research has shown that the presence of ethylene even during CA storage can exacerbate chilling injury symptoms.

Seafreight container systems. There is a range of commercial seafreight container systems available around the world which employ different technologies to achieve the CA atmosphere. The Maxtend® CA system is a distinctly different system using a more passive method of achieving lower O_2—akin to a MA system. The various systems may or may not include ethylene scrubbing, an important aspect for avocado where increased chilling injury can result. Because of differences in growing environments and storage potential, there may be differences in the effectiveness of the systems or atmospheres chosen by the various companies. It is recommended that careful assessment of out-turn quality (both at removal from storage and when ripe) be carried out using fruit from a range of growers/regions at a range of maturities (times in the season), and that careful measurement of temperatures and atmospheres be obtained.

Dynamic CA (DCA). This is a relatively new development which, to our knowledge, is only being examined experimentally. The system uses avocado skin chlorophyll fluorescence to determine low O_2 stress as a means to achieve a low, but not damaging O_2 concentrations. This can result in improved out-turn (reduced decay) over standard CA conditions and was found to achieve the beneficial results without the need for higher CO_2 concentrations which tend to increase shelf life (time to ripen after storage). In contrast, other treatments that improve storage life (such as 1-MCP) can increase shelf life and ripe fruit decay.

Ethylene considerations

- Avocados are sensitive to ethylene, which can be beneficial (when used to ripen fruit for sale—preripening or "ripe tonight" programs, similar to those for banana), or deleterious where they can induce softening/ripening during storage, which leads to internal chilling injury symptoms (diffuse flesh discoloration and vascular browning).
- Ethylene production by preclimacteric avocados is very low to undetectable, but once fruit commence ripening, ethylene production is very high, along with a high respiration rate and thus CO_2 production and O_2 consumption.
- Ripening avocados in a nonventilated space can lead to a significant health and safety risk, with atmospheres of $<5\%$ O_2 and over 10% CO_2. Such concentrations can be lethal.
- Under CA, other than managing temperature effectively, scrubbing of ethylene can have benefits.
- Sources of ethylene should be avoided in and around avocado storage facilities—other fruits that produce ethylene (e.g., apple, banana), combustion engines, fire, and cigarette smoke.

Application of 1-MCP after packing, and as an alternative or in conjunction to CA shipping, is used by some exporters. 1-MCP may reduce internal physiological disorders such as diffuse flesh discoloration which can occur especially toward the end of the harvest season. Although technically successful, the system is not widely used for the same reasons given for coatings—ripening time is delayed; associated increases in ripe fruit decay may also occur. This is especially important as ethylene ripening at destination market is deployed.

Problems under suboptimal conditions

Given CA is employed to extend storage life, the key disorders likely to be expressed are internal chilling symptoms (flesh discoloration and vascular browning) and/or increased decay (body or stem end rots).

Comments

- Avocado are a challenging crop in which to achieve optimum quality, and assessing the efficacy of any given storage regime requires careful assessment, and with trained, experienced assessors. The main point is to assess the fruit when they are *at the same stage of softening/ripening* (not at one time point after removal from storage).
- Overall, temperature management is the key postharvest factor, and controlled atmosphere storage is often used as an insurance against quality loss.

Further reading

Arpaia, M.L., Faubion, D., Mitchell, F.G., Mayer, G., 1991. The use of controlled atmospheres for long-term storage of 'Hass' avocados–progress report. Calif. Avocado Soc. Yearbook 74, 43–48.

Burdon, J., Lallu, N., Haynes, G., McDermott, K., Billing, D., 2008. The effect of delays in establishment of a static or dynamic controlled atmosphere on the quality of 'Hass' avocado fruit. Postharvest Biol. Technol. 49, 61–68.

Eaks, I.L., 1978. Ripening, respiration, and ethylene production of 'Hass' avocado fruits at 20°C to 40°C. J. Amer. Soc. Hort. Sci. 103, 576–578.

Everett, K.R., Hallett, I.C., Rees-George, J., Chynoweth, R.W., Pak, H.A., 2008. Avocado lenticel damage: The cause and the effect on fruit quality. Postharvest Biol. Technol. 48, 383–390.

Hertog, M.L.A.T.M., Nicholson, S.E., Whitmore, K., 2003. The effect of modified atmospheres on the rate of quality change in 'Hass' avocado. Postharvest Biol. Technol. 29, 41–53.

Hofman, P.J., Bower, J., Woolf, A., 2013. Harvesting, packing, postharvest technology, transport and processing, in: Schaffer B, Wolstenholme B.N., Whiley, A.W. (Eds). The Avocado: Botany, Production and Uses, 2nd Edition. CABI, Oxfordshire, UK. Pp 489–540.

Hopkirk, G., White, A., Beever, D.J., Forbes, S.K., 1994. Influence of postharvest temperatures and the rate of fruit ripening on internal postharvest rots and disorders of New Zealand 'Hass' avocado fruit. NZ J. Crop and Hortic. Sci. 22, 305–311.

Kader, A.A., Arpaia. M.L., 2001. Avocado. Recommendations for maintaining postharvest quality. http://postharvest.ucdavis.edu/produce/producefacts/fruit/avocado.html.

Lee, S.K., Young, R.E., Schiffman, P.M., Coggins, C.W., 1983. Maturity studies of avocado fruit based on picking dates and dry weight. J. Amer. Soc. Hort. Sci. 108, 390–394.

Lee, S.K., Young, R.E., 1984. Temperature sensitivity of avocado fruit in relation to C_2H_4 treatment. J. Amer. Soc. Hort. Sci. 109, 689–692.

Meir, S., Naiman, D., Hyman, J.Y., Akerman, M., Zauberman, G., Fuchs, Y., 1998. Modified atmosphere packaging enables prolonged storage of ‘Fuerte’ avocado fruit. Acta Hortic. (464), 397–402.

Obenland, D., Collin, S., Sievert, J., Negm, F., Arpaia, M.L., 2012. Influence of maturity and ripening on aroma volatiles and flavor in ‘Hass’ avocado. Postharvest Biol. Technol. 71, 41–50.

Pak, H.A., Dixon, J., Cutting, J.C., 2003. Influence of early season maturity on fruit quality in New Zealand ‘Hass’ avocados. In: Proceedings of the Fifth World Avocado Congress, pp. 635–640.

Yahia E.M., Woolf A., 2011. Avocado (*Persea americana* Mill), in: Yahia, E.M. (Ed.), Postharvest Biology & Technology of Tropical and Sub-Tropical Fruits, Volume 2. Woodhead Publishing, UK. pp 125–185.

Chapter 18.2

Subtropical fruits: Cherimoyas

Reinaldo Campos-Vargas and Patricio Olmedo
Centro de Biotecnologia Vegetal, Facultad Ciencias de la Vida, Universidad Andres Bello, Santiago, Chile

Quality characteristics

The Annonaceae family comprises more than 100 genera and 2400 species (Chatrou et al., 2012). The cherimoya (*Annona cherimola* Mill.) is the most cultivated species in the genera Annona and is an exotic subtropical fruit native to South America (Alique, 1995; Zonneveld et al., 2012). The cherimoya, within the group of edible Anonacaeas, presents the best organoleptic characteristics (Alique, 1995).

Other species are the atemoya (*Annona cherimoya* × *A. squamosa*), sweetsop (*Annona squamosa*), and custard apple (*Annona reticulata*).

Controlled and Modified Atmospheres for Fresh and Fresh-Cut Produce
https://doi.org/10.1016/B978-0-12-804599-2.00027-2

The primary quality characteristics are fruit size, color, firmness, and absence of visible defects and decay. Changes in skin color from dark-green to light-green or greenish-yellow are the primary maturity index (Kader and Arpaia, 1999).

During ripening of the cherimoya at room temperature, changes such as an increase in soluble sugars, reduction of starch and chlorophyll, softening, and cell wall disassembly occur (Cordeiro et al., 2013). Cherimoya, atemoya, and sweetsop when ripe have high concentrations of sugars (14%–15%) and moderate acidity (0.4%–0.7%). They are good sources of vitamin C (45–60 mg/100 g) and potassium (250–500 mg/100 g edible portion) (Kader and Arpaia, 1999).

Cherimoya has some peculiarities in its maturation physiology. It has been described as having a double increase in respiration (Brown et al., 1988; Alique and Oliveira, 1994), with an increase in ethylene after the first one (Alique et al., 1994). Also, after the initial rise in CO_2, there is a loss of firmness of the pulp, which is temperature dependent (Alique et al., 1994). Sanchez et al. (1998) describe that the softening of cherimoya is mainly due to the concerted action of cellulases and polygalacturonase, without discounting the possibility of the participation of other enzymes. Another characteristic that has been described as unusual in cherimoya is the increase of malic and citric acids during maturation that influences the taste of the fruit (Gonzalez-Agüero et al., 2016).

Lahoz et al. (1993) pointed out that the synthesis of ethylene in cherimoya depends on the storage temperature and that small concentrations of ethylene could be able to trigger the ripening process. When comparing cherimoya with other fruits it is considered as a high ethylene producer (Kader, 2002a). In relation to the synthesis of ethylene, the cherimoya has particularities.

Information on cherimoya biology is very scarce in the available literature; some available studies describe physiological characterization of the fruit, postharvest behavior, and use of storage strategies such as modified atmospheres and fresh-cut packaging. Further, there are excellent reviews that are relevant, although it would be helpful for them to be revised (Palma et al., 1993; Pareek et al., 2011).

Main causes of quality loss

Chilling injury symptoms include darkening and hardening of skin (Fig. 1), pitting, failure to develop full flavor, and mealy flesh.

Splitting as an increase in the receptacle diameter of the flesh and skin due to the turgor changes related to production of neutral sugars during ripening and the movement of water from the skin to the flesh.

Mechanical damage, mature fruits must be handled with care to minimize bruising. When ripe, fruits are very susceptible to mechanical damage (Fig. 2).

Pathological disorders include anthracnose, which appears as dark lesions and may produce pink spore masses under high humidity conditions; black

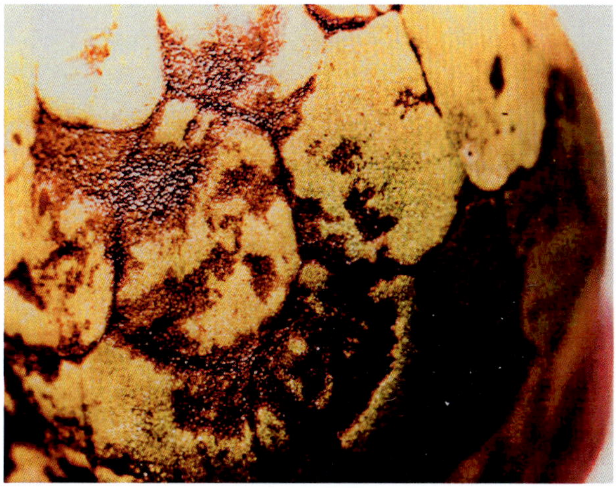

FIG. 1 Chilling injury symptoms of 'Concha Lisa' cherimoya.

FIG. 2 Mechanical damage of the mature fruits.

canker, which appears as purple spots on the fruit; and Botryodiplodia rot, which appears as purple areas in the skin and the flesh becomes brown and corky (Kader and Arpaia, 1999).

Optimum storage conditions

Temperature

Cherimoya is considered sensitive to low temperatures, therefore depending on cultivar, ripeness stage, and duration the optimum temperature for storage is 8–12°C (46–54°F) (Fig. 3).

FIG. 3 'Concha lisa' cherimoya stored for 3 days at 10°C (top image) and at 2°C (bottom image).

Thus the use of refrigeration brings risks of alterations in the physiology of the fruit. The alterations described in different species due to cold damage include alterations of the membrane, increase in respiration and ethylene synthesis, and enzymatic modification, among others (Wang, 1982). However, in the case of fruits sensitive to cold damage, the use of thermal conditions that induce damage has proved to be useful, given that the loss of quality due to cold damage was less than the increase in deterioration in conditions of higher temperature (Watada et al., 1996). Gutiérrez et al. (1994) in the introduction of his work refers to an unpublished experience in which it is described that half-season cherimoya fruits would be less sensitive to cold damage than those of the beginning of the season. Loyola (1988) exposed cherimoya storage temperatures of 5 and 7°C and described the development of physiological disorders such as pitting and browning, as well as structural changes in various fruit tissues.

In Chile, there are several cultivars of cherimoya, among which stands out the cultivar 'Concha Lisa' as the most resistant to low temperature (Gardiazábal and Rosenberg, 1993; Lahoz et al., 1993) (Fig. 4), withstanding a range of 7–8°C. Gutiérrez et al. (1994) found temperatures of 1 and 4°C on delayed cherimoya maturation in terms of hydrolysis of starch. This delay in starch hydrolysis could be reversed by warming the fruit, suggesting that the increase in soluble solids is not to be as sensitive to cold as injury symptom development. The same authors concluded that the site of damage is likely to be at the level of cellular membranes.

FIG. 4 'Concha Lisa' cherimoya at harvest.

Controlled atmosphere and ethylene

The optimum CA storage atmospheres for cherimoya and atemoyas range from 3% to 5% O_2 and 5% to 10% CO_2 (Kader and Arpaia, 1999). Benefits include delayed ripening, lower respiration and ethylene production rates, and firmness retention. Among the effects of CO_2 on the physiology of the fruit, it can be noted the reduction of senescence, a decrease in some physiological disorders, and reduced respiration and ethylene action (Kader, 2002b). Alique and Oliveira (1994) discuss the differential effect that the application of the CA may have on cherimoya depending on the variety. Additionally, they pointed out that high CO_2 would lower the organoleptic quality via a reduction of organic acids and so a compromise needs to be struck in terms of the combination of low O_2 and high CO_2 atmospheres used to optimize quality of the cherimoya after storage. Escribano et al. (1997) found high CO_2 inhibited maturation and senescence as determined by an improved maintenance of skin color and a decrease in rate of firmness loss.

Assis et al. (2001) described a decrease in the production of ethylene by CO_2 treatments in cherimoya fruits, accompanied by a greater firmness of the fruit pulp. Del Cura et al. (1996) described the positive effects of CO_2 treatments on the maintenance of firmness in cherimoya, possibly due to a reduction in polygalacturonase protein content in the fruit. These results, despite having been carried out under high temperature conditions (20°C) and on whole cherimoya fruits, could be useful in the reduction in quality loss in minimally processed custard apple. On the other hand, treatments with high CO_2 can bring about the activation of anaerobic respiration with the consequent synthesis of ethanol and acetaldehyde in cherimoya (Muñoz et al., 1997). Alique (1995) pointed out that, although exposure to a high CO_2 content produced a general slowing effect on ripening in certain combinations (9 days at 20% CO_2 at 8°C), it produced unpleasant flavors that reduced the organoleptic quality of the fruit.

Escribano et al. (1997), Merodio et al. (1998), and Maldonado et al. (2002) subjected cherimoya fruits to low temperature combined treatments with high CO_2 (6°C and 20%, respectively) and then they were kept at the same temperature in air only, describing a residual effect of CO_2 on the fruit which showed a greater resistance to cold damage, lower respiration, lower ethylene synthesis and acidity, retention of firmness, lower amount of proteins related to polygalacturonase, lower concentration of tannins and lignin, and retardation of senescence.

There are several studies that demonstrated that the low O_2 partial pressure acts in synergy with high CO_2 in CA storage to provide beneficial effects. Alique and Oliveira (1994) pointed out that changes in the respiration of cherimoya fruit subjected to CA would be explained by a reduction in O_2 rather than a high level of CO_2. Palma et al. (1993) point out that a reduction of O_2 produced a decrease in the loss of firmness in cherimoya fruits and a lower ethylene synthesis, situation that would be complemented with a lower sensitivity to ethylene in situations of less 8%O_2 (Kader, 2002b).

Cherimoya is a climacteric fruit sensitive to the action of ethylene, so factors that suppress ethylene action induce changes in its quality and storage capacity. Kader (1997) pointed out that the removal of ethylene could delay the advance of maturation. Several investigators have manipulated ethylene in the postharvest management of cherimoya. Pardo (1988), Toro (1989), and Jeria (1991) used ethylene absorbers (potassium permanganate) in storage of whole cherimoya of the Concha Lisa, Clavo, and Bronceada varieties, respectively, noting that the use of the absorbers did not induce important effects of modification of ripening. Recently, 1-methylcyclopropene (1-MCP) has been developed as a new tool to control ethylene responses. This gas binds the ethylene receptors for a long time (possibly permanently), inhibiting ethylene action completely (Sisler and Serek, 1997). This allows its use in the extension of postharvest life of various products and reduction of physiological disorders (Fan and Mattheis, 1999; Watkins et al., 2000). There is a little history of using 1-MCP in Annonaceae. Hofman et al. (2001) described the effect of 1-MCP (25 ppm) on custard apple (*Annona squamosa* × *Annona cherimola* L.). 1-MCP treatment increased the maturation period by 58% compared to the control, although some disorder development was reported. Likewise, Silva et al. (2016) using modified atmosphere in conjunction with 1-MCP (600 ppb) allowed storage for 25 days 15°C in atemoya with acceptable characteristics. In cherimoya, there are few published records although researchers associated with this review have studied the effectiveness of 1-MCP applications in quality preservation of unprocessed Concha Lisa cherimoya with promising results (unpublished).

Fresh-cut cherimoya

The organoleptic characteristics of the cherimoya, such as its sweetness, its creamy pulp, and its aroma (Manríquez et al., 2014), make it very attractive for its use as a fresh-cut product. However, cherimoya is extremely susceptible to browning, mainly through the activity of the polyphenol oxidase (PPO; Martinez-Cayuela et al., 1988; Prieto et al., 2007) on phenolics released in response to cell disruption.

The action of PPO could be responsible for the browning of the peel of the cherimoya after harvesting (Martínez et al., 1993). Gutiérrez et al. (1994) describe that cherimoya fruit subjected to 4°C underwent changes in coloration due to the action of PPO especially during the senescence of the fruit. Gardiazábal and Rosenberg (1993) pointed out that the use of heat is not viable to reduce the action of the enzymes present in the pulp due to the generation of off flavors. However, they also found the use of antioxidants such as ascorbic and citric acids and EDTA has shown good results. Abufom (1985) demonstrated the maintenance of PPO activity in custard apple even after processing of the fruits and their preservation by freezing (−38°C), storage (−18°C), and subsequent thawing. Therefore, although the use of antioxidants may mask the effect of PPO and help obtain a technical solution permitting the development of fresh-cut, the physiology associated with PPO would not be stopped even by freezing and is only slowed by low temperature.

Although browning is one of the most notorious characteristics of changes in fresh-cut because it is easily detectable, there are a number of other processes that are triggered by the preparation of fresh-cut products such as mechanical damage and deterioration, in which compounds associated with the cell wall and repair processes (i.e., lignin), phenylpropanoids, and aromas (Brecht, 1995) among others are involved. The development of events between the application of the stimulus (e.g., cutting) and the generated response (e.g., browning) are issues of interest in the development of fresh-cut products.

The use of temperature in the handling of minimally processed products has proven to be a great tool to reduce the deterioration produced by the stress of mechanical damage at the time of processing. Assays on cut cherimoya stored under CA of 20%–10% CO_2 showed acceptable quality levels after 60 days at 2°C (Gardiazábal and Rosenberg, 1993). Abufom (1985), Flores (1991), and Riveros (1991) explored the possibility of using cherimoya (var. 'Concha Lisa' and 'Bronceada') in different modalities as frozen products, using antioxidants in all of them, among which the use of mixtures of ascorbic and citric acids was a common factor. Alessandrini (1991), Molfino (1991), and Arze (1992) studied the use of cherimoya var. 'Local Serena,' 'Concha Usa,' and 'Tanned,' respectively, processed in rooms with the application of antioxidants and stored at 0°C under modified atmosphere in conditions of very low O_2 and high CO_2. These works show promising results in conditions for fresh-cut fruit, although problems in acceptability have been described. However, these investigations have not commented on the potential of the atmospheric conditions used to induce fermentation. The low concentration of oxygen and the low temperatures applied in this study are normally related to the production of compounds that cause off flavors.

Strategies to control browning

Some strategies have been described to reduce the development of browning in fruits or cherimoya pieces. Campos-Vargas et al. (2008) studied the effect of L-cysteine in the quality of fresh-cut cherimoya cv. 'Shell Lisa.' The results indicated that 0.5% of L-cysteine at the time of cutting reduced browning for 6 days at 0°C but did not affect PPO activity. Liu et al. (2016) evaluated the use of citric acid and chitosan coating in whole fruits in immersion treatments, finding that the treatments provided a better fruit quality and a reduction in PPO activity compared to the control. Analogously, Yonemoto et al. (2002) evaluated the application of waxes in cherimoya fruits, obtaining a reduction of the browning and increase of the postharvest life in the varieties 'El Bumpo' and 'Libby.'

References

Abufom, J.C. 1985. Efectos de aditivos, mondados y tiempo de almacenaje en la calidad de pulpa y rodelas congeladas de chirimoya (*Annona cherimola* Mill.) cv. Concha Lisa y Bronceada. Tesis. Universidad Católica de Valparaíso. Facultad de Agronomía. 117 pp.

Alessandrini, G.A. 1991. Efecto de diferentes niveles de anhídrido carbónico y nitrógeno sobre la calidad de cuartos de chirimoya (*Annona cherimola* Mill.) cv. local Serena, en almacenaje refrigerado. Tesis Universidad Católica de Valparaíso. Fac. de Agronomía 84p.

Alique, R., 1995. Residual effects of short-term treatments with high-CO_2 on the ripening of cherimoya (*Annona cherimola* Mill.) fruit. J. Hortic. Sci. 70, 609–615.

Alique, R., Oliveira, G.S., 1994. Changes in sugar and organic acids in cherimoya (*Annona cherimola* Mill.) fruit under controlled-atmosphere storage. J. Agric. Food Chem. 42, 799–803.

Alique, R., Zamorano, J.P., Calvo, M.L., Merodio, C., De la Plaza, J.L., 1994. Tolerance of cherimoya (*Annona cherimola* Mill.) to cold storage. J. Amer. Soc. Hort. Sci. 119, 524–528.

Arze, S.A. 1992. Efecto de tres niveles de atmósfera modificada sobre la calidad de cuartos de chirimoyas (*An- nona cherimola* Mill.) cv. Bronceada en almacenaje refrigerado. Tesis. Universidad Católica de Valparaíso. Facultad de Agronomía 92p.

Assis, J.S., Maldonado, R., Muñoz, T., Escribano, M.I., Merodio, C., 2001. Effect of high carbon dioxide concentration on PAL activity and phenolic contents in ripening cherimoya fruit. Postharvest Biol. Technol. 23, 33–39.

Brecht, J.K., 1995. Physiology of lightly processed fruits and vegetables. HortScience 30, 18–22.

Brown, B.I., Wong, L.S., George, A.P., Nissen, R.J., 1988. Comparative studies on the postharvest physiology of fruit from different species of Annona (custard apple). J. Hortic. Sci. 63, 521–528.

Campos-Vargas, R., Defilippi, B.G., Romero, P., Valdés, H., Robledo, P., Prieto, H., 2008. Effect of harvest time and L-cysteine as an antioxidant on flesh browning of fresh-cut cherimoya (*Annona cherimola* Mill.). Chilean J. Agric. Res. 68, 217–227.

Chatrou, L.W., Pirie, M.D., Erkens, R.H., Couvreur, T.L., Neubig, K.M., Abbott, J.R., Mols, J.B., Maas, J.W., Saunders, R.M., Chase, M.W., 2012. A new subfamilial and tribal classification of the pantropical flowering plant family Annonaceae informed by molecular phylogenetics. Bot. J. Linnean Soc. 169, 5–40.

Cordeiro, N., Sousa, L., Freitas, N., Gouveia, M., 2013. Changes in the mesocarp of *Annona cherimola* Mill. 'Madeira' during postharvest ripening. Postharvest Biol. Technol. 85, 179–184.

Del Cura, B., Escribano, M.I., Zamorano, J.P., Merodio, C., 1996. High carbon dioxide delays postharvest changes in RubCase and polygalacturonase-related protein in cherimoya peel. J. Amer. Soc. Hort. Sci. 121, 735–739.

Escribano, M.I., Del Cura, B., Muñoz, T., Merodio, C., 1997. The effect of high carbon dioxide at low temperature on ribulose 1,5-biphosphate carboxylase and polygalacturonase protein levels in cherimoya fruit. J. Amer. Soc. Hort. Sci. 122, 258–262.

Fan, X.T., Mattheis, J.P., 1999. Development of apple superficial scald, soft scald, core flush, and greasiness is reduced by MCP. J. Agric. Food Chem. 47, 3063–3068.

Flores, M. 1991. Efecto de dos estados de ablandamiento y dos presentaciones (rodelas y cuartos) en un medio de cobertura de jugo de naranjas, sobre la calidad y estabilidad en el almacenaje de la fruta congelada de chirimoya (*Annona cherimola* Mill.) cv. Bronceada. Tesis. Universidad Católica de Valparaíso. Facultad de Agronomía. 102 pp.

Gardiazábal, F.L., Rosenberg, M., 1993. El cultivo del chirimoyo. Universidad Católica de Valparaíso Fac. de Agronomía. 152 pp.

Gonzalez-Agüero, M., Tejerina Pardo, L., Zamudio, M.S., Contreras, C., Undurraga, P., Defilippi, B.G., 2016. The unusual acid-accumulating behavior during ripening of cherimoya (*Annona cherimola* Mill.) is linked to changes in transcription and enzyme activity related to citric and malic acid metabolism. Molecules 21, 398–413.

Gutiérrez, M., Lahoz, J.M., Sola, M.M., Pascual, L., Vargas, A.M., 1994. Postharvest changes in total soluble solids and tissue pH of cherimoya fruit stored at chilling and non-chilling temperatures. J. Hortic. Sci. 69, 459–463.

Hofman, P.J., Jobin-Decor, M., Meiburg, G.F., Macnish, A.J., Joyce, D.C., 2001. Ripening and quality responses of avocado, custard apple, mango and papaya fruit to 1-methylcyclopropene. Aust. J. Exp. Agric. 41, 567–572.

Jeria, L.A. 1991. Efecto del permanganato de potasio como absorbedor de etileno sobre la evolución de la madurez en frutos de chirimoya (*Annona cherimola* Mill.) cv. Bronceada en almacenaje refrigerado. Tesis. Universidad Católica de Valparaíso, Facultad de Agronomía. 81pp.

Kader, A.A., 1997. A summary of CA requirements and recommendations for fruits other than apples and pears. In: CA '97 Proceedings volume 3: Fruits other than apples and pears. 7th Controlled Atmosphere research conference. July 13-18. University of California, Davis, pp. 1–34.

Kader, A.A., 2002a. Postharvest biology and technology: An overview. In: Kader, A.A. (Ed.), Postharvest technology of horticultural crops. University of California. Agriculture and Natural Resources. Publication 3311. University of California, Oakland, CA, pp. 39–47.

Kader, A.A., 2002b. Modified atmosphere during transport and storage. In: Postharvest Technology of Horticultural Crops. University of California, Agriculture and Natural Resources. Publication 3311. 535 pp. pp. 135–145.

Kader, A.A., Arpaia, M.L. 1999. Cherimoya (Atemoya & Sweetsop). http://postharvest.ucdavis.edu/Commodity_Resources/Fact_Sheets/Datastores/Fruit_English/?uid=15&ds=798 (Accessed May, 2018).

Lahoz, J.M., Gutiérrez, M., Sola, M., Del, M., Salto, R., Pascual, L., Martínez-Cayuela, M., Vargas, A.M., 1993. Ethylene in cherimoya fruit (*Annona cherimola* Mill.) under different conditions. J. Agric. Food Chem. 41, 72–723.

Liu, K., Liu, J., Li, H., Yuan, C., Zhong, J., Chen, Y., 2016. Influence of postharvest citric acid and chitosan coating treatment on ripening attributes and expression of cell wall related genes in cherimoya (*Annona cherimola* Mill.) fruit. Sci. Hortic. 198(1-11).

Loyola, N.E. 1988. Identificación y caracterización morfológica de los desórdenes fisiológicos: pitting, moteado, empalado y cristalización de la pulpa en chirimoyo cv. Concha Lisa. Universidad de Valparaíso. Fac. de Agronomía. 137 pp.

Maldonado, R., Molina-Garcia, A.D., Sanchez-Ballesta, M.T., Escribano, M.I., Merodio, C., 2002. High CO_2 atmosphere modulating the phenolic response associated with cell adhesion and hardening of Annona cherimola fruit stored at chilling temperature. J. Agric. Food Chem. 50, 7564–7569.

Manríquez, D.A., Muñoz-Robredo, P., Gudenschwager, O., Robledo, P., Defilippi, B.G., 2014. Development of flavor-related metabolites in cherimoya (Annona cherimola Mill.) fruit and their relationship with ripening physiology. Postharvest Biol. Technol. 94, 58–65.

Martínez, G., Pretel, M.T., Serrano, M., Riquelme, F., 1993. Evolution of resistence to gas difussion during cherimoya (*Annona cherimola* Mill.) ripening. HortScience 28, 204–206.

Martinez-Cayuela, M., Medina, L.S., Faus, M.J., Gil, A., 1988. Cherimoya (*Annona cherimola* Mill) polyphenoloxidase: Monophenolase and dihydroxyphenolase activities. J. Food Sci. 53, 1191–1194.

Merodio, C., Muñoz, M.T., Del Cura, B., Buitrago, D., Escribano, M.I., 1998. Effect of high CO_2 level on the titres of y-aminobutyric acid, total polyamines and some pathogenesis-related proteins in cherimoya fruit stored at low temperature. J. Exp. Bot. 49, 1339–1347.

Molfino, P.A. 1991. Comportamiento durante el tiempo de almacenaje a ooc de chirimoyas (*Annona cherimola* Mili.) cv. Concha Lisa procesadas en cuartos con diferentes medios de cobertura en atmósfera modificada. Tesis. Universidad Católica de Valparaíso. Facultad de Agronomía. 81pp.

Muñoz, M.T., Escribano, M.I., Merodio, C., 1997. Ethanolic metabolism in cherimoya fruít during storage at ambient and under high C02 atmosphere. J. Hortic. Sci. 72, 363–370.

Palma, T., Aguilera, J.M., Stanley, D.W., 1993. A review of postharvest events in cherimoya. Postharvest Biol. Technol. 2, 187–208.

Pardo, H.A. 1988. Almacenaje refrigerado de chirimoya (*Annona cherimola* Mill.) cv. Concha Lisa. Efectos de índices de cosecha y un absorbedor de etileno. Tesis, Universidad Católica de Valparaíso, Facultad de Agronomía. 75 pp.

Pareek, S., Yahia, E.M., Parrek, O.P., Kaushik, R.A., 2011. Postharvest physiology and technology of Annona fruits. Food Res. Int. 44, 1741–1751.

Prieto, H., Utz, D., Castro, A., Aguirre, C., Gonzalez-Aguero, M., Valdes, H., Cifuentes, N., Defilippi, B.G., Zamora, P., Zuniga, G., Campos-Vargas, R., 2007. Browning in *Annona cherimola* fruit: role of polyphenol oxidase and characterization of a coding sequence of the enzyme. J. Agric. Food Chem. 55, 9208–9218.

Riveros, G.E. 1991. Influencia de dos estados de ablandamiento y dos formas de presentación (cuartos y rodelas), sobre la calidad y estabilidad en almacenaje de la fruta congelada de chirimoya (*Annona cherimola* Mill.) cv. Bronceada. Tesis. Universidad Católica de Valparaíso. Facultad de Agronomía. 90 pp.

Sanchez, J.A., Zamorano, J.P., Alique, R., 1998. Polygalacturonase, cellulase and invertase activities during cherimoya fruit ripening. J. Hortic Sci. Biotech. 73, 87–92.

Silva, G.M.C., Cordeiro, M.H.M., Mizobutsi, G.P., Sobral, R.R.S., Mizobutsi, E.H., Silva, N.M., Feitosa, F.M., Martins, J.C., Fonseca, S.N.A., 2016. 1-MCP extends atemoya shelf life. J. Food Agric. Environ. 14, 35–39.

Sisler, E.C., Serek, M., 1997. Inhibitors of ethylene responses in plants at the receptor level: recent developments. Physiol. Plant. 100, 577–582.

Toro, J.E. 1989. Efecto de tres dosis de ethysorb (permanganato de potasio) como absorbedores de etileno sobre la evolución de la madurez en frutos de chirimoyo (*Annona cherimola* Mill.) cv. Clavo en almacenaje refrigerado. Tesis. Universidad Católica de Valparaíso, Facultad de Agronomía, 76 pp.

Wang, Y.C., 1982. Physical and biochemical responses of plants to chilling stress. HortScience 17, 173–186.

Watada, A.E., Ko, N.P., Minott, D., 1996. Factors affecting quality of fresh-cut horticultural products. Postharvest Biol. Technol. 9, 115–125.

Watkins, C.B., Nock, J.F., Whitaker, B.D., 2000. Responses of early, mid and late season apple cultivars to postharvest application of 1-methylcyclopropene (1-MCP) under air and controlled atmosphere storage conditions. Postharvest Biol. Technol. 19, 17–32.

Yonemoto, Y., Higuchi, H., Kitano, Y., 2002. Effects of storage temperature and wax coating on ethylene production, respiration and shelf-life in cherimoya fruit. J. Japan. Soc. Hort. Sci. 71, 643–650.

Zonneveld, M., Scheldeman, X., Escribano, P., Viruel, M.A., Van Damme, P., Garcia, W., Tapia, C., Romero, J., Siguenas, M., Hormaza, J.I., 2012. Mapping genetic diversity of cherimoya (Annona cherimola Mill.): application of spatial analysis for conservation and use of plant genetic resources. PLoS One 7. e29845.

Further reading

Koussevitzky, S., Ne'eman, E., Peleg, S., Harel, E., 2008. Polyphenol oxidase can cross thylakoids by both the Tat and the Sec-dependent pathways: a putative role for two stromal processing sites. Physiol. Plant. 133, 266–277.

Muñoz, M.T., Aguado, P., Ortega, N., Escribano, M.I., Merodio, C., 1999. Regulation of ethylene and polyamine synthesis by elevated carbon dioxide in cherimoya fruit stored at ripening and chilling temperatures. Aust. J. Plant Physiol. 26, 201–209.

Chapter 18.3

Subtropical fruits: Citrus

María B. Pérez-Gago and Lluís Palou
Postharvest Technology Center (CTP), Valencian Institute of Agrarian Research (IVIA), Valencia, Spain

Quality characteristics

- Citrus are nonclimacteric fruits that are harvested when their commercial maturity index (MI = total soluble solids/acidity) has already been reached.
- Quality index also include rind color intensity (color index, $CI = 1000a/Lb$; Hunter parameters) and uniformity (especially in early citrus cultivars,

Controlled and Modified Atmospheres for Fresh and Fresh-Cut Produce
https://doi.org/10.1016/B978-0-12-804599-2.00028-4

where fully ripened fruit may still have green-colored rind), size, shape, firmness, juice yield, and freedom from decay and external defects.
- Recommendations for MI, color index, and firmness at harvest are specific for species and even cultivars.

Main causes of quality loss

- Water loss: Excessive water loss results in fruit shriveling and wilting, and triggers several physiological disorders that affect the peel. Some of these rind disorders that may appear under optimal, nonchilling temperatures include peel pitting, stem-end rind breakdown (SERB), and shriveling and collapse of the stem-end button.
- Storage below optimal temperatures causes chilling injury (CI) symptoms, which usually develop upon transfer the fruit from low to higher temperatures. CI occurrence and symptoms depend on citrus species, cultivar, maturity stage at harvest, cultural practices, weather conditions, and exposure conditions (temperature and duration) during storage. For instance, typical symptoms in oranges, mandarins, or grapefruits can be browning of the flavedo, appearance of dark sunken areas of collapsed tissue (pitting) (Fig. 1), or appearance of soft water-soaked areas (watery breakdown), while in lemons can be browning of the albedo or peteca (especial type of rind pitting). Other symptoms include scald, off-flavors, and increased decay incidence.
- Physical damage due to rind wounds or bruises caused during harvest, transportation, or postharvest handling in the packinghouse favors water loss and fungal infection. Furthermore, mechanical damage triggers many physiological disorders in the peel, including 'brush burn,' 'zebra skin,' and 'oleocellosis' caused by rind abrasion, rough handling, or thorn-branch punctures. Harvesting and handling turgid (e.g., early in the morning, after

FIG. 1 Severe rind pitting of mandarin caused by chilling injury.

irrigation or rain fall) or overmatured citrus fruits can boost some disorders such as oleocellosis or rind staining.

- Postharvest decay is typically caused by filamentous fungi and can cause significant economic losses. In general, there are two types of fungal pathogens. Wound pathogens infect the fruit through rind injuries inflicted during harvest, transportation, or postharvest handling and include *Penicillium digitatum* and *Penicillium italicum* causing green and blue mold, respectively, *Geotrichum citri-aurantii* causing sour rot, and *Rhizopus stolonifer* and *Mucor* spp. causing soft rot. Latent pathogens infect the fruit in the field during the growing season and remain latent or quiescent until they resume growth after harvest because of significant changes in fruit characteristics and environmental conditions. Main latent pathogens include *Lasiodiplodia theobromae* and *Phomopsis citri* causing stem-end rots, *Colletotrichum gloeosporioides* causing anthracnose, *Phytophthora citrophthora* causing brown rot, *Alternaria* spp. causing black rot, and *Botrytis cinerea* causing gray mold.
- Postharvest quarantine treatments are needed for citrus fruits shipped to pest-free export markets. One of the world's most damaging citrus fruit pests is the Mediterranean fruit fly, *Ceratitis capitata* (Diptera: Tephritidae), against which countries such as the United States, Japan, China, or Australia maintain strict quarantine measures.

Optimum storage conditions

Optimum temperature

- Postharvest storage of citrus fruits at low temperatures is the most efficient and widely used technology to maintain fruit quality, extend the market season mainly of late varieties, delay decay incidence, and apply quarantine treatments for insect control to fruit destined to pest-free export countries.
- Optimum temperature management is critical and depends on the species and cultivar. Grapefruits and lemons are more susceptible to CI than mandarins and oranges; whereas, hybrids (tangelos, tangors, and mandarin hybrids) display moderate susceptibility to CI. In general, grapefruit can be stored at 10–12°C for up to 2–3 months, lemons at 7–12°C for up to 4–6 months, limes at 10–13°C for up to 6–8 weeks, mandarins at 5–10°C for up to 1–2 months, and oranges at 2–5°C for up to 3 months. The renewal of air during storage is also important to prevent ethylene or CO_2 buildup.
- For the same cultivar, fruit grown and harvested in humid and warmer climates have a lower threshold of temperature tolerance to CI than fruit grown in cooler climates. Thus, for instance, optimum temperatures for 'Valencia' oranges are 4–5°C and 10–12°C in fruit grown in Mediterranean and tropical countries, respectively.

- The sensitivity to cold of some citrus cultivars can be reduced with the use of postharvest techniques such as fruit conditioning at medium temperature (15–18°C for several days prior to cold storage), curing (hot humid air at 30–37°C for up to 3 days; more difficult to implement at commercial scale), hot water dips (45–50°C for a few minutes), intermittent warming (alternance of cold storage with exposure to moderate temperatures of about 20°C), and commercial coatings (conventional waxes and edible coatings).
- The application of cold, as a quarantine treatment for export, requires a meticulous control of the harvesting operations, chemical treatments, degreening, prestorage, and transport so that the fruit reach their destination with the required quality. For shipments of mandarins and oranges to the United States, maximum quarantine temperatures of 1.1°C, 1.7°C, and 2.2°C are maintained for periods of fruit permanence of 14, 16, and 18 days, respectively. For export to Japan, the regulations require temperatures below 2°C in the center of the fruit for a minimum period of 16 days (lemons) or 17 days (oranges and mandarins).

Optimum relative humidity

In cold rooms, air movement around the fruit should be minimized to avoid transpiration, but ambient relative humidity should be as high as possible (90%–95%) to reduce water loss and skin-related disorders.

CA/MA considerations

- Commercial application of CA/MA technology to citrus fruit is very limited due to the low tolerance of most of species and cultivars to low O_2 and/or high CO_2, which stimulates the fermentative metabolism with the development of off-flavors caused by a buildup of ethanol, acetaldehyde, and other volatiles associated to anaerobic respiration.
- Although the use of packaging films for MA storage is not a commercial practice in postharvest handling of citrus fruit, the application of commercial waxes and coatings is widespread in the citrus industry. These commercial waxes create a MA within the fruit, reduce weight loss, improve appearance, and might reduce decay (if formulated with chemical fungicides or antifungal agents), all of which translate into improved fruit quality. Furthermore, appropriate coatings have been demonstrated to reduce CI and other citrus rind disorders due to the creation of gas and water barriers. However, if the coatings excessively restrict gaseous exchange during storage, they can adversely affect fruit flavor due to an overproduction of volatiles associated with anaerobic conditions (Fig. 2). Therefore the correct selection of a wax or coating is crucial to avoid anaerobic respiration. Citrus commercial waxes typically contain resins and/or waxes such as shellac, wood rosin, candelilla wax, carnauba wax, beeswax, polyethylene, and other

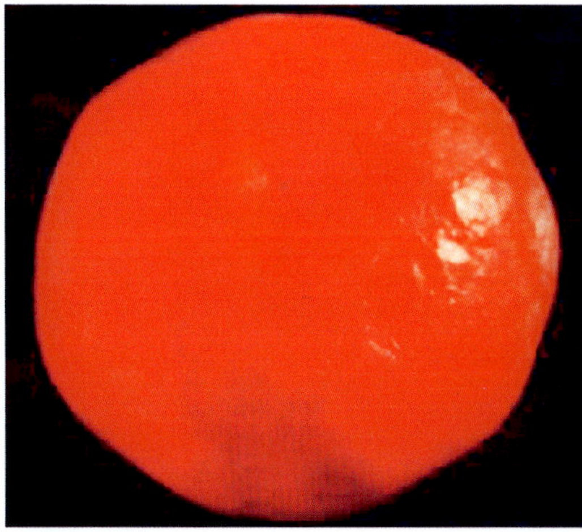

FIG. 2 Mandarin with deficient postharvest wax application that can cause irregular water and gas exchange and development of off-flavors.

petroleum waxes. In general, wax-based coatings are more suitable for citrus coating than resin-based coatings due to the more permeable nature of waxes.

- Composite edible coatings obtained by combination of biopolymers, such as proteins and polysaccharides, with lipids also provide an excellent alternative to commercial waxes as they allow combinations that match the product requirements. Edible coatings formulated with natural or food-grade antifungal agents are a nonpolluting alternative to conventional waxes amended with chemical fungicides such as imazalil, thiabendazole, pyrimethanil, and so on.
- The effect of commercial waxes and edible coatings on citrus quality greatly depends on the species and cultivar, coating composition, thickness, and environmental conditions. Furthermore, coating application should be done on dried surface fruit to ensure an adequate coating application.
- The application of ethylene for degreening of early season citrus fruits that have optimum sensory quality but not the required external rind color is a common practice in packinghouses to meet market requirements and consumers' expectations (Fig. 3). Commercial systems use a continuous ethylene flow in close rooms with appropriate ventilation to avoid CO_2 buildup. Environmental conditions in the room should be carefully monitored. Although the volatile profile of some cultivars may change, degreening does not affect overall internal fruit quality.
- Although cold quarantine treatments are currently mandatory, research has shown that insecticidal atmospheres consisting of short-term (20–24 h) CA treatments ($> 90\%$ CO_2) at around 20°C are interesting nonchemical alternatives for postharvest insect control that leave no residues on/in the

FIG. 3 Rind color index ($CI = 1000a/Lb$, Hunter parameters) values of early season mandarins subjected to degreening with exogenous ethylene.

commodity. These CA treatments, applied for longer times or at high temperatures (> 25°C), can be phytotoxic and/or adversely affect fruit quality. Moreover, brief gaseous shocks of CO_2 and O_2 at curing temperatures (30–35°C) can effectively control decay caused by *Penicillium* spp. Limiting factors are the cost and limited persistence of these treatments and the general low tolerance of citrus fruits to gas exposure and heat.

Tolerances

- Tolerance of citrus fruit to low O_2 and/or high CO_2 depends on the species and cultivar. In general, mandarins have a higher sensitivity than oranges, lemons, and grapefruits. This is related to major aldehyde dehydrogenase activity and to a more impermeable cuticle to gas diffusion of mandarins. Some works establish a tolerance for mandarins up to 5%–10% O_2 and 0–2% CO_2, for oranges up to 5%–10% O_2 and 0%–5% CO_2; for lemons up to 5%–10% O_2 and 0–10% CO_2, and for grapefruits up to 3%–10% O_2 and 5%–12% CO_2.
- Although difficult to establish, certain critical limits for internal gas concentration that adversely affect the flavor of different coated mandarin hybrids are internal O_2 concentration below 4% and internal CO_2 above 14%. Other works determined that the ethanol threshold before off-flavor detection in certain citrus cultivars was 1500–2000 ppm.
- Suitable conditions for citrus degreening depend on species and cultivar and are in the range: temperature of 20-22°C for California and Mediterranean countries fruits and 28–29°C for Florida and Brazil fruits, relative humidity higher than 90%, ethylene concentration of 2–5 ppm, and CO_2 concentration

lower than 0.2%. The duration of the process depends on the peel chlorophyll content (rind initial CI), but should not be longer than 120h to avoid blackening and falling of chalices (stem buttons), as well as other physiological alterations associated with rind senescence. In general, longer exposure times are preferred to higher ethylene concentrations. The rate of degreening is reduced if the CO_2 concentration reaches 0.1% and will nearly stop it above 1%.

Injuries

- High CO_2 concentrations may induce severe peel browning, especially if applied at high temperatures (Fig. 4).
- Low O_2 results in buildup of volatile components associated with anaerobic conditions and, therefore, in off-flavor development.
- Ethylene accumulation in cold rooms or excessive exogenous ethylene concentration during the degreening process can cause blackening and falling of chalices and peel injuries related to senescence (e.g., rind staining, pitting, SERB, etc.). Typically, the symptoms of these aging disorders are visible only after few days following ethylene exposure and their intensity is particularly high after mechanical fruit sorting and grading in the packing line (Fig. 5).

Problems under suboptimal conditions

Given that CA/MA are not commercially applied to extend storage life of citrus fruits, the key problems under suboptimal conditions are chilling injury causing peel pitting, watery breakdown, or albedo browning; rind disorders under nonchilling temperatures caused by water stress and mechanical damage (SERB, shriveling, collapse of the stem-end button, oleocellosis, etc.); and postharvest diseases caused by fungal pathogens (green mold, blue mold, stem-end rots, anthracnose, sour rot, brown rot, etc.).

FIG. 4 Rind phytotoxicity in mandarins exposed to 95% CO_2 for 20h at 25 (A) and 35°C (B).

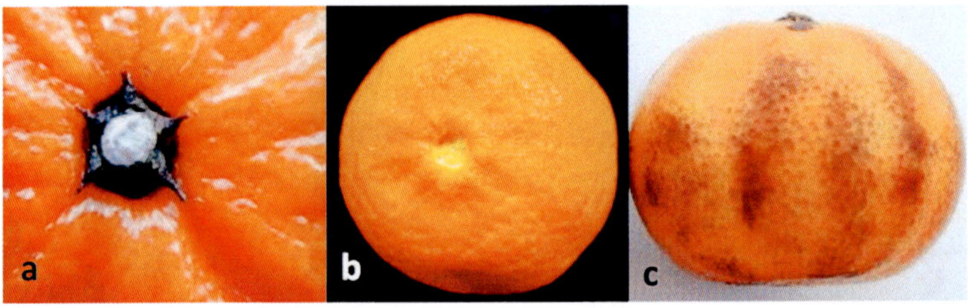

FIG. 5 Blackening (A) and falling (B) of chalices, and aging rind disorders in mandarins degreened with excessive ethylene followed by handling and processing in the packing line (C).

Comments

Although some reports show that CA/MA delayed senescence and decay and retained firmness of some oranges and mandarins, commercial application of CA/MA technology to citrus fruit is very limited because it is not cost effective and it can stimulate anaerobic respiration and physiological stress, with the buildup of volatile components that leads to off-flavor development.

Overall, optimum cold storage temperature and relative humidity higher than 90% are the key postharvest factors to maintain fruit quality and extend the market season of citrus fruits. At commercial scale and depending on the destination market, optimum storage conditions need to be accompanied by the use of waxes, edible coatings, chemical fungicides, or alternative antifungal treatments in order to reduce weight loss, physiological disorders, and decay.

Further reading

El-Otmani, M., Ait-Oubahou, A., Zacarías, L., 2011. *Citrus* spp.: orange, mandarin, tangerine, clementine, grapefruit, pomelo, lemon and lime. In: Yahia, E.M. (Ed.), Postharvest Biology and Technology of Tropical and Sub-tropical Fruits. In: 2, Woodhead Publishing Limited, Sawston, Cambridge, UK, pp. 436–514. Açai to Citrus.

Hagenmaier, R.D., 2002. The flavor of mandarin hybrids with different coatings. Postharvest Biol. Technol. 24, 79–87.

Ke, D., Kader, A.A., 1990. Tolerance of 'Valencia' oranges to controlled atmosphere as determined by physiological responses and quality attributes. J. Amer. Soc. Hort. Sci. 12, 779–783.

Landaniya, M.S., 2008. Citrus Fruit: Biology, Technology and Evaluation, Academic Press, Elsevier Inc. USA, San Diego, CA, p. 558.

Montesinos-Herrero, C., del Río, M.A., Rojas-Argudo, C., Palou, L., 2012. Short exposure to high CO_2 and O_2 at curing temperature to control postharvest diseases of citrus fruit. Plant Dis. 96, 423–430.

Navarro-Tarazaga, M.L., Pérez-Gago, M.B., 2006. Effect of edible coatings on quality of mandarins cv. Clemenules. Proc. Fla. State Hortic. Soc. 119, 350–352.

Palou, L., Jacas, J.A., Marcilla, A., Alonso, M., del Río, M.A., 2008. Physico-chemical and sensory quality of 'Clemenules' mandarins and survival of the Mediterranean fruit fly as affected by complementary cold and carbon dioxide quarantine treatments. Postharvest Biol. Technol. 48, 443–450.

Palou, L., Smilanick, J.L., Montesinos-Herrero, C., Valencia-Chamorro, S., Pérez-Gago, M.B., 2011. Novel approaches for postharvest preservation of fresh citrus fruits. In: Slaker, D.A. (Ed.), Citrus Fruits: Properties. Consumption and Nutrition. Nova Science Publishers, Inc., NY, USA, pp. 1–45.

Palou, L., Valencia-Chamorro, S.A., Pérez-Gago, M.B., 2015. Antifungal edible coatings for fresh citrus fruit: a review. Coatings 5, 962–986.

Petracek, P.D., Kelsey, D.F., Grierson, W., 2006. Physiological peel disorders. In: Wardowski, W.F., Miller, W.M., Hall, D.J., Grierson, W. (Eds.), Fresh Citrus Fruits, 2nd Ed, Florida Science Source Inc., Longboat Key, FL, USA, pp. 397–419.

Porat, R., 2008. Degreening of citrus fruit. Tree Forestry Sci. Biotechnol. 2 (Special Issue 1), 71–76.

Porat, R., Weiss, B., Cohen, L., Daus, A., Aharoni, N., 2004. Reduction of postharvest rind disorders in citrus fruit by modified atmosphere packaging. Postharvest Biol. Technol. 33, 35–43.

Ritenour, M.A., Miller, W.M., Wardowski, W.W., 2015. Recommendations for degreening Florida fresh citrus fruits. Cir 1170—Series of the Horticultural Sciences Department, UF/IFAS Extension. https://edis.ifas.ufl.edu/pdffiles/HS/HS19500.pdf (accessed 09.01.2018).

Rojas-Argudo, C., del Río, M.A., Montesinos-Herrero, C., Palou, L., 2010. Effects of CO_2 and O_2 shocks at high temperature on postharvest quality of cold-stored citrus fruit. Int. J. Food Sci. Technol. 45, 2062–2070.

Sdiri, S., Rambla, J.L., Besada, C., Granell, A., Salvador, A., 2017. Changes in the volatile profile of citrus fruit submitted to postharvest degreening treatment. Postharvest Biol. Technol. 133, 48–56.

Singh, Z., Singh, S.P., Yahia, E.M., 2009. Subtropical fruits. In: Yahia, E.M. (Ed.), Modified and Controlled Atmospheres for the Storage, Transportation, and Packaging of Horticultural Commodities. CRC Press. USA, Boca Raton, FL, pp. 317–361.

Chapter 18.4

Subtropical fruits: Dates

Nihad Alsmairat
Horticulture and Crop Science Department, University of Jordan, Amman, Jordan

Quality characteristics

The maturity of dates passes through five stages; hababouk, kimri, khalal (balah or bisr), rutab, and tamar. The quality characteristics of the fruit at harvest are based on sugar, tannin, and moisture contents while for fresh fruit consumption or storing are fruit appearance and texture. Some cultivars are acceptable to consumers upon harvest at khalal stage; these include Barhi, Hayani, Samany, and Zaghlel cultivars. The shelf life of khalal stage date fruit is very short for successful marketing.

The khalal stage is recommended for the harvest of fresh dates with top initial quality characteristics (Fig. 1). Expected characteristics include: (1) full mature fruits; (2) regular appearance, yellow, orange, or red in color depending on the cultivar; (3) no sign of sugar spots on the skin; (4) freedom from bruises, damage, or any cuts; (5) no fungal, mold, or insect damage; (6) smooth skin; (7) sweet flavor with low astringency; (8) free from defects such sunburn, insect damage, or dirt on the skin; and (9) appropriate oblate fruit size and shape with a medium-sized pit.

Once in the market, the top quality characteristics of fresh fruits include: (1) fruit skin with a rich and shiny color representative of the cultivar, (2) no sign of sugar spots on the fruit, (3) crispy fruit texture, (4) no sign of shrivel, (5) fruit clean from any blemishes, (6) thick flesh with rich flavor, and (7) fruit with no chilling injury or other damage symptoms.

Main causes of fruit quality loss

- Compaction of fruits due to improper fruit thinning
- Stressed trees resulting in compromised fruit quality
- Poor cultural practices before harvesting
- Improper harvest time

Controlled and Modified Atmospheres for Fresh and Fresh-Cut Produce
https://doi.org/10.1016/B978-0-12-804599-2.00029-6

FIG. 1 Visual quality of fresh dates at harvest at khalal stage of 'Barhi' cultivar.

- Lack of fruit bunch netting
- Sunburn and related defects
- Physical damage
- Improper handling after harvest
- Spoilage caused by yeast, fungi, or bacteria
- Insect damage
- Separation of the fruits from the strand
- Improper packaging in the room storage or during handling
- Lingering water on fruit surface following hydro-cooling
- Delay in fruit cooling after harvest
- Improper ventilation of stored fruit
- Exposure of fruit to a temperature less than 0°C
- Exposure of fruit to relative humidity under 85%
- Storage of fruit at high (i.e., ambient) temperatures
- Fluctuation in temperature during storage
- Exposure of date fruit to high than maximum recommended CO_2 concentration
- Improper temperatures and/or inappropriate handling procedures used during transport to the destination market

Optimum storage conditions

Optimum temperature

Forced-air cooling can be used to cool khalal fruits to near 5°C. Fruits should be stored at 0°C to: (1) Delay fruit ripening to the rutab stage, (2) maintain texture, and (3) maintain flavor quality.

Optimum relative humidity

To avoid water loss from the fruits, relative humidity of the atmosphere around the fruits should be 85–95%.

CA/MA consideration

Dates at khalal stage can benefit from controlled atmosphere (CA) or modified atmosphere (MA) storage. In cold storage, fruits can be stored about 7 weeks; however, when using CA, fruits at khalal stage can be stored approximately 24 weeks in 20% CO_2 enriched air at 0°C and 90-95% relative humidity (Fig. 2).

Benefits of CA/MA

- Extended storability.
- Maintained total phenolic content, SSC%, total sugar content, and total tannins.
- Maintained color brightness.
- Retarded decay and weight loss percentages.
- Retarded fruit softening.
- Prevented fruit darkening
- Reduced antioxidant losses.
- Allows shipping to distant markets with acceptable quality.
- Reduced chilling injury symptoms of low temperature (0°C).

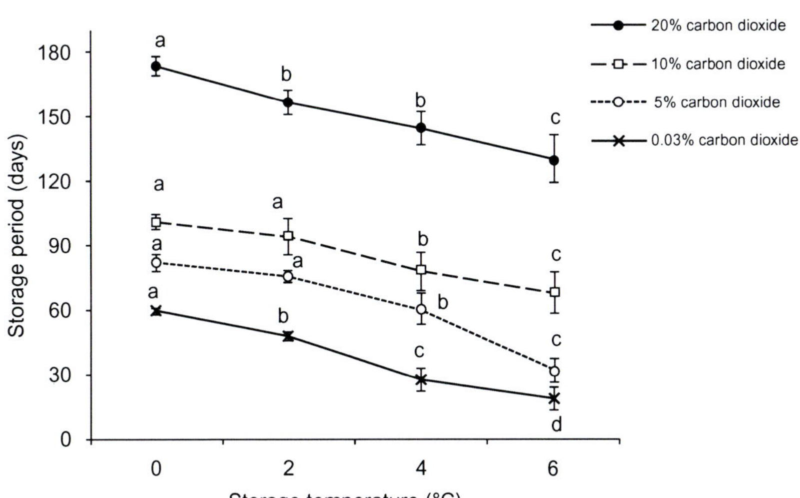

FIG. 2 The postharvest life of 'Barhi' date fruit at khalal stage is a function of CO_2 concentration and temperature.

Tolerances

Dates fruits eaten at khalal maturity stage could tolerate 5-20% CO_2 under MA or CA at temperatures between 0 and 5°C with little internal or external incidence of injury or chilling symptoms of the fruit.

Injuries

Exposing date fruits at khalal stage to high CO_2 concentration >30% CO_2 at ambient room temperature (i.e., about 20°C) could result in: (1) Advanced fruit ripening; (2) sugar spots on the fruit skin; (3) increased softening, especially when ambient temperature exceeds 20°C (Fig. 3); (4) loss in 'fresh' appearance; and (5) reduced shelf life and a foreshortened marketing window.

However, a CO_2 concentration >30% at low temperature (about 0°C) could result in fruit injury during storage (Fig. 4).

Recommendations

Carbon dioxide at 20% balanced with air at 0°C extends the storability and maintains fruit quality at khalal stage. It is recommended that the fruits remain attached to the strand when the strand is separated from its bunch. Appropriate cultural practice before harvesting and a suitable handling should be a complement to CA/MA storage conditions. Further research should be conducted on fresh dates at khalal stage to investigate the effect of different atmosphere and temperature combinations to allow shipping and storage of the fruits with acceptable quality.

FIG. 3 Exposure of fresh dates to CO_2 concentrations >30% balance with O_2 at ambient room temperature accelerates fruit ripening and softening.

FIG. 4 Storage of "Barhi" fresh dates at full mature stage (bisr) under high CO_2 concentrations (> 30% CO_2) could result in CO_2 injury (El-Rayes, 2009).

FIG. 5 An optimal combination of harvest maturity and storage temperature (0°C) can extend the storage of 'Barhi' dates at khalal stage. Dates stored at 0°C for 21 days (left) and for 28 days (right).

Ethylene considerations

At khalal stage, 'Barhi' dates were not affected when exposure to 100 ppm ethylene for up to 48 h at 20°C and 85–90% relative humidity. However, at high temperature, e.g., 30–35°C, they may respond to ethylene action.

Problems under suboptimal conditions

Fruits at khalal stage should be harvested at full mature harvesting time else the shelf life or duration of balah stage will be reduced (Fig. 5). Fruits are sensitive to any handling and holding procedure that may reduce the marketing storage period or fruit quality that reflects on the freshness and the potential for storing some dates cultivar at full mature to have at the end acceptable fruit quality.

References

El-Rayes, D.A., 2009. Effect of carbon dioxide-enriched atmosphere during cold storage on limiting antioxidant losses and maintaining quality of 'Barhy' date fruits. Env. Arid Land Agric. Sci. 20, 3–22.

Further reading

Al-Redhaiman, K.N., 2004. Modified atmosphere improves storage ability, controls decay, and maintains quality and antioxidant contents of barhi date fruits. Food. Agriculture and Environment 2, 25–32.

Al-Redhaiman, K.N., 2005. Chemical changes during storage of 'Barhi' dates under controlled atmosphere conditions. Hort Science 40, 1413–1414.

AlSaed, A.K., Mehyar, G.F., Arar, A., 2013. Effect of harvesting time and storage temperature on the duration of Balah stage of 'Barhi' dates. Ital. J. Food Sci. 25, 1–9.

Alsmairat, N.G., El-Assi, N.M., Al-Abdallat, A.M., Mehyar, G.F., 2013. Enhancement of edibility of 'Barhi' and 'Medjool' date palm cultivars at khalal mature stage. Int. J. Bot. 9, 123–132.

Lobo, N.G., Yahia, E.M., Kader, A.A., 2013. Biology and postharvest physiology of date fruit. In: Siddiq, M., Aleid, S.M., Kader, A.A. (Eds.), Dates: Postharvest Science, Processing Technology and Health Benefits. Wiley-Blackwell, New York, pp. 57–80.

Mehyar, G.F., El Asssi, N.M., Alsmairat, N.G., Holley, R.A., 2014. Effect of edible coating on fruit maturity and fungal growth on Berhi dates. Int. J. Food Sci. Technol. 49, 2409–2417.

Mortazavi, S.M.H., Arazani, K., Arujalian, A.A., 2010. Modified atmosphere packaging of date fruit (*Phoenix dactylifera* L.) cultivar 'Barhee' in khalal stage. Acta Hort. (882), 1063–1069.

Chapter 18.5

Subtropical fruits: Figs

Giancarlo Colelli and Maria Luisa Amodio
Department of Science of Agriculture, Food and Environment, University of Foggia, Foggia, Italy

Quality characteristics

The edible fruit of figs is actually a big infructescence that derives from a complex inflorescence called *syconium*. The syconium is a fleshy, involuted hollow, and almost closed receptacle with multiple ovaries on the internal surface. Fig shape may vary from pyriform to globose. Fruit skin turns from dark green to light green-yellow or to red-violet when fully ripe. Fruit flesh is juicy and sweet when ripe because of high amounts of sugars and a low organic acid content and gummy when unripe. Flesh color ranges from white to yellow, amber, pink, red, and purple.

Figs for fresh market are harvested when almost fully ripe and firm, so to be ready to eat and good for the market (Fig. 1); eating quality is enhanced with the ripening but fruits became very perishable (prompt market) (Fig. 2). Skin color and flesh firmness are strongly related to product quality (Fig. 3). Figs must be firm, with intact stem ends and without cracks and splitting (Fig. 4). Other

Controlled and Modified Atmospheres for Fresh and Fresh-Cut Produce
https://doi.org/10.1016/B978-0-12-804599-2.00030-2

FIG. 1 Fresh figs ready for the distribution to the market.

FIG. 2 Ripening stages and marketability of fresh figs.

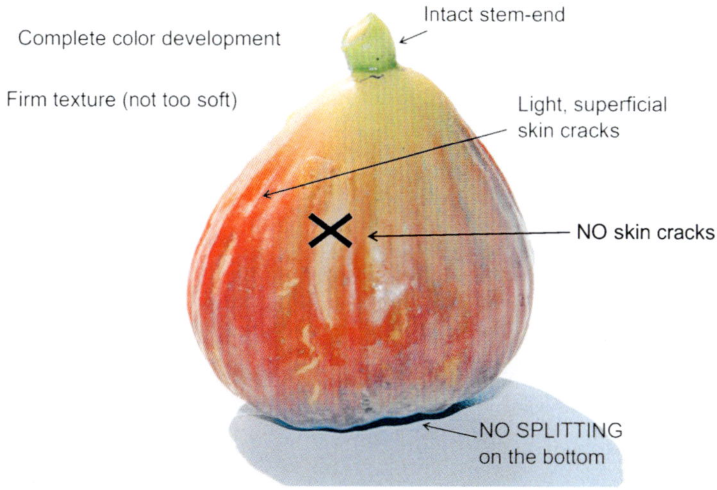

FIG. 3 Quality characteristics of the figs.

FIG. 4 Example of fig splitting.

quality indices include absence of external defects (sunburn, scab, skin break), insects, and decay (Crisosto et al., 1998).

In temperate climate, fig trees can produce two crops per year, depending on the cultivar type and growing conditions. The first crop, called the *breba crop*, is produced from flower buds on the previous year wood that was started during the previous summer. The main crop is instead produced from syconia initiated in the axils of the current season shoot. This crop ripens in late summer and usually produces fruit of better quality than the breba crop (Stover et al., 2007).

Main causes of quality loss

Figs are very perishable products and storage at ambient temperature leads to early senescence, fermentation, and decay occurrence which limit their shelf life (Karabulut et al., 2009). The main causes of produce quality losses are fungal decay (Michailides et al., 2008) (Fig. 5), mechanical damages, and weight loss (Chessa, 1997; Piga et al., 1998; Venditti et al., 2005).

The five major fig diseases in postharvest storage are: smut, Alternaria rot (caused by *Alternaria tenuis*, appears as small, round, brown-to-black spots over the fruit surface), black mold rot (caused by *Aspergillus niger*, appears as dark or yellowish spots in the flesh with no external symptoms), gray mold or Botrytis rot, and fig endosepsis (caused by *Fusarium moniliforme*).

The most sensitive part of figs to fungal decay is the ostiole; the natural opening of fig fruit could represent the access point through which the pathogens can reach the internal cavity of fruit (Cantín et al., 2011).

The severity of diseases caused by microbial molds and rots increase due to the easily damaged external surface and the high sugar content of fruits (Colelli et al., 1991), and the presence of cracks on the skin and fruit splitting make the fruit more susceptible to the rot.

Optimum storage conditions

Optimum temperature

Storage of fresh figs at −1°C to 0°C and maintaining the cold chain from farm to fork is important and recommended to preserve fruit quality. Prompt cooling

FIG. 5 Example of severe fungal decay on fresh figs.

to 0°C with forced air is strongly suggested in order to prevent weight losses (Crisosto and Kader, 2016). Figs exposed to 6h of delayed cooling showed an initial weight loss of approximately 6% compared to 0.6% if there is no delay in cooling (Dollahite et al., 2007).

Optimum relative humidity

The optimum relative humidity for fresh fig storage is between 90% and 95%. Postharvest life at optimum temperature and relative humidity depends upon cultivar and ripeness stage at harvest and ranges from 1 to 2 weeks in air and from 3 to 4 weeks in controlled atmosphere for California-grown 'Black Mission' and 'Calimyrna' figs.

CA/MA considerations

CA storage has been reported to extend the storage life and maintain quality of fresh figs. Particularly, high-CO_2 exposure reduced the ethylene production, softening, decay incidence, and maintained bright external appearance of fruit. Fig postharvest life can be extended by 2–3 weeks at 0–5°C in atmospheres enriched with 15%–20% CO_2, but off-flavors can develop by the fourth week of storage (Colelli et al., 1991). CA combinations of 5%–10% O_2 and 15%–20% CO_2 are effective in decay control, firmness retention, and reduction of respiration and ethylene production rates that prolong product shelf life (Chessa, 1997; Crisosto and Kader, 2007)

Benefits

Decay incidence and microbial contamination reduction
Maintaining of bright external appearance of the fruits
Respiration rate and ethylene production rate decrease
Reduction of firmness losses

Low O_2

Fully ripe 'Mavra Markopoulou' figs stored at −1°C in CA containing 2kPa O_2 (balance N_2) for 29days showed decreased softening, weight loss, ethylene production, and retention of bright skin color (Tsantili et al., 2003).

High CO_2

Storage of fresh figs in high CO_2 atmospheres reduced mold growth without greatly affecting the flavor of the fruit.

High CO_2 exposure is generally beneficial for fresh figs, reducing ethylene and CO_2 productions, decay incidence and softening.

'Mission' (Colelli et al., 1991) and 'Calymirna' (Colelli and Kader, 1993) figs could be stored for up to 4 weeks at 0, 2.2, or 5°C in atmospheres enriched with 15/20kPa CO_2. An example of the effect of storage temperature and high

CO_2 exposure for 4 weeks on 'Black Missions' figs is shown in Fig. 6. Ethanol content of the CO_2-treated fruit increased slightly during the first 3 weeks and moderately during the 4th week, while acetaldehyde concentration increased during the first week, and then decreased.

Claypool and Ozbek (1952) reported that storage in atmospheres containing up to 60 kPa CO_2 at 20°C was not effective, but pretreatment with 100 kPa CO_2 for 36 h at 5°C and 10°C delayed microbial growth.

CA combinations of 5–10 kPa O_2 and 15–20 kPa CO_2 are effective in decay control, firmness retention, and reduction of respiration and ethylene production for 3–4 weeks for California-grown 'Black Mission' and 'Calimyrna' figs (Crisosto and Kader, 2004).

Beneficial effect of high-CO_2 active modified atmosphere packaging (starting from 20 kPa of CO_2) can also be observed on "Petrelli" breba figs stored for 16 days between 0 and 5°C, as shown in Fig. 7.

Injuries

Prolonged exposition to controlled atmosphere can result in flavor losses.

Storage with less than 2 kPa O_2 and/or more than 25 kPa CO_2 develops off-flavors and can become undesirable due to fermentative metabolism by-products (Crisosto and Kader, 2016).

Recommendations

Gentle handling, strict sanitation of picking and transporting containers, as well as rapid cooling are highly recommended to minimize physical damages and

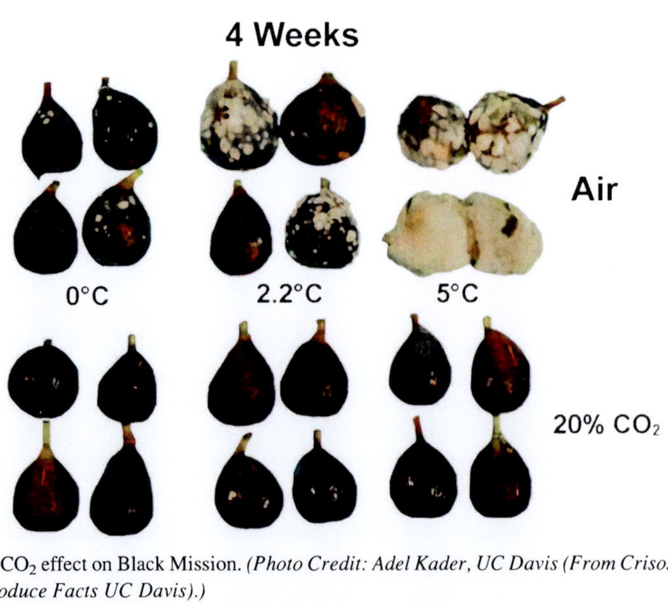

FIG. 6 CO_2 effect on Black Mission. *(Photo Credit: Adel Kader, UC Davis (From Crisosto et al., 1998. Produce Facts UC Davis).)*

FIG. 7 "Petrelli" breba figs stored for 13 days in active modified atmosphere packaging (initial 20 kPa CO_2), compared with samples stored in air. *(Photo credits: Giancarlo Colelli.)*

subsequent decay during storage; atmospheres enriched with 15%–20% CO_2 (CA or in MAP) can extend storage life of figs up to 3–4 weeks at 0°C.

Problems under suboptimal conditions

Suboptimal controlled atmosphere composition should be avoided in order to prevent product over ripening and fermentation occurrence. Overripe figs can become undesirable due to fermentative products (Crisosto et al., 1998).

Comments

Heat treatment did not extend shelf life of 'Petrelli' breba figs when stored at low temperature. On the other hand, when fruits were stored at room temperature a pretreatment at 50°C for 30 min contributed to reduce decay incidence, but only for a few days (Piazzolla et al., in press) (Fig. 8).

FIG. 8 Effect of different hot air treatments on quality and decay incidence after 3 days of storage at 25°C.

References

Cantín, C.M., Palou, L., Bremer, V., Michailides, T.J., Crisosto, C.H., 2011. Evaluation of the use of sulfur dioxide to reduce postharvest losses on dark and green figs. Postharvest Biol. Technol. 59, 150–158.

Chessa, I., 1997. Fig. In: Mitra, S. (Ed.), Postharvest Physiology and Storage of Tropical and Subtropical Fruits. CAB International, Walligford, UK, pp. 245–268.

Claypool, L.L., Ozbek, S., 1952. Some influences of temperature and carbon dioxide on the respiration and storage life of the Mission fig. Proc. Am. Soc. Hort. Sci. 60, 226–230.

Colelli, G., Kader, A.A., 1993. CO_2-enriched atmospheres reduce postharvest decay and maintain good quality in highly perishable fruits. Proc. COST 94 - 3rd Workshop on CA Storage. Milano, Italy, pp. 137–148.

Colelli, G., Mitchell, F.G., Kader, A.A., 1991. Extension of postharvest life of 'Mission' figs by CO_2-enriched atmospheres. Hort. Sci. 26, 1193–1195.

Crisosto, C.H., Kader, A.A., 2004. The Commercial Storage of Fruits, Vegetables, Florist, and Nursery Stocks. ARS-USDA Handb Beltsville, USA p. 66.

Crisosto, C., Kader, A.A., 2007. Figs. Postharvest quality maintenance guidelines. Postharvest information for fruits and nuts. http://www.uckac.edu/postharv.

Crisosto, C.H., Kader, A.A., 2016. Gross, K.C., Wang, C.Y., Saltveit, M.E. (Eds.), The commercial storage of fruits, vegetables, and florist and nursery stocks. U.S. Department of Agriculture, Washington, D.C Agriculture Handbook Number 66. http://ucanr.edu/datastoreFiles/234-2927.pdf.

Crisosto, C.H., Mitcham, E.J., Kader, A.A., 1998. Fig: Recommendations for Maintaining Postharvest Quality. http://postharvest.ucdavis.edu/Commodity_Resources/Fact_Sheets/Datastores/Fruit_English/?uid=23&ds=798 (Accessed January 16, 2017)

Dollahite, S., Bremer, V., Crisosto, G.M., Crisosto, C.H., Stover, E., Ferguson, L., 2007. Effects of delayed cooling on two fresh fig cultivars. http://groups.ucanr.org/freshfig/index.cfm.

Karabulut, O.A., Ilhan, K., Arslan, U., Vardar, C., 2009. Evaluation of the use of chlorine dioxide by fogging for decreasing postharvest decay of fig. Postharvest Biol. Technol. 52, 313–315.

Michailides, T.J., Morgan, D.P., Felts, D., Doster, M.A., 2008. Control of decay in Caprifigs and Calimyrna figs with fungicides. Acta Hortic. (798), 269–275.

Piazzolla, F., Amodio, M.L., Colelli, G., in press. Effects of thermal treatments on quality of "Petrelli" figs during storage. Proceedings of the 'VIII International Postharvest Symposium', Cartagena, Spain. 21–24 June 2016.

Piga, A., D'Aquino, S., Agabbio, M., Papoff, C.M., 1998. Short-term nitrose atmosphere exposure extends shelf life of fresh "Niedda longa" fig fruits'. In: Proc. First Intl. Symp. Fig. ISHS Acta Hort, 480pp. 295–299.

Stover, E., Aradhya, M., Ferguson, L., Crisosto, C.H., 2007. The fig: overview of an ancient fruit. HortScience 42, 1083–1087.

Tsantili, E., Karaiskos, G., Pontikis, C., 2003. Storage of fresh figs in low oxygen atmosphere. J. Hort. Sci. Biotechnol. 78, 56–60.

Venditti, T., Molinu, M.G., Dore, A., Hallewin, G.D., Fiori, P., Tedde, M., Agabbio, M., 2005. Treatments with gras compounds to keep fig fruit (*Ficus carica* L.) quality during cold storage. Commun. Agric. Appl. Biol. Sci. 70, 339–343.

Chapter 18.6

Subtropical fruits: Guavas

Gustavo Henrique de Almeida Teixeira
Universidade Estadual Paulista (UNESP), Faculdade de Ciências Agrárias e Veterinárias (FCAV), Departamento de Produção Vegetal, Jaboticabal, SP, Brazil

Quality characteristics

The guava fruit (*Psidium guajava* L.) quality characteristics are dependent on the destination of fruit production, table or processing fruit (Fig. 1). For processing it is used pink guava fruit cultivars and fruit must be fully ripe at harvest with 8–12°Brix, and 0.35%–0.63% total acidity (Pereira and Martinez Júnior, 1986).

Kader (1999) described various guava fruit quality indexes, such as: (i) skin color as a good indicator of ripeness stage; (ii) size and shape might be important in some markets; (iii) fruit must be free from defects, insects, and decay; (iv) firmness and extent of gritty texture due to the presence of stone cells (sclereids); (v) flesh color depends on cultivar and can be white, yellow, pink, or red; (vi) amount of seeds in the flesh (the fewer the better); (vii) aroma intensity; (viii) soluble solids and acidity; (ix) guava as a sources of vitamin C (200–400 mg.100g^{-1}) and vitamin A.

Fruit must be physiologically mature at harvest to achieve proper fruit ripening. Immature fruit will either fail to ripen or become senescent before reaching marketable quality (Singh, 2011).

Main causes of quality loss

The main causes of guava fruit quality loss are associated with its rapid rate of ripening (Pantastico et al., 1979), high susceptibility to diseases (Ko and Kunimoto, 1980), physical damage, chilling injury, diseases, and insects (Singh, 2011).

Guava fruit are considered climacteric fruit, which means the fruit present a sharp increment in respiration rate and ethylene production during maturation phase. Guava has a fast rate of ripening and senesces rapidly (Akamine and

Controlled and Modified Atmospheres for Fresh and Fresh-Cut Produce
https://doi.org/10.1016/B978-0-12-804599-2.00031-4

Goo, 1979; Pantastico et al., 1979). All together the fruit became inadequate to be consumed in a few days. The ripening rate of physiologically mature fruit occurs in a period of 24–48 h if guavas are maintained in ambient. After ripening, the senescence is also rapid, happening in one or two days (Castro and Sigrist, 1988).

According to Botelho (1996), guava climacteric behavior is a varietal trend. Many authors reported this fruit as climacteric (Akamine and Goo, 1979; Brown and Wills, 1983; Mercado-Silva et al., 1998), but others have stated a nonclimacteric pattern (Biale and Barcus, 1970; Medina et al., 1988; Chitarra and Chitarra, 1990; Azzolini et al., 2005), which might be a cultivar trend.

Singh (2011) reported that mechanical damage is one of the major causes of economic losses of guava fruit. Table fruit are always hand-harvested at the mature-green stage and carefully handled to avoid injury, processing fruit should be picked at the firm yellow to half-ripe stage (Paull and Duarte, 2012). Guava fruit skin is very delicate and do not resist to rough handling during harvest and postharvest operations (Singh, 2011). Therefore, guava fruit skin is susceptible to cuts, punctures, bruising, impact, and compression damage.

Guava is susceptible to many postharvest fungal decays and anthracnose infection is the most severe problem encountered during storage (Jacomino, 1999). It is caused by the fungus *Colletotrichum gloeosporioides* (Penz.) Penz. & Saac. which is capable of infecting a wide range of plant species both during their growth and after harvest (Piccinin and Pascholati, 1997).

As guava fruit is sensitive to low temperatures, during cold storage chilling injury can occur in a range of 0°C to 3°C (Bleinroth, 1996). Castro and Sigrist (1988) reported that white guavas stored at 0°C for 4 days and transferred to ambient (20°C) presented brown skin pitting and endocarp browning (chilling injury). After 10 days at 0°C it was observed a complete skin browning. On the other hand, at 5°C the chilling injury symptoms were observed after 16 days storage. Mature-green 'Beautmont' guava fruit stored at 10°C for 11 days did not develop full color and it was observed skin bronzing on some fruit (Reyes and Paull, 1995). Therefore, the severity of chilling injury depends on the storage temperature, the storage period, and the sensitivity of the cultivar to chilling temperatures (Paull, 1999).

Fresh weight loss and/or water loss is another cause of quality loss. During storage for up to 2 to 3 weeks at 7-10°C, and 85%–90% (RH), the fresh weight loss can be as high as 14% (Pantastico et al., 1979). According to Singh (2011), desiccation might result in shrivelling and loss of texture rendering the fruit unacceptable to the consumer.

Optimum storage conditions

As refrigeration is the most effective method to extend shelf-life of various products by slowing down physiological process such as respiration and ethylene production (Kader, 2003), it has been used during guava fruit storage. Guava fruit can be stored for up to 2–3 weeks at 7–10°C, and 85%–90% relative humidity (RH) (Salunkhe and Desai, 1984; Reyes and Paull, 1995; Bleinroth, 1996; Yamashita and Benassi, 2000). However, at such condition the fresh weight loss can be as high as 14% (Pantastico et al., 1979). The effects of low-temperature storage are also dependent on cultivar, harvest season, maturity stages, and storage period (Singh, 2011).

Kader (1999) reported the optimum temperatures for guava fruit storage, 8–10°C for mature-green and partially ripe guavas (storage potential of 2–3 weeks), and 5–8°C for fully ripe guavas (storage potential of 1 week), at a RH of 90%–95%.

CA/MA considerations

The beneficial effects of CA with low O_2 and/or high CO_2 concentrations on the postharvest shelf-life of guava fruit include reduction in respiration rates and ethylene production (Singh and Pal, 2008a; Teixeira and Durigan, 2010), color change, and softening (Teixeira and Durigan, 2010), retention of vitamins, sugars, and organic acids, the inhibition of some physiological disorders (Singh and Paul, 2008), and diseases (Teixeira et al., 2007). However, few studies have been conducted to evaluate the effects of CA storage on guavas fruit (Broughton and Leong, 1979; Castro and Sigrist, 1988; Kader, 1999; Kader, 2003; Singh and Pal, 2008a; Teixeira and Durigan, 2010; Brackmann et al., 2012; Teixeira et al., 2016).

Kader (1999) indicated that 2%–5% oxygen (O_2) levels might delay ripening of mature-green and partially ripe guavas kept at 10°C, and the tolerance to elevated carbon dioxide (CO_2) was not determined (Fig. 2). On the other hand, Kader (2003) recommended atmospheres with 2%–5% O_2 and 0%–1% CO_2 for guavas stored at 5–15°C (see Fig. 2).

Teixeira at al., (2016) reported that 'Pedro Sato' guava fruit stored at 12.2°C for 28 days in atmospheres with 5% O_2 and increasing levels of CO_2 (1%, 5%, 10%, 15%, and 20%) did not present differences in respiration rates, but it was possible to identify detrimental quality modifications in high-CO_2 atmospheres, especially 10%, 15%, and 20% (Fig. 3). Guava fruit stored in 5% O_2 + 20% CO_2 showed a sharp reduction in firmness and increase in soluble pectin especially after 14 days in this atmosphere.

Broughton and Leong (1979) reported that guava fruit do not tolerate CO_2 levels higher than 10%, Castro and Sigrist (1988) reported symptoms of damage caused by high CO_2 during guava fruit storage in atmospheres with 20% CO_2 after three weeks at 12°C. Prolonged CA storage in low O_2 and/or high CO_2 can cause adverse effects such as the accumulation of ethanol and acetaldehyde, development of off-odors and off-flavors, failure to ripen after removal from CA storage, and development of injuries by low O_2 and/or high CO_2 level(s) (Ke et al., 1991).

FIG. 1 Overall aspect of 'Paluma' guava fruit (*Psidium guajava* L.).

FIG. 2 External quality of CA-stored guava fruit.

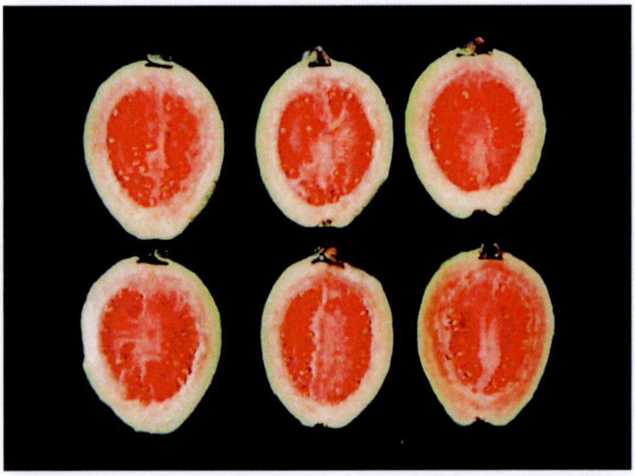

FIG. 3 Internal quality of CA-stored guava fruit.

FIG. 4 Water loss and decay in guava fruit.

CA storage can also be used to control another major cause of guava fruit quality loss which is the high susceptibility to diseases (Ko and Kunimoto, 1980). Teixeira et al. (2007) reported the low oxygen (5%) was the main factor for controlling of postharvest fungal development, especially anthracnose and stile rots (Fig. 4). It was not observed a synergistic effect of CO_2, though. Brackmann et al. (2012) also reported that the 'Paluma' guava fruit stored in different CA conditions presented less decay than normal atmosphere. Prestorage treatment with nitrous oxide (N_2O) has also been reported to inhibited decay development in guava (Qadir and Hashinaga, 2001). These authors reported that guava fruit inoculated with *Rhizopus stolonifer* and held in 80% N_2O for 6 days did not show any disease symptoms.

Guava fruit infecting pests can also be controlled by using CA storage (Yahia, 1997). Singh and Pal (2007) reported that a short-term exposure of guava fruit to very low oxygen (<1%) and high CO_2 (40%) at 40°C for 12 hours had extended the shelf-life by 2–3 days and might be useful for postharvest insect-pest disinfestation purposes.

Modified atmosphere (MA) generated in plastic packages and coatings reduce fresh weight loss and extend the shelf-life of guava fruit (Paull and Duarte, 2011). The responses of guava fruit to short- and long-term MA storage have been studied by many authors, but the majority of the studies related to modified atmosphere packaging (MAP) in guava fruit were done empirically and important information were not provided, such as plastic film thickness, the permeability to O_2, CO_2, and water vapor. Jacomino et al. (2001), Jacomino et al. (2005), and Sahoo et al. (2015) provided such information.

Although many studies can be found regarding the use of MAP in guava fruit, only few modeled the biochemical and physiological aspects (Mangaraj and Goswami, 2011; Mangaraj et al., 2014). Mangaraj and Goswami (2011) developed a respiration rate model based on enzyme kinetics and the Arrhenius equation of 'Baraipur' guava fruit as a function of O_2 and CO_2 concentrations and storage temperature. The respiration rates were found to decrease with storage time due to decrease in O_2 levels and increase in CO_2 in the respirometer. In the enzyme kinetics model, the dependence of respiration rate on O_2 and CO_2 was found to follow the uncompetitive inhibition (Mangaraj and Goswami, 2011).

In order to choose a plastic film, a systematic theoretical design and modeling is needed to establish conditions for the success and benefit of MAP for a particular produce (Mangaraj et al., 2014). By modeling the prediction of quality parameters of guava fruit under various MAP systems and control storage conditions using mathematical model equations, Mangaraj et al. (2014) showed that MA with 5% O_2 and 4% CO_2 was found to be suitable for preservation of 'Baruipur' guava fruit till 26, 19, 13, and 7 days of storage at 10, 15, 20, and 25°C, respectively.

An inadequate MA design can result in an accentuated reduction of O_2 levels and/or increase of CO_2 levels due to the respiration rate process. These adverse aspects will cause the accumulation of ethanol, acetaldehyde, and other volatile organic compounds in the atmosphere (Ke et al., 1991). In order to assess the freshness of guava fruit, Kuswandi et al. (2013) developed a novel on-package color indicator based on bromophenol blue immobilized onto bacterial cellulose membrane. This color indicator works based on pH decrease as the volatile organic compounds are produced gradually in the package headspace. As a consequence, the color of the indicator will change from blue to green as guava fruit ripens.

As water loss is one of the main causes of guava fruit quality loss, coating can be used to reduce fresh weight loss during short- and long-term storage. Numerous studies have been conducted using different coating in guava fruit storage. Carnauba wax has been effective in reducing fresh weight loss (Jacomino et al., 2003; Ribeiro et al., 2005; Zambrano-Zaragoza et al., 2013), but other coating such as chitosan (Oshiro et al., 2012; Aquino et al., 2015; Hong et al., 2012; Soares et al., 2011), gelatin (Oshiro et al., 2012), cassava starch (Aquino et al., 2015; Soares et al., 2011), fatty acid sucrose ester (Combrink et al., 1990), and solid lipid nanoparticles (Zambrano-Zaragoza et al., 2013) have been tested and presented the same good results.

Other postharvest treatments affecting fruit quality rather than CA/MA storage

The ethylene action inhibitor 1-methylcyclopropene (1-MCP) has been used to extend guavas' shelf-life. Cerqueira et al. (2009) observed a decrease in respiration and ethylene production rates in 'Kumagai' guava fruit treated with $300\eta LL^{-1}$ and $900\eta LL^{-1}$ of 1-MCP stored at 22°C for 12 days. The response to 1-MCP might be related to temperature as when 1-MCP was applied to light-dark green 'Pedro Sato' guava fruit stored at 25°C it was observed a reduction in the respiratory activity (900 ηL L^{-1}), but at 10°C this response was just observed after 3 and 5 days of storage (Basseto et al., 2005). During cold storage (10°C) of "Allahabad Safeda' guava fruit treated with 1-MCP (600 ηL L^{-1}) it was not observed a detectable ethylene production rate, and the 1-MCP treatment also inhibited ethylene production and delayed its climacteric peak during ripening at 25°C (Singh and Pal, 2008b).

The guava fruit response to 1-MCP is also dependent on cultivar. Porat et al. (2009) reported that the climacteric cultivar Ben Dov required a high 1-MCP concentration (500 ηL L^{-1}) to delay ripening and to extend postharvest shelf-life. On the other hand, the climacteric cultivar King required a moderate concentration (250 ηL L^{-1}) to inhibit ripening, and the nonclimacteric cultivar Omri required a low concentration of just 100 ηL L^{-1} to inhibit ripening.

Problems under suboptimal conditions

Under suboptimal CA/MA storage conditions the main causes of guava fruit quality loss can be exacerbated, due to the susceptibility to chilling injury and undesirable modifications in the respiration process, ethylene production and consequently guava fruit rapid rate of ripening.

Although CA/MA storage can alleviate the chilling injury symptoms, if guava fruit is exposed to long periods in low temperatures chilling injury can occus in the range of 0°C to 3°C (Bleinroth, 1996). However, the severity of chilling injury will depend on the storage temperature, the CA/MA storage period, and the sensitivity of the cultivar to low temperatures (Paull, 1999).

Regarding atmosphere composition, it seems that guava fruit do not tolerate high levels of CO_2 (>10%), (Broughton and Leong, 1979; Castro and Sigrist, 1988; Teixeira et al., 2016), and prolonged CA/MA storage in low O_2 and/or high CO_2 can cause adverse effects such as the accumulation of ethanol and acetaldehyde, development of off-odors and off-flavors, failure to ripen after removal from CA/MA. Therefore, guava fruit should be stored in atmosphere with 2%–5% O_2 and 0%–1% CO_2 (Kader, 2003).

As guava fruit is susceptible to postharvest fungal decays (Jacomino, 1999) inadequate MAP can favor the development of decay as water vapor can accumulate in the packaging. On the other hand, CA storage has been reported to be very effective in controlling decays (Usall et al., 2000; Fonseca et al., 2004), but the percentage of infected fruit is still elevated (Teixeira et al., 2007; Brackmann et al., 2012).

Comments

The application of CA/MA storage of guava fruit can extend the shelf-life and maintain the quality of the fruit.

Although some studies can be found regarding CA storage, many biochemical responses remain unclear and more investigation is required to fully understand how O_2 and CO_2 affect respiration process and ethylene production. Further investigation is necessary to identify the metabolites responsible for the onset of responses to high CO_2 in guava fruit, and the observed variations of guava cultivars to CA storage also need to be elucidated.

The uses of ethylene antagonists such as 1-methylcyclopropene (1-MCP), and inhibitors of ethylene synthesis, e.g., [S]-trans-2-amino-4-(2-aminoethoxy)-3-butenoic acid hydrochloride (AVG), also have to be evaluated during CA storage as they might contribute to extend guava fruit shelf-life.

Technologies already used during CA storage in other fruits, such as dynamic controlled atmosphere (DCA), ultra-low-oxygen (ULO), and so on, provide the possibility to maintain a much low O_2 level and retard the postharvest changes. As the skin of guava fruit contains abundant chlorophyll, these technologies can be tested with potentialities.

Regarding MA storage, modeling should be used to recommend the ideal film instead of empirical tests.

References

Akamine, E.K., Goo, T., 1979. Respiration and ethylene production in fruits of species and cultivars of Psidium and Eugenia. J. Am. Soc. Hort. Sci. 104, 632–635.

Aquino, A.B., Blank, A.F., Santana, L.C.L.A., 2015. Impact of edible chitosan–cassava starch coatings enriched with *Lippia gracilis* Schauer genotype mixtures on the shelf life of guavas (*Psidium guajava* L.) during storage at room temperature. Food Chem. 171, 108–116.

Azzolini, M., Jacomino, A.P., Bron, I.U., Kluge, R.A., Schiavinato, M.A., 2005. Ripening of 'Pedro Sato' guava: study on its climacteric or non-climacteric nature. Brazilian J. Plant Physiol. (3), 299–306.

Basseto, E., Jacomino, A.P., Pinheiro, A.L., Kluge, R.A., 2005. Delay of ripening of 'Pedro Sato' guava with 1-methylcyclopropene. Postharvest Biol. Technol. 35, 303–308.

Biale, J.B., Barcus, D.E., 1970. Respiratory patterns in tropical fruits of the Amazon basin. Tropical Sci. 12, 93–104.

Bleinroth, E.W., 1996. Colheita e beneficiamento. In: Neto, Á.G. et al., (Ed.), Goiaba para exportação: procedimentos de colheita e pós-colheita. Embrapa-SPI, Brasília, p. 35.

Brackmann, A., Anese, R.O., Both, V., Thewes, F.R., Fronza, D., 2012. Atmosfera controlada para o armazenamento de goiaba cultivar 'Paluma'. Rev. Ceres. 59, 151–156.

Broughton, W.J., Leong, S.F., 1979. Maturation of Malaysian fruits. III. Storage conditions and ripening of guava (*Psidium guajaba* L. var. GU3 and GU4). Mardi Res. Bull. 7, 12–26.

Brown, B.I., Wills, R.B.H., 1983. Postharvest changes in guava fruits of different maturity. Scientia Hort. 19, 237–243.

Castro, J.V., Sigrist, J.M.M., 1988. Matéria prima. In: Medina, J.C. et al., (Ed.), Goiaba: cultivo, matéria-prima, processamento e aspectos econômicos. ITAL, Campinas, pp. 121–140.

Cerqueira, T.S., Jacomino, A.P., Sasaki, F.F., Amorim, L., 2009. Controle do amadurecimento de goiabas 'Kumagai' tratadas com 1-metilciclopropeno. Rev. Bras. Frutic. 31, 687–692.

Chitarra, M.I.F., Chitarra, A.B., 1990. Pós-colheita de frutos e hortaliças: fisiologia e manuseio. ESAL/FAEPE, Lavras. 289p.

Combrink, J.C., Kock, S.L., van Eeden, C.J., 1990. Effect of post-harvest treatment and packaging on the keeping quality of fresh guava fruit. Acta Hort. (275), 639–645.

Fonseca, M.J.O., Leal, N.R., Cenci, S.A., 2004. Padrão de ocorrência de doenças em mamão armazenado sob atmosfera controlada. Rev. Bras. Frutic. 26, 547–549.

Hong, K., Xie, J., Zhang, L., Sun, D., Gong, D., 2012. Effects of chitosan coating on postharvest life and quality of guava (*Psidium guajava* L.) fruit during cold storage. Scientia Hort. 144, 172–178.

Jacomino, A.P., 1999. Conservação de goiabas 'Kumagai' em diferentes temperaturas e materiais de embalagem. PhD Thesis, Escola Superior de Agricultura Luiz de Queiroz, Universidade de São Paulo, Piracicaba: 90p.

Jacomino, A.P., Bron, I.U., Sarantópoulos, I.G.L., Sigrist, J.M.M., 2005. Preservation of cold-stored guavas influenced by package materials. Packag. Technol. Sci. 18, 71–76.

Jacomino, A.P., Kluge, R.A., Sarantópoulos, I.G.L., Sigrist, J.M.M., 2001. Evaluation of plastic packages for guava refrigerated preservation. Packaging Technol. Sci. 14, 11–19.

Jacomino, A.P., Ojeda, R.M., Kluge, R.A., Scarpare Filho, J.A., 2003. Conservação de goiabas tratadas com emulsões de cera de carnaúba. Revista Brasileira de Fruticultura. 25, 401–405.

Kader, A.A., 2003. A summary of CA requirements and recommendations for fruits other than apples and pears. Acta Hort. (600), 737–740.

Ke, D., Rodriguez-Sinobas, L., Kader, A.A., 1991. Physiology and prediction of fruit tolerance to low-oxygen atmospheres. J. Amer. Soc. Hort. Sci. 116, 253–260.

Ko, W.H., Kunimoto, R.K., 1980. Guava fruit firm rot induced by brusing. Hort Science 15, 722–723.

Kuswandi, B., Jayus, C.M., Abdullah, A., Heng, L.K., 2013. Real time on-package freshness indicator for guavas packaging. J. Food Measurement Characterization. 6, 1–4.

Mangaraj, S., Goswami, T.K., 2011. Measurement and modeling of respiration rate of guava (cv. baruipur) for modified atmosphere packaging. International J. Food Properties. 14, 609–628.

Mangaraj, S., Goswami, T.K., Giri, S.K., Joshy, C.C., 2014. Design and development of modified atmosphere packaging system for guava (cv. Baruipur). J. Food Sci. Technol. 51, 2925–2946.

Mercado-Silva, E., Benito-Bautista, P., Garcia-Velasco, M.A., 1998. Fruit development, harvest index and ripening changes of guavas produced in central Mexico. Postharvest Biol. Technol. 13, 143–150.

Oshiro, A.M., Dresch, D.M., Scalon, S.P.Q., 2012. Preservação de goiabas 'Pedro Sato' armazenadas sob atmosfera modificada em refrigeração. Rev. Ciências Agrárias 35, 213–221.

Pantastico, E.B., Chattopadhyay, T.K., Subramanyam, H., 1979. Almacenamiento y operaciones comerciales de almacenaje. In: Pantastico, E.B. (Ed.), Fisiologia de la postrecoleccion, manejo y utilizacion de frutas y hortalizas tropicales y subtropicales. Compañía Editorial Continental, S.A., Mexico, pp. 375–405

Paull, R.E., 1999. Effect of temperature and relative humidity on fresh commodity quality. Postharvest Biol. Technol. 15, 263–277.

Paull, R.E., Duarte, O., 2011. Tropical fruits, 2nd ed. vol. II. CAB International, Wallingford, pp. 91–122.

Paull, R.E., Duarte, O., 2012. Tropical Fruits. vol. 2. CAB International, Cambridge.

Pereira, F.M., Martinez Júnior, H., 1986. Goiabas para industrialização. UNESP, Jaboticabal. 142p.

Piccinin, E., Pascholati, S.F., 1997. Doenças da goiabeira (*Psidium guajava*). In: Kimati, H., Amorin, L., Bergamin Filho, A., Camargo, L.E.A., Rezende, J.A.M. (Eds.), Manual de fitopatologia: volume 2—doenças das plantas cultivadas. Editora Agronômica Ceres Ltda, São Paulo, pp. 450–454.

Qadir, A., Hashinaga, F., 2001. Inhibition of postharvest decay of fruits by nitrous oxide. Postharvest Biol. Technol. 22, 279–283.

Reyes, M.U., Paull, R.E., 1995. Effect of storage temperature and ethylene treatment on guava (*Psidium guajava* L.) fruit ripening. Postharvest Biol. Technol. 6, 357–365.

Ribeiro, V.G., Simão de Assis, J., Silva, F.F., Siqueira, P.P.X., Vilaronga, C.P., 2005. Armazenamento de goiaba 'Paluma' sob refrigeração e em condições ambiente, com e sem tratamento com cera de carnaúba. Rev. Bras. Frutic. 27, 203–206.

Sahoo, N.R., Pandaa, M.K., Balb, L.M., Pala, U.S., Sahoo, D., 2015. Comparative study of MAP and shrink wrap packaging techniques forshelf life extension of fresh guava. Scientia Hort. 182 (1–7).

Salunkhe, D.K., Desai, B.B., 1984. Postharvest Biotechnology of Fruits. CRC Press, Boca Raton, pp. 39–46.

Singh, S.P., 2011. Guava (*Psidium guajava* L.). In: Yahia, E.M. (Ed.), Postharvest biology and technology of tropical and subtropical fruits. In: Cocona to mango, vol. 3. Woodhead Publishing Limited, Cambridge, pp. 213–245.

Singh, S.P., Pal, R.K., 2007. Postharvest fruit fly disinfestation strategies in rainy season guava crop. Acta Hort. (735), 591–596.

Singh, S.P., Pal, R.K., 2008a. Controlled atmosphere storage of guava (*Psidium guajava* L.) fruit. Postharvest Biol. Technol. 47, 296–306.

Singh, S.P., Pal, R.K., 2008b. Response of climacteric-type guava (*Psidium guajava* L.) to postharvest treatment with 1-MCP. Postharvest Biol. Technol. 47, 307–314.

Soares, N.F.F., Silva, D.F.P., Camilloto, G.P., Oliveira, C.P., Pinheiro, N.M., Medeiros, E.A.A., 2011. Antimicrobial edible coating in post-harvest conservation of guava. Rev. Bras. Frutic. E. 281–289.

Teixeira, G.H.A., Cunha Júnior, L.C., Ferraudo, A.S., Durigan, J.F., 2016. Quality of guava (*Psidium guajava* L. cv. Pedro Sato) fruit stored in low-O_2 controlled atmosphere is negatively affected by increasing levels of CO_2. Postharvest Biol. Technol. 111, 62–668.

Teixeira, G.H.A., Durigan, J.E., 2010. Effect of controlled atmospheres with low oxygen levels on extended storage of guava fruit (*Psidium guajava* L. 'Pedro Sato'). Hort Science 45, 918–924.

Teixeira, G.H.A., Durigan, J.F., Santos, L.O., Ogassavara, F.O., Martins, R.N., Cunha Júnior, L.C., Donadon, J.R., 2007. Effect of controlled atmosphere with reducing levels of oxygen on incidence of postharvest diseases in guava (Psidium guajava L. cv. 'Pedro Sato'). In: Int. Congr. Novel Approaches Control Postharvest Dis. Disorders. p. 59.

Usall, J., Teixido, N., Fons, E., Vinas, I., 2000. Biological control of blue mold on apple by a strain of Candida sake under several controlled atmosphere conditions. Inter. J. Food Microbiology. 58, 83–92.

Yamashita, F., Benassi, M.T., 2000. Influência da embalagem de atmosfera modificada e do tratamento com cálcio na cinética de degradação de ácido ascórbico e perda de massa em goiabas (*Psidium guajava* L.). Ciênc. Tecnol. Aliment. 20, 27–31.

Zambrano-Zaragoza, M.L., Mercado-Silva, E., Ramirez-Zamorano, P., Cornejo-Villegas, M.A., 2013. Use of solid lipid nanoparticles (SLNs) in edible coatings to increase guava (*Psidium guajava* L.) shelf-life. Food Research International 51, 946–953.

Botelho, R.V., 1996. Efeito do tratamento pós-colheita com cálcio na ocorrência de antracnose (Colletotrichum gloeosporioides Penz) e no amadurecimento de goiabas (Psidium guajaba L.) Branca de Kumagaia. MSc Thesis, Universidade Estadual Paulista, Botucatu.

Medina, J.C., Castro, J.V., Sigrist, J.M.M., de Martin, Z.J., Kato, K.K., Maia, M.L., Garcia, A.E.B., Fernandes, R.S.S., 1988. Goiaba: cultura, matéria-prima, processamento e aspectos econômicos, second ed. Instituto de Tecnologia de Alimentos, Campinas Série Frutas Tropicais 6.

Kader, A.A., 1999. Guava produce facts. University of California, Davis, Perishables Handling Quarterly 97, 19–20.

Porat, R., Weiss, B., Zipori, I., Dag, A., 2009. Postharvest longevity and responsiveness of guava varieties with distinctive climacteric behaviors to 1-methylcyclopropene. Hor Technology 19, 580–585.

Chapter 18.7

Subtropical fruits: Kiwifruit

Jeremy Burdon
The New Zealand Institute for Plant & Food Research Limited, Auckland, New Zealand

Quality characteristics

Typical of cultivar: flesh color, shape, size, flavor.
Main cultivar is "Hayward"—green-flesh.
Other cultivars: "Zesy002," "Jintao"—yellow-flesh.
"Hongyang"—yellow/red-flesh.
Harvest maturity—specific to meet marketing requirement.
Firmness—appropriate to position in the supply chain.
Free from major defects—including cuts, rots.
Cosmetic quality to meet the commercial grade class.

Controlled and Modified Atmospheres for Fresh and Fresh-Cut Produce
https://doi.org/10.1016/B978-0-12-804599-2.00032-6

Kiwifruit in international trade are larger-fruited species of *Actinidia chinensis* var. *deliciosa* ("Hayward") and *A. chinensis* var. *chinensis* ("Zesy002," "Jintao," "Hongyang").

Kiwifruit are harvested in a mature but unripe state.

The hard, unripe fruit at harvest must ripen to a soft, melting texture when eating ripe.

The harvest maturity should match the marketing requirement—with immediate sale and long storage (4–5 months) being the extremes.

For early harvested fruit, ripening may require the use of ethylene which helps to induce softening and coordinates the batch of fruit. These fruit may be chilling sensitive and should not be long stored.

Long storage is achieved through matching a more advanced fruit at harvest (but not over-mature) with temperature management and either air or controlled atmosphere storage conditions.

A key element of maturation is the reduction in chilling susceptibility of the fruit as it matures.

Fruit firmness, measured by penetrometer (commonly with a 7.9-mm Effegi^{TM} probe), is the main check on fruit quality during commercial storage and distribution, along with visual inspection for defects.

Main causes of quality loss

Physical damage

The susceptibility to physical damage depends on fruit firmness and may be cultivar specific:

Impact:

external symptoms may include a puncture or lifted fruit skin, which may lead to a rot. In "Hort16A," skin discoloration may occur around the point of impact.

internal symptoms may include a white area in the flesh where cells are broken and starch does not break down; may have water soaking (Fig. 1).

(A) (B)

FIG. 1 Physical damage in "Hayward": (A) impact, (B) compression.

Compression—symptoms are water-soaked areas of the outer pericarp, usually on opposite sides of the fruit. Commonly caused by fruit being allowed to soften too far in bulk storage, or fruit poorly packed into trays/packs.

Physiological disorders

Chilling—all cultivars appear susceptible. Symptoms can include a granular appearance to the outer pericarp and/or water soaking to any of the outer pericarp, inner pericarp, and core. In some cultivars, external symptoms may include skin darkening at the stylar end.

Hard cores—generally in less mature fruit, including fruit harvested early and ripened without ethylene.

Over-soft—caused by incorrect harvest maturity, temperature management or storage conditions (in particular the presence of ethylene), fruit stored for too long. The exact definition of an over-soft fruit depends on position in the supply chain and any commercial specification.

Shrivel—caused by excessive water loss, usually as the fruit softens during prolonged storage.

Other—dependent on vine management/growing conditions, e.g., physiological pitting, lenticel spots which may be exacerbated, but not created by, low temperatures (Fig. 2).

Rots

Off orchard: may include *Botryosphaeria* sp., *Pseudocercospora* sp., *Phomopsis* sp.—depending on seasonal/environmental conditions, any spray program and whether fruit ripen on the vine.

In storage: generally categorized initially by position on fruit:

Stem end rot—particularly *Botrytis cinerea* in "Hayward," others include *Cylindrocarpon* sp., *Phomopsis* sp.

(A) (B) (C)

FIG. 2 Physiological damage: (A) "Hayward" chilling injury, (B) "Zesy002" chilling injury, (C) "Hort16A" lenticel spots.

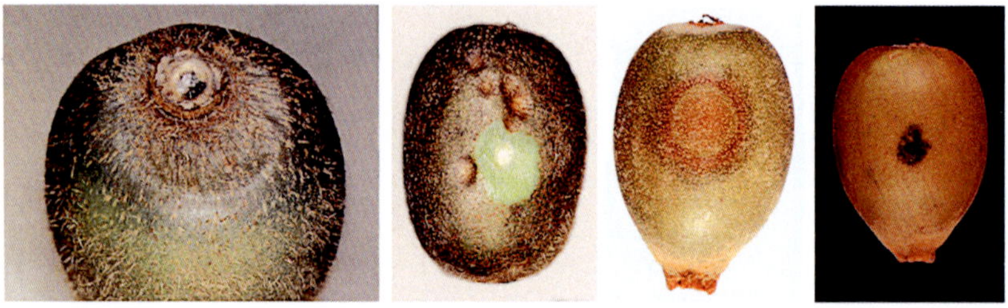

FIG. 3 Rots. (A) "Hayward" Stem-end rot, (B) "Hayward" *Cryptosporiopsis actinidiae* (fungal pitting), (C) "Hort16A" *Botryosphaeria*, (D) "Hort16A" storage stain.

Body rots—largely ripe rots including *Cryptosporiopsis actinidiae* (*Neofabraea actinidiae*), *Botryosphaeria* sp., *Phomopsis* sp.

Opportunistic rots (*B. cinerea*, *Phoma* sp., *Penicillium*) indicative of damage to the fruit and sometimes referred to as wound rots. Also commonly rots on chill-damaged fruit.

Storage stain—growth of rots such as *Alternaria* sp. on the fruit surface (Fig. 3).

Flavor

Consumer liking of kiwifruit is largely associated with the ripe fruit soluble solids content (SSC)—too low dry matter (starch) at harvest leaves the fruit with a low SSC when ripe. Ripening fruit too quickly with ethylene can leave fruit soft but without full starch conversion to SSC.

Flesh colour

Fruit marketed as yellow-fleshed fruit need to degreen before harvest, or at least have started and have an off-vine degreening program. Degreening may be temperature sensitive depending on cultivar, and fruit harvested before fully degreened may therefore remain green at storage temperatures.

The occurrence of red pigmentation in fruit of some cultivars depends on growing conditions and may be inhibited by high temperatures. A lack of red pigmentation cannot be recovered after harvest.

Optimum storage conditions

Optimum temperature

Cultivar-specific, general guidelines: "Hayward" ~ 0°C, "Zesy002" and "Jintao" ~ 1°C.

An optimum temperature depends on cultivar and harvest maturity.

Maturity influences the susceptibility to chilling damage—more mature fruit are less susceptible.
Temperature and storage duration determine the expression of chilling symptoms.
Temperature management has two components—cooling rate and final storage temperature.
Faster cooling (e.g., immediately after harvest, forced air and directly to storage temperature) and lower storage temperatures tend to exacerbate chilling injury.
Storage temperature can be reduced slightly as fruit ripen in storage.

Relative humidity

Store relative humidity (RH) is often recommended as 90-95% to minimize water loss from the fruit.
For fruit stored in bulk and exposed to the store atmosphere, the store RH should be as high as possible to reduce the water loss. To minimize water loss from the fruit, and therefore to minimize the risk of shrivel, other aspects to be considered include: having a full room, air flow just sufficient to maintain fruit temperature, and refrigeration equipment which will extract little water—a large evaporator with a small temperature differential.
A high store RH is not essential for packed fruit (in a polybag in a box) since the high in-pack humidity maintains the fruit condition. More importantly, packages may soften excessively in a high RH environment. Thus for packed fruit, a store atmosphere of 85-90% may be acceptable.

Controlled atmosphere (CA)/modified atmosphere (MA) considerations

Kiwifruit may be stored for several months and CA may be used both before and during export; therefore both land-based stores and CA-capable shipping containers are relevant.
Land-based CA stores operate to maintain predetermined atmospheres whereby both oxygen and carbon dioxide concentrations are monitored and maintained.
Shipping containers are available from a range of companies and operate with different degrees of sophistication—either actively maintaining both O_2 and CO_2, or less precise control with ventilation on demand dependent on exceeding predetermined low O_2 or high CO_2 thresholds.
Kiwifruit soften in response to ethylene. CA stores should be monitored for ethylene and scrubbers should be fitted to remove any exogenous ethylene from the store atmosphere. Fruit may also be treated with the ethylene action inhibitor SmartFresh™ (AgroFresh) at the start of storage.

Any benefits from newer dynamic CA technologies, in which oxygen concentrations are established based on a fruit response, are still to be proven for kiwifruit.
To maximize the benefits of CA, the CA conditions should be established as soon as practical after harvest, as delays greater than one week may reduce the benefit to firmness retention. The reality is that it may take 3-4 days to fill a CA room and the fruit should be cooled before CA establishment.
It is essential to use well-matured fruit, with a low risk of chilling injury, for long-term storage in CA.
The use of MA depends on the development of a film with an appropriate permeability to match the fruit respiration at a given temperature to the required atmosphere.
Successful use of MA is dependent on good temperature management.
MA packs may also need an ethylene absorbent in the pack to limit the risk from ethylene (disordered and over-soft fruit produce ethylene).

Benefit

Low O_2 and high CO_2 concentrations in CA or MA can slow ripening/improve firmness retention.
There is a synergistic effect of low O_2 and high CO_2 on firmness retention.
There is a synergistic effect of CA with SmartFresh on firmness retention.
For green-fleshed fruit, CA can improve color retention during prolonged storage.
CA may reduce rots—delaying fruit ripening will delay the expression of ripe rots.
Disinfestation under CA has been researched but is not used commercially.

Tolerances

CA/MA regimes often recommended for kiwifruit are in the region of 2% O_2 and 5% CO_2, or given as a range of 1%–3% O_2 and 2%–5% CO_2. These have been determined largely for "Hayward" fruit, although similar atmospheres seem to work for newer cultivars (including "Hort16A," "Zesy002," "Hongyang").

Injuries

Too low O_2/too high CO_2 can lead to abnormal softening, flesh discoloration, and off-flavors.
Other injuries tend to be associated with the way in which CA is used, e.g., increased chilling damage if fruit of incorrect maturity are cooled too quickly before CA establishment.
Growing region-specific problems: in New Zealand high CO_2 can increase the incidence of the ripe rot *Cryptosporiopsis* (also known as fungal pitting) which may be managed by using a lower CO_2 concentration or by slowing the accumulation of CO_2 in storage.

Problems under suboptimal conditions

Suboptimal CA/MA conditions may include:

Too high O_2/too low CO_2

Reduced efficacy in maintaining firmness. May lead to compression damage, as CA fruit are generally stored in bulk.

May result in fruit too soft for grading and/or marketing and riper fruit are more prone to rots.

Too low O_2

Fruit become anaerobic—causing abnormal softening, discoloration, and off-flavors.

Too high CO_2

May stimulate anaerobic metabolism.

May exacerbate some types of rots, e.g., *C. actinidiae*.

May cause water soaking of the inner pericarp and a very firm core.

In addition to O_2/CO_2 conditions:

Ethylene in the store atmosphere will cause fruit to soften rapidly.

Fruit harvested too early, harshly cooled and stored for too long may develop chilling injury.

Poorly designed refrigeration, high air flows, and only partially filled rooms may lead to high water loss and an increased incidence of shrivel in exposed bulk-stored fruit.

Comments

CA storage of kiwifruit is an established part of commercial practice in some regions.

Until recently, global trade was limited to fruit of the "Hayward" cultivar.

Fruit of *A. chinensis* var. *chinensis* cultivars seem to be slightly more chill sensitive than "Hayward" fruit.

Commercial coolstore operators have become used to the long storage life of "Hayward" kiwifruit. In new cultivars storage life may be shorter, and CA more relevant to extend the marketing window.

CA is used to prolong fruit storage and it is essential to use only well-matured fruit. Much of the published harvest recommendations for kiwifruit refer only to "Hayward" and then tend to go no further than the 6.2°Brix SSC harvest index introduced about 30 years ago in New Zealand. This was only ever a minimum index, and more mature fruit (e.g., 8-10°Brix) should be used for CA storage. For yellow-fleshed cultivars, harvest indices should also factor in the requirement for flesh color.

After CA storage, softening may resume at a rate equivalent to fruit at the same firmness in air storage. However, some CA operators claim an even more rapid softening may occur.

The recent spread of *Pseudomonas syringae* pv. *actinidiae* (Psa) in the major kiwifruit-growing regions is a serious impediment to fruit quality. Psa may affect vines to different degrees—from asymptomatic to dead vines. The effect of Psa on fruit may be from nothing noticeable to there being no fruit. Fruit from affected vines should not be put into the commercial supply chain, including for CA storage.

There is an increasing interest in kiwiberry fruit. These are small-fruited (5–15 g) cultivars mostly of *A. arguta* which are marketed alongside berryfruit. Their biology and commercial possibilities differ from those of the main large-fruited kiwifruit, yet they seem to respond well to CA storage conditions.

Further reading

Arpaia, M.L., Mitchell, F.G., Kader, A.A., Mayer, G., 1985. Effects of 2% O_2 and varying concentrations of CO_2 with or without C_2H_4 on the storage performance of kiwifruit. J. Amer. Soc. Hortic. Sci. 110, 200–203.

Arpaia, M.L., Mitchell, F.G., Mayer, G., Kader, A.A., 1984. Effects of delays in establishing controlled atmospheres on kiwifruit softening during and following storage. J. Amer. Soc. Hortic. Sci. 109, 768–770.

Beever, D.J., Hopkirk, G., 1990. Fruit development and fruit physiology. In: Warrington, I.J., Weston, G.C. (Eds.), Kiwifruit: Science and Management. Ray Richards Publisher/New Zealand Society for Horticultural Science, pp. 97–126.

Brigati, S., Pratella, G.C., Bassi, R., 1989. CA and low oxygen storage of kiwifruit: effects on ripening and diseases. In: Proceedings of the fifth international controlled atmosphere research conference, Wenatchee, Washington, USA, 14–16 June, 1989, publ. 1991, Vol. 2, 41–48.

Burdon, J., 2015. Soluble solids revisited: a maturity or harvest index for kiwifruit. Acta Hortic. 1096, 257–266.

Burdon, J., Lallu, N., 2011. Kiwifruit (*Actinidia* spp.). In: Yahia, E.M. (Ed.), Postharvest biology and technology of tropical and subtropical fruit. Volume 3: Cocona to mango. Woodhead Publishing Limited, Sawston, Cambridge, UK, pp. 326–360.

Lallu, N., 1997. Low temperature breakdown in kiwifruit. Acta Hortic. 444, 579–585.

Lallu, N., Burdon, J., Yearsley, C.W., Billing, D., 2003. Commercial practices used for controlled atmosphere storage of 'Hayward' kiwifruit. Acta Hortic. 610, 245–251.

McDonald, B., 1990. Precooling, storage, and transport of kiwifruit. In: Warrington, I.J., Weston, G.C. (Eds.), Kiwifruit: Science and Management. Ray Richards Publisher/New Zealand Society for Horticultural Science, pp. 429–459.

McDonald, B., Harman, J.E., 1982. Controlled-atmosphere storage of kiwifruit. I. Effect on fruit firmness and storage life. Sci. Hortic. 17, 113–123.

Manning, M.A., Lallu, N., 1997. Fungal diseases of kiwifruit stored in controlled atmosphere conditions in New Zealand. Acta Hortic. 444, 725–732.

Organisation for Economic Co-operation and Development, 1992. International Standardisation of Fruit and Vegetables. Kiwifruit. OECD Publications Service, Paris, France.

Regiroli, G., Vriends, P., 2007. SmartFresh^{SM} (1-methylcyclopropene) benefits for kiwifruit. Acta Hortic. 753 (753), 745.

Thompson, A.K., 2010. Controlled Atmosphere Storage of Fruits & Vegetables, second ed. CAB International, Wallingford, Oxford, UK.

Tonini, G., Barberini, K., Bassi, F., Proni, R., 1999. Effects of new curing and controlled atmosphere storage technology on Botrytis rots and flesh firmness in kiwifruit. Acta Hortic. 498, 285–291.

Chapter 18.8

Subtropical fruits: Melons

Manuel Madrid
Fruit Profits, Valencia, Spain

Quality characteristics

There are many cultivars of melon (*Cucumis melo L.*) and availability is dependent upon the geographical region and historical preferences.

The image shows examples of melon types marketed in Europe and North America. There are many different types of melons with markedly differing flavor, texture, and appearance characteristics, and the popularity of any given variety is highly dependent on country of origin. Consumers prefer local types in general.

Controlled and Modified Atmospheres for Fresh and Fresh-Cut Produce
https://doi.org/10.1016/B978-0-12-804599-2.00033-8

In Europe, the most common types of melons consumed are as follows:

- Galia (netted skin and green flesh, round),
- Yellow (yellow skin and white flesh, oval),
- Piel de sapo (green shell and white flesh, oval),
- Sugar Baby (yellow skin and orange flesh, oval).

In North America, the most common types of melons are as follows:

- Honeydew (green skin and green flesh, round),
- Cantaloupe (netted skin and orange flesh, round),
- Orange flesh (green striped skin and orange flesh, round).

Each of these types behaves very differently in terms of shelf life, recommended temperature, and ideal storage conditions. It is important to observe the individual recommendations per melon type.

Main quality characteristics:

- Skin: In nonnetted melon types, the skin must be smooth and shiny, without bumpy texture. In striped varieties, the stripe must be clearly defined and uniform on the whole surface of the fruit.
- Net: In the netted varieties (North American cantaloupe and Galia) the net must be well formed, clean, not damaged, light in color. The space between the net must be clean, without brown or dark areas.
- Shape: In the round types, the shape must be round without bumps or sunken areas. In the oval types, the shape must be oval and symmetrical.
- External color: Surface should be clean without dark areas. Areas between nets should be clean and light in color, never brown or dark.

Type of melon	Skin color
Yellow	Intense yellow with no green reflections
Galia	Pale yellow, with no green tonality
EU Cantaloupe	Grayish green background, dark green stripes
Piel de sapo	Dark green background, with black reflections
Sugar Baby	Bright yellow, no dark areas
Honeydew	Light green, no dark or brownish spots
N. Am. Cantaloupe	Well developed net, pale yellow or orange
Orange flesh	Grayish green background, dark green stripes

- Internal quality: Juicy flesh that gives off juice when bitten, crunchy, and not jellylike. Aromatic. No off-flavors. Not under ripe (cucumber taste).

Type of melon	Minimum sugar (°Brix)	Flesh colour
Yellow	10	Pale yellow/ whitish
Sugar Baby Gold	13	Light orange

Continued

Type of melon	Minimum sugar (°Brix)	Flesh colour
Galia	10 (extra sweet ≥ 12)	Light green at the border to yellow next to the seeds (center)
EU Cantaloupe	12 (extra sweet ≥ 14)	Deep orange
Piel de sapo	12	White / Very pale green/light mber
Honeydew	10	Light green, waxy
N. Am. Cantaloupe	11	Light orange
Orange flesh	11	Deep orange

- Appearance: Free from pests, diseases, cuts, and scars. Free from moisture and foreign smell and taste. Free from any foreign matter. Absence of finger marks.
- Peduncle: The peduncle or stem-end should be green, with fresh appearance and without dehydrations, and not longer than 1 cm.
- Taste: Melons should be picked ripe; failure to do so will result in dull tasting fruit. The choice of variety is important when considering taste.

Main causes of quality loss

There are several causes of quality loss in melons:

- Physical damage during picking and handling in the field: Commercial melons are hand-picked in the field and directly packed into crates or lugs. Handling should be very careful. Too much pressure while picking will result in damaged skin. Drops and bumps should be assiduously avoided. Use padded materials on the sides and bottom of bins. Avoid piling too high in field container when collecting melons and transporting to the packing station.
- Exposure to sunlight: Melon should never be exposed to direct sunlight after picking. Keep bins in a shaded place while waiting for transport to the cold storage facility.
- Wet fruit from the field (from dew or rain): Melon should always be harvested dry. Sometimes melons are wet due to dew or recent rain. In that case, harvest should be postponed until melons are dry. In some countries, melons are grown under plastic tunnels (a.k.a. 'hoops'). In this case, dryness of the fruit and sanitation are improved as well as earliness.
- Poor field sanitation: The level of decay pathogen inoculum in the field should be kept low by removing damaged and decayed fruit, and spraying with authorized fungicides only, respecting reentry periods and manufacturer labels. Failure to do so will result in high incidence of postharvest diseases.

- Delays in cooling: Melon should be cooled to their optimum temperature within 4-6 hours of picking. Failure to do so will result in reduced shelf life and higher presence of mold in storage.
- Cold chain interruptions: The cold chain of melons should be maintained at all times. Failure to maintain cold chain in precooled melons will result in condensation of ambient moisture on the skin, causing severe proliferation of fungi and skin stains.
- Chilling injury: Melons should never be stored below the recommended temperature (see below). They are very sensitive to chilling injury. Symptoms of chilling injury in melons are sunken areas, dark stains in skin, and fungal growth.
- Fertilization and its effect on melon quality: Fertilization has a significant impact on melon quality and shelf life.
  - Calcium content: Applying weekly soluble, systemic calcium applications during the growth of the melon helps improve skin and flesh firmness, shelf life, and shine.
  - Nitrogen content: Avoid high nitrogen fertilization during flowering and fruiting as it is detrimental to fruit quality and shelf life.
  - Potassium content: Potassium fertilization during fruit growth helps to improve fruit quality and sweetness.

Optimum storage conditions

Optimum storage temperature (°C) varies depending on melon type:

European melon types:

- Galia: 7°C
- Yellow: 8–10°C
- Cantaloupe: 8–10°C
- Piel de sapo: 14–16°C
- Sugar Baby: 8–10°C

North American melon types:

- Honeydew: 8–10°C
- Cantaloupe: 3–4°C
- Orange flesh: 7°C

Melon should be cooled down to the optimum temperature within 4–6h after harvest. Hydrocooling or forced air cooling is the most common cooling method. To avoid risk of damaging the fruit, cooling tunnel temperature setpoint should always be set at the recommended temperature, because thermostats usually present a temperature oscillation.

If hydrocooling is used, melons should be free of moisture after cooling, using cold air fans to remove all moisture

Relative humidity should be 85%–90%. Avoid condensation and free moisture on the surface of the melon, free moisture on the skin stains the melon and promotes fungus growth.

CA/MA considerations

Melons react well to modified atmospheres for shelf life extension. Atmospheres of 2%–3% O_2 help extend the shelf life of the product. Additionally to low O_2, an atmosphere rich in CO_2 ($\sim$10%) helps prevent fungal growth. However, CA is not commonly used for short travel distances because the benefits do not justify the costs when the duration of transport is short. Also MA bags have not been extensively used due to condensation problems. MA is used as a standard procedure in South America for shipment melons to Europe. The most common bags are nylon-based microperforated bags. They provide a fair moisture barrier and minimize condensation. However, they do not provide all the benefits of low O_2 or high CO_2.

This MAP system is not used in Europe or other continents on a regular basis, due to the proximity of the production areas to the market and the short postharvest interval.

Benefits and injuries of CA/MA control

Melons rarely develop CA injury, because CA is not extensively used. Disease, dehydration, and overmaturity are more common problems and related symptoms typically show up before CA injury can be observed.

Problems under suboptimal conditions

Mold: Melons being grown on the ground are very susceptible to a variety of fungal infections. The most common postharvest disease organisms causing decay of melon are *Fusarium*, *Rhizopus*, *Botrytis*, and *Alternaria* (Snowdon, 1990). It is essential to keep good sanitation and low inoculum in the field. It is also essential to avoid condensation on the fruit following harvest.

Blossom-end rot: softening of the blossom end of the melon. It is caused by a calcium deficiency and is exacerbated by hot and humid growing conditions. Regular use of foliar calcium sprays prevents this problem.

Chilling injury: Melons are very susceptible to chilling injury. Storage at temperatures below the recommended temperatures induces rots, sunken areas, and brownish skin spots. These symptoms appear more rapidly when the melons are displayed at elevated temperatures (i.e., ambient or room) at retail.

Sunscald: Exposing the melons to direct sunlight for a long period causes bronzing or browning of skin. It is very important to maintain adequate canopy foliage to protect growing melons. Harvested melons should be held in the shade prior to transport from the field.

Comments

It is key to harvest melons at the proper maturity to achieve high sugars and flavor. Do not harvest immature fruit. Gentle handling at harvest, prompt cooling, transportation at optimum temperature per type of melon, and sanitation in the field are the keys to successful storage and distribution. Avoiding moisture on the melon surface is key for shelf life and quality preservation.

References

Snowdon, A., 1990. A color atlas of post-harvest diseases. Volume 1. General introduction and fruits. CRC Press, Boca Raton, FL.

Further reading

Gross, K.C., Wang, C.Y., Saltveit, M.E., 2016. The commercial storage of fruits, vegetables, and florist and nursery stocks. Agriculture Handbook Number 66, U.S. Department of Agriculture, Washington, DC. http://ucanr.edu/datastoreFiles/234-2927.pdf.

Photos

Open air melon stand in Costa Rica—subject to high ambient temperatures. Melons should be stored and displayed in a cool environment.

Galia melons. For good quality melons, and especially netted melons, sanitation in the field is essential. Good field drainage and preventing contact with surfaces having free moisture are keys to disease- and defect-free melons. A good balance between foliage and fruit avoids sunburn.

Chapter 18.9

Subtropical fruits: Persimmons

Ruth Ben-Arie

Emeritus, The Institute of Postharvest and Food Sciences, ARO, Volcani Center, Bet Dagan, Israel

Quality characteristics

Persimmon cultivars have been classified into four categories based on two characteristics, namely, pollination type and astringency of mature fruit (Woolf and Ben-Arie, 2011). Astringency disappears naturally only when the fruit become excessively soft. It can be removed artificially by hypoxia.

Controlled and Modified Atmospheres for Fresh and Fresh-Cut Produce
https://doi.org/10.1016/B978-0-12-804599-2.00034-X

Main causes of quality loss

- Decay caused primarily by *Alternaria alternata*.
- Chilling injury in the form of pulp jellification of certain cvs., when stored at 0°C.
- Excessive softening.
- Flesh discoloration.

Optimum storage conditions

Optimum temperature

Temperature between −1°C and +2°C, depending on sensitivity to chilling injury.

CA/MA considerations

Depend upon the group to which the cultivar belongs. For nonastringent cvs. the important factors are softening, chilling injury, and decay. For astringent cultivars, a major factor is also removal of astringency, for which acetaldehyde production is required, with the added danger of internal injury.

Nonastringent cvs. (e.g. Fuyu)

Since the turn of the century, R&D have focused on MAP, testing different films to obtain various gas combinations, and the results have supported the earlier findings summarized by Testoni (2002), although the current recommended O_2 levels are at the lower end of the range: 3%–8% O_2, 3%–10% CO_2 (Bae et al., 2010; Cia et al., 2006; Kim et al., 2010). More recently, attempts have been made to solve the problems arising after the fruit is removed from cold storage. When the fruit remains in the package, the rise in temperature causes increased respiration accompanied by rapid changes in the composition of the atmosphere (reduced O_2 and increased CO_2), leading to the accumulation of ethanol and the formation of off-flavors. On the other hand, when the fruit is removed from the package, there is a rapid loss of firmness and deterioration in quality. Two directions have been investigated: (a) prestorage treatment with 1-MCP (Choi et al., 2013); (b) the use of two layers of packaging consisting of a micro-perforated inner film for individual fruit wrapping and an outer bulk pack, each with different permeability characteristics at the low and high temperatures examined (Jeong et al., 2013). Both methods enable the retention of fruit firmness during shelf life after approximately 3 months' cold storage.

- Benefits—delayed softening, prevention of peel browning, flesh discoloration and chilling injury, reduction of decay.
- Tolerances—Minimum O_2—0.5%, maximum CO_2—15%,
- Injuries—Low O_2 induces stem-end browning, high CO_2 causes internal browning.

Astringent cvs. (e.g., Triumph, Rojo Brillante)

A major advantage of CA for astringent persimmon cvs. is that prolonged cold storage at O_2 levels below 5% enables loss of astringency without the need to treat the fruit upon removal from storage. Additional benefits are maintenance of fruit firmness and suspension of decay development. Thus CA at 1.5%–3% of both O_2 and CO_2 can prolong the storage life of Triumph persimmons by ca. 2 months at −1°C (Guelfat-Reich and Ben-Arie, 1976). The optimum gas composition for Rojo Brillante has yet to be determined, as flesh browning occurred at 4%–5% O_2 (Arnal et al., 2008; Besada et al., 2014). Whereas CO_2 is beneficial for retarding decay development on Triumph (Guelfat-Reich et al., 1975), it appears to be detrimental to Rojo Brillante (Arnal et al., 2008). After removal of the fruit from CA it may soften rapidly with an immediate burst in decay development, so that the shelf life under ambient conditions can be very short. The solution to fruit softening is the application of 500–600 $nl\,L^{-1}$ 1-MCP for 24 h as a prestorage or poststorage treatment (Tsviling et al., 2003; Besada et al., 2014). However, 1-MCP has no beneficial effect on the development of Alternaria black spot decay at ambient temperature. This may be achieved by maintaining a modified atmosphere around the fruit until it is marketed. MAP has the advantage over CA, when combined with 1-MCP prior to storage, of preserving fruit quality during shelf life, in addition to the removal of astringency (Oz, 2011).

The following protocol for long-term storage of 'Triumph' persimmons has been developed in Israel (Ben-Arie et al., 2011):

1. 1-MCP treatment—600 $nl\,L^{-1}$ at 20°C for 24 h.
2. Sort fruit and seal (3–30 fruit) in 0.06 mm LDPE bags, in either N_2 or in vacuum.
3. Hold at 20°C or at ambient temperature for 48 h. Although this step is to ensure total removal of astringency, it also enables accumulation of CO_2 to a level that inhibits the development of Alternaria decay in storage.
4. Store at −1°C for up to 6 months.
5. Examine upon removal from storage. If the fruit is free of decay it can be marketed in the pack, as is. If there is an occasional decayed fruit, the intact fruit should be repacked in sealed LDPE and can be held at ambient temperature for at least one week.

Problems under suboptimal conditions

There are two problems that tend to arise causing suboptimal atmospheric conditions to develop that may result in loss of fruit quality.

1. Incomplete sealing or punctures in the LDPE bags allowing increases in O_2 levels, which allow fruit to soften and decay while still in storage. If the bags

are packed under vacuum instead of nitrogen, the broken seals can be detected at the end of the 48-h period prior to cold storage.

2. Temperature fluctuations during storage. An increase in temperature inducing anaerobiosis and/or CO_2 accumulation may induce internal and external browning (Figs. 1–3).

FIG. 1 'Triumph' persimmons at removal from storage after 4 months at −1°C.

FIG. 2 1-MCP treated 'Triumph', held for 48h in MAP at 20°C before storage for 5 months at −1°C, after 1 week at 20°C, with (left) or without (right) the pack.

FIG. 3 'Triumph' persimmons upon removal from MAP storage at −1°C for 22 weeks.

Comments

The permeability of the LDPE is critical. If the film is too thick and the CO_2 exceeds 10%, there will be both external and internal damage.

References

Arnal, L., Besada, C., Navarro, P., Salvador, A., 2008. Effect of controlled atmospheres on maintaining quality of persimmon fruit cv. "Rojo Brillante" J. Food Sci. 73, S26–S30.

Bae, M.S., Kim, H.J., An, D.S., Ahn, G.H., Lee, S.C., Lee, D.S., 2010. Comparison of two packaging conditions in preserving antioxidant activity of sweet persimmons. J. Food, Agric. Environ. 8, 192–195.

Ben-Arie, R., Tsviling, A., Nerya, O., Gizis, A., Gamrasni, D., Azani, L., 2011. Extending the storage life of 'Triumph' persimmons. Alon HaNotea 65, 34–38 (in Hebrew).

Besada, C., Novillo, P., Navarro, P., Salvador, A., 2014. Effect of a low oxygen atmosphere combined with 1-MCP pretreatment on preserving the quality of 'Rojo Brillante' and 'Triumph' persimmon during cold storage. Sci. Hortic. 179, 51–58.

Choi, H.S., Jung, S.K., Kim, Y.K., 2013. Storage ability of non-astringent 'Fuyu' persimmon fruit is affected by various concentrations of 1-methylcyclopropene and/or modified atmosphere packaging. J. Hortic Sci. Biotech. 88, 195–200.

Cia, P., Benato, E.A., Sigrist, J.M.M., Sarantopóulos, C., Oliveira, L.M., Padula, M., 2006. Modified atmosphere packaging for extending the storage life of 'Fuyu' persimmon. Postharv. Biol. Technol. 42, 228–234.

Guelfat-Reich, S., Ben-Arie, R., 1976. CA storage of 'Triumph' persimmons. In: Proceedings of the XIV Congress of the International Institute of, Melbourne Refrigeration, pp. 59–63.

Guelfat-Reich, S., Ben-Arie, R., Metal, N., 1975. Effect of CO_2 during and following storage on removal of astringency and keeping quality of 'Triumph' persimmons. J. Am. Soc. Horticult. Sci. 100, 95–98.

Jeong, M., An, D.S., Ahn, G.H., Lee, D.S., 2013. Master packaging system for sweet persimmon applicable to produce supply chains. Postharv. Biol. Technol. 86, 141–146.

Kim, H.J., An, D.S., Ahn, G.H., LEE, D.S., 2010. Respiration rate of sweet persimmon fruit depending on cultivar, harvest date and temperature. J. Food, Agric. Environ. 8, 74–76.

Oz, A.T., 2011. Combined effects of 1-methyl cyclopropene (1-MCP) and modified atmosphere packaging (MAP) on different ripening stages of persimmon fruit during storage. Afr. J. Biotechnol. 10, 807–814.

Testoni, A., 2002, Postharvest and processing of persimmon fruit. Proceedings of the First Mediterranean Symposium on persimmon, 2002 CIHEAM, International Centre for Advanced Mediterranean Agronomic studies.

Tsviling, A., Nerya, O., Gizis, A., Sharabi-Nov, A., Ben-Arie, R., 2003. Extending the shelf-life of 'Triumph' persimmons after storage, with 1-MCP. Acta Horticult. (599), 53–58.

Woolf, A.B., Ben-Arie, R., 2011. Persimmon (Diospyros kaki L.). In: Yahia, E.M. (Ed.), Postharvest Biology and Technology of Tropical and Subtropical Fruits. In: 4, Woodhead Publishing, pp. 166–193.

Chapter 18.10

Subtropical fruits: Prickly pear

José Angel Guerrero-Beltrán
Food and Chemical Engineering Department, Universidad de las Américas Puebla, Cholula, Puebla, Mexico

Quality characteristics

Opuntia is a genus of the cactus family, Cactaceae that grows in dry and hot environments. Its fruits are variously referred to as tuna, prickly pear, cactus pear, or Indian fig and are harvested around the world in the summer season for a very short period. The quality of the fruits may change during storage at different rates, depending upon the storage temperature. There are prickly pears cultivars having a peel of pale-green, light-green, green, or dark-green colors (Fig. 1) as well as yellow, orange, magenta (pink), red, red-purple, and purple colors. However, even though the color of the peel of some cultivars may change during storage, external quality loss is not clearly observed in all

Controlled and Modified Atmospheres for Fresh and Fresh-Cut Produce
https://doi.org/10.1016/B978-0-12-804599-2.00035-1

FIG. 1 Fresh prickly pears with optimum edible characteristics.

types of fruits due to their color, which masks degreening. Without taking into account the color, fruits should not have prickles (glochids) and be moderately glossy, turgid, without peduncle damage, and be free from brown/black spots. Moreover, no molds growth should be observed on any part of the fruit surface, mainly on peduncle. Wrinkles may be observed on the surface. Fruits should not be overripe as they may be overly soft and the characteristic flavor may be lost because they start to ferment.

Main causes of quality loss

Perishability of prickly pears depends on the cultivar, and firmness and pericarp thickness of the cultivar are two of the most important factors in storability. The thickness of pericarp depends on the cultivar and is associated with skin color; therefore the handling of fruits should be different according to the cultivar. In general, some green and dark-green prickly pears are more resistant to handling because their pericarp is thick or medium-thick. The pericarp of prickly pears having different colors is typically less thick; therefore they are more susceptible to be damaged when handling.

As with other fresh produce, prickly pears continue their biological metabolism after harvest. In general, the respiration rate (CO_2 production) of most cultivars of prickly pears ranges from 20.8 to 32.8 $\text{mL}\,\text{kg}^{-1}\,\text{h}^{-1}$. However, prickly pears are not climacteric fruits and so suppression of ethylene action has minimal impact on their storability. Therefore their storability primarily depends on the storage conditions such as relative humidity, temperature, and packaging types.

As with other fruit, relative humidity (RH) and temperature govern quality loss and chemical and physical damages markedly impact microbiological decays during storage.

Physiological damage can take place during storage when temperature is too high. The peel starts to generate small brown spots and these spots will continue growing until they cover almost the whole surface of the fruit (Fig. 2). Since the peduncle of prickly pears is cut from the paddles, the peel is susceptible to tearing. If the peel is torn, deterioration (browning) starts very soon and it expands along the peel of the whole fruit (Fig. 3). Peel browning can start there; however, sometimes the browning spots on the surface of pericarp grow faster than the browning of the peduncle. Oxygen from the air drives an oxidation reaction dependent upon polyphenol oxidase activity and is necessary for the browning of the damaged peduncle and the fruit surface.

FIG. 2 Superficial brown spots on prickly pear.

FIG. 3 Prickly pear decay at peduncle end of the fruit following wounding.

FIG. 4 Prickly pears start to lose humidity and may have wrinkles and changes in color during storage.

If prickly pears are harvested correctly, the physical damage during storage is mainly due to drying of pericarp and some physiological changes in color (Fig. 4). Excessive loss of weight may occur if relative humidity is lower than the water vapor equilibrium of the fruit. Also, wrinkling of the skin (shrivel) appears due to water loss from the fruit.

However, even if there is some shrivel, the fruit pulp is typically of sufficient quality to be consumed if decay is not present. Similarly, the presence of some brown spots on the surface does not always limit edibility. The presence of surface defects still permits fruit making processed products [e.g., to ready-to-eat prickly pear (with or without treatments such as sanitizing or coating with some edible films), prickly pear preserves, and prickly pear liqueur) once the skin has been removed].

Microbiological damage can be promoted if temperature and relative humidity are not adequate; the growth of several opportunistic fungal decay organisms (*Fusarium* spp., *Alternaria* spp., and *Penicillium* spp.) and bacteria may cause rot of injured fruits (Fig. 3).

Packaging may also affect the prickly pears' shelf life. For example, plastic bags may create a modified atmosphere (MA) inside after packaging in passive (Fig. 5) or active MA packaging (MAP). Even though beneficial passive or active modified atmospheres can be generated inside the bags, high relative humidity and the presence of free water (condensation) may increase risk of rot due to molds and bacteria (Fig. 6). The respiration of the fruits, which is affected by the O_2 and CO_2 concentrations in the package, can generate enough heat to create conditions conducive to condensation.

Chilling injuries may also occur when fruits are stored at very low temperature, mainly below 8°C. Sensitivity to chilling depends on the cultivar. Hard and dry tissue can be formed during storage at very low temperature.

FIG. 5 Prickly pears stored at room temperature (20.4 ± 2.4°C) in passive MAP.

FIG. 6 Prickly pears stored at room temperature (20.4 ± 2.4°C) in passive MAP (59.7 ± 6.9% RH) for 70 days (53 days without MA and 17 days under MA) without condensation (left) and with condensation (right).

Optimum storage conditions

Optimum temperature

The optimal storing temperature of prickle pears may depend on the type of cultivar; however, it is recommended that the range is between 5°C and 9°C, but the relative humidity of the storage room is also important. The recommended RH is between 85% and 90%. It is important to mention again that relative humidity below the water activity (RH/100) of the fruit might dehydrate the fruit peel, forming wrinkles (Fig. 4). Relative humidity above the water activity of the fruit may create an ideal environment for growing, mainly molds or yeasts and eventually bacteria. However, rot is frequently caused by molds in high relative humidity environments. It is also important to change the air in the storage environment due to the increase in CO_2 and ethylene in the environment that promotes maturation.

CA/MA considerations

Benefits

CA/MA can be used to optimize the O_2 and CO_2 in the system and thereby increase the shelf life of prickly pears by controlling their metabolic activity. An initial MA is sometimes created in the package at the time of package sealing (active packaging); however, the respiration rate of the fruit continues, driving passive changes in package atmosphere. Eventually, a steady state is established for the respiratory gases in the package the fruit; however, if the package design is not adequate, O_2 can be depleted, causing the fruit to change its respiratory metabolism to an anaerobic one (fermentation) and unpleasant flavors and aromas may be created. Therefore an adequate modification of gases is required as well as low temperature for lowering the respiration rate. Controlled atmosphere (CA) systems can be used to maintain target concentration of gases (O_2, CO_2, and N_2) in the storage room during the storage period. Typically, fruit respiration continues more or less normally if the concentration of O_2 is not too low and the concentration of CO_2 is not too high. An adequate low temperature in the storage room is also required, in the range of 5–9°C, depending on the cultivar.

Tolerances

Prickly pears do not tolerate concentrations of O_2 lower than 3% and concentrations of CO_2 higher than 5%. Temperatures above than 20°C are not recommended, although fruits may last up to 20 days at this temperature (Fig. 7). However, some cultivars of prickly pear may be maintained in good conditions for up to three months at this temperature if properly handled and stored.

Injuries

The ideal way for increasing the shelf life of prickly pears is detach them very carefully from the paddle without damaging the peduncle. The risk of rotting may occur if peduncle is torn. Prickly pears may form brown spots during storage in ambient atmospheres or in MA packages. Molds may quickly develop on prickly pears in MA if storage temperatures are higher than 20°C.

Adequate temperate in the range of 5–9°C and MA having 3% or higher concentration of O_2 and no more than 5% CO_2 may prolong the shelf life of prickly pears for more than 40 days, depending on the cultivar.

FIG. 7 Prickly pears stored at room temperature ($20.4 \pm 2.4°C$) in passive MA ($59.7 \pm 6.9\%$ RH) for 85 days (53 days without MA and 32 days under MA).

- No insect control is required for stored prickly pears. Growth of cochineal (an insect generated material used to make carmine dye) may be observed on some paddles of some prickly pears.

Problems under suboptimal conditions

Problems under suboptimal conditions have been mentioned. However, it is important to point out that each cultivar of *Opuntia* may behave different under different environments (gas concentrations and temperatures) in a manner affected by skin color and thickness of pericarp. Some cultivars may last less than other cultivars.

Comments

It is notable that prickly pears may be stored longer if they are packaged individually.

Further reading

Guerrero-Beltrán, J.Á., Ochoa-Velasco, C.E., 2018. Figo da India—*Opuntia* spp. In: Rodrigues, S., Silva, E., de Brito, E. (Eds.), Exotic Fruits Reference Guide. Elsevier/Academic Press, Chennai, India, pp. 187–201.

Ochoa-Velasco, C.E., Guerrero-Beltrán, J.A., 2016. The effects of modified atmospheres on prickly pear (*Opuntia albicarpa*) stored at different temperatures. Postharvest Biol. Technol. 111, 314–321.

Piga, A., Aquino, S.D., Agabbio, M., Schirra, M., 1996. Storage life and quality attributes of cactus pears cv Gialla as affected by packaging. Agric Mediterr. 126, 423–427.

Thompson, A.K., 2010. Controlled Atmosphere Storage of Fruits and Vegetables, second ed. MPG Books Group, Preston, UK.

Chapter 18.11

Subtropical fruits: Pomegranates

Mustafa Erkan and Adem Dogan
Department of Horticulture, Akdeniz University, Antalya, Turkey

Quality characteristics

Pomegranate fruit do not ripen off the tree and should be harvested when they reach fully harvest maturity to ensure their best eating quality. Quality of pomegranate fruit is determined by both internal and external attributes of the fruit. The acceptability of pomegranates to the consumer depends on a combination of several quality characteristics.

Controlled and Modified Atmospheres for Fresh and Fresh-Cut Produce
https://doi.org/10.1016/B978-0-12-804599-2.00036-3

These properties are as follows:

- Free from internal and external deterioration symptoms
- Free from preharvest and harvest defects such as cracking, splitting, and sunburn
- Free from packing house defects such as bruises, wounds, and calyx damages
- Free from excessive water loss from the skin, shriveling, and hardening
- Aril color (dark red) and uniformity
- Fruit size, skin color (red or reddish), smoothness, and calyx opening
- Taste; soluble solids, acidity, and flavor (Prasad et al., 2010; Kader, 2006; Erkan and Kader, 2011).

Main causes of quality loss

Chilling injury (CI) is a physiological disorder affecting quality and marketing of pomegranate fruit. The minimum safe temperature for postharvest handling of pomegranates ranges between 5°C and 8°C (Elyatem and Kader, 1984). If the fruit are stored under these temperatures, CI occurs on the skin of pomegranates. External symptoms of CI include rind pitting, brown discoloration of the skin, and increased susceptibility to decay. Internal symptoms of CI were manifested as pale color of the arils and brown discoloration of the white segments separating the arils (Elyatem and Kader, 1984; Pekmezci and Erkan, 2004; Erkan and Kader, 2011). To reduce the occurrence of CI in pomegranate fruit, several techniques have been applied including controlled atmosphere storage (CA) (Artés et al., 1996), modified atmosphere packaging (MAP) (Selcuk and Erkan, 2015), intermittent warming (Artés et al., 1998), polyamine (Mirdehghan et al., 2007a,b), salicylic acid (Sayyari et al., 2009), and methyl jasmonate treatments (Ranjbar et al., 2007; Sayyari et al., 2011). Among these techniques, the most successful method reducing CI in pomegranate fruit was to store the fruit in CA.

Fruit cracking is one of the most important physiological disorder causing fruit loss and reducing marketability and quality of pomegranates. Cracking susceptibility among the cultivars varies in different parts of the world and this disorder can be reduced by regular irrigation (Prasad et al., 2003). This injury can reach up to 20%–30% if they are overripen and not harvested at full maturity stage. Heavy rain before harvest, excessive irrigation on mature pomegranates, and delayed harvest increase the amount and severity of the injury. Spraying fruit with gibberellic acid (GA_3) at 150ppm or with benzyl adenine (BA) at 40ppm could significantly reduce fruit cracking (Yılmaz and Özgüven, 2006). Similarly, application of boron (Singh et al., 2003) and paclobutrazol at 300 ppm (Khalil and Aly, 2013) may also reduce this disorder.

Husk scald is another important physiological disorder limiting the storage potential of the pomegranates. This symptom appears on the fruit surface as superficial skin browning initiating from stem end of the fruit without affecting the internal tissues and spreading toward the blossom end as the severity increases (Defilippi et al., 2006). Ben-Arie and Ore (1986) reported that scald symptoms may be caused by the enzymatic oxidation of phenolic compounds in the husk when stored at temperatures above 5°C. CA storage (1 kPa O_2 +15 kPa CO_2 or 5 kPa O_2 +15 kPa CO_2) significantly reduced scald incidence on pomegranates for up to 6 months at 7°C (Defilippi et al., 2006).

Sunburn is another physiological disorder which may negatively affect quality and trading of pomegranate fruit. The causes of sunburn are the combined action of high solar radiation, low humidity, and high temperatures. The incidence of sunburn damage can vary depending on the years, cultivation areas, and cultivars and may exceed 20%–25% of fruit loss. Yazici and Kaynak (2009) reported that fruit surface temperatures that cause sunburn damage vary between 41°C and 47.5°C and kaolin treatment proved to be one of the most successful method to reduce this injury in Turkish 'Hicaznar' cultivar. Holland et al. (2009) reported that in Israel, late cultivars such as 'Wonderful' is much more susceptible to sunburn than early cultivars such as 'Akko' and 'Shani-Yonay'.

Another important quality loss in pomegranate fruit is water loss from the skin. Pomegranate fruit are very susceptible to water loss which may result in hardening and shriveling of the skin and browning of the rind and arils. In pomegranates, if the weight loss exceeds more than 10%, the fruit turn to diagonal shape that affects market quality of pomegranates negatively. Storing fruit in MAP and waxing can reduce water loss, especially under conditions of lower relative humidity in storage (Artés et al., 2000).

Aril browning is another physiological disorder characterized by soft, light creamy, brown, dark blackish or brown, and slightly flattened arils which are deformed and possessing an unpleasant odor when the fruit is cut open. The disorder is accompanied by desiccation, wrinkling, and development of internal spaces in the arils. The affected fruits exhibit poor dessert quality and are unfit for consumption in the advanced stages of the disorder. In many cases, the intensity of the disorder in mature ripe fruits could be more than 50% causing severe loss of quality (Shivashankar et al., 2012).

The most common economically important postharvest disease resulting in heavy fruit loss in pomegranates is gray mold caused by *Botrytis cinerea* (Kinay, 2015). Gray mold usually starts at the calyx of the fruit and as it progresses, the skin becomes light-brown, tough, and leathery. Other fungi causing fruit rot include *Aspergillus niger*, *Penicillium* spp., *Alternaria* spp., *Nematospora* spp., *Coniella granati*, or *Pestalotiopsis versicolor*. The application of fludioxonil after harvest has reduced postharvest decay caused by *B. cinerea* (Palou et al., 2007; D'Aquino et al., 2010). Pomegranates have benefited from CA or MAP for decay control. The combination of CA storage (5 kPa O_2+15 kPa CO_2) with antifungal treatment (potassium sorbate) reduced *Botrytis* spp decay in pomegranates during storage (Palou et al., 2007).

Optimum storage conditions

Pomegranate fruit can be stored at 5°C for up to 2 months, but longer storage should be at 7°C to avoid CI (Kader et al., 1984; Erkan and Kader, 2011). Spanish 'Mollar' pomegranates can suffer CI if they are stored for more than 2 months at temperatures below 5°C (Artés et al., 2000). However, some cultivars of pomegranate such as 'Wonderful' can be stored without CI symptoms for 2 months at 5°C. Similarly, Hicaznar, a popular Turkish cultivar, can be stored up to 6 months at 6°C in MAP without any CI symptoms (Selcuk and Erkan, 2015). Therefore the minimum safe temperatures for storing pomegranates vary among cultivars and growing regions and should be 5–7°C. The relative humidity (RH) is the second important factor after temperature in maintaining postharvest quality of pomegranates. The objective of storage in pomegranate fruit is minimizing the weight loss without increasing the microbial development and decay and temperature and relative humidity are the key factors to fulfill this objective.

CA/MA considerations

CA storage with different O_2 and CO_2 levels has reduced weight loss, fungal decay, and physiological disorders during storage of different varieties of pomegranate. Similarly, husk scald development was inhibited at between 2kPa and 4kPa O_2 and 2–6°C as compared with regular atmosphere (Hess-Pierce and Kader, 2003). Artés et al. (1996) reported that CA storage was effective in maintaining fruit quality by reducing CI, weight loss, and fungal decay in Spanish 'Mollar de Elche' pomegranates. This cultivar could be stored at 5°C in a CA of 5 kPa O_2 + 5 kPa CO_2 for up to 2 months and at 7.5°C in 5 kPa O_2 + 15 kPa CO_2 for up to 5months. In another study, 'Hicaznar' pomegranates stored in 1 kPa CO_2 + 3 kPa O_2, 3 kPa CO_2 + 3 kPa O_2, or 6 kPa CO_2 + 3 kPa O_2 had a storage life of 130 days (Küpper et al., 1996) (Fig. 1).

FIG. 1 CA storage of pomegranate fruit.

Defilippi et al. (2006) have found that the most effective postharvest decay control on 'Wonderful' pomegranates was the combination of prestorage treatment with potassium sorbate followed by storage in an atmosphere of 5 kPa O_2 + 15 kPa CO_2.

MAP is a simple and low-cost method that has been proven to maintain postharvest quality of pomegranates up to 6 months after harvest (Selcuk and Erkan, 2015) (Fig. 2). It has been proven to be a successful means of reducing water loss, shrinkage, scald development, and decay and can facilitate maintenance of fruit quality for three months after harvest of 'Mollar de Elche' (Artés et al., 2000), 'Ganesh' (Nanda et al., 2001), 'Shlefy' (Ghafir et al., 2010), 'Primosole' (D'Aquino et al., 2010), 'Mollar de Elche' (Laribi et al., 2012), and 'Hicrannar' (Selcuk and Erkan, 2014) pomegranates. Porat et al. (2008) reported that crop-specific MAP for pomegranate fruit have different gas permeability results in the accumulation of 5 kPa O_2 + 12–14 kPa CO_2 within the bags at the end of prolonged storage. These types of bags reduced weight loss from 7% to 3.5% and husk scald development from 38% to 21% in "Wonderful" pomegranates. 'Hicaznar' fruit packed in MAP and stored at 6.0°C for 6 months were better conditions in fruit quality criteria than those unpacked fruit (Selcuk and Erkan, 2015). Furthermore, MAP reduces the moisture loss by creating a high RH within the environment, surrounding the product during handling, storage, and transportation of pomegranate fruit. However, if the permeability of the MAP is insufficient, it can create a saturated environment within the package that can promote the development of fungal decay. Therefore it is vitally important to correlate the permeability properties of the packing films with the respiration rate of pomegranates, in order to avoid anaerobic conditions and accumulation of ethanol.

FIG. 2 MA storage of pomegranate fruit.

Several postharvest techniques have been tested for long-term storage of pomegranates, including low-temperature storage (Ben-Arie and Ore, 1986), CA storage (Artés et al., 1996; Küpper et al., 1996; Defilippi et al., 2006), MA storage (Artés et al., 2000; Caleb et al., 2012; Selcuk and Erkan, 2014, 2015), film wrapping (D'Aquino et al., 2010), and intermittent warming (Artés et al., 1998). Among these techniques, the most successful method for reducing decay and physiological disorders is the use of CA storage. However, CA storage facilities are not always economical and available in many pomegranate-producing countries. For that reason, the use of MAP is a simple, economical, and effective method in delaying the postharvest deterioration and maintaining the postharvest quality of pomegranates.

The tolerance limit for CO_2 and O_2 levels could be different for different cultivars of intact pomegranates and extracted arils. For example, Hess-Pierce and Kader (2003) reported that in intact fruit 10kPa of CO_2 resulted in higher concentrations of fermentative metabolites, including acetaldehyde and ethanol, especially after 4 and 5 months at 5°C, 7.5°C, or 10°C. In another study, Holcroft et al. (1998) reported that it is possible to store intact pomegranates at 7.5°C in 5 kPa O_2 + 15 kPa CO_2 for 5 months. Furthermore, Holcroft et al. (1998) found that arils stored in regular atmosphere enriched with 10kPa CO_2 had a lower anthocyanin concentration than that found in regular atmosphere-stored arils and with 20kPa CO_2 had even much lower level anthocyanin concentration, possibly due to suppressed anthocyanin biosynthesis. In general, the tolerance atmosphere limit for pomegranate fruit is 5kPa O_2 + 15kPa CO_2 and at 7.5°C fruit can be stored for up to 5 months.

Other postharvest treatments affecting fruit quality rather than CA/MA storage

Another technology used with success in pomegranate storage is the preconditioning of the fruits at moderate temperature (30–40°C) and high relative humidity of 90%–95% for a short period of time (1–4 days). Pretreatment of pomegranates at 35°C and 90%–95% RH for 3 days before 80-day storage at 5°C or 2°C and 90%–95% RH reduced the pitting and husk scald considerably compared to that observed in control fruit (Artés and Tomás-Barberán, 2000). Heat treatments have also been used to extend the storability of pomegranates. Mirdehghan et al. (2007a) reported that arils from heat-treated pomegranates (at 45°C for 4min) exhibited higher total antioxidant activity than that found on arils of control fruit. Intermittent warming has proved useful in the prevention of CI symptoms in pomegranates (Artés et al., 1998). Pretreatment of pomegranates with intermittent warming and hot water was shown to reduce CI and electrolyte and K leakage and in maintaining pomegranate quality during storage, in terms of retention of anthocyanin and titratable acidity, and reduction of decay (Artés et al., 1998; Mirdehghan and Rahemi, 2005).

Problems under suboptimal conditions

One of the major problems associated with pomegranate fruit under suboptimal conditions is excessive water loss which may result in weight loss, hardening of the husk, and browning of the rind and arils (Caleb et al., 2012). Even in the absence of shriveling, water loss can cause undesirable textural and flavor changes, ultimately resulting in loss of visual appeal (Fig. 3). Pomegranate fruit are very susceptible to chilling injury (CI) if they are stored longer than one month at temperatures between their freezing point −3°C and 5°C or longer than two months at 5°C (Elyatem and Kader, 1984; Kader et al., 1984).

Comments

Pomegranate fruit should be handled with as much care as other perishable horticultural crops during preharvest and postharvest handling to minimize mechanical damage and to prolong storage duration. Proper management of approved preharvest and postharvest fungicides is critical to reduce decay and to avoid the development of fungal resistance for pomegranates. Use of MAP and/or CA storage as supplements to low-temperature storage and proper RH are effective in prolonging postharvest life of pomegranates. Further research, such as ozone and other nonchemical methods, is needed to evaluate their efficacy on preventing spread of decay in different pomegranate fruit having different acidity levels. Increasing mechanization of aril extraction with minimal damage and subsequent preparation steps of arils as a ready-to-eat, value-added product will increase marketability. Further development of nutraceuticals derived from the edible and nonedible portions of pomegranates will increase the popularity and profitability of pomegranates.

FIG. 3 Water loss from pomegranate fruit.

References

Artés, F., Tomás-Barberán, F.A., 2000. Postharvest technological treatments of pomegranate and preparation of derived products. In: Melgarejo, P., Martinez-Nicolas, J.J., Martinez-Tome, J. (Eds.), Production, Processing and Marketing of Pomegranate in the Mediterranean Region: Advances in Research and Technology. CIHEAM, Zaragoza, pp. 199–204.

Artés, F., Marin, G.J.J.A., Martinez, J.A., 1996. Controlled atmosphere storage of pomegranate. Lebensm Unters Forsch. 203, 33–37.

Artés, F., Tudela, J.A., Gil, M.I., 1998. Improving the keeping quality of pomegranate fruit by intermittent warming. Eur. Food Res. Technol. 207, 316–321.

Artés, F., Villaescusa, R., Tudela, J.A., 2000. Modified atmosphere packaging of pomegranates. J. Food Sci. 65, 1112–1116.

Ben-Arie, R., Ore, E., 1986. Development and control of husk scald on 'Wonderful' pomegranate fruit during storage. J. Am. Soc. Hortic. Sci. 111, 395–399.

Caleb, O.J., Opara, U.L., Witthuhn, C.R., 2012. Modified atmosphere packaging of pomegranate fruit and arils: a review. Food Bioprocess Technol. 5, 15–30.

D'Aquino, S., Palma, A., Schirra, M., Continella, A., Tribulato, E., La Malfa, S., 2010. Influence of film wrapping and fludioxonil application on quality of pomegranate fruit. Postharvest Biol. Technol. 55, 121–128.

Defilippi, B.G., Whitaker, B.D., Hess-Pierce, B.M., Kader, A.A., 2006. Development and control of scald on Wonderful pomegranates during long-term storage. Postharvest Biol. Technol. 41, 234–243.

Elyatem, S.M., Kader, A.A., 1984. Postharvest physiology and storage behaviour of pomegranate fruits. Scientia Hortic. 24, 287–298.

Erkan, M., Kader, A.A., 2011. Pomegranate (*Punica granatum* L.). In: Yahia, E.M. (Ed.), Postharvest Biology and Technology of Tropical and Subtropical Fruits. Woodhead Publishing, New York, pp. 287–311.

Ghafir, S.A.M., Ibrahim, I.Z., Zaied, S.A., 2010. Response of local variety 'shlefy' pomegranate fruits to packaging and cold storage. Acta Hort. (877), 427–432.

Hess-Pierce, B., Kader, A.A., 2003. Responses of 'Wonderful' pomegranates to controlled atmospheres. Acta Hort. 600, 751–757.

Holcroft, D.M., Gil, M.I., Kader, A.A., 1998. Effect of carbon dioxide on anthocyanins, phenylalanine ammonia lyase and glucosyltransferase in the arils of stored pomegranates. J. Amer. Soc. Hort. Sci. 123, 136–140.

Holland, D., Hatib, K., Bar-Ya'akov, I., 2009. Pomegranate: Botany, horticulture, breeding. Hort. Rev. 35, 127–191.

Kader, A.A., 2006. Postharvest biology and technology of pomegranates. In: Seeram, N.P., Schulman, R.N., Heber, D. (Eds.), Pomegranates Ancient Roots to Modern Medicine. Taylor and Francis, New York, pp. 211–222.

Kader, A.A., Chardas, A., Elyatem, S., 1984. Responses of pomegranate to ethylene treatment and storage temperature. Calif. Agr. 38, 14–15.

Khalil, H.A., Aly, H.S.H., 2013. Cracking and fruit quality of pomegranate (*Punica granatum* L.) as affected by pre-harvest sprays of some growth regulators and mineral nutrients. J. Hortic. Sci. Ornamental Plants 5 (2), 71–76.

Kinay, T.P., 2015. Alternative technologies to control postharvest diseases of pomegranate. Stewart Postharvest Rev. 4 (3), 1–7.

Küpper, W., Pekmezci, M., Henze, J., 1996. Studies on CA-storage of pomegranate fruit (*Punica Granatum* L., cv. Hicaz). Acta Hort. 398, 101–108.

Laribi, A.I., Palou, L., Taberner, V., Pérez-Gago, M.B., 2012. Modified atmosphere packaging to extend cold storage of pomegranate cv. 'Mollar de Elche'. http://www.academia.edu/2500799/ (Accessed 5.5.2016).

Mirdehghan, S.H., Rahemi, M., 2005. Effects of hot water treatment on reducing chilling injury of pomegranate (*Punica granatum* L.) fruit during storage. Acta Hort. (682), 887–892.

Mirdehghan, S.H., Rahemi, M., Martinez-Romero, D., Guillen, F., Valverde, J.M., Zapata, P.J., Serrano, M., Valero, D., 2007a. Reduction of pomegranate chilling injury during storage after heat treatment: role of polyamines. Postharvest Biol. Technol. 44, 19–25.

Mirdehghan, S.H., Rahemi, M., Serrano, M., Guillén, F., Martínez-Romero, D., Valero, D., 2007b. The application of polyamines by pressure or immersion as a tool to maintain functional properties in stored pomegranate arils. J. Agric. Food Chem. 55, 755–760.

Nanda, S., Sudhakar Rao, D.V., Krishnamurthy, S., 2001. Effects of shrink film wrapping and storage temperature on the shelf life and quality of pomegranate fruits cv. Ganesh. Postharvest Biol. Tecnol. 22, 61–69.

Palou, L., Crisosto, C.H., Garner, D., 2007. Combination of postharvest antifungal chemical treatments and controlled atmosphere storage to control gray mold and improve storability of 'Wonderful' pomegranates. Postharvest Biol. Technol. 43, 133–142.

Pekmezci, M., Erkan, M., 2004. Pomegranate. Postharvest quality maintenance guidelines. USDA, United States Department of Agriculture, Agricultural Research Service. Agricultural Handbook Number-66.

Porat, R., Weiss, B., Fuchs, Y., Sandman, A., Ward, G., Kosto, I., 2008. Keeping quality of pomegranate fruit during prolonged storage and transport by MAP: New Developments and Commercial Applications. Acta Hort. (804), 115–120.

Prasad, R.N., Bankar, G.J., Vashishtha, B.B., 2003. Effect of drip irrigation on growth, yield and quality of pomegranate in arid region. Indian J. Hort. 60, 140–142.

Prasad, R.N., Chandra, R., Teixeira da Silva JA, 2010. Postharvest handling and processing of pomegranate. In: Chandra, R. (Ed.), Pomegranate. Fruit Veg. Cereal Sci. Biotechnol., vol. 4, Special Issue 2, pp. 88–95.

Ranjbar, H., Zou Alfaghari Nasab, R., Ghasemnezhad, M., Sarkhoush, A., 2007. Effect of methyl jasmonate on inducing chilling tolerance in pomegranate fruits (Malas Save). Pajouhes Va-Sazandegi Agron Hortic. 20, 43–49.

Sayyari, M., Babalar, M., Kalantari, S., Serrano, M., Valero, D., 2009. Effect of salicylic acid treatment on reducing chilling injury in stored pomegranates. Postharvest Biol. Technol. 53, 152–154.

Sayyari, M., Babalar, M., Kalantari, S., Martínez-Romero, D., Guillén, F., Serrano, M., Valero, D., 2011. Vapour treatments with methyl salicylate or methyl jasmonate alleviated chilling injury and enhanced antioxidant potential during postharvest storage of pomegranates. Food Chem. 124, 964–970.

Selcuk, N., Erkan, M., 2014. Changes in antioxidant ac tivity and postharvest quality of sweet pomegranates cv. Hicrannar under modified atmosphere packaging. Postharvest Biol. Technol. 92, 29–36.

Selcuk, N., Erkan, M., 2015. Changes in phenolic compounds and antioxidant activity of sour–sweet pomegranates cv. 'Hicaznar' during long-term storage under modified atmosphere packaging. Postharvest Biol. Technol. 109, 30–39.

Shivashankar, S., Sing, H., Sumathi, M., 2012. Aril browning in pomegranate (*Punica granatum* L.) is caused by the seed. Current Sci. 103, 26–28.

Singh, D.B., Sarma, B.D., Bhargava, R., 2003. Effect of boron and GA_3 to control fruit cracking in pomegranate (*Punica granatum*). Current Agric. 27, 125–127.

Yazici, K., Kaynak, L., 2009. Effects of kaolin and shading treatments on sunburn in fruit of hicaznar cultivar of pomegranate (Punica granatum L.cv. Hicaznar). Acta Hortic. (818), 167–173.

Yılmaz, M., Özgüven, A.I., 2006. The effect of some plant nutrients, gibberellic acid and pinolate treatments on the yield, fruit quality and cracking in pomegranate, p. 59. In: ISHS. 1st Int. Symp., Pomegranate and Minor Mediterranean Fruits, Abstract contributed papers, 16–19 Oct., Adana, Turkey.

Chapter 19.1

Fresh-cut fruits: Apples and pears

Robert Soliva-Fortuny and Olga Martín-Belloso
Department of Food Technology, University of Lleida—Agrotecnio Center, Lleida, Spain

Quality characteristics

Visual appearance and texture are two fundamental characteristics determining the acceptance of fresh-cut apples and pears. The quality and shelf life potential of the cut produce is greatly determined by the characteristics of the intact fruit, which are in turn dependent on many different aspects, including growing and postharvest conditions, variety, maturity at harvest, ripeness stage at processing. The flavor of the product has a strong varietal component and is determined by the soluble solids/acidity ratio at the stage of processing.

Controlled and Modified Atmospheres for Fresh and Fresh-Cut Produce
https://doi.org/10.1016/B978-0-12-804599-2.00037-5

In fact, although fresh apples and pears are generally graded according to size; shape; color; and freedom from defects, blemishes, and diseases, the sugar/acid ratio is the main parameter determining the sensory characteristics of the processed product. Although the ratio is highly variety dependent, those fruits with sugar content below 10.5 °Brix should not be considered for minimal processing. Size and shape do not significantly affect the quality of the end product. However, small size and irregular shapes might significantly decrease the overall yield or coring and peeling operations, especially when these are performed mechanically. Diameters below 70 mm for apples and below 60 mm for pears will easily lead to yields falling below 50%.

General provisions concerning the quality of raw materials must be in accordance with the requirements of the Codex Standard for fresh Apples (CODEX STAN 299-2010) (CODEX Alimentarius, 2010). Therefore the fruits must be: (i) whole, without any skin damage, although the stalk (stem) may be missing, provided the adjacent tissues are not damaged; (ii) sound, without any rotten or deteriorated tissue that makes the fruit unfit for fresh consumption; (iii) as firm as possible, but with the organoleptical traits that make them fit for consumption; (iv) clean, practically free of foreign matters; (v) free of damage caused by pests affecting the product general appearance; (vi) free of abnormal external moisture, excluding condensation following removal from cold storage; (vii) free of any foreign smell and/or taste; (viii) without damage caused by chilling or high temperatures; (ix) without signs of dehydration.

Main causes of quality loss

Apples and pears are seasonal cultivars and some varieties are available all year round thanks to cold storage. Fruits that allow long-term refrigerated storage without severe quality decay are good candidates for fresh-cut processing.

Fruits stored in CA environments are more likely to exhibit good quality traits and susceptibility to processing. In order to satisfy the requirements of a fresh-cut product, fresh apples and pears need to be free of internal disorders such as bitter pit, water core, browning disorders, or internal breakdown. Superficial disorders such as storage scald and light sunscald may be acceptable when the peel is going to be removed during processing.

Regarding microbiological quality, fungi are the main microorganisms present on the skin of the intact product. Therefore dips in chlorinated water are commonly recommended for pome fruits before processing and also after peeling or cutting operations (Soliva-Fortuny and Martín-Belloso, 2003). As regular pH values for apples and pears do not generally exceed 4.0 and 4.5, respectively, both spoilage and foodborne pathogenic bacteria are not likely to be a significant thread in apple and pear fresh-cut products.

An important quality trait is the susceptibility to brown. In that regard, a previous evaluation must be carried out, as it may significantly differ from cultivar to cultivar, or even from field to field, in both apples and pears depending on the amount of polyphenol oxidase (PPO) enzyme activity and, to a lesser extent, on the amount and types of phenolic compounds that act as substrates of the browning reaction. Nevertheless, these are not the only relevant factors. Although phenolic levels tend to decrease in more ripened fruits or with delayed harvest times, phenolic levels do not always correlate with the susceptibility to browning (Amiot et al., 1995). In pear and apple (Fig. 1), PPO activity is generally higher in the core tissue than in the peel and the flesh, in this order (Macheix and Fleuriet, 1990). It is important to point out that the qualitative aspects of the phenolic composition are determined as well, as browning rates and overall browning susceptibility are affected by the relative content of individual phenolics (Amiot et al., 1992). Cultivar selection for fresh-cut processing should therefore consider raw materials with either low PPO levels or low phenolic substrates, or both.

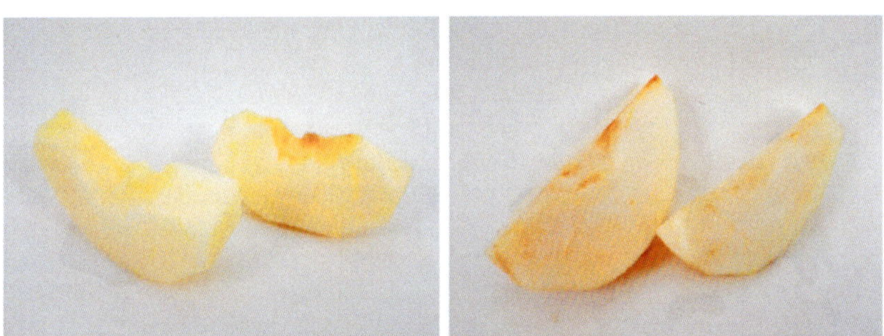

FIG. 1 Browning evidence in remaining core tissues of apple slices (left) and tissue browning in core and cut surfaces due to insufficient antioxidant treatment of pear slices (right).

Softening is another undesirable quality change which considerably curbs the acceptance of fresh-cut apples and pears by consumers. Mechanical stress may lead to ethylene production and the subsequent triggering of several metabolic disorders that result into the hydrolysis of cell wall pectic substances (Varoquaux and Wiley, 1997). However, not only enzymatic pathways are involved but also physical changes associated with turgor pressure of the cut tissues and temperature, among other factors. The complexity of these phenomena makes it difficult to prevent fresh-cut apple and pear products from softening just by modulating the MA environments. However, dipping in calcium solutions and 1-MCP treatments have been successfully been used to delay softening of cut produce (Perera et al., 2003; Lu et al., 2009).

Off-flavor generation has been related to the development of undesirable fermentative metabolites in the cut tissues. However, this evidence of physiological disorders usually appears after prolonged storage under abusive conditions and when other major quality traits have already decayed significantly.

Some important sources of quality loss in fresh-cut apple and pear products are listed as follows:

- Improper ripeness stage at processing (immature or too aged fruits).
- Improper handling of raw materials before processing leading to mechanical injuries.
- Delayed cooling and improper cold/CA storage of the raw materials.
- Mechanical injuries due to unacceptable conditions during processing (blunt blades, excessive falls, or excessive vacuum during the packaging operation).
- Inadequate cooling of the cut product before packaging.
- Improper use of packaging materials (permeability does allow reaching equilibrium with O_2 consumption and CO_2 production).
- Improper temperatures over storage, thus leading to O_2 concentrations below the fermentative threshold or two high CO_2 levels.
- Exposure of the packaged product to sudden temperature fluctuations.

Optimum storage conditions

The benefits of MA storage regarding the physiological response of the cut products are relatively similar to those of the fresh fruits. Hence, MA packaging allows controlling the respiratory burst of the cut tissue, ethylene production rates, and softening of the product, while delaying microbial spoilage. MA packaging may also significantly reduce other ripeness-related changes in the cut produce. However, as the shelf life of cut products is significantly shorter than that of their fresh counterparts, these benefits are attenuated.

Temperature range for appropriate storage should be 1–4°C (34–39°F). Higher temperatures will significantly reduce the product shelf life, which can be directly related to the increase in the respiratory activity of the cut product. However, respiration rates of fresh-cut apples and pears are greatly dependent on many other factors including cultivar, physiological state, harvest and postharvest handling and atmosphere conditions. Gorny et al. (2000) reported respiration rates for fresh-cut 'd'Anjou,' 'Bartlett,' 'Red Anjou,' and 'Bosc' cultivars ranging between 0.15 and 0.43, 0.53 and 1.04, and 1.70 and 3.21 mmol $CO_2\ kg^{-1}\ h^{-1}$ at 0°C, 10°C, and 20°C, respectively. Similarly, ethylene production dramatically increased at higher temperature values ranging between 0.029 and 0.979, 0.230 and 3.092, and 0.374 and 5.098 μmol $C_2H_4\ kg^{-1}\ h^{-1}$. These values illustrate a great temperature and cultivar dependency of the respiratory response.

As well, the composition of the package atmosphere may also significantly affect respiration although in a much lesser extent than temperature. Gomes et al. (2010) reported significant increases of the O_2 uptakes rate of 'Rocha' pears with O_2 partial pressures within the range between 0.02 and 0.95 mmol O_2. The increases were especially noticeable at temperatures above 5°C. Hence, from the point of view of metabolic activity, very low oxygen atmospheres are not shown to offer major benefits without the danger of inducing anaerobic respiration. In addition, it is far more detrimental to the cut fruit quality to expose the product to temperature fluctuations, especially once packaged, as the formation of in-package condensates is favored, thus promoting microbial growth and greatly reducing shelf life. On the other hand, the use of superatmospheric O_2 environments is not generally acceptable as, beyond the positive effects regarding microbial spoilage and the competitive inhibition of PPO enzyme (O'Beirne et al., 2011) a significant depletion of the antioxidant content has been reported in pears (Oms-Oliu et al., 2008; Abreu et al., 2012). The use of argon-enriched atmospheres has also some potential for maintaining the quality of fresh-cut apples and pears. The major advantage of enriching low-O_2 atmospheres with argon is the reduction of enzymatic browning in those products that develop browning even at relatively low oxygen levels (O'Beirne et al., 2011). However, a cost–benefit analysis should be performed.

Atmospheres with low O_2 and elevated CO_2 concentrations reduce ethylene biosynthesis and related side effects. This effect may have limited consequences from the point of view of quality of the fresh-cut products, as most metabolic processes involved in ripening have already been upregulated. However, some beneficial effects regarding softening delay over storage can be associated with MA environments. Hence, apart from the reduction of respiration rates, the main reason to utilize MA for fresh-cut apples and pears is the reduction of browning phenomena, water loss, and the inhibition of the growth of spoilage microorganisms. As yeasts and molds are the main microorganisms growing on fresh-cut apples and pears, due to the low pH values of these fruits, MA environments may significantly contribute to control their proliferation. Thus microbial quality should not be a major quality-limiting factor especially when optimal storage temperatures are observed.

FIG. 2 Development of translucency as a sign of tissue injury caused by excessive in-package accumulation of CO_2.

Taking into account these constraints, optimal in-package concentrations for both fresh-cut pears and apples under steady conditions should be 0.5–1 kPa O_2 and 10–15 kPa CO_2. Further accumulation of CO_2, beyond 30 kPa may lead to significant injury of the cut tissue. This is especially relevant in the case of pears (Fig. 2).

Problems under suboptimal conditions

Cut tissues are much more tolerant to high and low CO_2 and O_2 concentrations, respectively. Hence, CO_2 partial pressures of up to 25–30 kPa can be tolerated by the cut tissues without triggering the development of undesirable flavors and odors, compared to the rather lower values of 2–5 kPa for the fresh fruits. Similarly, fresh-cut apples and pears can withstand environments with O_2 partial pressures that would be injurious to whole produce because the removal of the cuticle significantly reduces the resistance to gas diffusion (Gunes et al., 2001; Beaudry, 2000). In fact, several research works report that storage under anaerobic conditions may even contribute to increase the shelf life of fresh-cut apples, leading to the accumulation of fermentative metabolites such as ethanol and acetaldehyde but without significant sensory impairment (Gil et al., 1998). Ultralow oxygen atmospheres may be therefore applicable in certain cases but will generally induce production of fermentative metabolites that will impair the flavor and aroma of the fruits.

Further studies have suggested that high CO_2 partial pressures of up to 30 kPa may have a positive effect on shelf life of sliced apples by increasing the tissue tolerance to low O_2 environments and by reducing the production of ethylene as well as the activity of several enzymes involved in cellular respiration (Gunes et al., 2001). Nevertheless, taking into account that greater CO_2 concentrations may lead to important imbalances of the physiological activity of the tissue and that respiration is strongly affected by temperature, it must be advised to design package systems that do not favor the generation of strict anaerobic in-package conditions.

Comments

Edible coatings may be used to extend the shelf life of fresh-cut apple and pear products by controlling respiration, transpiration, or even physiological disorders, thus reducing overall quality decay. Because of their potential to carry active ingredients such as antibrowning agents, antimicrobial compounds, or calcium salts, shelf life may be significantly extended. However, as they provide an additional protective layer, gas diffusion properties need to be carefully considered with regard to their impact on internal gas concentrations in order to prevent the generation of anoxic conditions at the tissue level.

Acknowledgments

This work was supported by the Ministry of Economy and Competitiveness of Spain, Project AGL2013-44851-R.

References

Abreu, M., Alegria, C., Gonçalves, E.M., Pinheiro, J., Moldao-martins, M., Empis, J., 2012. Effects of superatmospheric oxygen pre-treatment on fresh-cut 'Rocha' pear quality. Acta Horticulturae (934), 567–574.

Amiot, M.J., Tacchini, M., Aubert, S., Nicolas, J., 1992. Phenolic composition and browning susceptibility of various apple cutivars at maturity. J. Food Sci. 57, 958–962.

Amiot, M.J., Tacchini, M., Aubert, S., Nicholas, J., 1995. Influence of cultivar, maturity stage, and storage conditions on phenolic composition and enzymatic browning on pear fruit. J. Agri. Food. Chem. 43, 1132–1137.

Beaudry, R.M., 2000. Response of horticultural commodities to low oxygen: limits to the expanded use of MAP. HortTechnol. 10, 491–500.

CODEX Alimentarius, 2010. Codex Standard for Apples. CODEX STAN 299-2010. Food And Agriculture Organization of the United Nations. World Health Organization, Rome.

Gil, M.I., Gorny, J.R., Kader, A.A., 1998. Responses of "Fuji" apple slices to ascorbic acid treatments and low-oxygen atmospheres. Hortscience 33, 305–309.

Gomes, M.H., Beaudry, R.M., Almeida, D.P.F., Malcata, F.X., 2010. Modelling respiration of packaged fresh-cut 'Rocha' pear as affected by oxygen concentration and temperature. J. Food Eng. 9681, 74–79.

Gorny, J.R., Cifuentes, R.A., Hess-Pierce, B., Kader, A.A., 2000. Quality changes in fresh-cut pear slices as affected by cultivar, ripeness stage, fruit size, and storage regime. J. Food Sci. 65, 541–544.

Gunes, G., Watkins, C.B., Hotchkiss, J.H., 2001. Physiological responses of fresh-cut apple slices under high CO_2 and low O_2 partial pressures. Postharvest Biol. Technol. 22, 197–204.

Lu, C., Cureatz, V., Toivonen, P.M.A., 2009. Improved quality retention of packaged 'Anjou' pear slices using a 1-methylcyclopropene (1-MCP) co-release technology. Postharvest Biol. Technol. 51, 378–383.

Macheix, J.J., Fleuriet, A., 1990. In: Billot, J. (Ed.), Fruit Phenolics. CRC Press, Boca Raton.

O'Beirne, D., Murphy, E., Ni Eidhin, D., 2011. Effects of argon enriched low-oxygen atmospheres and of high-oxygen atmospheres on the kinetics of polyphenoloxidase (PPO). J. Food Sci. 76, E73–E77.

Oms-Oliu, G., Odriozola-Serrano, I., Soliva-Fortuny, R., Martín-Belloso, O., 2008. Antioxidant content of fresh-cut pears stored in high-O_2 active packages compared with conventional low-O_2 active and passive modified atmosphere packaging. J. Agric. Food Chem. 56, 932–940.

Perera, C.O., Balchin, L., Baldwin, E., Stanley, R., Tian, M., 2003. Effect of 1-methylcyclopropene on the quality of fresh-cut Apple slices. J. Food Sci. 68, 1910–1914.

Soliva-Fortuny, R.C., Martín-Belloso, O., 2003. New advances in extending the shelf-life of fresh-cut fruits: a review. Trends Food Sci. Technol. 14, 341–353.

Varoquaux, P., Wiley, R.C., 1997. Biological and biochemical changes in minimally processed refrigerated fruits and vegetables. In: Wiley, R.C. (Ed.), Minimally Processed Refrigerated Fruits and Vegetables. Chapman & Hall, New York, pp. 226–268.

Chapter 19.2

Fresh-cut fruits: Mangoes

Jeffrey K. Brecht
Horticultural Sciences Department, University of Florida, Gainesville, FL, United States

Quality characteristics

Fresh-cut mango should be ripe and ready to eat, with characteristic color, texture, taste, and aroma as in whole fruit. Absence of tissue browning and translucency, as well as off-flavors, mealy tissue breakdown, and drying of the cut surfaces is also important.

Main causes of quality loss

Oxidative browning and tissue translucency (Fig. 1) are issues with riper fruit that are related to senescence (Gonzalez-Aguilar et al., 2008; Plotto et al., 2010; de Souza et al., 2006); overripening also leads to fermentation and other off-odors and off-flavors (Dea et al., 2013). Failure to remove the fruit peel to a sufficient depth during processing leads to browning (Fig. 2). Polyphenol oxidase (PPO) activity is most intense near the endocarp (pit), and the phenolic compounds that are PPO substrates are concentrated near the endocarp and subtending the epicarp (peel) (Dea et al., 2007). Use of insufficiently

Controlled and Modified Atmospheres for Fresh and Fresh-Cut Produce
https://doi.org/10.1016/B978-0-12-804599-2.00038-7

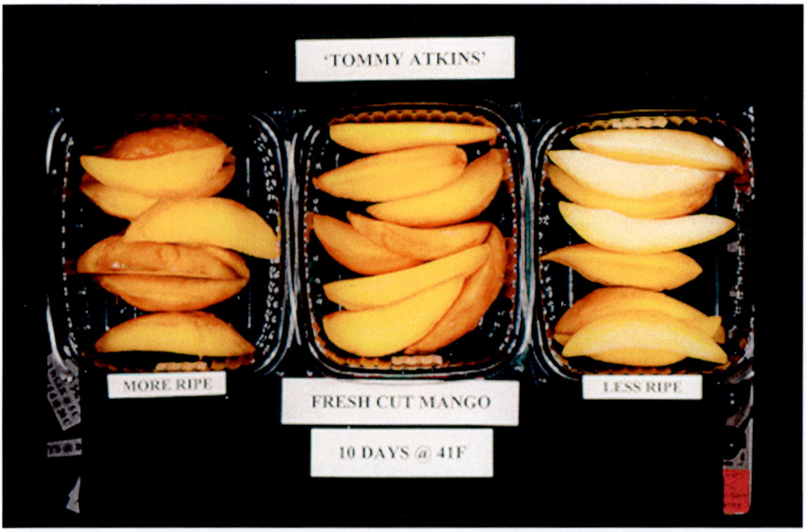

FIG. 1 Oxidative browning and tissue translucency on fresh-cut 'Tommy Atkins' mango stored for 10 days at 41°F (5°C) showing the greater severity in slices from more ripe fruit.

FIG. 2 A peeled mango showing where the depth of peeling was insufficient to fully remove the epidermal and subepidermal tissue that is most prone to oxidative browning.

sharp cutting and peeling implements leads to tissue breakdown. Desiccation of the cut surface may also limit shelf life. With proper sanitation and temperature management practices, spoilage due to microorganisms is not usually encountered.

Optimum storage conditions

Optimum temperature

As with all fresh-cut products, fresh-cut mango must be maintained below 5°C for food safety reasons. The optimum temperature for handling fresh-cut mango is 2–5°C. Concerns that such low temperatures may cause chilling injury, given that 13°C is the chilling threshold temperature for whole mangos, have proven to be unfounded. Although indications of chilling stress have been found, the exposure times at the low temperatures appear to be too brief to cause serious injury, in contrast to the rapid quality deterioration that occurs at nonchilling temperatures (Chantanawarangoon, 2000; Dea et al., 2010a).

CA/MA considerations

Benefits

Modified atmosphere packaging (MAP) is used to reduce the quality changes in fresh-cut mango related to ripening and senescence, such as softening, browning, and tissue translucency (Fig. 3), as well as to reduce loss of sugars and organic acids due to respiration. The high humidity maintained in MAP alleviates tissue drying, which was found to be an important factor in the ability of antibrowning treatments to prevent browning of fresh-cut mango (Gonzalez-Aguilar et al., 2000). An atmosphere of 2%–5% O_2 plus 10% CO_2 reduced yeast and total mesophilic aerobic counts on fresh-cut "Tommy Atkins" and "Kent" mango during temperature abuse at 10°C, but had no effect at 5°C (Rattanapanone et al., 2001). Similarly, 10% CO_2 reduced bacterial counts on fresh-cut "Carabao" and "Nam Dokmai" mango at 13°C, but not at 5°C (Poubol and Izumi, 2005).

Tolerances

Fresh-cut mango tolerates similar O_2 and CO_2 as the intact fruit at the same ripeness stage despite the lower temperature used for the fresh-cut product. Atmospheres of 2%–5% O_2 plus 10% CO_2 have been reported to be ideal for maintaining the quality and composition of fresh-cut "Haden," "Keitt," "Kent," and "Tommy Atkins" mangos at 5°C (Chantanawarangoon, 2000; Dea 2009; Rattanapanone et al., 2001). Polysaccharide-based materials that are used as edible coatings limit gas exchange, thus creating a modified internal atmosphere in the fresh-cut fruit (Baldwin et al., 1995). Caution should be used when

FIG. 3 Symptoms of ripening and senescence during storage of fresh-cut mango are reduced by proper use of modified atmosphere packaging (MAP).

combining edible coatings and MAP to avoid development of potentially injurious modified internal atmospheres.

Injuries

Fresh-cut mango exposed to extreme atmosphere conditions transition to fermentative metabolism, resulting in alcoholic off-flavors and off-odors (Rattanapanone and Watada, 2000). Exposure to CO_2 greater than 10% promotes abnormal softening (Dea et al., 2010b; de Souza et al., 2006).

Problems under suboptimal conditions

When fresh-cut mangos are handled at above optimum temperature and/or in too extreme atmospheres, the consequences are typically rapid softening and browning, and development of translucency and off-odors. If microorganisms are present, proliferation is extremely rapid under suboptimal conditions (Fig. 4).

FIG. 4 If present, proliferation of microorganisms is rapid when temperature and atmosphere conditions are not optimal.

Comments

Because the duration of postharvest handling for fresh-cut mango is usually short, 4 to 5 days is typical, proper ripeness selection and temperature control are probably more important than modified atmospheres for obtaining the required shelf life. However, the importance of surface tissue desiccation as a shelf life limiting factor for fresh-cut mango argues for the use of moisture barrier packaging and hygroscopic edible coatings to maintain a moist surface.

References

Baldwin, E.A., Nisperos-Carriedo, M.O., Baker, R.A., 1995. Edible coatings for lightly processed fruits and vegetables. HortScience 30, 35–38.

Chantanawarangoon, S. 2000. Quality maintenance of fresh-cut mango cubes. M.S. thesis in Food Science, University of California at Davis, 72 p.

de Souza, B.S., O'Hare, T.J., Durigan, J.F., de Souza, P.S., 2006. Impact of atmosphere, organic acids, and calcium on quality of fresh-cut 'Kensington' mango. Postharv. Biol. Technol. 42, 161–167.

Dea, S., 2009. Establishment of favorable physical and environmental conditions for the optimization of the total product quality of fresh-cut 'Kent' mangoes. University of Florida, Ph.D. dissertation in Horticultural Science 251 p.

Dea, S., Brecht, J.K., Nunes, M.C., 2007. Visualization of polyphenoloxidase and phenolics distribution in mesocarp of fresh-cut mango (cv. Kent) during storage. Proc. Florida State Hort. Soc. 120, 263–266.

Dea, S., Brecht, J.K., Nunes, M.C.N., Baldwin, E.A., 2010a. Occurrence of chilling injury in fresh-cut 'Kent' mangoes. Postharvest Biol. Technol. 57, 61–71.

Dea, S., Brecht, J.K., Nunes, C.N., Émond, J.P., 2010b. Modified atmosphere packaging for fresh-cut 'Kent' mango under common retail display conditions. Proc. Fla. State. Hort. Soc. 123, 278–289.

Dea, S., Brecht, J.K., Nunes, M.C.N., Baldwin, E.A., 2013. Optimal ripeness stage for processing mangoes into fresh-cut slices. HortTechnology 23, 12–23.

Gonzalez-Aguilar, G.A., Celis, J., Sotelo-Mundo, R.R., de la Rosa, L.A., Rodrigo-Garcia, J., Alvarez-Parrilla, E., 2008. Physiological and biochemical changes of different fresh-cut mango cultivars stored at 5 C. Intl. J. Food Sci. Technol. 43, 91–101.

Gonzalez-Aguilar, G.A., Wang, C.Y., Buta, J.G., 2000. Maintaining quality of fresh-cut mangoes using antibrowning agents and modified atmosphere packaging. J. Agric. Food Chem. 48, 4204–4208.

Plotto, A., Narciso, J.A., Rattanapanoneb, N., Baldwin, E.A., 2010. Surface treatments and coatings to maintain fresh-cut mango quality in storage. J. Sci. Food Agric. 90, 2333–2341.

Poubol, J., Izumi, H., 2005. Shelf life and microbial quality of fresh-cut mango cubes stored in high CO_2 atmospheres. J. Food Sci. 70, M69–M74.

Rattanapanone, N., Lee, Y., Wu, T., Watada, A.E., 2001. Quality and microbial changes of fresh-cut mango cubes held in controlled atmosphere. HortScience 36, 1091–1095.

Rattanapanone, N., Watada, A.E., 2000. Respiration rate and respiration quotient of fresh-cut mango (*Mangifera indica* L.) in low oxygen atmosphere. Acta Hort. (509), 471–478.

Chapter 19.3

Fresh-cut fruits: Melons

Victor Rodov* and Ranjeet Shinde†

*Agricultural Research Organization, The Volcani Center, Rishon LeTsiyon, Israel, †Department of Plant Agriculture, University of Guelph, Guelph, ON, Canada

Quality characteristics

Fresh-cut melon products are prepared from fruit of two physiologically distinct horticultural groups: *Cucumis melo* var. *cantalupensis* bearing climacteric fruit with typical aroma (e.g., cantaloupe, charentais, Galia, ananas types), and *C. melo* var. *inodorus* bearing nonclimacteric fruit having weak or no aroma (e.g., honeydew, Piel de Sapo, Yellow Canary types) (Robinson and Decker-Walters, 1997). Each type comprises numerous cultivars.

Due to the natural and breeding-related genetic variability, both *Cantalupensis* and *Inodorus* groups nowadays include cultivars of orange, green, and creamy to off-white flesh shades. Major quality attributes of fresh-cut melons include cultivar-specific flavor, aroma (especially for *Cantalupensis* cultivars), color, and texture. Fruit pieces (chunks) should be of more or less uniform size and shape, with clean-cut edges, free of microbial decay and physiological

Controlled and Modified Atmospheres for Fresh and Fresh-Cut Produce
https://doi.org/10.1016/B978-0-12-804599-2.00039-9

disorders (e.g., translucency), without noticeable liquid accumulation on container bottom. Besides genotype, fresh-cut melon characteristics are strongly affected by fruit maturity at harvest. To attain optimal combination of sensory and keeping quality, a melon typically should be harvested and processed slightly prior to its full maturity.

Main causes of quality loss

Microbial decay

Cut melon surface provides a rich nutrient substrate for the growth of bacteria, yeast, and molds. Fig. 1A presents early stages of microbial development on fresh-cut Galia-type melons visible as dark points (molds) and moist areas (bacteria or yeast). At advanced stage, visible microbial colonies may be observed on the spoiled product surface (Fig. 1B). Accumulation of leachate (typically turbid) in the container and off-odors often accompany the microbial spoilage. Efficient precut surface decontamination of whole melons intended for fresh-cut processing reduces the risk of microbial spoilage. Storage temperature is the most important environmental factor affecting the rate of microbial colonization of fresh-cut melon surface.

Tissue degradation

Mechanical damage inflicted to melon tissues by fresh-cut processing may result in tissue degradation that further progresses during storage. One of the most common manifestations of this degradation is appearance of water-soaked translucent areas, typically on the periphery of the piece (Fig. 2). The translucency is aggravated by using a dull instrument for melon cutting (Portela and Cantwell, 2001). Although no in-depth research of mechanisms underlying this phenomenon has been conducted by now, presumably it may be associated with

(A) (B)

FIG. 1 Microbial spoilage of fresh-cut melons. (A) early stage (Galia-type melon), (B) advanced stage (orange-fleshed ananas-type melon).

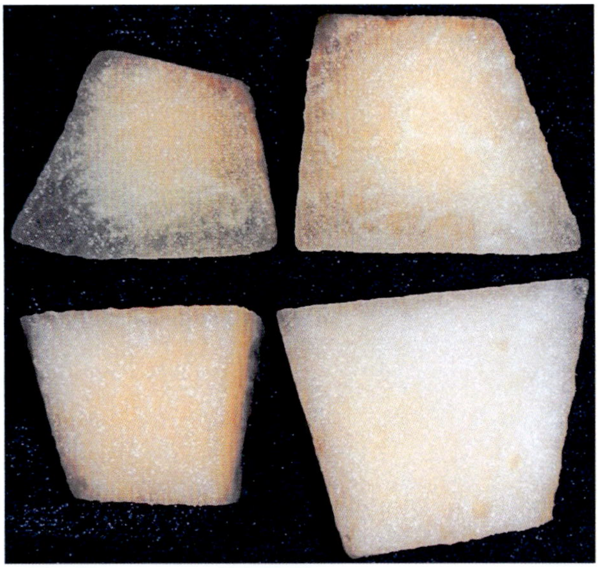

FIG. 2 Translucency in fresh-cut melons (upper row) in comparison with nontranslucent pieces (lower row).

disorders in cell wall integrity. The translucency is alleviated by treating the fresh-cut melons with calcium ions (Saftner et al., 2003) that strengthen the plant cell walls via cross-linking the pectin chains. In addition to translucent appearance, tissue degradation results in shape changes of fresh-cut melon chunks, loss of their clean-cut edges, and visual attractiveness. Note the rounded edges of the melon chunk presented in Fig. 1A.

Senescence

Similar to physiological changes taking place in whole melons in course of their overripening, fresh-cut melons during storage undergo senescence-associated changes resulting in their quality decline. Prevalent trends among these phenomena include softening and textural changes (tissue mushiness, translucency), color fading, decline in sugar content, and flavor deterioration. The senescence-related phenomena are especially characteristic to fresh-cut melons of climacteric *Cantalupensis* genotypes and are associated with ethylene action. Treating whole Galia-type melons prior to processing with ethylene antagonist 1-MCP delayed softening and translucent appearance, extending the commercial life of fresh-cut products by 2–3 days (Ergun et al., 2007). Similarly, fresh-cut product prepared from transgenic charentais-type melon with blocked ethylene biosynthesis demonstrated improved storage performance in terms of maintaining firmness, soluble solids content, flavor, sweetness, texture, and visual quality (Fonseca et al., 2001).

Off-flavor and off-odor

In fresh-cut melons (especially of *Cantalupensis* genotypes), quality loss is often primarily manifested as flavor and/or aroma deterioration. The off-flavors and off-odors may stem from various causes: microbial spoilage, senescence, inappropriate packaging and storage conditions, and so on. On the other hand, maintenance of desirable product quality under optimal storage conditions is usually reflected by persistence of cultivar-specific flavor and aroma.

It should be noted that deterioration symptoms causing quality loss of fresh-cut melons are typically manifested in complex. For example, microbiological spoilage presented in Fig. 3 results in physiological degradation such as color fading and softening. On the other hand, senescent tissues may be especially prone to microbial colonization.

Optimum storage conditions

Optimum temperature

Similar to most fresh-cut products, temperature range between 0 and 5°C is most suitable for preservation of fresh-cut melons. However, this temperature is not always attainable under real-life supply chain conditions. During marketing on cooled display at least part of the produce is exposed to temperatures about 6–8°C or even higher (Chen and Yuan, 2005; Nunes et al., 2009). These situations are too widespread to be ignored as "temperature abuse." Furthermore, such temperature regime has to be taken into account while evaluating the potential of various postharvest techniques for extending the produce shelf life.

FIG. 3 Effect of MA packaging on the appearance of fresh-cut melons after 14 days of storage at 6–8°C. (A) pink-fleshed Piel-de-Sapo-type melon cv. Sorbeto; left—no MA, right—passive MA package (nonperforated). (B) Galia-type melon cv. Raanan; left—passive MA package (microperforated), right—no MA.

CA/MA considerations

Atmosphere containing optimal concentrations of oxygen and carbon dioxide can slow down the deterioration of fresh-cut melons and help preserving their quality. Controlled atmosphere (CA) technique is helpful for experimental determination of beneficial and injurious O_2 and CO_2 levels, but usually is not applicable in real-life supply chain of fresh-cut produce. On the other hand, modified atmosphere (MA) packaging has good applied potential and is used commercially for preserving quality of fresh-cut melons. The MA packaging (MAP) can be implemented in two versions: so-called passive (or equilibrium) MAP using air as initial atmosphere, and active MAP when air in a package is initially substituted by flushing with a gas mix of desirable composition. In both cases, the initial atmosphere is further modified by the interaction of produce respiration with gas transmission through packaging material. The obvious advantage of passive MAP is its simplicity and lower cost. The advantage of the active MAP is in shortening or omitting a transient period of atmosphere equilibration when the product is exposed to nonoptimal O_2 and CO_2 levels and keeps deteriorating. On the other hand, active MA may sometimes aggravate the translucency in some melon types (e.g., Piel de Sapo) as compared with passive MA (Oms-Oliu et al., 2007; Rodov, nonpublished). The effect of vacuum applied during gas flushing on turgescent and fragile melon tissues might be involved in this phenomenon.

Benefits

Inhibiting growth of bacteria, yeast, and molds and controlling microbial spoilage is one of the major benefits of MA/CA storage to fresh-cut melons. Fig. 3 illustrates the advantages of MA packaging for preservation of fresh-cut melons of both *Inodorus* and *Cantalupensis* groups. Carbon dioxide plays the major role in the antimicrobial effect of modified atmospheres. Keeping fresh-cut cantaloupes in CA of air supplemented with 7.5% or 15% CO_2 diminished their aerobic plate counts by 4–5 log CFU/g at 5°C and by almost 3 log CFU/g at 10°C (Portela et al., 1997). Similar results were reported by Aguayo et al. (2007) for Amarillo melon of *Inodorus* group. Without the CO_2 supplementation, the inhibitory effect of low-oxygen CA (1.5% or 3% O_2) on microbial load of fresh-cut cantaloupe was limited, but combination of low oxygen with enhanced CO_2 levels provided the most efficient antimicrobial CA composition (Portela et al., 1997). In the same way, the population of yeast and molds on fresh-cut cantaloupe was inhibited by MA packaging, especially by active MA containing 3–5 kPa O_2 and 9–12 kPa CO_2 (Bai et al., 2001).

Another benefit of CA/MA storage is inhibition of physiological deterioration of fresh-cut melons (softening, translucency, discoloration, sugar loss). Low oxygen was the major factor that improved firmness preservation in CA-stored fresh-cut Amarillo melons (Aguayo et al., 2007). In fresh-cut cantaloupes kept at 10°C, softening was postponed by both reduced oxygen and enhanced carbon dioxide levels, either alone or in combination (Portela et al., 1997). Reduced translucency was the major positive physiological effect of MA packaging of fresh-cut cantaloupe (Bai et al., 2003) and honeydew (Bai et al., 2001). With both varieties, active MAP was much more efficient for translucency control than the passive one. The translucency in fresh-cut honeydew was reduced by CA storage in CO_2-enriched air (Portela and Cantwell, 1998) indicating that enhanced carbon dioxide rather than reduced oxygen plays a prevalent role in translucency control.

Tolerances/injuries

Beneficial CA compositions for preserving fresh-cut cantaloupes at 4.5°C contained 3.5%–6% oxygen and 6%–15% carbon dioxide, with the highest overall quality maintained at 4.5%–6% O_2 and 10%–15% CO_2. Carbon dioxide levels of $\geq$20% caused excessive softening, texture degradation, and off-flavor (O'Connor-Shaw et al., 1996). Atmosphere compositions with $\leq$3% O_2 resulted in accumulation of fermentative volatiles (ethanol and acetaldehyde) correlating with off-odor development in fresh-cut cantaloupe (Portela et al., 1997). In MA packaging, oxygen depletion occurs due to the imbalance between package oxygen transmission rate and produce respiration, typically at superoptimal storage temperature. In addition to off-odor and off-flavor, severe hypoxic fermentation may cause bloating of nonperforated MA packages resulting from anaerobic metabolism when carbon dioxide production is not balanced by oxygen consumption (Fig. 4). The risk of hypoxia can be omitted by using packaging materials with sufficiently high oxygen transmission rate, e.g., microperforated plastic films.

No big differences were found between *Cantalupensis* and *Inodorus* fresh-cut melons in their response to carbon dioxide level. Similar to cantaloupe melons, CA containing 15% CO_2 was helpful for preserving visual quality

FIG. 4 MA packages of fruit salad including fresh-cut orange-fleshed ananas-type melon. Left: nonperforated package bloated due to the severe hypoxic fermentation; right: microperforated package. The pencils were put on top of the packages in order to better visualize the difference in their height and shape.

and flavor of fresh-cut honeydew (Portela and Cantwell, 1998) and Amarillo melons (Aguayo et al., 2007).

At the same time, nonclimacteric *Inodorus* genotypes showed higher tolerance to low oxygen than cantaloupe-type melons. Storage at 2% O_2 and 10% CO_2 was beneficial for retaining quality of fresh-cut honeydew at 5°C. Even though respiratory quotient indicated the presence of anaerobic metabolic pathway in the fresh-cut honeydews kept under 2% O_2, it had no deleterious effect on the product quality (Qi et al., 1999). The quality of fresh-cut Piel de Sapo melons was better preserved in active MA packages with oxygen level declining from 2.5 kPa to <1 kPa than in passive MA maintaining the O_2 level well above 3 kPa. Minimum of 1 kPa O_2 and maximum of 15 kPa CO_2 were suggested as safe limits for MA storage of fresh-cut Piel de Sapo melons at 5°C. Crossing this low-oxygen damage threshold resulted in the onset of headspace ethanol accumulation (Oms-Oliu et al., 2007). Similarly, high-barrier MA package effectively preserved the quality of fresh-cut Barattiere melons (*Inodorus* group) in spite of the clearly hypoxic in-package atmosphere. Though, also in this case after oxygen depletion the product eventually underwent fast degradation (Conte et al., 2009). Using active MA with superatmospheric oxygen level (70 kPa) was demonstrated as means to preserve the quality of fresh-cut Piel de Sapo melons without a risk of hypoxic damage (Oms-Oliu et al., 2008).

In addition to product quality degradation, oxygen depletion in MA packages is allegedly associated with the health hazard of development of toxigenic anaerobic microorganisms such as *Clostridium botulinum*. However, no botulinal toxin was revealed in *C. botulinum*-inoculated fresh-cut cantaloupe stored for up to 21 days in passive MA packages at 7°C and 15°C, and in honeydew samples stored at 7°C. The toxin was detected in some honeydew samples stored at 15°C, but only at the stage when the product was severely spoiled and clearly unacceptable for consumption (Larson and Johnson, 1999).

CA/MA-temperature intersections

Similar to other types of fresh produce, oxygen consumption by fresh-cut melons and consequently chances of hypoxia development directly depend on temperature. Therefore while 2% O_2 was sufficient for quality maintenance of CA-stored fresh-cut honeydew at 5°C, the level of 4% O_2 was required at 10°C (Qi et al., 1999). Without such atmosphere adjustment, elevated storage temperature may increase the incidence of fermented off-flavor (Bett-Garber et al., 2003).

On the other hand, the benefits of CA/MA may be accentuated under the challenging conditions of elevated storage temperatures while adherence to optimal temperature is often sufficient for maintaining the product quality even without atmosphere manipulations. For example, CA storage was helpful for firmness maintenance of fresh-cut cantaloupe (Portela et al., 1997) and honeydew (Qi et al., 1999) stored at 10°C but not at 5°C because at 5°C the products

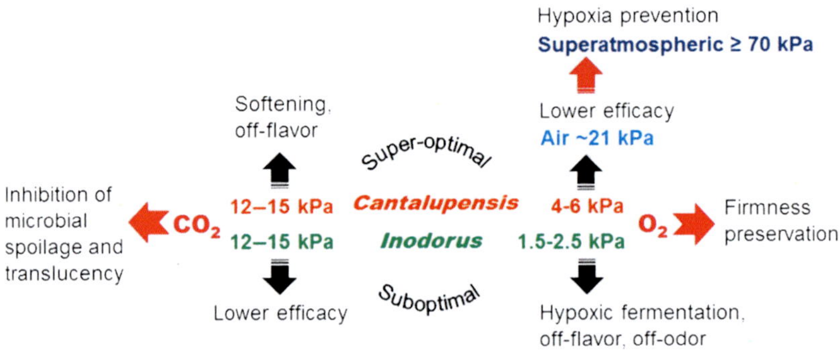

FIG. 5 Effects of CA/MA conditions on fresh-cut *Cantalupensis* and *Inodorus* melons. Arrow directions designate increase or decrease of O_2 or CO_2 levels compared with the optimal ones. Positive effects on the product quality are marked by red arrows, negative effects by the black ones.

did not undergo significant softening. Similarly, the improvement of visual quality of fresh-cut honeydew melons by passive or active MA packaging was more evident in samples kept at least part of the storage period at 10°C than in the samples continuously stored at 5°C (Bai et al., 2003).

Summary: Problems under suboptimal conditions

Fig. 5 summarizes optimal CA/MA compositions for the preservation of fresh-cut *Cantalupensis* and *Inodorus* melons and consequences of products' exposure to optimal, suboptimal, and excessive O_2 and CO_2 concentrations.

References

Aguayo, E., Escalona, V.H., Artés, F., 2007. Quality of minimally processed *Cucumis melo* var. *saccharinus* as improved by controlled atmosphere. Eur. J. Hort. Sci. 72, 39–45.

Bai, J.H., Saftner, R.A., Watada, A.E., Lee, Y.S., 2001. Modified atmosphere maintains quality of fresh-cut cantaloupe (*Cucumis melo* L.). J. Food Sci. 66, 1207–1211.

Bai, J.H., Saftner, R.A., Watada, A.E., 2003. Characteristics of fresh-cut honeydew (*Cucumis* x *melo* L.) available to processors in winter and summer and its quality maintenance by modified atmosphere packaging. Postharvest Biol. Technol. 28, 349–359.

Bett-Garber, K.L., Beaulieu, J.C., Ingram, D.A., 2003. Effect of storage on sensory properties of fresh-cut cantaloupe varieties. J. Food Qual. 26, 323–335.

Chen, Y.G., Yuan, X.L., 2005. Experimental study of the performance of single-band air curtains for a multi-deck refrigerated display cabinet. J. Food Eng. 69, 261–267.

Conte, A., Scrocco, C., Brescia, I., Del Nobile, M.A., 2009. Different packaging strategies for fresh-cut 'Barattiere' melon cultivar (*Cucumis melo* L.). Int. J. Food Sci. Technol. 44, 1422–1428.

Ergun, M., Jeong, J., Huber, D.J., Cantliffe, D.J., 2007. Physiology of fresh-cut 'Galia' (*Cucumis melo* var. *reticulatus*) from ripe fruit treated with 1-methylcyclopropene. Postharvest Biol. Technol. 44, 286–292.

Fonseca, R.M., Goularte, M.A., Silva, J.A., Lucchetta, L., Marini, L., Zanuzo, M.R., Antunes, P.L., Rombaldi, C.V., 2001. Conservabilidade de meloes transgenicos, cv. Vedrantais, minimamente processados e refrigerados. Rev. Bras. Agrocienc. 7, 149–151.

Larson, A.E., Johnson, E.A., 1999. Evaluation of botulinal toxin production in packaged fresh-cut cantaloupe and honeydew melons. J. Food Prot. 62, 948–952.

Nunes, M.C.N., Emond, J.P., Rauth, M., Dea, S., Chau, K.V., 2009. Environmental conditions encountered during typical consumer retail display affect fruit and vegetable quality and waste. Postharvest Biol. Technol. 51, 232–241.

O'Connor-Shaw, R.E., Roberts, R., Ford, A.L., Nottingham, S.M., 1996. Changes in sensory quality of sterile cantaloupe dice stored in controlled atmospheres. J. Food Sci. 61, 847–851.

Oms-Oliu, G., Soliva-Fortuny, R., Martín-Belloso, O., 2007. Effect of ripeness on the shelf-life of fresh-cut melon preserved by modified atmosphere packaging. Eur. Food Res. Technol. 225, 301–311.

Oms-Oliu, G., Martínez, R.R.M., Soliva-Fortuny, R., Martín-Belloso, O., 2008. Effect of superatmospheric and low oxygen modified atmospheres on shelf-life extension of fresh-cut melon. Food Control 19, 191–199.

Portela, S.I., Cantwell, M.I., 1998. Quality changes of minimally processed honeydew melons stored in air or controlled atmosphere. Postharvest Biol. Technol. 14, 351–357.

Portela, S.I., Cantwell, M.I., 2001. Cutting blade sharpness affects appearance and other quality attributes of fresh-cut cantaloupe melon. J. Food Sci. 66, 1265–1270.

Portela, S., Nie, X., Suslow, T., Cantwell, M., 1997. Changes in sensory quality and fermentative volatile concentrations of minimally processed cantaloupe stored in controlled atmospheres. In: Gorny, J.R. (Ed.), CA '97, Proceedings Volume 5: Fresh-cut Fruits and Vegetables and MAP. Postharvest Horticulture Series No. 19. Department of Pomology, University of California, Davis, CA, pp. 123–129.

Qi, L., Wu, T., Watada, A.E., 1999. Quality changes of fresh-cut honeydew melons during controlled atmosphere storage. J. Food Qual. 22, 513–521.

Robinson, R.W., Decker-Walters, D.S., 1997. Cucurbits. CAB International, Wallingford, UK.

Saftner, R.A., Bai, J., Abbott, J.A., Lee, Y.S., 2003. Sanitary dips with calcium propionate, calcium chloride, or a calcium amino acid chelate maintain quality and shelf stability of fresh-cut honeydew chunks. Postharvest Biol. Technol. 29, 257–269.

Chapter 19.4

Fresh-cut fruits: Pineapple

Marta Montero-Calderón*, Olga Martín-Belloso† and Robert Soliva-Fortuny†

*Postharvest Technology Lab., Center for Agronomic Research, University of Costa Rica, San Jose, Costa Rica, †Department of Food Technology, University of Lleida—Agrotecnio Center, Lleida, Spain

Quality characteristics

Fresh-cut pineapple is a tasty and exotic fruit, with high nutritional value. Because of the large size of the whole fruit, fresh-cut presentations are optimal to allow consumers enjoying its juiciness, sweetness, acidity, texture, flavor, and aroma, without any further preparation required.

Controlled and Modified Atmospheres for Fresh and Fresh-Cut Produce
https://doi.org/10.1016/B978-0-12-804599-2.00040-5

Fresh-cut pineapple has an important share of the fresh-cut market. In 2010, market share in the United States reached 14% and keeps growing (Rojas-Graü et al., 2010). Fresh-cut apples, watermelons, pineapples, and mixed fruits represent more than 80% of the traded volume in this market.

Quality and shelf life of the fresh-cut products depend on the raw materials and the processes used to prepare them. Proper selection of pineapples considering origin, cultivar, climate and growing conditions, season, stage of maturity, as well as postharvest handling practices prior to processing is key factor for success, together with adequate peeling, cutting, handling, packaging, cooling, and storage procedures and conditions. The Codex standard for pineapples (CODEX Alimentarius, 1993) establishes that fruit to be supplied fresh to the consumers need to be whole, clean, free of various damages, and practically free of pests; the fruit must be physiologically ripe without signs of unripening or over ripening; it must be able to withstand transport and handling; the fruit flesh total soluble solids content (SSC) should be at least 12° Brix; and sizes between 700 and 2750 g. For fresh-cut processing pineapple, large, crownless, cylindrical, and firm physiologically ripe fruit with bright flesh and nontranslucent tissues, with 12 °Brix or above, can lead to larger yields and better quality of the fresh-cut products.

Fresh-cut pineapple is offered in a wide variety of presentations: completely peeled and sliced fruit, slices, chunks, longitudinal wedges, sticks, cubes or bits in small plastic trays or pouches, or larger containers for the foodservice industry.

Several cultivars including Josephine, Spanish red, Queen Victoria, Pérola, Phuket, and Smooth cayenne, are used to produce fresh-cut pineapple depending on fresh fruit availability and consumer preferences, but gold cultivar is generally preferred because of the flesh color, texture, sweetness, and acidity, together with its long shelf life potential (Singh et al., 2007; Montero-Calderón et al., 2008; Bierhals et al., 2011; Finnegan et al., 2013).

Pineapple flesh quality attributes vary along the fruit, because its rather large size and unique morphology composed of multiple fruitlets which show a ripening pattern from the bottom of the fruit to the crown; thus differences among individual fruit pieces are endogenous. In fact, some authors recommend discarding 3 cm of the apical extreme of the fruit when preparing fresh-cut products (Antoniolli et al., 2007). For Gold cultivar, Montero-Calderón et al. (2010) reported changes along the fruit, from the base to the crown, for water content (81.2%–85.7%), ratio of soluble solids content to titratable acidity (28.9–17.9), POD activity (6.7–6.0 UA/min/mL), juiciness (10.4–12.1 g/100 g fresh weight), color (L*: 66.8–70.5; b*: 46.0–49.0), total phenolic content (50.8–40.3 mg gallic acid/100 g fresh weight), vitamin C (305–351 mg/kg fresh weight), and antioxidant capacity (43.1%–45.6%DPPH inhibition). Texture parameters and aroma volatile compounds (methyl-butanoate, methyl 2-methyl butanoate, and methyl hexanoate) change as well, due to both position inside each fruit and differences among fruits.

Main causes of quality loss

Juice leakage, changes in fresh appearance given by color, gloss and texture retention, odd-flavors and odd-odors, microbial growth, and, for some cultivars, tissue browning are the major causes of quality loss.

Juice leakage and water vapor losses are favored by the skin removal, handling, and cutting operations. Skin and core removal causes wounding tissue and increases the surface/mass ratio, thus increasing gas exchange between the flesh and the surrounding atmosphere. In turn, wounding triggers metabolic processes within the fresh-cut fruit. The lack of a protective barrier also favors juice drainage and water evaporation from the product surface.

Dripping of pineapple juice increases during handling and along storage and distribution. It accumulates within the package, and part of the product may eventually sink in it (Fig. 1). As the fruit pieces are immersed in pineapple juice, they become flooded and translucent. Respiration rate can be affected because air transfer through the tissues becomes harder, eventually leading to anaerobic respiration. Under these conditions, the loss of quality attributes and microbial spoilage is likely to be accelerated, thus reducing the product shelf life. Even though no information is available on either how fast overall quality deterioration of fresh-cut pineapple occurs, or how the presence of juice within the package affects the consumers' acceptance, it may denote lack of freshness. Juice accumulation inside fresh-cut packages can be partially controlled with a proper selection of the ripeness stage at processing, but also, with proper handling, which requires the use of sharp knives for peeling and cutting, the use of trays or other rigid packages instead of bags, and/or the application of edible coatings and temperature control (Montero-Calderón et al., 2008; Bierhals et al., 2011).

The fresh glossy appearance of the fruit flesh is lost when fruit packages allow considerable water vapor exchange through the package walls or lids, which favors fresh-cut pineapple pieces to shrink and look dry. On the other hand, color changes may occur due to enzymatic browning in some sensible cultivars like Smooth Cayenne, which has high PPO activity, but also due to

FIG. 1 Juice loss in fresh-cut pineapple throughout storage: juice drained from the cut fruit (glass) (left) and juice accumulation and product soaking in a commercial package (right).

an increase in the degree of translucency, which occurs in overripened fruit, water soaked tissues, or after the application of vacuum pressures during the packaging steps, especially in some cultivars such as MD2 (González-Aguilar et al., 2005; Montero-Calderón et al., 2008; Tavarini et al., 2010).

Even though respiration rate of fresh-cut pineapple is relatively low, it increases after preparation for a few hours. Finnegan et al. (2013) reported that at 4°C, the respiration rate of ripe pineapple increased along processing to 8.52 ± 4.68 mL CO_2/kg/h and equilibrated to an average value of 2.64 ± 0.68 mL CO_2/kg/h after 12–20h. Physiological age, origin, piece size, and sharpness of the blade to prepare the cut fruit also affect respiration rate. As the fruit pieces respire, they consume O_2 and produce carbon dioxide, hence modifying the internal atmosphere. When the headspace composition reaches levels of O_2 below 2% and/or high levels of CO_2, anaerobic respiration may occur, with the production of ethanol, acetaldehyde, and other aromatic compounds, which leads to the appearance of off-flavors and odors. Such changes, which can occur even within few days, greatly depend on the design of the packaging system, which involves a proper selection of the packaging material and initial in-package composition (Marrero and Kader, 2006; Montero-Calderón et al., 2010; Manzocco et al., 2016).

Selection of packaging material with the desired permeability to O_2, CO_2, and water vapor is needed to grant adequate headspace concentration, but packages size, surface area exposed to inner atmosphere, free space, wall thickness, quantity of packed fresh-cut product inside the package, product/volume ratio, sealing closing characteristics, and storage conditions through the cold chain can also affect the composition changes inside the package and the mass transfer to the surrounding atmosphere. Ideal permeability characteristics should consider these factors, and adjustments might be needed depending on the size of the pieces and the quantity inside the packages.

Finnegan et al. (2013) modeled changes in the atmosphere composition along storage and recommended some O_2 and CO_2 permeability coefficients to achieve optimum concentrations for specific serving size of fresh-cut pineapple. In addition, Chonhenchob et al. (2007) found that fresh-cut pineapple packed in polyethylene terephthalate trays under initial air conditions reached optimum O_2 and CO_2 concentration in one week; however, equilibrium was not reached, and concentration continued to change thereafter.

On the other hand, microbial growth of yeasts, molds, psychrophilic and mesophilic bacteria highly limit the shelf life of fresh-cut pineapple, because of the microbial load coming from the field in concomitance with preparation procedures. Rohrbach and Johnson (2003) reported that microorganisms can become trapped between fruitlets during pineapple development, thus leading to rather large microbial counts in the fresh-cut fruit compared to other fruit. Despite most of such organisms do not compromise the consumers' safety, sanitation of the fruit is jeopardized, which dramatically impacts the shelf life of the cut product.

Optimum storage conditions

Parameter	Range/description	Comments
Temperature, T (°C)	0–5	No chilling injuries have been reported for fresh-cut pineapple within two weeks of storage (Marrero and Kader, 2006, Montero-Calderón et al., 2008)
Relative Humidity, RH (%)	85%–90% or higher	Internal relative humidity in trays or pouches is likely to be near saturation; however, it has not been reported up to date. Water vapor losses could occur through films with high permeability or through pressure closure lids when fresh-cut pineapple is stored under low relative humidity conditions along the cold chain
Headspace gases concentration	(4%–8% O_2)/(5%–10% CO_2)[a] 50% O_2/50% CO_2[b]	Reported benefits: inhibition or reduction of microbial growth, reduced respiration rate, better color preservation and overall sensory attributes. Some authors report better tissue firmness control and reduced incidence of translucent tissues
Packaging materials	PET trays, microperforated films, PA/PE pouches, PP trays wrapped with PVC or PP film with selected permeability[c]. Finnegan et al., 2013 recommended O_2 and CO_2 transmission rates of 7300–12,500 and 13,900–23,500 ml m^{-2} day^{-1} atm^{-1} for optimal modified atmosphere packaging	Authors have reported good results with different types of packaging materials, with little differences in reported shelf life of fresh-cut pineapple
Maximum reported shelf life	15 days	Microbial growth and development of off-flavors and off-odors
Adjuvant treatments to MAP	Edible coatings (single and multiple layers, alone or as carriers of antioxidant, antimicrobial (lemongrass, trans-cinnamaldehyde), or antisoftening agents); High pressure Argon, Argon/Xenon treatment, high O_2/high CO_2 atmospheres; 1-MCP, UV-C light treatment and enriched MAP with N_2O.	Edible coatings act as a barrier to control internal composition within fruit pieces and as carriers of adjuvant treatments to reduce microbial growth and delay quality deterioration and juice drainage[d] High pressure Ar or AR/Xe, high O_2/high CO_2, 1-MCP help to reduce microbial decay[e]. UV-C light treatment and MAP supplemented with N_2O reduce respiration rate and ethylene production and delay softening[f].

[a]*Chonhenchob et al. (2007), Singh et al. (2007), Finnegan et al. (2013), Finnegan and O'Beirne (2015), Wu et al. (2012a,b, 2013), Mantilla et al. (2013), Rocculi et al. (2009), Liu et al. (2007) and Pan and Zu (2012).*

[b]*Zhang et al. (2013) to inhibit* Candida sake *and* Candida argentea *growth, but did not control* Leuconostoc citreum *bacteria.*

[c]*Singh et al. (2007), Paull and Lobo (2012), Manzocco et al. (2016) and Finnegan et al. (2013).*

[d]*Azarakhsh et al. (2014) and Mantilla et al. (2013).*

[e]*Zhang et al. (2013) and Wu et al. (2012a,b, 2013).*

[f]*Manzocco et al. (2016) and Rocculi et al. (2009).*

The main benefits of MAP include delay in microbial growth of yeasts, molds, and bacteria (both psychrophilic and mesophilic), which directly affects the product shelf life. It also reduces fruit respiration and ethylene production rate, and better preserves color, texture, and overall sensory attributes, though it does not have an effect on juice leakage and fermentation processes (Marrero and Kader, 2006; Rocculi et al., 2009). Montero-Calderón et al. (2010) also reported better preservation of aroma compounds in MAP with initial O_2 concentrations between 12% and 38% for two weeks; their results suggested that fresh-cut pineapple was able to tolerate low levels of O_2 and high levels of carbon dioxide for several days, before switching to anaerobic respiration. The same authors did not find any difference in soluble solids content, titratable acidity, pH, and color among fresh-cut pineapple stored with MAP with initial O_2 concentration of 12%, 21%, and 38% but vitamin C content and antioxidant capacity decreased when a gas mixture containing 38% O_2 was used.

Maximum claimed shelf life for fresh cut pineapple under MAP is 15 days, but several authors report one week or even a few days. Shelf life reduction is frequently caused by microbial growth or by the formation of off-odors and flavors due to the reduction of the in-package O_2 content to levels below 2% (Manzocco et al., 2016, Marrero and Kader, 2006). Several authors have proposed the use of MAP in combination with other treatments, such as UV-C light, high argon pressures, the use of mixtures containing Argon and Xenon, 1-methyl cyclopropane (1-MCP), single or multiple layer edible coatings, and the addition of N_2O to the gas mixture. Some of them can help to better control microbial growth; better preserve texture, color, and other quality attributes; or even favor product wound healing; however, none have yield shelf lives beyond two weeks.

Problems under suboptimal conditions

Higher temperatures: faster changes in product quality, respiration rates, headspace composition, and microbial growth, thus reducing shelf life.

Lower temperatures: storage below 0°C might result in freezing injuries.

Low oxygen or high carbon dioxide concentration of package headspace: Anoxia conditions in the headspace of the packages (less than 2% O_2) or elevated CO_2 concentrations lead to fermentation processes, ethanol and acetaldehyde presence, and development of off-odors and off-flavors.

Inadequate selection of packaging materials: internal atmosphere can reach undesirable levels of gas composition. Accelerated transpiration can cause dehydration and loss of fresh-like appearance.

Comments

Proper selection of fruit to be processed, adequate packaging materials, and storage conditions are the key elements to preserve product quality and extend its shelf life.

References

Antoniolli, L.R., Benedetti, B.C., Sigrist, J.M.M., Silveira, N.F.A., 2007. Quality evaluation of fresh-cut 'Perola' pineapple stored in controlled atmosphere. Food Sci. Technol. 27, 530–534.

Azarakhsh, N., Osman, A., Ghazali, H.M., Tan, C.P., Mohd Adzahan, N., 2014. Effects of gellan-based edible coating on the quality of fresh-cut pineapple during storage. Food Bioprocess Technol. 7, 2144–2151.

Bierhals, V.S., Chiumarelli, M., Hubinger, M.D., 2011. Effect of cassava starch coating on quality and shelf life of fresh-cut pineapple (*Ananas comosus* L. Merril cv 'Pérola'). J. Food Sci. 76, E62–E72.

Chonhenchob, V., Chantarasomboon, Y., Singh, S.P., 2007. Quality changes of treated fresh-cut tropical fruits in rigid modified atmosphere packaging containers. Packaging Technol. Sci. 20, 27–37.

CODEX Alimentarius. 1993. Codex Standard for Pineapples. CODEX STAN 182-1993. CODEX Alimentarius. Volume 5B. Tropical Fresh Fruits and Vegetables. Food and Agriculture Organization of the United Nations. World Health Organization. Rome.

Finnegan, E., O'Beirne, D., 2015. Characterizing deterioration patterns of fresh-cut fruit using principal component analysis. II: Effects of ripeness stage, seasonality, processing and packaging. Postharvest Biol. Technol. 100, 91–98.

Finnegan, E., Mahajan, P., O'Connell, M., Francis, G., O'Beirne, D., 2013. Modeling respiration in fresh-cut pineapple and prediction of gas permeability needs for optimal modified atmosphere packaging. Postharvest Biol. Technol. 79, 47–53.

González-Aguilar, G.A., Ruiz-Cruz, S., Soto-Valdez, H., Vázquez-Ortiz, F., Pacheco-Aguilar, R., Wang, C.Y., 2005. Biochemical changes of fresh-cut pineapple slices treated with antibrowning agents. nternational J. Food Sci. Technol. 40, 377–383.

Liu, C.L., Hsu, C.K., Hsu, M.M., 2007. Improving the quality of fresh-cut pineapples with ascorbic acid/sucrose pretreatment and modified atmosphere packaging. Packaging Technol. Sci. 20, 337–343.

Mantilla, N., Castell-Perez, M.S., Gomes, C., Moreira, R.G., 2013. Multilayered antimicrobial edible coating and its effect on quality and shelf-life of fresh-cut pineapple. LWT-Food Sci. Technol. 51, 37–43.

Manzocco, L., Plazzotta, S., Maifreni, M., Calligaris, S., Anese, M., Nicoli, M.C., 2016. Impact of UV-C light on storage quality of fresh-cut pineapple in two different packages. LWT-Food Sci. Technol. 65, 1138–1143.

Marrero, A., Kader, A.A., 2006. Optimal temperature and modified atmosphere for keeping quality of fresh-cut pineapple. Postharvest Biol. Technol. 39, 163–168.

Montero-Calderón, M., Rojas-Graü, M.A., Martín-Belloso, O., 2008. Effect of packaging conditions on quality and shelf-life of fresh-cut pineapple (*Ananas comosus*). Postharvest Biol. Technol. 50 (2/3), 182–189.

Montero-Calderón, M., Rojas-Gräu, M.A., Aguiló-Aguayo, I., Soliva-Fortuny, R., Martín-Belloso, O., 2010. Influence of modified atmosphere packaging on volatile compounds and physicochemical and antioxidant attributes of fresh-cut pineapple (*Ananas comosus*). J. Agric. Food Chem. 58, 5042–5049.

Pan, Y.G., Zu, H., 2012. Effect of UV-C radiation on the quality of fresh-cut pineapples. In: Second SREE conference on Engineering modeling and simulation (CEMS 2012), Hong Kong 37, 113–119.

Paull, R.E., Lobo, M.G., 2012. Pineapple. In: Siddiq, M. (Ed.), Tropical and Subtropical Fruits: Postharvest Physiology, Processing and Packaging. John Wiley & Sons, Inc., Malaysia, pp. 333–357

Rocculi, P., Cocci, E., Romani, S., Sacchetti, G., Rosa, M.D., 2009. Effect of 1-MCP treatment and N_2O MAP on physiological and quality changes of fresh-cut pineapple. Postharvest Biol. Technol. 51, 371–377.

Rohrbach, K.G., Johnson, M.W., 2003. Pests, Diseases and Weeds. In: Bartholomew, D.P., Paull, R.E., Rohrbach, K.G. (Eds.), The Pineapple: Botany, Production and Uses. CABI Publishing, New York, pp. 203–251.

Rojas-Graü, M.A., Garner, E., Martín-Belloso, O., 2010. Martín-Belloso, O., Soliva-Fortuny, R.J. (Eds.), The fresh-cut fruit and vegetables industry. Current Situation and market trends. 1st ed. CRC Press, Taylor & Francis Group, Boca Raton, pp. 1–11.

Singh, S.P., Chonhenchob, V., Chantarasomboon, Y., Singh, J., 2007. Testing and evaluation of quality changes of treated fresh-cut tropical fruits packaged in thermoformed plastic containers. J. Testing Evaluation 35, 1–7.

Tavarini, S., Trinci, L., Degl'inocenti, E., Guidi, L., 2010. Different sensitivity to browning in fresh-cut pineapple, apple and pear: the role of endogenous Vitamin C. Italian J. Food Sci. 22, 171–179.

Wu, Z., Zhang, M., Adhikari, B., 2012a. Application of high pressure Argon treatment to maintain quality of fresh-cut pineapples during cold storage. J. Food Engineering 110, 395–404.

Wu, Z., Zhang, M., Adhikari, B., 2012b. Application Effects of high-pressure argon and nitrogen treatments on respiration, browning and antioxidant potential of minimally processed pineapples during shelf life. J. Sci. Food Agric. 92, 2250–2259.

Wu, Z., Zhang, M., Adhikari, B., 2013. Effects of high pressure Argon and Xenon mixed treatment on wound healing and resistance against the growth of Escherichia coli or *Saccharomyces cerevisiae* in fresh-cut apples and pineapples. Food Control 30, 265–271.

Zhang, B., Samapundo, S., Pothakos, V., de Baenst, I., Sürengil, G., Noseda, B., Devlieghere, F., 2013. Effect of atmospheres combining high oxygen and carbon dioxide levels on microbial spoilage and sensory quality of fresh-cut pineapple. Postharvest Biol. Technol. 86, 73–84.

Part III

CA/MA requirements and recommendations for fresh and fresh-cut vegetables

Chapter 20.1

Fresh and fresh-cut fruit vegetables: Peppers

Maria Isabel Gil and Juan A. Tudela
Group on Quality and Safety, Department of Food Science and Technology, CEBAS-CSIC, Murcia, Spain

Quality characteristics

Bell peppers should have uniform shape with the size and color typical of the cultivar. The appearance should be glossy and the color green in immature stage but it can turn red, gold, purple, orange, or brown when ripening (Fig. 1). Flavor can be sweet, mild, or strongly pungent. The texture of the flesh should be firm and the thickness can vary depending on the cultivar. Maturity is assessed by size and color. When the color criterion is used, more than 50% of fruit surface should have the selected color. Fruit with cracks, decay, and sunburn must be sorted out and discarded before storage.

Controlled and Modified Atmospheres for Fresh and Fresh-Cut Produce
https://doi.org/10.1016/B978-0-12-804599-2.00041-7

FIG. 1 Bags composed by three different color peppers.

Main causes of quality loss

- Water losses and dehydration.
- Mechanical damage such as crushing, stem punctures, and cracks.
- Chilling injury.
- Accelerated ripening.
- Decay, bacterial soft rot.

Optimum storage conditions

To prolong storage life, rapid cooling is recommended.

Optimum temperature

The best temperature for storing whole peppers is 7.5°C, which can permit storage for 3–5 weeks. However, peppers can be stored also at 5°C, but only for a maximum of 2 weeks. After that, chilling injury will be manifested. A storage temperature of 5°C is recommended for preserving the quality of fresh-cut peppers.

Optimum relative humidity

Besides a low temperature, a RH of 95% will reduce dehydration during storage. Excessive RH and the presence of free water can promote bacterial soft rot.

CA/MA considerations

Benefits

Atmospheres with low O_2 (2%–5%) delay ripening during storage of whole peppers. CA of 3% O_2 + 5% CO_2 preserves the quality of red peppers better than that of green peppers for 3–4 weeks of storage at 5–10°C. Modified atmosphere packaging (MAP) in low density polyethylene (LDPE) film generating a gas composition of 2%–5% O_2 + 2%–5% CO_2 maintains green bell pepper quality for 2 weeks at 5°C. Quality of whole green, yellow, and red peppers has been preserved up to 14 days at 8°C using a microperforated film that maintained an atmosphere of 20% O_2 + 2% CO_2. Under these conditions, weight loss was only 0.2% when compared with a 4% weight loss for peppers stored in packages composed of macroperforated film.

For fresh-cut, CA of 3% O_2 + 10% CO_2 can preserve quality of green and red pepper slices up to 12 days at 5°C. A moderate CA of 5% O_2 + 5% CO_2 at 5°C has proved to reduce respiration rate and to maintain organoleptic, nutritional, and microbiological quality. A less stringent MA of 13%–14% O_2 + 7% CO_2 can maintain visual quality for 9 days at higher temperature (8°C) (Fig. 2).

However, these recommendations are impacted by the stage of development. The responses of fresh-cut peppers to CA/MA are impacted by ripening stage in that ripe fresh-cut peppers are sensitive to O_2 concentrations lower than 5% and CO_2 levels higher than 5%–10%.

Atmospheres with high O_2 (50%–80%) combined with 15% CO_2 have been reported to maintain visual quality and inhibited the growth of *Enterobacteriaceae* in fresh-cut peppers.

Additionally, a pretreatment with pressurized argon at 4 MPa (about 40 atmospheres of pressure), applied before storage in MA with 5% O_2 + 8% CO_2 at 4°C, can permit a shelf life of up to 12 days.

FIG. 2 Slices of fresh-cut red pepper packaged in LDPE film.

FIG. 3 Skin pitting in yellow and red bell peppers caused by elevated levels of CO_2 in the storage atmosphere.

Injuries

For whole fruits, a concentration of CO_2 higher than 5% can cause calyx discoloration, skin pitting, and softening during storage at 10°C. Levels higher than 10% CO_2 control growth of *Botrytis*, but reduce pepper quality.

For fresh-cut peppers, physiological injuries can be caused when using atmospheres with CO_2 higher than 15% (Fig. 3).

Problems under suboptimal conditions

A storage temperature lower than 7°C can induce chilling injury. Chilling injury symptoms include surface pitting, water-soaked areas, decay, discoloration of the seed cavity, and softening. Sensitivity to chilling injuries depends on maturity stage and cultivar. Immature peppers (green) are more sensitive to chilling temperatures than the colored peppers.

Temperatures higher than 13°C promote ripening, bacterial soft rot (*Erwinia* sp.), and fungal decay (*Botrytis* and *Alternaria*) (Fig. 4).

FIG. 4 Bacterial soft rot in red bell pepper.

Comments

Mechanical damage promotes water loss, discoloration, and decay. Preventing drops from excessive heights to avoid impact damage and preventing undue pressure during harvest, handling, and packaging to avoid compression damage is essential for maximizing shelf life.

Further reading

Cantwell, M.I., 1996. Bell pepper: Recommendations for Maintaining Postharvest Quality. http://ucanr.edu/sites/Postharvest_Technology_Center_/Commodity_Resources/Fact_Sheets/Datastores/Vegetables_English/?uid=19&ds=799 (Accessed January 18, 2014).

Conesa, A., Artés-Hernández, F., Geysen, S., Nicolaï, B., Artés, F., 2007. High oxygen combined with high carbon dioxide improves microbial and sensory quality of fresh-cut peppers. Postharvest Biol. Technol. 43, 230–237.

González-Aguilar, G.A., Ayala-Zavala, J.F., Ruiz-Cruz, S., Acedo-Félix, E., Díaz-Cinco, M.E., 2004. Effect of temperature and modified atmosphere packaging on overall quality of fresh-cut bell peppers. LWT- Food Sci. Technol. 37, 817–826.

González-Aguilar, G.A., 2016. Pepper. In: Gross, K.C., Wang, C.Y., Saltveit, M.E. (Eds.), USDA Handbook 66—The Commercial Storage of Fruits, Vegetables, and Florist and Nursery Crops. pp. 481–484. http://www.ba.ars.usda.gov/hb66/contents.html.

Lopez-Galvez, G., El-Bassuoni, R., Nie, X., Cantwell, M.I., 1997. Quality of red and green fresh-cut peppers stored in controlled atmospheres. In: Seventh international controlled atmosphere research conference, CA '97, vol. 5. pp. 152–157.

Manolopoulou, H., Xanthopoulos, G., Douros, N., Lambrinos, G., 2010. Modified atmosphere packaging storage of green bell peppers: quality criteria. Biosyst. Eng. 106, 535–543.

Meng, X., Zhang, M., Adhikari, B., 2012. Extending shelf-life of fresh-cut green peppers using pressurized argon treatment. Postharvest Biol. Technol. 71, 13–20.

Ranjitha, K., Sudhakar-Rao, D.V., Shivashankara, K.S., Roy, T.K., 2015. Effect of pretreatments and modified atmosphere packaging on the shelf life and quality of fresh-cut green bell pepper. J. Food Sci. Technol. 52, 7872–7882.

Rodoni, L., Vicente, A., Azevedo, S., Concellón, A., Cunha, L.M., 2015. Quality retention of fresh-cut pepper as affected by atmosphere gas composition and ripening stage. LWT- Food Sci. Technol. 60, 109–114.

Saltveit, M.E., 2003. A summary of CA requirements and recommendations for vegetables. Proceedings of the 8th International Conference. Oosterhaven, J., Peppelembos, H.W. (Eds.), Acta Hort., 600, 723–727.

Chapter 21.1

Leafy vegetables: Baby leaves

Maria Isabel Gil and Yolanda Garrido
Group on Quality and Safety, Department of Food Science and Technology, CEBAS-CSIC, Murcia, Spain

Quality characteristics

Baby leaves are marketed as a mixture of fresh young specialty greens also called salad mix, spring mix, California salad or mezclum. They are assorted small young leaves that contain a combination of red and green oak leaf, Romaine and Lollo Rosso lettuces, tatsoi, bok choy, arugula (rocket or roquette), Swiss chard, spinach, endive, frisée (curly endive), mustard, mizuna (Japanese mustard), dandelion, watercress, lamb's lettuce, radicchio, sorrel, or other leaf vegetables, depending on the season. A mix of baby leaves might include different types and varieties with a wide range of leaf shapes, colors, textures, and tastes and strong flavors such as those from rocket and watercress.

Baby leaves must be fresh, tender, and turgid with no yellowing, dehydration, and wilting and with the typical color of the variety. Leaves must not show any symptoms of discoloration, decay, mechanical damage, and physiological

Controlled and Modified Atmospheres for Fresh and Fresh-Cut Produce
https://doi.org/10.1016/B978-0-12-804599-2.00066-1

FIG. 1 Symptoms of quality loss such as wilting, yellowing, petiole cut edge browning and decay from left to right.

FIG. 2 Baby leaves package in trays to reduce physical damage and maintain high RH to prevent wilting.

disorders (Fig. 1). Quality characteristics include the maturity stage based on leaf and petiole length and in some cases, leaf width as well (Gil et al., 2012). Leaf length can be from 5 to 12 cm with an optimum size of 8–10 cm with a leaf width of 7–8 cm. The petiole must be present with a maximum length of 35% of leaf length, but it depends on species and variety. In the case of fresh-cut spinach and Swiss chard, the pieces should have a uniform length of 2.5–3 cm.

Baby leaves are sold unwashed or washed, in plastic trays in air, or in bags under MA (Fig. 2). The new trend is to sell unwashed baby leaves, packaged in trays or pouches to keep physical damage to a minimum and to prevent wilting from loss of water (Løkke et al., 2012).

Main causes of quality loss

- Mechanical damage can occur during handing and processing, particularly when the leaves are more turgid (Fig. 3).

FIG. 3 Quality categories (leaf damage) for commercial packaged spinach (*Spinacia oleracea* L.): from left to right, no damage, slight damage, moderate damage, severe damage, and leaf pieces.

- Water loss when stored under low RH or under light causes wilting.
- Development of undesirable off-odors is one of the major problems when stored under low O_2 and high CO_2, particularly in baby spinach with the production of ammonia and in rocket with the degradation of glucosinolates.
- Stem cut edge browning.
- Loss of green color (yellowing, chlorophyll loss) and presence of discoloration and color changes.
- Decay as a result of bacterial soft rot caused by *Pseudomonas* spp. and *Erwinia carotovora*, first on the cut stems or broken leaf areas and in some cases on edge margins of delicate leaves such as Lollo Rosso lettuce.

Optimum storage conditions

Baby leaves are highly perishable with mid-high respiration and water loss rates. In general, they should be precooled promptly after harvest either by hydrocooling or vacuum-cooling but specific conditions for time and pressure reduction rates are needed to prevent tissue damage (Garrido et al., 2015). Contact with melting ice and water sprays help in preventing dehydration. However, careful handling is necessary to avoid crushing or bruising the delicate leaves (Morris and Jobling, 2016).

Optimum temperature

Temperature is the single most important factor determining the postharvest quality of baby leaves. Shelf life of baby leaves is better maintained at 0°C. Temperature of 5°C is insufficient to maintain very good quality beyond 7 days. At 10°C, shelf life is limited to 3–4 days because of loss of quality and yellowing. Cantwell et al. (1998) observed that an average 90% of product volume was still marketable after 21 days at 0°C while less than 70% of the product was marketable after 7 days at 10°C. In general, baby lettuce has lower quality scores for the same storage period and temperature than other baby greens. Among the Brassicaceae, rocket salad typically lasts 7–10 days while other leafy greens last 10–14 days (Cantwell, 1997; Cantwell et al., 1998). It has been reported that rocket leaves can be stored successfully at 0°C for 16 days, while shelf life was 13 days at 5°C and only 8 days at 10°C (Koukounaras et al., 2007). In the

Chenopodiaceae family, spinach and chards have similar postharvest behavior. In particular, shelf life of fresh-cut spinach packaged in perforated bags is about 8 days at 8°C (Kou et al., 2014). For baby red leaves, beet leaves (*Beta vulgaris*) had a shelf life similar to that of green chard while red orach (*Atriplex hortensis*) had a shelf life similar to that of spinach (Cantwell et al., 1998). These authors also studied several kale cultivars including white and red feather kale and white and Nagoya Red and observed a long shelf life (4 weeks storage in air at 2.5°C) compared to rocket and mustard greens. In the case of watercress, it can be stored for 2–3 weeks at 0°C with >95% RH (Hruschka and Wang, 1979).

Optimum relative humidity

The general recommendation is that baby leaves should be stored with 95%–100% RH to prevent water loss (Cantwell et al., 2010). In general, shelf life is reduced to 2–3 days if stored under low RH. However, when leaves are going through minimal processing, lower RH (70%–80%) is recommended as leaves can be rehydrated during washing. It has been reported that when baby spinach was exposed to very high RH (98%) before processing, more decay and loss of quality during storage were observed because leaves were more fragile, increasing the damage during processing and the susceptibility to microbial colonization (Medina et al., 2012). After packaging and while waiting for dispatch, temperature has been observed to affect the package headspace atmosphere. Indeed, an increase in storage temperature from 3°C to 7°C led to a significant increase in respiration. The best combination of conditions in the high-care room was 3°C and 74% RH for a maximum of 1 day before dispatch (Oliveira et al., 2016).

MA considerations for lettuce, chard, and Brassicaceae

Modified atmospheres are generally not beneficial for baby leaves. MA is mostly beneficial for controlling RH because of the complex combination of a mixture of baby leaves of diverse botanical origins with differences in the recommendations of O_2 and CO_2 levels and their injurious levels (Fig. 4). Maturity stage affects respiration rate, being 60% higher for 2–3 cm leaves, on average, than the rates for 6–8 cm leaves (Cantwell et al., 1998). The ethylene production rate for baby leaves has been reported to be double the rates typically reported for mature leafy vegetables (0.4 and 1.20 $nL\,g^{-1}\,h^{-1}$ at 0° and 10°C, respectively, during the first 5 days after harvest) (Cantwell et al., 1998). For lettuce and chards, the average respiration rates (2–5 days after harvest) were similar, averaging 4.5 and 21 $\mu L\,CO_2\,kg^{-1}\,h^{-1}$ at 0° and 10°C, respectively, while for other baby Brassicaceae leaves, the average was 60% higher. Partial pressures of 15.8 kPa O_2 and 4.8 kPa CO_2 (1% ≈ 1 kPa) at equilibrium have been recommended for baby leaves including red chard stored at 5°C (Tomás-Callejas et al., 2011). Baby leaf lettuces such as green leaf, red leaf, and Lollo Rosso achieved a gas composition within an

FIG. 4 Some types of leaves of Brassicaceae family included in green salads.

MA bag of about 5% O_2 and 10% CO_2 and were able to retain acceptable quality after 12 days of storage (3 days at 4°C and then transferred at 7°C) (Martínez-Sánchez et al., 2012). The use of "novel" gases has also received some attention such as He-enriched MAs for the preservation of fresh-cut red chard quality (Tomás-Callejas et al., 2011). MA using super-atmospheric O_2 concentrations also provided a beneficial effect by inhibiting the aerobic mesophilic growth in baby spinach and baby Swiss chard (Allende et al., 2004; Tomás-Callejas et al., 2011).

MA considerations for baby spinach and fresh-cut spinach

Spinach can be marketed as fresh, small, young leaves washed and packaged for salads and also as cut leaves washed and packaged for cooking.

The vast majority of research published on CA/MA of baby spinach and fresh-cut has been studied using MA packing. Most of the atmospheres reported were very low concentration of O_2 in combination with elevated CO_2 to investigate the effects of MA on quality maintenance. Some of these publications described the benefits of low O_2 (0.8%–3%) and high CO_2 (8%–10%) as it related to a decrease in microbial proliferation (Babic and Watada, 1996; Gorny, 1997). Low O_2 atmospheres (0.8%) reduced the main spoilage agent *Pseudomonas*, but the control of microbial development in fresh-cut spinach is possible if the temperature does not exceed 5°C (Babic and Watada, 1996). In addition, low O_2 atmospheres have been shown to be beneficial in reducing respiration, weight loss, and antioxidant loss such as ascorbic acid and flavonoids (Ko et al., 1996; Gil et al., 1999). However, any beneficial effects of low O_2 atmospheres on reducing respiration and ascorbic acid loss were not

FIG. 5 Visual quality of baby spinach stored in air (left) and CA conditions of 0.5% O_2 + 10% of CO_2 (right) after 10 days at 7°C.

noted when leaves were held at 0°C (Izumi et al., 1997). Several authors also reported enriched CO_2 atmospheres benefitted storability by inhibiting growth of aerobic microorganisms in fresh-cut spinach (Babic and Watada, 1996). We observed that low O_2 (0.2%) and high CO_2 (10%) prevented leaf senescence measured as cell damage, lipid peroxidation, and chlorophyll degradation, but this MA was detrimental because of the intense off-odor development (Tudela et al., 2013; Fig. 5). On the contrary, atmospheres with moderate concentrations of O_2 and CO_2 (7%–10% O_2 and 5%–10% CO_2) controlled oxidative damage while avoiding the negative effect of air storage or the combination of low O_2 with injurious high CO_2 levels (Garrido et al., 2016), but moderate concentrations of O_2 and CO_2 also offered moderate benefit in delaying yellowing (Suslow and Cantwell, 1998).

In addition, Gorny (1997) reported that chlorophyll loss in spinach could be reduced by 50% when using C_2H_4 scrubbers.

The first recommended beneficial levels of 0.8%–3% O_2 + 8%–10% CO_2 made over the years have been refined. Modifications resulting from current research have reached that the recommended optimum range of MA for baby spinach and fresh-cut spinach leaves is a more moderate level of O_2 (7%–10%) and a similar, but broader range in CO_2 (5%–10% CO_2).

Problems under suboptimal conditions for baby spinach and cut leaves

If O_2 is reduced below a critical level (<0.8%), undesirable anaerobic metabolism of spinach can be induced (Ko et al., 1996). Low oxygen injury symptoms have been described as the development of strong off-odors (considered the major problem for baby spinach), decay associated with tissue breakage, and tissue softening (Gorny, 1997; Tudela et al., 2013). Low O_2 and high CO_2 were also detrimental because of the increase in pH and electrolyte leakage and the reduction in chlorophyll fluorescence (Tudela et al., 2013; Garrido et al., 2016). Moreover, increasing CO_2 concentrations up to 13% caused rapid tissue breakdown (Gorny, 1997) and the development of off-odors, resulting in quality rejection after 1 week at 7°C (McGill et al., 1966). Baby spinach exposed to CO_2 accumulates ammonia (Cantwell et al., 2010; Tudela et al., 2013), which can be toxic to plant cells (Toivonen, 1997). Ammonia has been reported as a good indicator of CO_2 injury in spinach (Rodriguez and Cantwell, 2002). Therefore the control of the concentration of O_2 and CO_2 gases combined with refrigeration is critical for the maintenance of product quality, avoiding the development of off-odors.

Postharvest lighting for leafy greens

Our recent results showed the strong impact of light on MA generated when packages of baby spinach were stored under different light conditions (Garrido et al., 2016). The impact of light on MA of fresh and fresh-cut vegetables is covered in a specific chapter in this book. For baby spinach, it is worth mentioning that under light, in MAP, the atmosphere has a higher O_2 and lower CO_2 compared to packages stored under darkness which has a more pronounced shift in the atmosphere was observed with lower O_2 and higher CO_2. Differences in these headspace gas compositions were due to photosynthesis and respiration reactions and they strongly affected the quality characteristics of baby spinach. Storage conditions under light are detrimental as the tissue senescence progresses with all characteristics of oxidative stress (damaged cells, lipid peroxidation, and chlorophyll degradation) and bacterial growth. On the other hand, under darkness, the low O_2 and high CO_2 were also detrimental because of the intense off-odor developments, the increase in pH and electrolyte leakage, and the reduction in chlorophyll fluorescence (Fig. 6).

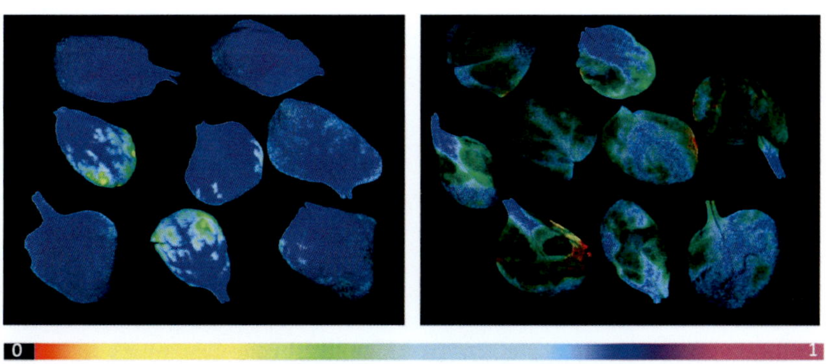

FIG. 6 Changes in chlorophyll fluorescence parameters of baby spinach stored in air (left) and CA of 0.5% O_2 + 10% CO_2 (right) in darkness for 10 days at 7°C. Images of the quantum yield of photosystem II (PSII) (Fv/Fm) are depicted using a false-color code ranging from 0 (black) to 1 (purple).

MA considerations for rocket

Rocket leaves are also known as rocket salad, roquette, arugula, rucola, or rugula (*Eruca sativa* Mill.) and wild rocket (*Diplotaxis tenuifolia* (L.) DC. CA storage of wild rocket leaves at 5 kPa O_2 + 10 kPa CO_2 showed a beneficial effect preserving the sensory and microbiological quality as well as the content of health-promoting phytonutrients compared to air (Martínez-Sánchez et al., 2006a). However, extremely low O_2 and high CO_2 levels promote visual quality deterioration followed by significant nutrient loss (Martínez-Sánchez et al., 2006b). Endelembos et al. (2015) studied the discrepancies between visual and overall freshness of rocket leaves stored in films with too low or too high O_2 transmission rate (OTR). In low OTR packages, they observed that a discrepancy between visual and overall freshness occurred as the packaged looked fresh but they were perceived spoiled after opening. Off-odor formation and change in color occurred at 0.5 kPa O_2 (Løkke et al., 2012; Endelembos et al., 2015). At low OTR, the O_2 concentration decreased to $\leq$0.5 kPa, and the leaves developed a "smoked odor," lost leaf integrity and texture, and turned olive-brown. At high OTR, the O_2 concentrations inside the package remained at 10–18 kPa, and leaves showed signs of senescence, i.e. light-green to yellow leaves and a heterogeneous appearance of leaves in the tray. Therefore recommendations for wild rocket must be MA packaged in an OTR film that permits sufficient levels of O_2 for aerobic respiration when storage temperature cannot be controlled and kept low during distribution and retail.

MA considerations for watercress and lamb's lettuce

There is limited information on watercress. The rate of yellowing has been reported to be reduced by storing watercress in atmospheres with >7% CO_2 (Aharoni et al., 1989). They suggested that the O_2 levels in the packages (4%–20.5%) were not impactful.

Lamb's lettuce retains acceptable quality after 28 days in sealed plastic bags with reduced O_2 and elevated CO_2 at <4°C (Wright, 2016). MAP reduces yellowing and decay of sorrel (Aharoni et al., 1993).

References

Aharoni, N., Reuveni, A., Dvir, O., 1989. Modified atmospheres in film packages delay senescence and decay of fresh herbs. Acta Hort. 258, 255–263.

Aharoni, N., Dvir, O., Chalupowicz, D., Aharon, Z., 1993. Coping with postharvest physiology of fresh culinary herbs. Acta Hort. 344, 69–78.

Allende, A., Luo, Y., McEvoy, J.L., Artés, F., Wang, C.Y., 2004. Microbial and quality changes in minimally processed baby spinach leaves stored under super atmospheric oxygen and modified atmosphere conditions. Postharvest Biol. Technol. 33, 51–59.

Babic, I., Watada, A.E., 1996. Microbial populations of fresh-cut spinach leaves affected by controlled atmospheres. Postharvest Biol. Technol. 9, 187–193.

Cantwell, M. 1997. Properties and recommended conditions for storage of fresh fruits and vegetables. http://postharvest.ucdavis.edu/produce_information/storagefreshproduce.

Cantwell, M., Rovelo, J., Nie, X., Rubatzky, V., 1998. Specialty salad greens: postharvest physiology and shelf-life. Acta Hort. 467, 371–377.

Cantwell, M., Hong, G., Nie, X., 2010. Using tissue ammonia and fermentative volatile concentrations as indicators of beneficial and stressful modified atmospheres for leafy and floral vegetables. Acta Hort. 876, 165–172.

Endelembos, M., Seefeldt, H.F., Løkke, M.M., 2015. MA_Packaged fruit and vegetables—Discrepancy between visual and overall freshness of green produce and relations to product quality. In: Amodio, M.L., Colleli, G. (Eds.), Proceedings of the XIth International Controlled and Modified Atmosphere Research Conference, Acta Horticulturae 1071, pp. 421–428.

Garrido, Y., Tudela, J.A., Gil, M.I., 2015. Comparison of industrial precooling systems for minimally processed baby spinach. Postharvest Biol. Technol. 102, 1–8.

Garrido, Y., Tudela, J.A., Hernández, J.A., Gil, M.I., 2016. Modified atmosphere generated during storage under light conditions is the main factor responsible for the quality changes of baby spinach. Postharvest Biol. Technol. 114, 45–53.

Gil, M.I., Ferreres, F., Tomás-Barberán, F.A., 1999. Effect of postharvest storage and processing on the antioxidant constituents (flavonoids and vitamin C) of fresh-cut spinach. J. Agric. Food Chem. 47, 2213–2217.

Gil, M.I., Tudela, J.A., Martínez-Sánchez, A., Luna, M.C., 2012. Harvest maturity indicators of leafy vegetables. Stewart Postharvest Rev. 8, 1–9.

Gorny, J.R., 1997. A summary of CA and MA requirements and recommendations for fresh-cut (minimally processed) fruits and vegetables. In: Gorny, J.R. (Ed.), CA'97 Proceedings Volume 5: Fresh-cut Fruits and Vegetables and MAP. Postharvest Horticulture Series. University of California, Davis, No. 19, 53.

Hruschka, H.W., Wang, C.Y., 1979. Storage and shelf life of packaged watercress, parsley and mint. Market Research Report 1102, U.S. Department of Agriculture, Washington, DC.

Izumi, H., Nonaka, T., Muraoka, T., 1997. Physiology and quality of fresh-cut spinach stored in low O2 controlled atmospheres at various temperatures. In: Gorny, J.R. (Ed.), CA'97 Proceedings Volume 5: Fresh-cut Fruits and Vegetables and MAP. In: Postharvest Horticulture Series, University of California, Davis, No. 19. pp. 130–133.

Ko, N.P., Watada, A.E., Schlimme, S.V., Bouwkamp, J.C., 1996. Storage of spinach under low oxygen atmosphere above the extinction point. J. Food Sci. 61, 398–400.

Kou, L., Luo, Y., Park, E., Turner, E.R., Barczak, A., Jurick, W.M., 2014. Temperature abuse timing affects the rate of quality deterioration of commercially packaged ready-to-eat baby spinach. Part I: sensory analysis and selected quality attributes. Postharvest Biol. Technol. 91, 96–103.

Koukounaras, A., Siomos, A.S., Sfakiotakis, E., 2007. Postharvest CO_2 and ethylene production and quality of rocket (*Eruca sativa* Mill.) leaves as affected by leaf age and storage temperature. Postharvest Biol. Technol. 46, 167–173.

Løkke, M.M., Seefeldt, H.F., Edelenbos, M., 2012. Freshness and sensory quality of packaged wild rocket. Postharvest Biol. Technol. 73, 99–106.

Martínez-Sánchez, A., Marín, A., Llorach, R., Ferreres, F., Gil, M.I., 2006a. Controlled atmosphere preserve quality and antioxidant constituents in wild rocket (*Diplotaxis tenuifolia* L.). Postharvest Biol. Technol. 40, 26–33.

Martínez-Sánchez, A., Allende, A., Bennett, R.N., Ferreres, F., Gil, M.I., 2006b. Microbial, nutritional and sensory quality of rocket leaves as affected by different sanitizers. Postharvest Biol. Technol. 42, 86–97.

Martínez-Sánchez, A., Luna, M.C., Selma, M.V., Tudela, J.A., Abad, J., Gil, M.I., 2012. Baby-leaf and multi-leaf of green and red lettuces are suitable raw materials for the fresh-cut industry. Postharvest Biol. Technol. 63, 1–10.

McGill, J.N., Nelson, A.I., Steinberg, M.P., 1966. Effects of modified storage atmospheres on ascorbic acid and other quality characteristics of spinach. J. Food Sci. 31, 510–517.

Medina, M.S., Tudela, J.A., Marín, A., Allende, A., Gil, M.I., 2012. Short postharvest storage under low relative humidity improves quality and shelf life of minimally processed baby spinach (*Spinacia oleracea* L.). Postharvest Biol. Technol. 67, 1–9.

Morris, S., Jobling, J., 2016. Watercress. In: Gross, K.C., Wang, C.Y., Saltveit, M.E. (Eds.), USDA Handbook 66—The Commercial Storage of Fruits, Vegetables, and Florist and Nursery Crops, pp. 597–598. http://www.ba.ars.usda.gov/hb66/contents.html.

Oliveira, A., Castro, P.M., Amaro, A.L., de Sain, J., Pintado, M. 2016. Optimization of temperature, relative humidity and storage time before and after packaging of baby spinach leaves using response surface methodology. Food Bioproc. Tech., in press, 1–10.

Rodriguez, S.C., Cantwell, M. 2002. Efecto de atmósferas controladas en la conservación de espinacas mínimamente procesadas. 3er Cong. Iberoamericano de Tecnología Postcosecha y Agroexportaciones, Santiago de Chile, December, 2001.

Suslow, T., Cantwell, M., 1998. Spinach. Recommendations for Maintaining Postharvest Quality, http://postharvest.ucdavis.edu/Commodity_Resources/Fact_Sheets/Datastores/Vegetables_English/?uid=32&ds=799.

Toivonen, P.M.A., 1997. Non-ethylene, non-respiratory volatiles in harvested fruits and vegetables: their occurrence, biological activity and control. Postharvest Biol. Technol. 12, 109–125.

Tomás-Callejas, A., Boluda, M., Robles, P.A., Artés, F., Artés-Hernández, F., 2011. Innovative active modified atmosphere packaging improves overall quality of fresh-cut red chard baby leaves. LWT- Food Sci. Technol. 44, 1422–1428.

Tudela, J.A., Marín, A., Garrido, Y., Cantwell, M., Medina-Martínez, M.S., Gil, M.I., 2013. Off-odour development in modified atmosphere packaged baby spinach is an unresolved problem. Postharvest Biol. Technol. 75, 75–85.

Wright, K.P., 2016. Salad greens. In: Gross, K.C., Wang, C.Y., Saltveit, M.E. (Eds.), USDA Handbook 66—The Commercial Storage of Fruits, Vegetables, and Florist and Nursery Crops, pp. 538–540. http://www.ba.ars.usda.gov/hb66/contents.html.

Further reading

Barth, M.M., Zhuang, H., Saltveit, M.E., 2016. Fresh-cut vegetables. In: Gross, K.C., Wang, C.Y., Saltveit, M.E. (Eds.), USDA Handbook 66—The Commercial Storage of Fruits, Vegetables, and Florist and Nursery Crops, pp. 624–641. http://www.ba.ars.usda.gov/hb66/contents.html.

Nicola, S., Tibaldi, G., Fontana, E., 2009. Fresh-cut produce quality: implications for a systems approach. In: Florkowski, W.J., Shewfelt, R.L., Brueckner, B., Prussia, S.E. (Eds.), Postharvest Handling, second ed. Elsevier Inc, pp. 247–282.

Saltveit, M.E., 2001. A summary of CA requirements and recommendations for vegetables. In: Postharvest Horticulture Series. University of California, Davis, No. 22A. pp. 71–94.

Chapter 21.2

Leafy vegetables: Lettuce, escarole, and radicchio

Maria Isabel Gil and Juan A. Tudela

Group on Quality and Safety, Department of Food Science and Technology, CEBAS-CSIC, Murcia, Spain

Quality characteristics of Lettuce

The main lettuce (*Lactuca sativa* L.) types grown are crisphead or iceberg, butterhead, romaine (cos), and leaf lettuce. Iceberg produces closed and compact heads; butterhead and romaine produce loose, open heads; and leaf lettuce does not form any head. To achieve the maximum shelf life, these products should be harvested at optimal maturity stage based on size of the "head" and/or compactness. They must be harvested clean without soil, with no evidence of browning.

Main causes of quality loss

- Mechanical damages during harvest and manipulation that can cause browning and decay.

Controlled and Modified Atmospheres for Fresh and Fresh-Cut Produce
https://doi.org/10.1016/B978-0-12-804599-2.00043-0

- Bacterial soft rot that occurs in bruised leaves and causes the breakdown of the tissue, which can be caused by *Erwinia*, *Pseudomonas*, *Xanthomonas*, and *Sclerotinia* spp.
- Tipburn, a physiological problem originated in the field that can be more evident during storage. It is recognized by the presence of brown necrotic leaf edges that can develop decay.
- Pinkrib, a pink discoloration of the midribs more frequently found in over-mature heads and it is related to genetic and environmental factors.
- Russet spotting that causes dark brown, oval lesions mainly on the midrib and it is produced by the presence of ethylene.

Optimum storage conditions

Lettuces should be cooled as soon as possible after harvest. For heading types, it is recommended vacuum cooling and hydrovacuum cooling, while for not heading types is better hydrocooling.

Optimum temperature

The temperature that best preserves lettuce quality is 0°C. However, it needs a precise control because the freezing point is −0.2°C.

Optimum relative humidity

To reduce water losses, it is recommended to apply a not condensing but high humidity in the cold room (98%–100% RH).

CA/MA considerations

Lettuce is a produce that obtain great benefits of CA and its use with an optimum temperature and RH can maintain lettuce quality up to a 4 weeks.

Benefits

Russet spotting can be reduced if lettuces are stored under a CA of 1%–3% O_2 at 0–5°C.

An atmosphere of 2%–5% O_2 + 0% CO_2 reduces respiration rate, decreases detrimental effects of ethylene, inhibits pink rib and butt browning, and maintains visual quality. If the storage needs to be prolonged up to 1 month, 3% O_2 + 2% CO_2 can also control decay.

Tolerances

The lower limit of tolerance for O_2 in iceberg and romaine lettuce is between 0.5% and 1%, depending on the cultivar.

Levels higher than 2% CO_2 can cause brown stain in iceberg lettuce.

A super-atmospheric level of O_2 up to 80% can develop russet spotting.

Injuries

Heart leaf injury is a reddish brown discoloration in the lead edge or in the entire leave of interior leaves. It appears when CO_2 levels are higher than 2.5% for 20 days.

Brown stain is a physiological disorder that occurs in iceberg lettuce that causes oval brown lesions on the margins of inner leaves. It is caused by a not adequate CA with moderate levels of CO_2 (>5%) for 2–3 weeks. This physiological injury is intensified when O_2 level decreases until 2%–3% (depending of the cultivar).

Problems under suboptimal conditions

To avoid wilting, storage temperature must be very low and RH very high.

Comments

Mechanical damages induce water losses, discoloration, and decay.

Quality characteristics of escarole

Cichorium endivia L. has two forms, a narrow-leafed form called endive and a broad-leafed form called escarole. For escarole, there are also two varietal groups that differ in the shape of the leaves: variety Crispa, also named Frisée that has highly divided leaves and twisted leaves with serrated edges and variety Latifolia that has broad and wavy leaves without little serrated edges. The outer leaves are dark green with bitter taste while the inner leaves are light green to creamy white with mild flavor. Escarole heads should be clean, crisp, and bright green.

Main causes of quality loss

- Water loss, dehydration, and wilting.
- Mechanical damage during harvest, handling, and packing.
- Presence of ethylene can produce leaf yellowing and brown spots on the midribs.
- Decay for bacterial soft rot that begins on stem and lower leaves and induce water-soaked appearance, softening, and breakdown of tissues, sometimes with a putrid odor.
- Marginal browning on the tips of outer leaves.

Optimum storage conditions

After harvest it is recommended to apply a precooling using vacuum cooling and hydrocooling at 0°C.

Optimum temperature

At 0°C escarole quality can be preserved for 2–4 weeks. This shelf life is reduced by half when the storage temperature increases until 5°C. Freezing injuries can appear when the temperature decreases to −0.1°C.

Optimum relative humidity

To prevent wilting during storage, it is necessary to apply a very high RH (95%–100%). The recommendation for temperature and humidity can be achieved by using top ice or package ice.

CA/MA considerations

Although there are no recommended CA/MA levels specific for escarole, those used to lettuce can be also used (1%–3% O_2 + <2% CO_2).

Injuries

A CA of 25% CO_2 during storage can produce browning in the inner leaves.

Problems under suboptimal conditions

To avoid wilting, storage temperature must be very low and RH very high.

Comments

Mechanical damages induce water losses, discoloration, and decay.

Quality characteristics of radicchio

Radicchio is a leaf chicory (*Cichorium intybus*) also known as Italian chicory with variegated leaves or red leaves. Heads are harvested by size and compactness with a maximum of 4 cm of root with no splitting. The growing period can vary between 45 and 150 days, being the most common around 80 days. Leaves have a long conical shape and a bittersweet and spicy taste similar to Belgian endive and escarole and a crispy texture. They should have bright appearance without marginal leaf necrosis, insect damage, or mechanical damage. To induce tender leaves, some cultivars are "whitened" by holding them in darkness at 10–15°C for 1–2 weeks.

Main causes of quality loss

- Dehydration that can appear in a few days during storage at low RH and could also lead to necrosis of the leaf margins.
- Bacterial soft rot caused by *Erwinia carotovora* at warmer temperatures.
- Decay by *Pseudomonas* at lower temperatures.
- Decay by *Botrytis cinerea* even with good temperature control during storage.
- Leaf abscission could occur in some cultivars.
- Mechanical damages during harvest and manipulation that can generate browning of the white midribs.

- Marginal leaf browning and fungal decay that can be caused by exposure to ethylene. Presence of this gas during storage can induce the change of pink to purple of the white midribs.

Optimum storage conditions

After harvest it is recommended to apply a rapid precooling using vacuum cooling.

Optimum temperature

The more effective technology to preserve radicchio quality after harvest is to hold the heads at low temperatures. The best temperature for maintaining radicchio quality up to 16–21 days of storage is 0°C.

Optimum relative humidity

Besides a low temperature, a high RH of 95%–100% will reduce dehydration during storage.

CA/MA considerations

Benefits

Although there are not very clear benefits of CA/MA to maintain quality of radicchio, an atmosphere of 3% O_2+5% CO_2 can be adequate to control decay.

Injuries

Low O_2 can promote internal browning of head.

Further reading

Cantwell, M.I., 2002. Appendix: Summary Table of Optimal Handling Conditions for Fresh Produce. In: Kader, A.A. (Ed.), Postharvest Technology of Horticultural Crops. University of California, Agriculture and Natural Resources, Publication 3311, pp. 511–518.

Cantwell, M.I., Suslow, T.V., 2001. Lettuce, Romaine: Recommendations for Maintaining Postharvest Quality. http://ucanr.edu/sites/Postharvest_Technology_Center_/Commodity_Resources/Fact_Sheets/Datastores/Vegetables_English/?uid=19&ds=799.

Cantwell, M.I., Suslow, T.V., 2002. Lettuce, Crisphead: Recommendations for Maintaining Postharvest Quality. http://ucanr.edu/sites/Postharvest_Technology_Center_/Commodity_Resources/Fact_Sheets/Datastores/Vegetables_English/?uid=19&ds=799.

Gil, M.I., Tudela, J.A., Martínez-Sánchez, A., Luna, M.C., 2012. Harvest maturity indicators for leafy vegetables. Stewart Postharvest Rev. 1 (2), 1–9.

Mateos, M., Ke, D., Kader, A., Cantwell, M., 1992. Differential responses of intact and minimally processed lettuce to high-carbon dioxide atmospheres. Acta Horticult. 343, 171–174.

Mencarelli, F., 2016. Radichhio. In: Gross, K.C., Wang, C.Y., Saltveit, M.E. (Eds.), USDA Handbook 66—The Commercial Storage of Fruits, Vegetables, and Florist and Nursery Crops, pp. 521–523. http://www.ba.ars.usda.gov/hb66/contents.html.

Moline, H.E., Lipton, W.J., 1987. Market diseases of beets, chicory, endive, escarole, globe artichokes, lettuce, rhubarb, spinach and sweet potatoes. USDA Handbook. 155.

Saltveit, M.E., 2016a. Endive and Escarole. In: Gross, K.C., Wang, C.Y., Saltveit, M.E. (Eds.), USDA Handbook 66—The Commercial Storage of Fruits, Vegetables, and Florist and Nursery Crops, pp. 326–327. http://www.ba.ars.usda.gov/hb66/contents.html.

Saltveit, M.E., 2016b. Lettuce. In: Gross, K.C., Wang, C.Y., Saltveit, M.E. (Eds.), USDA Handbook 66—The Commercial Storage of Fruits, Vegetables, and Florist and Nursery Crops, pp. 386–389. http://www.ba.ars.usda.gov/hb66/contents.html.

Suslow, T.V., Cantwell, M.I. 1999. Radicchio: Recommendations for Maintaining Postharvest Quality. http://ucanr.edu/sites/Postharvest_Technology_Center_/Commodity_Resources/Fact_Sheets/Datastores/Vegetables_English/?uid=19&ds=799.

Thompson, J.F., Gordon Mitchell, F., Rumsey, T.R., Kasmire, R.F., Crisosto, C.H., 1998. Commercial Cooling of Fruits, Vegetables, and Flowers. University of California. Agriculture and Natural Resources, Publication 21567.

Chapter 21.3

Leafy vegetables: Fresh-cut lettuce

Juan A. Tudela and Maria Isabel Gil
Group on Quality and Safety, Department of Food Science and Technology, CEBAS-CSIC, Murcia, Spain

Quality characteristics

To obtain a good quality fresh-cut product, it is absolutely necessary to use high quality lettuce heads. Reducing irrigation intensity and/or frequency before harvest can extend shelf life of fresh-cut iceberg and romaine lettuce (Luna et al., 2012, 2013). Harvesting at optimal maturity stage can also assure a longer shelf life (Kader et al., 1973; Gil et al., 2012). In addition, it is recommended to avoid cultivars with a high respiration rate and low sugar content (Varoquaux et al., 1996). After harvest, lettuce heads should be rapidly cooled at $<2°C$ with high RH to preserve quality (Cantwell and Suslow, 2001). Also, delaying processing for >48 h can seriously detract from product quality, reducing shelf life. During processing, excess water must be removed to minimize the risk of product decay.

Controlled and Modified Atmospheres for Fresh and Fresh-Cut Produce
https://doi.org/10.1016/B978-0-12-804599-2.00044-2

Main causes of quality loss

Leaf edge browning.
Off-odor and flavor development.
Decay and tissue softening.
Brown stain.
Translucency.
Dehydration that can occur when high RH is not maintained within the bag.
Crispness loss.

Optimum storage conditions

Optimum temperature

The optimal temperature for maintaining fresh-cut lettuce quality is 0°C. However, freezing injuries can occur when the temperature control is not adequate. Therefore, the general temperature recommendation is between 1°C and 3°C.

Optimum relative humidity

To avoid weight loss, fresh-cut lettuce is packaged in bags made of plastic films with reduced permeability to water vapor pressure. Packaging maintains RH near saturation.

CA/MA considerations

Atmospheres with low O_2 ($<1\%$) and high CO_2 ($>10\%$) prevent browning of lettuce cultivars that are not sensitive to brown stain (Mateos et al., 1993b; Varoquaux et al., 1996) (Fig. 1). The control of browning under this gas composition has been correlated with the suppression of enzymes phenylalanine ammonia lyase (PAL) and polyphenol oxidase (PPO), thereby slowing the synthesis and accumulation of caffeic acid derivatives (Luna et al., 2016). Therefore, the control of browning under MA can be achieved by the modification of the concentrations of both gases, and it is not exclusively due to the O_2 as previously described (Mateos et al., 1993a) (Fig. 2).

FIG. 1 Quality losses in fresh-cut romaine lettuce. From left to right: midrib browning after storage when O_2 concentration $\geq 1\%$ (left) and moderate (middle) and severe (right) brown stain as a consequence of exposure to high CO_2.

FIG. 2 Visual appearance of fresh-cut iceberg lettuce after 11 days of storage under different CA compositions: left (1% O_2 + 5% CO_2), center (1% O_2 + 10% CO_2), and right (1% O_2 + 15% CO_2). *(Source: Gil, M.I.)*

Benefits and tolerances

Modified atmosphere (MA) packaging is used to avoid wilting and discoloration (Ballantyne et al., 1998). Atmospheres with low O_2 (0.5%–3%) are recommended to reduce browning of iceberg, romaine, green leaf, and red leaf lettuces, while other lettuce types such as butterhead lettuce do not tolerate levels lower than 1% O_2 (Gorny, 2003). However, this O_2 range is even considered too high to prevent browning in iceberg and in romaine lettuces and the recommended O_2 levels are slightly lower (0.2%–0.5%) (Smyth et al., 1998; Martínez-Sánchez et al., 2011). Beside application of low O_2 levels, the use of high levels of CO_2 between 5% and 10% is recommended for most of lettuce types or, in the case of iceberg, is even higher (up to 15%) (Mateos et al., 1993b; Varoquaux et al., 1996; Gorny, 2003). Thus, visual quality of fresh-cut iceberg and romaine can be maintained for 12 days at 5°C under CA with 3% O_2 + 10% CO_2 (Lopez-Galvez et al., 1996).

Commonly, bags of fresh-cut lettuce are flushed before sealing with 100% N_2 to reduce O_2 from the beginning and get 1%–2% of residual concentration (Varoquaux et al., 1996). However, other authors have recommended higher levels of O_2 (2.5%–21%) to reduce off-odors (Kim et al., 2005). In commercial operations, fresh-cut processors flush with a mix of N_2 + CO_2 to reach 1%–2% O_2 + 8%–12% CO_2 immediately after sealing, although this gas composition could increase off-odor development. A positive correlation between CO_2 concentration within the package and off-odor development has been described (Tudela et al., 2013). Other packaging parameters that contribute to increase in CO_2 in the package and influence off-odors are lower initial bag volume, higher mass of product, and lower film permeability.

Injuries

Off-odors and brown stain can appear at the end of shelf life if low levels of O_2 and high levels of CO_2 are established. These disorders can be associated with different pre- and postharvest factors. Within the preharvest factors, cultivar plays an important role as does maturity stage and seasonality. Immature heads and cold season induce the highest accumulation of off-odors in iceberg and in romaine lettuce (Tudela et al., 2013, 2017).

Problems under suboptimal conditions

The use of CA or MA techniques cannot substitute for the correct temperature management. The best tool for the preservation of overall visual quality (OVQ) of fresh-cut lettuce is low temperature (2°C). This is independent of the level of O_2 in the range of 0.1%–5%. However, at higher temperatures, for example, if the storage is between 4°C and 7°C, the lowest levels of O_2 (0.1%) are needed to maintain quality, but a high risk of fermentation would be assumed (Fig. 3).

Temperature abuse induces fermentation of fresh-cut lettuce. A package with a targeted concentration of 0.5 kPa O_2 at 5°C can cause fermentation, producing off-odors if temperature increases up to 10°C (Smyth et al., 1998). Moreover, temperature abuse when O_2 concentration is very low could be a potential hazard by enhancing the growth of *Listeria monocytogenes* (O'Beirne et al., 2015).

Comments

Even though it is well known the recommended gas composition for preserving quality of fresh-cut lettuce, it is not easy to design an optimal MA package using

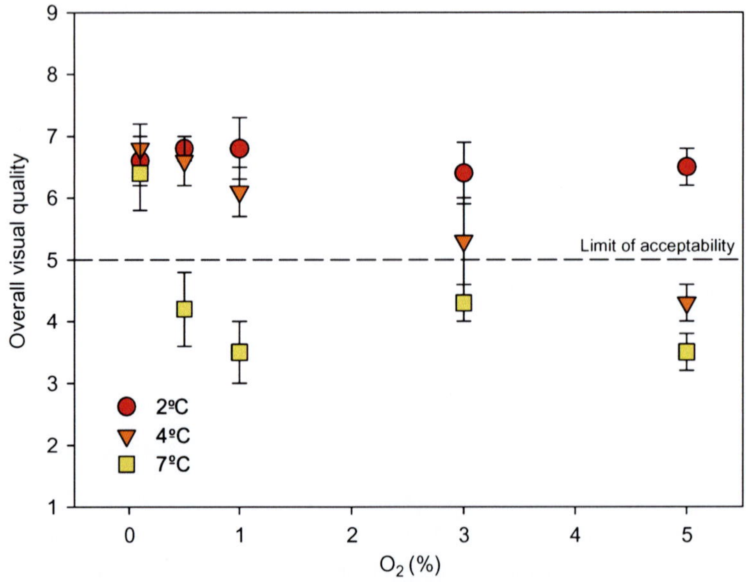

FIG. 3 Overall visual quality of fresh-cut romaine lettuce after 9 days of storage at 2°C, 4°C, and 7°C under CA with 10% CO_2 and different O_2 levels (Lopez-Galvez et al., 2019).

mathematical models such as that described by Mannapperuma et al. (1989) because these models are only valid under aerobic conditions and this is far from the situation that fresh-cut lettuce has in a bag (Gil and Tudela, 2013).

Moreover, light conditions during shelf life can also modify the O_2 concentrations within the bags and these can affect directly the lettuce visual appearance because of the increase in O_2 levels as a consequence of photosynthesis by the green tissue (Martínez-Sánchez et al., 2011).

Use of other gases such as argon has not shown a clear benefit in maintaining the quality of fresh-cut lettuce compared to N_2 (Baldassarre et al., 2015; Jaime and Saltveit, 2002).

References

Baldassarre, V., Amodio, M.L., De Chiara, M.L., Colelli, G., 2015. Effect of argon-enriched atmospheres on shelf life of fresh-cut "iceberg" lettuce. Acta Hort. (1071), 755–761.

Ballantyne, A., Stark, R., Selman, J.D., 1998. Modified atmosphere packaging of shredded lettuce. Int. J. Food Sci. Technol. 23, 267–274.

Cantwell, M., Suslow. T. 2001. Lettuce, Romaine. Recommendations for maintaining postharvest quality, 2001. http://postharvest.ucdavis.edu/pfvegetable/LettuceRomaine/ (Accessed December, 2016).

Gil, M.I., Tudela, J.A., 2013. Modified atmosphere packaging of fresh-cut lettuce: trial and error? In: XI International Controlled & Modified Atmosphere Research Conference. Trani.

Gil, M.I., Tudela, J.A., Martínez-Sánchez, A., Luna, M.C., 2012. Harvest maturity indicators of leafy vegetables. Stewart Postharvest Review 1 (2), 1–9.

Gorny, J.R., 2003. A summary of CA and MA requirements and recommendations for fresh-cut (minimally processed) fruits and vegetables. Acta Hort. (600), 609–614.

Jaime, P., Saltveit, M.E., 2002. Postharvest changes in broccoli and lettuce during storage in argon, helium, and nitrogen atmospheres containing 2% oxygen. Postharvest Biol. Technol. 26, 113–116.

Kader, A.A., Lipton, W.J., Morris, L., 1973. Systems for scoring quality of harvested lettuce. HortScience 8, 408–409.

Kim, J.G., Luo, Y., Tao, Y., Saftner, R.A., Gross, K.C., 2005. Effect of initial oxygen concentration and film oxygen transmission rate on the quality of fresh-cut romaine lettuce. J. Sci. Food Agric. 85, 1622–1630.

Lopez-Galvez, G., Saltveit, M.E., Cantwell, M., 1996. The visual quality of minimally processed lettuces stored in air or controlled atmosphere with emphasis on romaine and iceberg types. Postharvest Biol. Technol. 8, 179–190.

Lopez-Galvez, F., Vermeulen, A., Devlieghere, F., Tudela, J.A., Jacxsens, L., Gil, M.I., 2019. Modelling of quality loss of fresh-cut lettuce. In: Modeling in Horticultural Supply Chain. Molfetta, Italy, June.

Luna, M.C., Tudela, J.A., Martínez-Sánchez, A., Allende, A., Marín, A., Gil, M.I., 2012. Long-term deficit and excess of irrigation influences quality and browning related enzymes and phenolic metabolism of fresh-cut iceberg lettuce (*Lactuca sativa* L.). Postharvest Biol. Technol. 73, 37–45.

Luna, M.C., Tudela, J.A., Martínez-Sánchez, A., Allende, A., Gil, M.I., 2013. Optimizing water management to control respiration rate and reduce browning and microbial load of fresh-cut romaine lettuce. Postharvest Biol. Technol. 80, 9–17.

Luna, M.C., Tudela, J.A., Tomás-Barberán, F.A., Gil, M.I., 2016. Modified atmosphere (MA) prevents browning of fresh-cut romaine lettuce through multi-target effects related to phenolic metabolism. Postharvest Biol. Technol. 119, 84–93.

Mannapperuma, J.D., Zagory, D., Singh, R.P., Kader, A.A., 1989. Design of polymeric packages for modified atmosphere storage of fresh produce. In: Fellman, J.K. (Ed.), Proceedings of the Fifth International Controlled Atmosphere Research Conference, Vol. 2-Other Commodities and Storage Recommendations, pp. 225–233.

Martínez-Sánchez, A., Tudela, J.A., Luna, C., Allende, A., Gil, M.I., 2011. Low oxygen levels and light exposure affect quality of fresh-cut Romaine lettuce. Postharvest Biol. Technol. 59, 34–42.

Mateos, M., Ke, D., Cantwell, M., Kader, A.A., 1993a. Phenolic metabolism and ethanolic fermentation of intact and cut lettuce exposed to CO_2-enriched atmospheres. Postharvest Biol. Technol. 3, 225–233.

Mateos, M., Ke, D., Kader, A.A., Cantwell, M., 1993b. Differential responses of intact and minimally processed lettuce to high carbon dioxide atmospheres. Acta Hortic. (343), 171–174.

O'Beirne, D., Gomez-Lopez, V., Tudela, J.A., Allende, A., Gil, M.I., 2015. Effects of oxygen-depleted atmospheres on survival and growth of *Listeria monocytogenes* on fresh-cut iceberg lettuce stored at mild abuse commercial temperatures. Food Microbiol. 48, 17–21.

Smyth, A.B., Song, J., Cameron, A.C., 1998. Modified atmosphere packaged cut iceberg lettuce: effect of temperature and O_2 partial pressure on respiration and quality. J. Agric. Food Chem. 46, 4556–4562.

Tudela, J.A., Marín, A., Martínez-Sánchez, A., Luna, M.C., Gil, M.I., 2013. Preharvest and postharvest factors related to off-odours of fresh-cut iceberg lettuce. Postharvest Biol. Technol. 86, 463–471.

Tudela, J.A., Hernández, N., Pérez-Vicente, A., Gil, M.I., 2017. Growing season climates affect quality of fresh-cut lettuce. Postharvest Biol. Technol. 123, 60–68.

Varoquaux, P., Mazollier, J., Albagnac, G., 1996. The influence of raw material characteristics on the storage life of fresh-cut butterhead lettuce. Postharvest Biol. Technol. 9, 127–139.

Further reading

Cameron, A.C., 2003. Modified-atmosphere packaging of perishable horticultural commodities can be risky business. Acta Hort. (600), 305–310.

Cantwell, M.I., Kasmire, R.F., 2002. Postharvest handling systems: flowers, leafy and stem vegetables. In: Kader, A.A. (Ed.), Postharvest Technology of Horticultural Crops. University of California, Agriculture and Natural Resources, pp. 423–433. Publication 3311.

Chapter 21.4

Leafy vegetables: Fresh and fresh-cut mature spinach

Maria Isabel Gil and Yolanda Garrido
Group on Quality and Safety, Department of Food Science and Technology, CEBAS-CSIC, Murcia, Spain

Quality characteristics

Spinach leaves (*Spinacia oleracea* L.) covered in this chapter are slightly older, medium-sized leaves and should not be confused with baby or immature leaves. The information regarding baby spinach is detailed in the chapter related to baby leaves.

Controlled and Modified Atmospheres for Fresh and Fresh-Cut Produce
https://doi.org/10.1016/B978-0-12-804599-2.00046-6

Mature spinach leaves can be marketed as whole leaves (loose leaves or in bunches) and as cut leaves. Quality leaves should be uniformly dark green, fresh, crisp, and free from discoloration, yellowing, blemishes, and decay.

Leaves should be turgid, with minimum breakage and practically devoid of soil and roots as well as free of visible residues of fertilizers or other products.

Main causes of quality loss

- Water loss, wilting, and tissue softening.
- Mechanical damage during harvest, handling, and packing.
- Loss of chlorophyll and yellowing (Fig. 1)
- Freezing injury that darkens portions of the leaves or causes the tissue to become water-soaked and translucent.
- Symptoms of downy mildew infection (Bremia), presence of white rust *(Albugo officinalis*) and bacterial soft-rot (primarily *Erwinia* and *Pseudomonas*) associated with damaged leaves and stems.
- Presence of soil particularly after raining when fields are muddy.

Optimum storage conditions

Spinach is characterized as very perishable, with moderate rates of respiratory activity and water loss.

Optimum temperature

The maximum shelf life with the best retention of quality is achieved within the temperatures range of 0–5°C. Shelf life can be 3 weeks at 0°C, but only 2 weeks at 5°C (Cantwell and Kasmire, 2002).

FIG. 1 Symptoms of yellowing of mature spinach leaves and the yellowing rating scale for spinach (from 1 none to 5 severe).

Optimum relative humidity

The general recommendation is that RH must be high (90%–95%) to prevent water loss.

CA/MA considerations

Research on the CA/MA storage requirements for mature spinach leaves has been very limited (Fig. 2). In general, CA or MA is not used for the commercial storage of mature spinach leaves. Even though the CA/MA recommendations for spinach at the temperature range of 0–5°C is 7%–10% O_2 +5%–10% CO_2 (Saltveit, 1997), the potential application is very small (Saltveit, 2003).

Recommendations

Temperature between 0 and 5°C in macroperforated film to reduce water loss without modifying the gas composition.

Problems under suboptimal conditions

A combination of high humidity and high temperature favors development and spread of diseases in spinach and many other perishable leafy greens (Fig. 3). Mature spinach benefit very little by modified atmosphere because off-odors can be easily developed if O_2 concentration drops too low (Tudela et al., 2013).

FIG. 2 Experiments to test various combinations of O_2 and CO_2 for varying duration of storage for spinach are challenging but are required to find optimal gas combinations. Treatments can include cultivar, cultural practices, temperature, humidity, oxygen, carbon dioxide, growth regulators, plant characteristics, and growing environment.

FIG. 3 Visual quality of mature spinach stored in air (left) and in high CO_2 concentrations (8%–10%) (right) (Garrido et al., 2016).

Modified atmospheres must keep CO_2 at 5% or less as high CO_2 atmospheres are stressful to spinach (Cantwell and Kasmire, 2002). The concentration of ethylene in CA/MA should be minimal as spinach leaves are highly sensitive to exogenous ethylene that leads to yellowing (Saltveit, 2003). Exposure to light during storage can promote yellowing (Toledo et al., 2003). Chlorophyll content decreased after storage for 10 days at 7°C in air compared to the initial value, while the content was similar to the initial value in high CO_2 (Garrido et al., 2016) (Fig. 3).

Comments

The impact of specific atmosphere combinations can vary with different cultivars and growing conditions (e.g., seasons and locations), but CA and MAP are typically not recommended for mature spinach leaves. Performance under MA is very dependent on the maturity stage; the recommendations for baby spinach are different from those of mature leaves.

References

Cantwell, M.I., Kasmire, R.F., 2002. Postharvest handling systems: flower, leafy and stem vegetables. In: Kader, A.A. (Ed.), Postharvest Technology of Horticultural Crops. University of California, Agriculture and Natural Resources, Publication 3311. pp. 423–433.

Garrido, Y., Tudela, J.A., Hernández, J.A., Gil, M.I., 2016. Modified atmosphere generated during storage under light conditions is the main factor responsible for the quality changes of baby spinach. Postharvest Biol. Technol. 114, 45–53.

Saltveit, M.E., 1997. A summary of CA and MA requirements and recommendations for harvested vegetables. In: Seventh International Controlled Atmosphere Research Conference. Volume 4: Vegetables and Ornamentals. University of California, Davis, Postharvest Horticulture Series. No. 18. pp. 98–117.

Saltveit, M.E., 2003. A summary of CA requirements and recommendations for vegetables. In Proceedings of 8th International Controlled Atmosphere Research Conference. Oosterhaven, J., Peppelenbos, H.W. (Eds.), Acta Hort. 600, 723–727.

Toledo, M.E.A., Ueda, Y., Imahori, Y., Ayaki, M., 2003. L-Ascorbic acid metabolism in spinach (*Spinacia oleracea* L.) during postharvest storage in light and dark. Postharvest Biol. Technol. 28, 47–57.

Tudela, J.A., Marín, A., Garrido, Y., Cantwell, M., Medina-Martínez, M.S., Gil, M.I., 2013. Off-odour development in modified atmosphere packaged baby spinach is an unresolved problem. Postharvest Biol. Technol. 75, 75–85.

Further reading

Barth, M.M., Zhuang, H., Saltveit, M.E., 2016. Fresh-cut vegetables: Spinach (whole leaves, cut leaves). In: Gross, K.C., Wang, C.Y., Saltveit, M.E. (Eds.), USDA Handbook 66—The Commercial Storage of Fruits, Vegetables, and Florist and Nursery Crops, pp. 635–636. http://www.ba.ars.usda.gov/hb66/contents.html.

Moline, H.E., Lipton, W.J., 1987. Market diseases of beets, chicory, endive, escarole, globe artichokes, lettuce, rhubarb, spinach and sweet potatoes. In: USDA Handbook No. 155, pp. 39–43.

Suslow, T., Cantwell, M., 1998. Spinach. Recommendations for Maintaining Postharvest Quality. http://postharvest.ucdavis.edu/Commodity_Resources/Fact_Sheets/Datastores/Vegetables_English/?uid=32&ds=799.

U.S. Department of Agriculture, Agricultural Marketing Service, 1987a. Fruits and Vegetable Division. Fresh Products Branch, U.S. Standards for Grades of Bunched Spinach. https://www.ams.usda.gov/sites/default/files/media/Spinach_Bunched_Standard%5B1%5D.pdf.

U.S. Department of Agriculture, Agricultural Marketing Service, 1987b. Fruits and Vegetable Division. Fresh Products Branch, U.S. Standards for Grades of Spinach Leaves. https://www.ams.usda.gov/sites/default/files/media/Spinach_Leaves_Standard%5B1%5D.pdf.

Chapter 22.1

Floral vegetables: Artichokes

Maria Luisa Amodio, Michele la Zazzera and Giancarlo Colelli
Department of Science of Agriculture, Food and Environment, University of Foggia, Foggia, Italy

Quality characteristics

Artichoke (*Cynara cardunculus* var. scolymus) is cultivated for its large immature inflorescence, called "capitula" or "head" (Fig. 1). The edible part, that represents the 30%–40% of its fresh weight, is positioned in the central portion of the capitula, including the fleshy receptacle, on which are gathered the flowers reduced to a short hair, and inner tender bracts, protected by tough and fibrous inedible outer bracts (Fig. 2).

Each plant produces several heads, with the earliest formed at the apex of the central stem; they are harvested when the buds have grown to maximum size but before the bracts begin to spread open. The outer bracts of an artichoke ready for harvest should be tightly closed, firm, and turgid.

Controlled and Modified Atmospheres for Fresh and Fresh-Cut Produce
https://doi.org/10.1016/B978-0-12-804599-2.00049-1

FIG. 1 Artichoke "heads". *(Photo credits: Giancarlo Colelli.)*

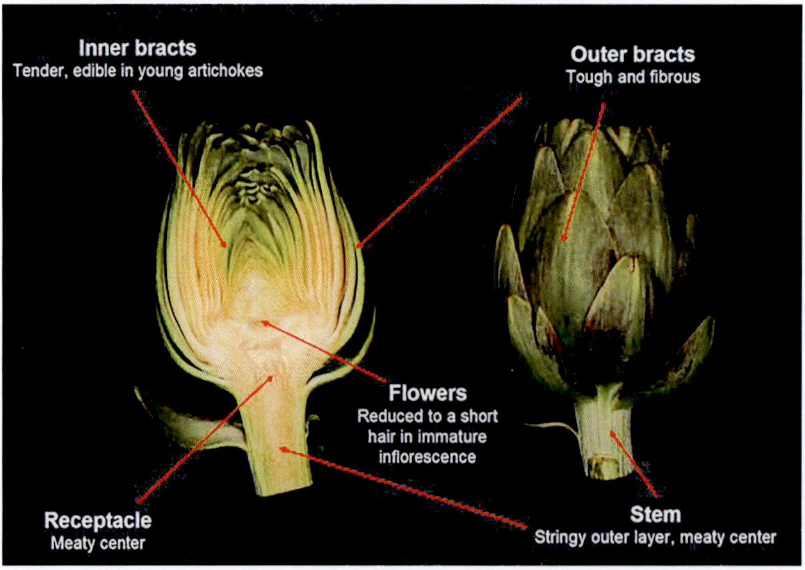

FIG. 2 Artichoke head and cross section. *(Photo credits: Michele la Zazzera.)*

According to last revised 2010 United Nations Economic Commission for Europe (UNEC) standards concerning the marketing and commercial quality control for fresh artichokes in 56 member States in Europe, North America, and Asia (72.5% of fresh artichokes global market), minimum quality requirements are the following:

- Artichokes must be (1) intact; (2) free from rotting or deterioration; (3) clean, free of any visible foreign matter; (4) fresh in appearance, with no sign of shrivel; (5) free from pests and damage caused by pests; (6) free of abnormal external moisture; (7) free of any foreign smell and/or taste; (8) stem not longer than 10 cm (not applicable to artichokes packed in bunches).

Moreover, the highest quality heads show no discoloration on the inner bracts, a short pappus, outer bracts tightly closed, and no blemishes. The content of each package must be uniform containing only artichokes of the same origin, variety or commercial type, quality and size, if sized. Artichokes must be packed in such a way as to protect the produce properly. Packages must be free of all foreign matter. Storage potential of artichoke is generally less than 21 days as visual and sensory quality deteriorate rapidly.

Main causes of quality loss

Freezing injury: Artichoke is quite resistant to low temperatures down to 0°C, but lower temperatures can cause serious damage to inflorescences eventually to the leaves. Freezing injury initiates at −1.2°C (29.9°F), with blistering and detachment of the cuticle and a bronzing of the outer bracts; at −4°C, cold damage may become more severe, resulting in watersoaked bracts with the heart becoming brown to black and gelatinous in appearance over time.

Chilling injury: Few studies indicated that some cultivars may show some chilling injuries when stored at 0°C, in the form of striped browning from the base of the external face of the inner bracts as reported for "Blanca de tudela" (Fig. 3), or as suggested by the higher susceptibility to browning of external bract, when processed (e.g., "Catanese" cultivar).

FIG. 3 Chilling injury in "Blanca de Tudela" artichoke bract. *(Photo credit: Mabel Gil.)*

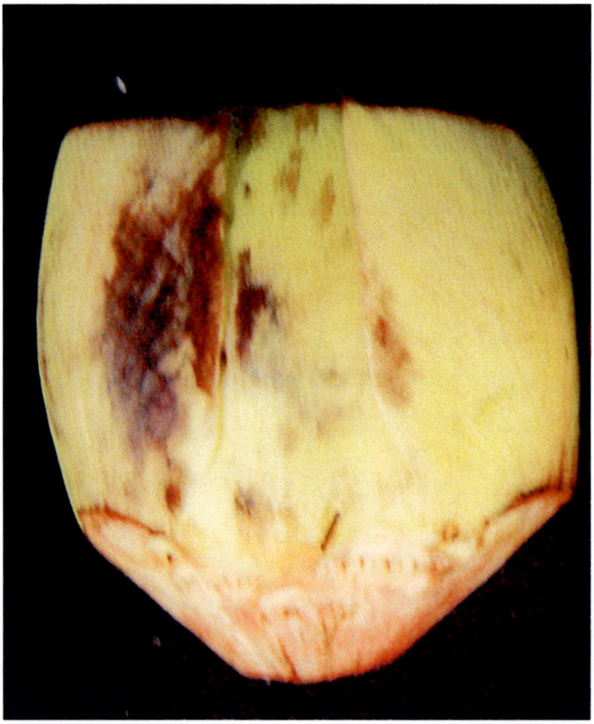

FIG. 4 Internal physical damage artichoke bract. *(Photo credit: Maria Luisa Amodio.)*

Weight loss: Due to the high artichoke respiration and transpiration rates, weight loss increases if the product is stored at temperatures higher than 25° C and relative humidity lower than 90%. Artichoke heads become dry and more fibrous as moisture loss increases.

Physical injury (bruising and compression) is common for artichoke when not handled with care. Rough handling during and after harvest also causes splitting of the bract tip. The abraded areas usually turn brown or black, providing a route through which microorganisms can enter, leading to decay. Moreover, internal browning can be detected, once the external bracts are removed (Fig. 4).

Bract opening and growth of floral primordia in the form of reddish tissues at the base of receptacle results from high temperature storage and senescence

Artichokes heads harvested late (after the commercial maturity) tend to harden because of an increase of fiber content, and the bracts begin to open.

Pathological disorders: Grey mold (*Botrytis cinerea*) and bacterial soft rot (*Erwinia carotovora*) can develop during storage and distribution, especially if the storage temperature is well above optimal.

Atmosphere oxygen levels below 2% may result in internal blackening of artichokes.

FIG. 5 "Blanca de Tudela" stored at 0, 2°C, 5°C, 7°C, and 10°C and high RH (90%–95%) for 14 days. *(Photo: Mabel Gil.)*

Atmospheres with high CO_2 (>10%) may result in internal blackening of artichokes.

Optimum storage conditions

Optimum temperature

The storage temperature should be kept between 0 and 2°C (32–35.6°F), for those cultivars not susceptible to chilling injury. Low temperatures slow metabolic processes, reduce the respiratory rate, minimize weight loss, and retard development of fungal and bacterial decay (Fig. 5). Temperatures below 0 °C must be avoided. The heads must be kept in a covered, cool, and ventilated place with appropriate humidity management to control moisture loss. Rapid cooling after the harvest to a target temperature of 2–4°C (35.6–39.2°F) is necessary especially in autumn and spring. Hydrocooling, forced air cooling, and package icing (e.g., top icing) are common methods of postharvest cooling of artichokes. Top icing and spraying with water the products are very effective in keeping the head temperature low and surface humidity high.

Optimum relative humidity

RH should be higher than 95% and never lower than 90% to avoid dehydration and, subsequently, weight loss.

CA/MA considerations

Some authors reported that controlled atmospheres (CA) or modified atmosphere packaging (MAP) offers moderate to little benefit to maintaining quality of fresh intact artichokes. The modest positive effects of CA or MAP depend on several factors in addition to gas mixture used, such as cultivar, storage temperature, and bud maturity. Therefore optimal CA conditions are reported as varying widely for this crop and a general recommendation cannot be made.

CA benefits

Reduction of transpiration rate.
Reduction of weight loss and water loss.
Retention of smoothness and firmness and, subsequently, reduction of wilting and shriveling.
Protection from mechanical damages.
Retention of nutritional content (e.g., phenolic compound).
Delay of bracts discoloration

Tolerances

While slight beneficial effects have been observed in response to storage atmospheres ranging between 1%–6% O_2 and 2%–7% CO_2 (Andre et al., 1980; Escriche et al., 1982; Ryall and Lipton, 1979) or restricted to 2%–3% O_2 and 2%–3% CO_2 (Beaudry, 1999; Saltveit, 2003; Sandhya, 2010), other studies report either no effect of atmosphere modification on shelf life of artichokes, or a response of some, but not all cultivars. According to Suslow and Cantwell (1997) conditions of 2%–3% O_2 and 3%–5% CO_2 at 5°C may delay discoloration of bracts and the onset of decay by a few days.

Injuries

Atmospheres of 3%–5% CO_2 may stimulate the growth of pappus on the receptacle (Miccolis and Saltveit, 1988). Atmospheres below 2% O_2, considered the fermentation threshold, may result in internal blackening of artichokes; similarly, injury will occur if CO_2 exceeds tolerable levels.

Particularly, atmosphere with high CO_2 (>10%), depending on the cultivar, may result in browning of internal bracts (Fig. 6) and internal core blackening. Among four popular cultivars, "Madrigal" has been found to be the most susceptible to internal core browning, but only when exposed to 20% CO_2 (Fig. 7), whereas some core browning was also observed in "Opal."

Problems under suboptimal conditions

Handling artichokes under suboptimal conditions will accelerate decay, wilting, and browning, reducing quality and storage potential. Artichoke of late harvests may be even more susceptible to suboptimal conditions.

FIG. 6 Effect of CO_2 concentration on browning of internal bracts of 4 artichoke cultivar. *(Photo credits: Maria Luisa Amodio.)*

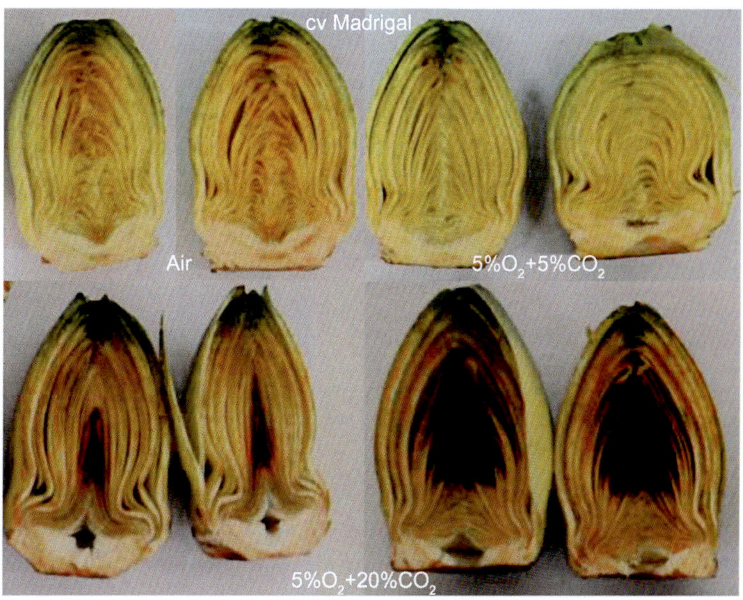

FIG. 7 Effect of CO_2 concentration on internal core blackening of artichoke cultivar "Madrigal." *(Photo credits: Maria Luisa Amodio.)*

Comments

Artichoke quality and shelf life is highly variable, depending on cultivar, growing location, and time of harvest.

References

Andre, P., Blanc, R., Buret, M., Chambroy, Y., Flanzy, C., Foury, C., Martin, F., Pelisse, C., 1980. Trails of preservation of artichokes by combined utilization of vacuum pre-cooling, controller atmospheres, and cold. Revue Hort. 211, 33–40.

Beaudry, R.M., 1999. Effect of O_2 and CO_2 partial pressure on selected phenomena affecting fruit and vegetable quality. Postharvest Biol. Technol. 15 (3), 293–303.

Escriche, A., Artes, F., Marin, J.C., 1982. Conservation d artichauts en atmosphere controlee. In: Progress in the design and operation of refrigerating equipment in the storage of fruits and vegetables by refrigeration. Inst. Intl. du Froid, Paris, France, pp. 336–341.

Miccolis, V., Saltveit, M.E., 1988. Influence of temperature and controlled atmosphere on storage of 'Green Globe' artichoke buds. HortScience 23, 736–741.

Ryall, A.L., Lipton, W.J., 1979. Handling, transportation and storage of fruits and vegetables. In: Vegetables and melons. Vol. 1. AVI, Westport, CT.

Saltveit, M.E., 2003. Is it possible to find an optimal controlled atmosphere? Postharvest Biol. Technol. 27, 3–13.

Sandhya, 2010. Modified atmosphere packaging of fresh produce: current status and future needs. LWT—Food Science and Technol. 43 (3), 381–392.

Suslow, T., Cantwell, M., 1997. Artichoke (globe): recommendations for Maintaining Postharvest Quality. http://postharvest.ucdavis.edu/Commodity_Resources/Fact_Sheets/Datastores/Vegetables_English/?uid=1&ds=799. Accessed 01/16.

Further reading

Aldoni, R., Girdini, L., 1989. Book chapter "Carciofo" on: Coltivazioni erbacee. Pàtron Editore: 745–761. ISBN 9788855597708.

Artichokes. How to manage pests. University of California. http://ipm.ucanr.edu/PMG/GARDEN/VEGES/artichokes.html (Accessed 01.16)

Bartz, J.A., Brecht, J.K., 2002. Postharvest physiology and pathology of vegetables. CRC Press. ISBN 9780824706876 (Accessed http://1.0.0.16).

Boriss, H., 2014. Artichoke profile. http://www.agmrc.org/commodities-products/vegetables/artichoke-profile/.

Cantore, V., Boari, F., 2009. Irrigazione e salinità. In: Calabrese, N. (Ed.), Il Carciofo e il cardo. Collana Coltura&Cultura. Edizioni Script, Bologna Italy, pp. 190–199. ISBN 978-88-96301-05-0.

Colelli, G., Calabrese, N., Calabrese, N., 2009. Trasformazione industriale. In: Il Carciofo e il cardo. Collana Coltura&Cultura. Edizioni Script, Bologna Italy, pp. 348–355. ISBN 978-88-96301-05-0.

Gil, M.I., Conesa, M.A., Artés, F., 2001. Effect of storage temperature on quality of artichokes (*Cynara scolymus* L.). Acta Hortic. (553), 555–557.

Gil-Izquierdo, A., Conesa, M.A., Ferreres, F., Gil, M.I., 2002. Influence of modified atmosphere packaging on quality, vitamin C and phenolic content of artichokes (*Cynara scolymus* L.). Eur Food Res Technol. 215, 21–27.

Gross, K.C., Wang, C.Y., Saltveit, M.E., 2016. The commercial storage of fruits, vegetables, and florist and nursery stocks, Agriculture Handbook Number 66, U.S. Department of Agriculture, Washington, DC http://ucanr.edu/datastoreFiles/234-2927.pdf (accessed 10.16).

Lattanzio, V., Kroon, P.A., Linsalata, V., Cardinali, A., 2009. Globe artichoke: a functional food and source of nutraceutical ingredients. Journal of Functional Foods 1, 131–144.

Mencarelli, F., Massantini, R., Casella, M., 1993. Influence of chemicals, stem lengthand plastic films on quality of artichoke buds. J. Hortic. Sci. 68, 597–603.

Piazza, R., Caccioni, D., 2009. Aspetti commerciali. In: Calabrese, N. (Ed.), Il Carciofo e il cardo. Collana Coltura&Cultura. Edizioni Script, Bologna Italy, pp. 431–439. ISBN 978-88-96301-05-0.

Rao, C.G., 2015. Engineering for Storage of Fruits and Vegetables, first ed. Elsevier. ISBN: 9780128033654.

Rappaport, L., Watada, A.E., 1958. Effect of temperature on artichoke quality. In: Proc. Conf. Transport Perishables, Univ. Calif., Davis CA, pp. 142–146.

Regione Puglia, 2008. Disciplinare regionale di produzione integrata—Carciofo- Anno 2008. In: Bollettino Ufficiale Regione Puglia No. 99 del 24-6-2008. http://www.regione.puglia.it/burp_doc/pdf/xxxix/N099_24_06_2008.pdf. (accessed 01.16).

Regulations (EU) No 1221/2008, http://eur-lex.europa.eu/LexUriServ/LexUriServ.do?uri=OJ:L:2008:336:0001:0080:en:PDF.

Regulations (EU) No 543/2011, http://eur-lex.europa.eu/legal content/EN/TXT/PDF/?uri=CELEX:32011R0543&from=it.

Ryder, E.L., DeVos, N.E., Bari, M.A., 1983. The globe artichoke (*Cynara scolymus* L). HortScience 18, 646–653.

UNECE Fresh fruit and vegetable—standards. Artichokes. http://www.unece.org/fileadmin/DAM/trade/agr/standard/fresh/FFV-Std/English/03Artichokes_2010.pdf. (Accessed 01.16).

Chapter 22.2

Floral vegetables: Fresh-cut artichokes

Maria Luisa Amodio, Michele la Zazzera and Giancarlo Colelli
Department of Science of Agriculture, Food and Environment, University of Foggia, Foggia, Italy

Quality characteristics

(Photo credits: Giancarlo Colelli)

Fresh-cut artichokes are highly perishable and suffer several degradative reactions, which strongly limit their marketability; the suitability of artichokes as a fresh-cut, 100% edible product is considered extremely difficult and therefore the R&D is needed to develop a high quality product. On the other hand, whole artichokes, with only the stem and the very external bracts removed and cut about 2 cm on top, are currently available on the market, packed in trays and wrapped in film (Fig. 1).

Controlled and Modified Atmospheres for Fresh and Fresh-Cut Produce
https://doi.org/10.1016/B978-0-12-804599-2.00050-8

FIG. 1 Fresh-cut artichokes. *(Photo credits: Maria Luisa Amodio.)*

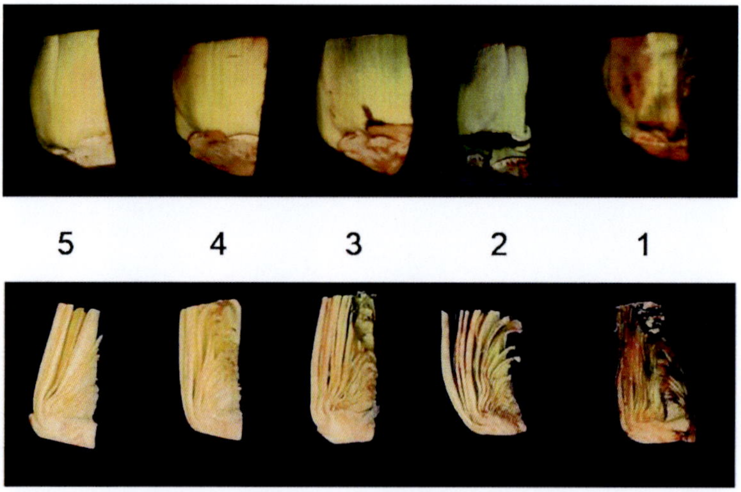

FIG. 2 Appearance score rating scale of outer bracts and cut surfaces of fresh-cut artichokes. *(Photo credits: Maria Luisa Amodio.)*

As for all the fresh-cut products, also for fresh-cut artichokes, appearance is the first quality attribute evaluated by consumers, even if other attributes such as organoleptic and nutritional characteristics are also very important. Fresh-cut produce should, in fact, appear clean, free of any visible foreign matter, as fresh as just cut, perfectly hydrated, free of abnormal external moisture, with bright color surface and with no sign of discoloration (browning, whitening, etc.), free from molds and rotten/deteriorated areas, pests, and any other defect; no off-odors must be perceived when the bag is open. In Fig. 2 an example of a laboratory photographic scale for appearance evaluation based on a 5-point scale is shown for outer bracts and cut surfaces. In this scale 5 = excellent appearance;

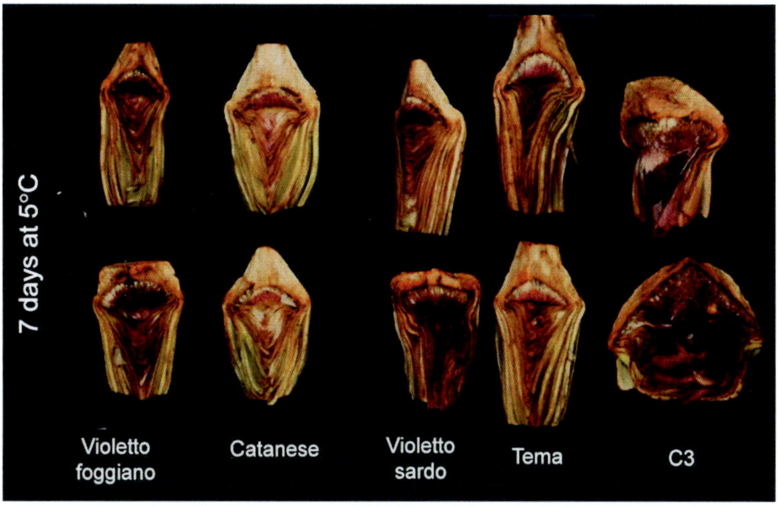

FIG. 3 Appearance of untreated fresh-cut artichokes of five different cultivars. *(Photo Credits: Maria Luisa Amodio; Results published in Cabezas-Serrano, A.B., Amodio, M.L., Cornacchia, R., Rinaldi, R., Colelli, G., 2009, Screening quality and browning susceptibility of 5 artichoke cultivars for fresh-cut processing. J. Sci. Food Agric., 89, 2588–2594.)*

4 = good; 3 = fair (limit of marketability); 2 = poor (limit of edibility); 1 = very bad, nonedible (Amodio et al., 2007; Cabezas-Serrano et al., 2009).

Main causes of quality loss

Enzymatic browning: Fresh-cut artichokes show a great tendency to enzymatic browning, due to the high content of phenolic compounds, mainly represented by chlorogenic acid, that constitute the substrate for oxidative reactions led by the enzyme polyphenol oxidase (PPO) (Lattanzio et al., 1994; Lattanzio, 2003; Shütz et al., 2004).

Selection of artichoke cultivars with specific physical, physiological, and biochemical characteristics, such as a low content of phenolic compounds that lead to enzymatic browning reactions, may be very critical for processing artichokes as fresh-cut products. Several studies pointed out this aspect (Cabezas-Serrano et al., 2009; Calabrese et al., 2012). An example of browning susceptibility of different cultivars is shown in Fig. 3.

Harvest season is very critical for postcutting performance of artichokes, showing that heads harvested in southern Italy during the coldest season (January–February) are generally more suitable for fresh-cut processing, compared to later harvest dates, in terms of browning, appearance score, and color parameters (Fig. 4).

Nonenzymatic browning: caused by iron-polyphenol complexing; these reactions may occur in cold stored plant tissues.

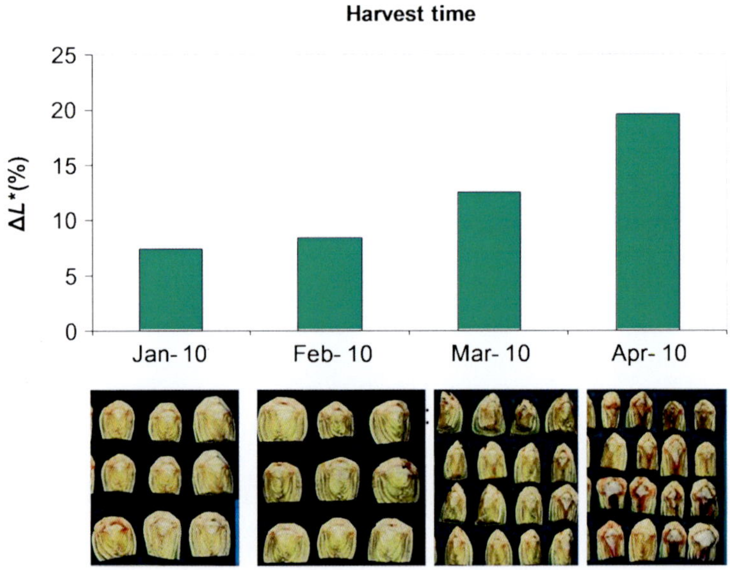

FIG. 4 Appearance and color variation (calculated as percentage variation from fresh samples) of untreated fresh-cut artichokes from different harvested times over the 2010 winter-spring. *(Photo credits: Maria Luisa Amodio; Results published in Ricci, I., Amodio, M.L., Colelli, G. 2013a. Influence of pre-cutting operations on quality of fresh-cut artichokes (Cynara scolymus L.): Effect of storage time and temperature before cutting. Postharvest Biol. Technol., 85, 124–131.)*

Growth of floral primordia in the form of fibrous pinkish tissue at the base of the receptacle can occur particularly for artichokes collected at the end of production season (late spring) when growth results much faster.

Weight loss and dehydration: high respiration and transpiration rates of artichokes increase with wounding, above all if the produce is not stored properly in terms of temperature and relative humidity.

Physical injury (bruising and compression): if the product is not handled with care. Damaged areas are more visible in the inner, more tender, bracts, which usually turn brown or black.

Optimum storage conditions

Optimum temperature

Similarly to other fresh-cut produce, minimally processed artichokes must be kept refrigerated at 0-4 °C (32-41 °F), and the cold chain must be maintained during all the postcutting life. Low temperatures allow to control respiration rate slowing down metabolic processes and reducing the weight loss; molds and bacterial development is also retarded. Temperature below 0°C must be avoided because of the risk of freezing. Temperatures above 4 °C cause dehydration and, subsequently, weight loss. Storage temperature of whole heads before cutting can affect fresh-cut artichoke quality; low temperatures (0°C) for 'Catanese' cv caused physiological injuries such as browning of bract surfaces, while precutting storage at 12°C resulted in the growth of floral primordia in the form of pinkish and fibrous tissue at the base of the receptacle, increasing the susceptibility to browning (Ricci et al., 2013b).

Optimum relative humidity

RH should be higher than 95% and never lower than 90%, to avoid dehydration and, subsequently, weight loss.

CA/MA considerations

Low O_2 concentrations in combination with increased levels of CO_2 are often used to maintain quality of several fresh-cut produce. However, differences between beneficial and harmful atmosphere combinations may be small when fresh-cut artichokes are stored in MA. In few studies, the effect of modified atmospheres on minimally processed artichokes was tested (Giménez et al., 2003; Del Nobile et al., 2009), without finding the safe atmosphere levels for this product. The presence of a very high CO_2 concentration (25%) in the storage atmosphere has been shown to be deleterious (la Zazzera et al., 2012), inducing the development of brown spots on the outer bracts, while only slight beneficial effects were observed at lower concentrations (5 and 15%), delaying browning on the cut surfaces. As for the effect of low O_2 concentrations, generally no or only slight beneficial effects were observed reducing the O_2 concentration in the range of 3-8%, while in some trials concentration of 1% resulted in a greater deterioration of the product, inducing dark spots on the outer bracts, similar to those observed for very high CO_2 concentrations (Fig. 5). Moreover the combination of 1 O_2% with 12% of CO_2 resulted in an increase of these kind of physiological damages (Fig. 6), as previously suggested by Kader (2002). Other authors, combining a soy protein isolate enriched with cysteine and different modified atmospheres (active with 5 kPa O_2 + 15 kPa CO_2; active with 80 kPa O_2 and passive), found that the MA did not increase shelf-life compared to coated samples, obtaining only 4 days of shelf-life (Ghidelli et al., 2015). Therefore, the avoidance of extreme conditions in terms of CO_2 and O_2 concentrations within the package should be the main objective when designing a modified atmosphere packaging (MAP) system for fresh-cut artichokes.

Benefits

- Reduction of transpiration rate.
- Delay of bract discoloration
- Reduction of weight loss and water loss.
- Protection from mechanical damages.
- Retention of nutritional content (e.g., phenolic compound).

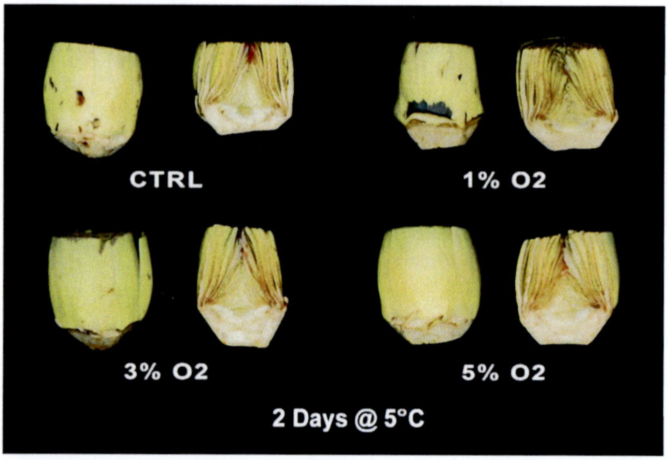

FIG. 5 Appearance of untreated fresh-cut artichokes stored in controlled atmosphere with different oxygen concentrations: example of blackening caused by low oxygen. *(Photo credits: Michele la Zazzera; Results from la Zazzera, doctoral thesis.)*

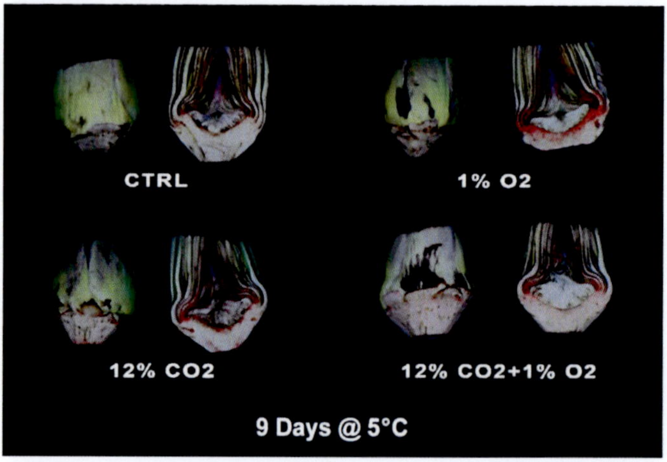

FIG. 6 Appearance of untreated fresh-cut artichokes stored in controlled atmosphere with different gas concentrations: example of blackening caused by low oxygen in combination of high CO_2. *(Photo credits: Michele la Zazzera; Results from la Zazzera, doctoral thesis.)*

Tolerances

A gas concentration of 5% O_2+10% CO_2 has been proposed as "target atmosphere," to be maintained inside the package in MAP conditions, and a "safety range" ($O_2 \geq 3\%$+$CO_2 \leq 15\%$) in which 3% O_2 was defined as the lowest level and 15% CO_2 as the highest one (la Zazzera et al., 2015).

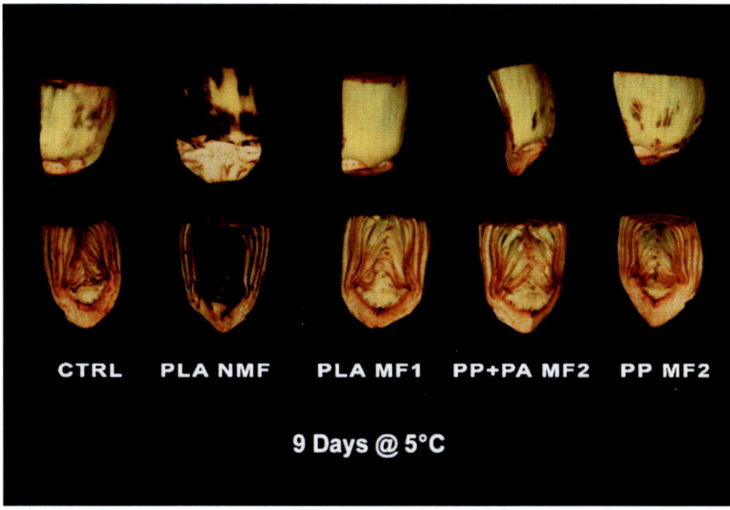

FIG. 7 Appearance of untreated fresh-cut artichokes stored in different packaging materials (PLA NMF: polylactic acid not microperforated; PLA MF1 and PLA MF2: polylactic acid with 1 or 2 lines of microperforation; PP MF2: polypropylene with microperforations. *(Results from la Zazzera, M., Amodio, M.L., Colelli, G. 2015. Designing a modified atmosphere packaging (MAP) for fresh-cut artichokes. Adv. Hort. Sci., 29(1), 24-29.)*

The use of microperforated films, allowing a greater permeability to O_2 and CO_2 than conventional films, should be recommended for fresh-cut produce with a very high respiration rate such as fresh-cut artichokes (Kader, 2002), since the accumulation of CO_2 can still be reached within the package but it avoids an extreme concentration and total O_2 depletion. In the optimal MAP, fresh-cut artichokes could be stored for 9 days (Fig. 7) in the absence of any stabilizing treatment (la Zazzera et al., 2015).

Injuries

Atmospheres below 3% O_2 may result in internal blackening of artichokes; similarly, injury will occur if CO_2 exceeds tolerable levels. Complete anaerobic condition (16% CO_2 and 0% O_2) developed in MAP caused a drastic quality reduction as soon as bags were opened (Fig. 8), with blackening of cut bracts and receptacle, while black spots appeared on outer bracts; also a significant accumulation of ethanol and acetaldehyde, potentially responsible for off odors, was detected in these samples. Level of O_2 below the fermentation threshold may also stimulate the growth of some anaerobic psychrotrophic pathogens.

Additional antibrowning treatments

Fresh-cut artichokes may be dipped in antioxidant solutions to delay browning of cut surfaces. Among several compounds, cysteine (0.5%–1% concentration) has

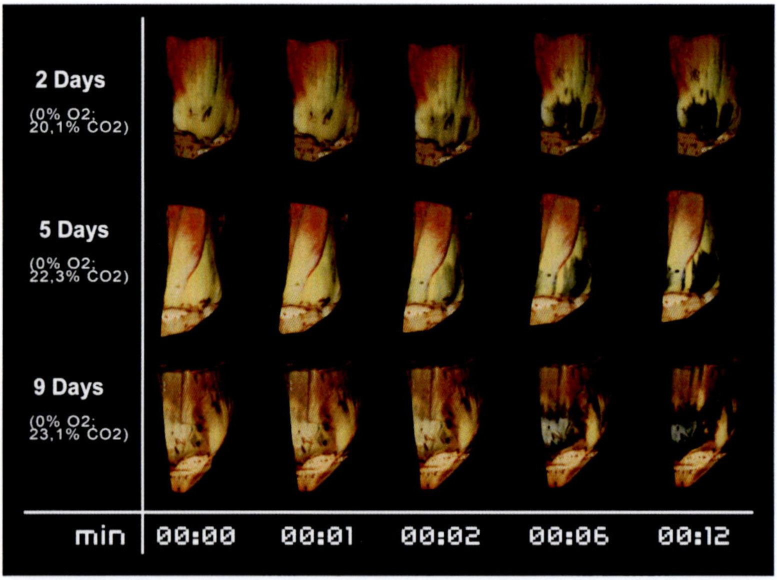

FIG. 8 Example of blackening of untreated fresh-cut artichokes occurring after opening bags under anoxia conditions and extreme CO_2 accumulation. *(Results from la Zazzera et al., doctoral thesis.)*

proved to be one of the most effective treatments to prevent browning of fresh-cut artichokes (Amodio et al., 2010; Cabezas-Serrano et al., 2013; Ghidelli et al. 2015). Cysteine is recognized as a GRAS substance, but not yet allowed as food additive. Other authors suggest the inclusion of 0.3 g/100 mL Cys in soy protein isolate (SPI) and beeswax (BW) edible (40 g/100 g BW dry basis, and 7 g/100 mL SPI). The use of this coating prevented browning of fresh-cut artichokes up to 4 days, whereas the combination with MAP did not extend the shelf-life, but gas conditions at equilibrium during storage were not optimal if compared to other findings. Dipping in 1% lactic acid has also shown to delay browning of artichoke hearts (Calabrese et al., 2012)

Ethylene considerations

Artichokes produce only very low amounts of ethylene and are not particularly sensitive to ethylene exposure.

Recommendations

Gentle handling avoiding compression of artichokes heads before processing; using appropriate cultivars and good quality heads; use permeable material for packaging.

References

Amodio, M.L., Cabezas, A., Rinaldi, R., Colelli, G., 2007. Implementation of rating scales for visual quality evaluation of various vegetable crops. In: Kader, A.A., Cantwell, M. (Eds.), Produce Quality Rating Scales and Color Charts. University of California, Davis, CA, p. 151. Postharvest Horticultural Series No. 23.

Amodio, M.L., Cabezas-Serrano, A.B., la Zazzera, M., Cibelli, F., Raimondo, M.L., Carlucci, A., Colelli, G., 2010. Effetto della cisteina sul controllo dell'imbrunimento e della crescita microbica in carciofo (*Cynara cardunculus* L. subsp. Scolymus (L.) Hayek) di IV gamma. Italus Hortus 17 (3), 87–91.

Cabezas-Serrano, A.B., Amodio, M.L., Cornacchia, R., Rinaldi, R., Colelli, G., 2009. Screening quality and browning susceptibility of 5 artichoke cultivars for fresh-cut processing. J. Sci. Food Agric. 89, 2588–2594.

Cabezas-Serrano, A.B., Amodio, M.L., Colelli, G., 2013. Effect of solution pH of cysteine-based pre-treatments to prevent browning of fresh-cut artichokes. Postharvest Biol. Technol. 75, 17–23.

Calabrese, N., Cefola, M., Carito, A., Pace, B., Baruzzi, F., Vanadia, S., 2012. Effects of dipping treatments on quality of fresh cut artichoke. Acta Hortic. 942, 407–412.

Del Nobile, M.A., Conte, A., Scrocco, C., Laverse, J., Brescia, I., Conversa, G., Elia, A., 2009. New packaging strategies to preserve fresh-cut artichoke quality during refrigerated storage. Innov Food Sci Emerg Technol. 10, 128–133.

Ghidelli, C., Mateos, M., Rojas-Argudo, C., Pérez Gago, M.B., 2015. Novel approaches to control browning of fresh-cut artichoke: Effect of a soy protein-based coating and modified atmosphere packaging. Postharvest Biol. Technol. 99, 105–113.

Giménez, M., Olarte, C., Sanz, S., Lomas, C., Echávarri, J.F., Ayala, F., 2003. Relation between spoilage and microbiological quality in minimally processed artichoke packaged with different films. Food Microbiol. 20, 231–242.

Kader, A.A., 2002. Postharvest biology and technology: an overview. In: Kader, A.A. (Ed.), Postharvest Technology of Horticultural Crops, third ed. University of California, ANR Publication, USA, p. 40.

la Zazzera, M., Rinaldi, R., Amodio, M.L., Colelli, G., 2012. Influence of high CO_2 atmosphere composition on fresh-cut artichoke quality attributes. Acta Hortic. (934), 633–640.

la Zazzera, M., Amodio, M.L., Colelli, G., 2015. Designing a modified atmosphere packaging (MAP) for fresh-cut artichokes. Adv. Hort. Sci. 29 (1), 24–29.

Lattanzio, V., 2003. The role of plant phenolics in the postharvest physiology and quality of fruit and vegetables. Advances in Phytochemistry, 49–83. ISBN: 81-271-0018-8.

Lattanzio, V., Cardinali, A., Di Venere, D., Linsalata, V., Palmieri, S., 1994. Browning phenomena in stored artichoke (*Cynara scolymus* L.) heads: Enzymic or chemical reactions? Food Chem. 50, 1–7.

Ricci, I., Amodio, M.L., Colelli, G., 2013b. Influence of pre-cutting operations on quality of fresh-cut artichokes (*Cynara scolymus* L.): Effect of harvest dates. Postharvest Biol. Technol 83, 90–96.

Shütz, K., Kammerer, D., Carle, R., Schieber, A., 2004. Identification ad quantification of caffeoylquinic acids and flavonoids from artichoke (*Cynara scolymus* L.) heads, juice and pomace by HPLC-DAD-ESI/MSn. J. Agric. Food Chem. 52, 4090–4096.

Further reading

la Zazzera, M., 2011. Management of gas composition for fresh-cut artichokes (Cynara scolymus L.). Doctoral thesis, University of Foggia.

Ricci, I., Amodio, M.L., Colelli, G., 2013a. Influence of pre-cutting operations on quality of fresh-cut artichokes (*Cynara scolymus* L.): Effect of storage time and temperature before cutting. Postharvest Biol. Technol. 85, 124–131.

Chapter 23.1

Mushrooms

Maria Isabel Gil and Juan A. Tudela
Group on Quality and Safety, Department of Food Science and Technology, CEBAS-CSIC, Murcia, Spain

Quality characteristics

Quality button mushrooms *(Agaricus bisporus* Lange) have a rounded cap, intact veils, and are free of darkening/browning. The cap is typically white, but there are brown-capped strains. Mushrooms are harvested based on the maturity stage, but not by the size of the caps. Optimal maturity is reached when the caps are well rounded, the veil is completely intact, and the stipe length is sufficient to permit trimming (Fig. 1). *Pleurotus ostreatus*, the oyster mushroom, is a common edible mushroom that has a white to gray cap color. It is lighter in the spring and darker in the fall.

Controlled and Modified Atmospheres for Fresh and Fresh-Cut Produce
https://doi.org/10.1016/B978-0-12-804599-2.00052-1

FIG. 1 White mushrooms after harvest without trimming the stipe.

Main causes of quality loss

- Upward bending of caps.
- Opening of the veil.
- Brown discoloration due to mechanical damage.
- Stipe elongation.
- Decay caused by bacterial blotch mainly by *Pseudomonas* spp.

Optimum storage conditions

Immediately after harvest, mushrooms should be precooled to 2–4°C using vacuum-cooling, hydrocooling, or forced-air cooling.

Optimum temperature

When mushrooms are stored at 0–1°C and 95% RH, the shelf life can reach 7–9 days. If the storage temperature increases to 2°C, then the shelf life can be reduced by half because of the development of surface browning, stipe elongation, and veil opening.

Optimum relative humidity

Low RH during storage causes loss of brightness, dehydration, stipe blackening, and veil opening. Therefore, it is necessary to store mushrooms at high RH, but avoiding condensation (Fig. 2).

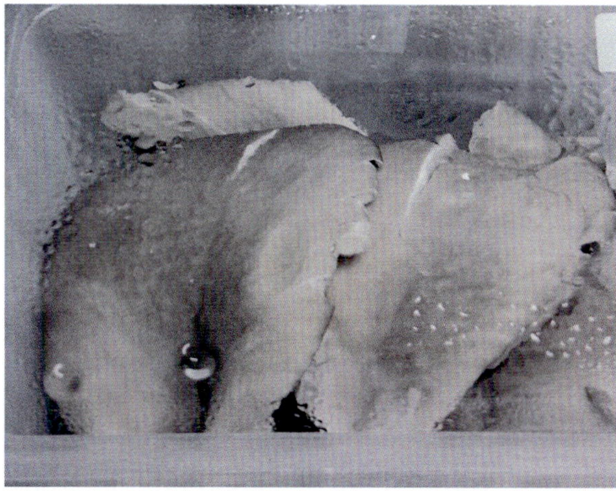

FIG. 2 Condensation on packaging film for *Pleurotus ostreatus*.

CA/MA considerations

CA/MA technology can benefit quality of mushrooms. However, very low levels of O_2 can compromise quality.

Benefits

In general, mushroom quality can be preserved using CA/MA with 3%–21% O_2 + 5%–15% CO_2. The impact of different O_2 and CO_2 ranges and storage period has been reported.

- A CA of 3% O_2 + 10% CO_2 can prolong mushroom shelf life, permitting up to 12–15 days storage at 0°C.

For *Pleurotus*, an atmosphere of 15% O_2 + 5% CO_2 was beneficial to preserve quality for 7 days at 4°C.

- Mushroom visual quality can be maintained up to 14 days at 4°C using 39 μm-thick polyethylene (PE) film with an initial gas composition of 50% O_2 + 50% N_2.
- A pretreatment of 95%–100% CO_2 for 12 h at the beginning of storage reduced browning and maintained flavor.
- Using microperforated packages with an initial atmosphere of 5% O_2 + 10% CO_2 preserved aroma during 27 days at 0°C.
- Application of MAP combined with high argon pressure allowed a storage life of mushrooms to 15 days at 4°C.

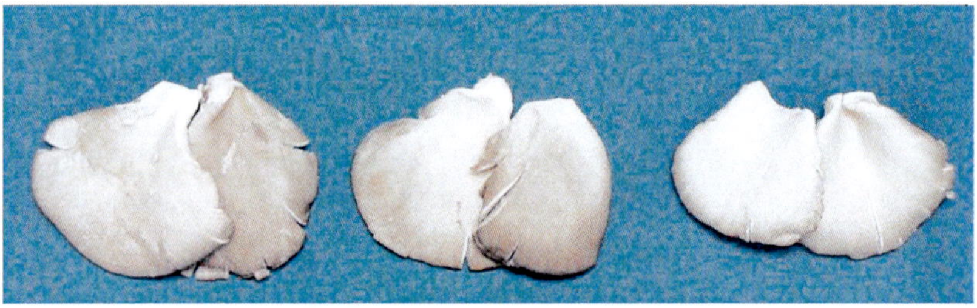

FIG. 3 Appearance of *Pleurotus ostreatus* under air (left), 2% O_2 + 10% CO_2 (center), and 15% O_2 + 5% CO_2 (right) after 7 days at 4°C. Note the reduction of discoloration by the higher CO_2 levels.

Injuries

Atmospheres with low O_2 and high CO_2 can cause cap yellowing (Fig. 3). Levels below 1% O_2 bring about the development of off odors and off flavors, cap opening, and stipe elongation and could stimulate the growth of *Clostridium botulinum* and its toxin.

Problems under suboptimal conditions

The freezing point of mushrooms is −0.6°C. When temperature decreases below that point, freezing injuries appear as water-soaked areas and soft caps (Fig. 4).

Storing mushrooms at elevated temperatures can induce growth of bacterial blotch caused by *Pseudomonas* spp.

Mushrooms are ethylene sensitive and can develop browning on the caps in its presence.

FIG. 4 Freezing injuries in mushrooms during transport.

Further reading

Adamicki, F., 2016. Mushroom. In: Gross, K.C., Wang, C.Y., Saltveit, M.E. (Eds.), USDA Handbook 66—The Commercial Storage of Fruits, Vegetables, and Florist and Nursery Crops, pp. 415–417. http://www.ba.ars.usda.gov/hb66/contents.html.

An, D.S., 2016. Effect of active master packaging system on preservation of fresh shiitake mushrooms in supply chain. J. Korean Soc. Food Sci. Nutr. 45, 402–408.

Capotorto, I., Amodio, M.L., Colelli, G., 2015. Influence of modified atmosphere packaging on shelf-life of whole and sliced "Cardoncello" mushroom (*Pleurotus eryngii*). Acta Hort. (1079), 553–559.

Dhalsamant, K., Dash, S.K., Bal, L.M., Panda, M.K., 2015. Effect of perforation mediated MAP on shelf life of mushroom (*Volvariella volvacea*). Sci. Hortic. 189, 41–50.

Gantner, M., Guzek, D., Pogorzelska, E., Brodowska, M., Wojtasik-Kalinowska, I., Godziszewska, J., 2016. The effect of film type and modified atmosphere packaging with different initial gas composition on the shelf life of white mushrooms (*Agaricus bisporus* L.). J. Food Process. Preserv. http://doi:10.111/jfpp.13083.

Han, L., Qin, Y., Liu, D., Chen, H., Li, H., Yuan, M., 2015. Evaluation of biodegradable film packaging to improve the shelf-life of *Boletus edulis* wild edible mushrooms. Innov. Food Sci. Emerg. Technol. 29, 288–294.

Lagnika, C., Zhang, M., Wang, S., 2011. Effect of high argon pressure and modified atmosphere packaging on the white mushroom (*Agaricus bisporus*) physico-chemical and microbiological properties. J. Food. Nutr. Res. 50, 167–176.

Li, T., Zhang, M., Wang, S., 2008. Effects of temperature on *Agrocybe chaxingu* quality stored in modified atmosphere packages with silicon gum film windows. LWT—Food Sci. Technol. 41, 965–973.

Lin, Q., Lu, Y., Zhang, J., Liu, W., Guan, W., Wang, Z., 2017. Effects of high CO_2 in-package treatment on flavor, quality and antioxidant activity of button mushroom (*Agaricus bisporus*) during postharvest storage. Postharvest Biol. Technol. 123, 112–118.

Parentelli, C., Ares, G., Corona, M., Lareo, C., Gámbaro, A., Soubes, M., Lema, P., 2007. Sensory and microbiological quality of shiitake mushrooms in modified-atmosphere packages. J. Sci. Food Agric. 87, 1645–1652.

Suslow, T.V., Cantwell, M.I. 2002. Mushroom: Recommendations for Maintaining Postharvest Quality. http://postharvest.ucdavis.edu/Commodity_Resources/Fact_Sheets/Datastores/Vegetables_English/?uid=21&ds=799 (Accessed June, 2019).

Villaescusa, R., Gil, M.I., 2003. Quality improvement of *Pleurotus* mushrooms by modified atmosphere packaging and moisture absorbers. Postharvest Biol. Technol. 28, 169–179.

Chapter 23.2

Mushrooms: Fresh-cut mushrooms

Juan A. Tudela and Maria Isabel Gil
Group on Quality and Safety, Department of Food Science and Technology, CEBAS-CSIC, Murcia, Spain

Quality characteristics

When processing fresh-cut mushrooms, washing is not recommended after cutting because the contact of the cut surfaces with water induces browning.

Packaging is also a critical step in mushroom processing. The general recommendation for preserving quality is to maintain an air (21% O_2) gas composition with high RH. For that reason, it is common to pack fresh-cut mushrooms in trays sealed with a macroperforated film that allows water vapor

Controlled and Modified Atmospheres for Fresh and Fresh-Cut Produce
https://doi.org/10.1016/B978-0-12-804599-2.00067-3

FIG. 1 Cut surface and cap browning of cut mushrooms after storage.

exchange, reduces the risk of *Clostridium botulinum* growth, and permits vacuum cooling.

Most of the literature on fresh-cut mushrooms describes different preservative treatments to control browning such as the use of sodium metabisulfite. Treatment of mushrooms with sulfite, however, is not permitted in many countries. Commercial application of browning inhibitor treatments might enable producers to market washed mushrooms having superior quality and shelf life (Sapers et al., 1994).

Main causes of quality loss

- Browning development on the cut surface (Fig. 1).
- Bacterial growth.
- Tissue dehydration at the cut surface.

Optimum storage conditions

Optimum temperature

The recommended storage temperature for mushroom slices is 0°C.

Optimum relative humidity

Fresh-cut mushrooms should be stored under moderate RH and excessive high RH should be avoided to prevent condensation and free water accumulation.

FIG. 2 Fresh-cut mushrooms stored under different controlled atmospheres.

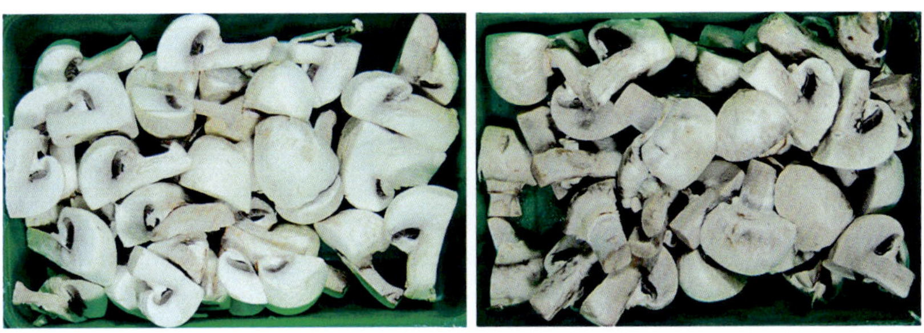

FIG. 3 White mushrooms under MAP with 0.1% O_2 + 20% CO_2 (left) and 5% O_2 + 14% CO_2 (right) after 3 days at 4°C + 9 days at 8°C.

CA/MA considerations

Under CA, fresh-cut mushroom quality can be maintained if the O_2 level is reduced to 0.1% and CO_2 increased up to 20% (Figs. 2 and 3) (M.I. Gil, personal communication, 2006). However, levels of O_2 <1% can favor growth of *Clostridium botulinum* as happens with whole mushrooms and therefore CA is not used for fresh-cut mushrooms.

Several reports suggest that storage of fresh-cut mushrooms can be improved through the use of MA. Amodio et al. (2003) reported that fresh-cut mushroom quality characteristics can be preserved for 24 days after cutting when stored at 0°C with a CA of 3% O_2 + 20% CO_2. A shelf life of almost 8 days was achieved when sliced mushrooms were stored at 0°C in MA with a gas composition of approximately 10% O_2 + 12% CO_2 (Oliveira et al., 2012a).

However, shelf was dramatically reduced around 3 days when stored at 10°C with a MA of 4% O_2 + 12% CO_2. Under these conditions, if CO_2 is removed with a scavenger, deterioration of mushrooms increases (Oliveira et al., 2012b).

At 5°C, a CA of 3% O_2 + 10% CO_2 can reduce tissue darkening and maintain quality (Barth et al., 2016).

No differences were found between CA of 3% O_2 + 10% CO_2 or 10% O_2 + 10% CO_2 and air if temperature is not well controlled as it was the case of simulated commercial conditions of 3 days at 4°C followed by 6 days at 9° C (M.I. Gil, personal communication, 2006).

Injuries

More than 20% CO_2 accelerates tissue darkening.

Problems under suboptimal conditions

O_2 levels below 3% O_2 are not recommended as it increases food safety risk because *C. botulinum* can sporulate and produce endotoxin with O_2 levels between 0.9% and 2% after 6 days at 25°C.

References

Amodio, M.L., Colelli, G., De Cillis, F.M., Lovino, R., Massignan, L., 2003. Controlled-atmosphere storage of fresh-cut 'Cardoncello' mushrooms (*Pleurotus eryngii*). Acta Hortic. 599, 731–735.

Barth, M.M., Zhuang, H., Salveit, M.E., 2016. Fresh-cut vegetables. In: Gross, K.C., Wang, C.Y., Saltveit, M.E. (Eds.), The Commercial Storage of Fruits, Vegetables, and Florist and Nursery Stocks. USDA Agriculture Handbook Number 66. U.S. Department of Agriculture, Washington, DC, pp. 632. https://www.ars.usda.gov/is/np/CommercialStorage/CommercialStorage.pdf.

Oliveira, F., Sousa-Gallagher, M.J., Mahajan, P.V., Teixeira, J.A., 2012a. Evaluation of MAP engineering design parameters on quality of fresh-sliced mushrooms. J. Food Eng. 108, 507–514.

Oliveira, F., Sousa-Gallagher, M.J., Mahajan, P.V., Teixeira, J.A., 2012b. Development of shelf-life kinetic model for modified atmosphere packaging of fresh sliced mushrooms. J. Food Eng. 111, 466–473.

Sapers, G.M., Miller, R.L., Miller, F.C., Cooke, P.H., Choi, S.-W., 1994. Enzymatic browning control in minimally processed mushrooms. J. Food Sci. 59, 1042–1047.

Chapter 24.1

Roots: Beetroots

Merete Edelenbos*, Anne-Berit Wold†, Justyna Wieczynska* and Alexandru Luca*

*Department of Food Science, Aarhus University, Aarhus, Denmark, †Department of Plant Sciences, Faculty of Biosciences, Norwegian University of Life Sciences, As, Norway

Quality characteristics

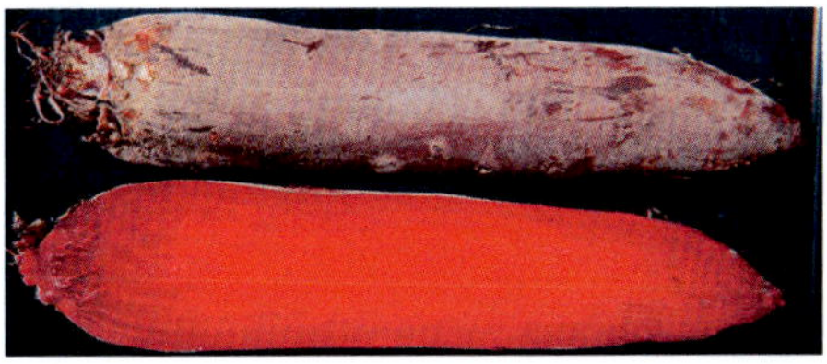

Photo: Jens Michael Madsen, Department of Food Science, Aarhus University, Denmark.

Beetroots (*Beta vulgaris* L.) are classified based on root size, shape, uniformity, surface smoothness, lack of sprouts, damages, wounds and decay, surface color, and a minimum of internal zoning of the color, firmness, sweet taste, lack of fusty and earthy odors and flavors, and a bitter taste (Bach et al., 2015; Adamicki, 2016).

Main causes of quality loss

(i) Moisture loss and desiccation
(ii) Spoilage caused by fungi and bacteria
(iii) Fusty and earthy odors and flavors
(iv) Sprouting
(v) Suberinization and development of white blush on surfaces

Controlled and Modified Atmospheres for Fresh and Fresh-Cut Produce
https://doi.org/10.1016/B978-0-12-804599-2.00054-5

Optimum storage conditions

- Optimum temperature and relative humidity: 1–2°C and 98%–100% RH
- CA/MA considerations:
  - Benefits: >17% O_2 and <4% CO_2
  - Tolerances: >10% O_2 and < 10% CO_2 for 1 month
  - Injuries: <10% O_2 and > 10% CO_2

Topped beetroots can be stored for 8–10 months at 1–2°C and 98%–100% RH (Cantwell, 2002; Adamicki, 2016). Lower temperatures enhance fungal growth and development of black spots caused by *Phoma betae* Frank (Adamicki, 2016; Henze and Baumann, 1979; Schouten and van Schaik, 1980; Tucker et al., 1980).

CA is not recommended for long-term storage of beetroots, as the benefits are limited (Stoll and Weichmann, 1987; Cantwell, 2002). However, beetroots can tolerate 3%–5% CO_2 for 4.5 months at 1°C (Henze and Baumann, 1979). More than 10% CO_2 in the storage atmosphere is not recommended, as it increases fungal growth (Stoll, 1974) and promotes a fusty and earthy odor and flavor of geosmin (Murray et al., 1975; Lu et al., 2003; Freidig and Goldman, 2014). Packaging in films with low water vapor transmission rates such as polyethylene films helps to maintain root texture and turgidity (Caleb et al., 2013).

Problems under suboptimal conditions

Variable temperature and relative humidity

Low RH (<95%) for a prolonged time has a negative effect on beetroot quality (Fig. 1) as the cell turgor decreases, the tissue loses water, and the roots dehydrate and lose weight (van den Berg and Lentz, 1973; Afek and Kays, 2004). Beetroot can tolerate losing up to 7% of its original weight and still be marketable (Robinson et al., 1975). During prolonged storage in air at low RH, beetroots may lose roughly one-fifth of their weight and roots will desiccate, wilt, shrivel, and become rubbery with a white blush appearance (Fig. 1A and B). Storage in high RH (>95%) prevents desiccation and preserves the root turgidity and firmness, but it cannot prevent the development of white blush (Fig. 1C–F). Smaller roots lose moisture more rapidly than larger roots (Adamicki, 2016) as the former have less suberin deposited between the cell walls compared to the latter (Bernards, 2002), where suberin functions as a water-impermeable barrier against desiccation (Artschwager, 1926; Fugate et al., 2016).

Beetroots are relatively resistant to spoilage and decay after harvest (Figs. 1 and 2). Sprouting may be observed after storage in air at variable temperatures and high RH, but not in CA/MA (Fig. 1). White blush and suberinization develop regardless of storage conditions, and they gives the beetroot a corky

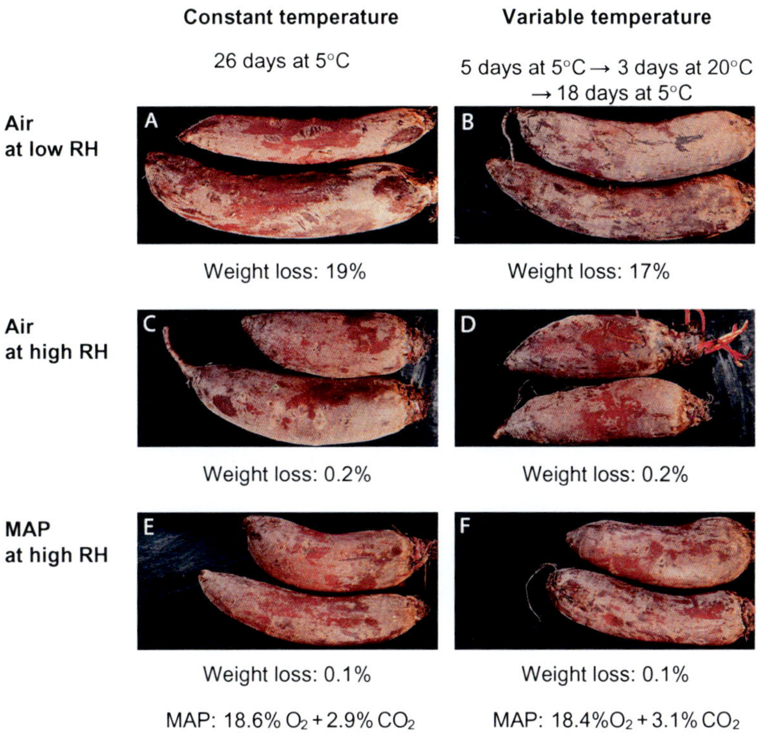

FIG. 1 Visual quality of beetroots stored for 26 days at constant or variable temperatures in air or in modified atmosphere packaging (MAP). Low RH: <95%; high RH: >95%. *(Photo: Jens Michael Madsen, Department of Food Science, Aarhus University, Denmark.)*

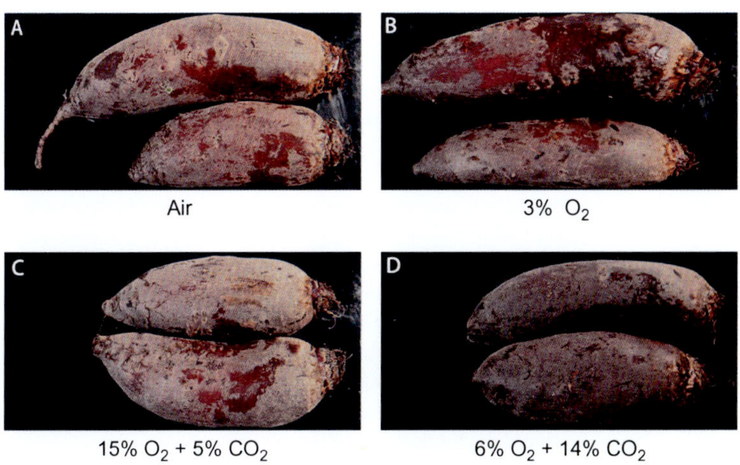

FIG. 2 Visual quality of beetroots stored for 26 days in air and controlled atmosphere at 5°C and >95% RH. *(Photo: Jens Michael Madsen, Department of Food Science, Aarhus University, Denmark.)*

and whitish appearance during prolonged storage (Figs. 1 and 2) (Avena-Bustillos et al., 1993; Simõnes et al., 2010). High RH and MAP cannot prevent the development of white blush and a corky appearance on washed roots (Fig. 1C and D).

Gas composition

Spoilage and microbial decay increase during storage in low O_2 (3%) and decrease in high CO_2 ($\geq$3%) (Figs. 1E, F and 2B–D). The respiration rate of beetroots is influenced by the gas composition. The respiration rate of the beetroots shown in Fig. 2 was 0.5 times lower in 3% O_2 and 2.5–3.5 times higher in 15% O_2 + 5% CO_2 and 6% O_2 + 14% CO_2, respectively, than in air. A higher respiration rate indicates that the beetroots were responding to elevated CO_2 concentrations (Watkins, 2000), but not to such an extent that acetaldehyde, ethanol, and other fermentative off-odors and flavors were emitted (Edelenbos et al., 2010). Similar findings were observed with fresh radishes stored in CA/MA (Edelenbos et al., 2019).

Comments

Beetroots are relatively resistant to spoilage during postharvest storage; however, MAP and high RH in storage cannot prevent white blush and a corky appearance. Elevated CO_2 ($\geq$3%) prevents sprouting and postpones microbial decay while low O_2 (3%) increases decay.

References

Adamicki, F., 2016. The Commercial Storage of Fruits, Vegetables, and Florist and Nursery Stocks. In: Gross, K.C., Wang, C.Y., Saltveit, M.E. (Eds.), In: Agriculture Handbook 66, U.S. Department of Agriculture, Beltsville, MD, pp. 234–236.

Afek, U., Kays, S., 2004. Postharvest physiology and storage of widely used root and tuber crops. Hortic. Rev. 30, 253–316.

Artschwager, E., 1926. Anatomy of the vegetative organs of the sugar beet. J. Agric. Res. 33, 143–176.

Avena-Bustillos, R.J., Cisneros-Zevallos, L.A., Salveit, M.E., 1993. Application of casein-lipid edible film emulsions to reduce white blush on minimally processed carrots. Postharvest Biol. Technol. 4, 319–329.

Bach, V., Mikkelsen, L., Kidmose, U., Edelenbos, M., 2015. Culinary preparation of beetroot (*Beta vulgaris* L.): The impact on sensory quality and appropriateness. J. Sci. Food Agr. 95, 1852–1859.

Bernards, M.A., 2002. Demystifying suberin. Can. J. Bot. 80, 227–240.

Caleb, O.J., Mahajan, P.V., Fahad, A.A., Opara, U.L., 2013. Modified atmosphere packaging technology of fresh and fresh-cut produce and the microbial consequences—a review. Food Bioprocess. Technol. 6, 303–329.

Cantwell, M., 2002. Appendix: Summary table of optimal handling conditions for fresh produce. In: Kader, A.A. (Ed.), Postharvest Technology of Horticultural Crops. University of California, Agriculture and Natural Resources, pp. 511–518 Publication 3311.

Edelenbos, M., Balasubramaniam, M., Pedersen, H.T., 2010. Effects of minimal processing and packaging on volatile compounds and other sensory aspects in carrots. Acta Hortic 876, 269–277.

Edelenbos, M., Wold, A.-B., Wieczynska, J., Luca, A., 2019. Radish: requirements and recommendations. In: Gil, M., Beaudry, R. (Eds.), Controlled and Modified Atmospheres for Fresh and Fresh-Cut Produce. first ed. Elsevier, Amsterdam, The Netherlands.

Freidig, A.K., Goldman, I.L., 2014. Geosmin (2β,6α-dimethylbicyclo[4.4.0]decan-1β-ol) production associated with *Beta vulgaris* ssp. *vulgaris* is cultivar specific. J. Agric. Food Chem. 62, 2031–2036.

Fugate, K.K., Ribeiro, W.S., Lulai, E.M., Deckard, E.L., Finger, F.L., 2016. Cold temperature delays wound healing in postharvest sugarbeet roots. Frontiers Plant Sci. 7, 1–14.

Henze, J., Baumann, H., 1979. Quality of red beet (*Beta vulgaris* L.) as affected by storage conditions. Acta Hortic. 93, 59–65.

Lu, G., Edwards, C.G., Fellmann, J.K., Mattinson, D.S., Navazio, J., 2003. Biosynthetic origin of geosmin in red beets (*Beta vulgaris* L.). J. Agric. Food Chem. 51, 1026–1029.

Murray, K.E., Bannister, P.A., Buttery, R.G., 1975. Geosmin: an important volatile constituent of beetroot (*Beta vulgaris*). Chem. Ind. London 22, 973–974.

Robinson, J.E., Browne, K.M., Burton, W.G., 1975. Storage characteristics of some vegetables and soft fruits. Ann. Appl. Biol. 81, 399–408.

Schouten, S.P., van Schaik, A.C.R., 1980. Storage of red beets. Acta Hortic. 116, 25–29.

Simões, A.N., Ventrella, M.C., Moretti, C.L., Carnelossi, M.A.G., Puschmann, R., 2010. Anatomical and physiological evidence of white blush on baby carrot surfaces. Postharvest Biol. Technol. 55, 45–52.

Stoll, K., 1974. Storage of vegetables in modified atmospheres. Acta Hortic. 38, 13–22.

Stoll, K., Weichmann, J., 1987. Root vegetables. In: Weichmann, J. (Ed.), Postharvest Physiology of Vegetables. Marcel Dekker, New York, pp. 541–553.

Tucker, W.G., Drew, R.L.K., Ward, C.M., Rutherford, P.P., 1980. The effects of lifting date and storage temperature on the keeping quality of beetroots. J. Hort. Sci. 55, 409–414.

van den Berg, L., Lentz, C.P., 1973. High humidity storage of carrots, parsnips, rutabagas and cabbage. J. Amer. Soc. Hort. Sci. 98, 129–132.

Watkins, C.B., 2000. Responses of horticultural commodities to high carbon dioxide as related to modified atmosphere packaging. HortTechnology 10, 501–506.

Chapter 24.2

Roots: Fresh-cut beetroots

Merete Edelenbos*, Anne-Berit Wold†, Justyna Wieczynska* and Alexandru Luca*

*Department of Food Science, Aarhus University, Aarhus, Denmark, †Department of Plant Sciences, Faculty of Biosciences, Norwegian University of Life Sciences, As, Norway

Quality characteristics

Photo: Jens Michael Madsen, Department of Food Science, Aarhus University, Denmark.

Fresh-cut beetroot (*Beta vulgaris* L.) includes peeled and fresh-cut products that are diced, shredded, or cut into sticks. Important quality characteristics are surface color, juiciness, moistness, crunchiness, and sweet taste, lack of brown-pigmented vascular bundles, fusty, earthy, and fermentative odors and flavors, and slimy surfaces (Osornio and Chaves, 1998; Bach et al., 2015).

Main causes of quality loss

(i) White blush
(ii) Slimy surfaces and fermentative odors and flavors
(iii) Fusty and earthy odors and flavors

Controlled and Modified Atmospheres for Fresh and Fresh-Cut Produce
https://doi.org/10.1016/B978-0-12-804599-2.00055-7

Optimum storage conditions

- Optimum temperature and relative humidity:
  - 0–5°C and 98%–100% RH to minimize white blush
  - 0–5°C and 94%–96% RH to minimize slimy surfaces
- CA/MA considerations:
  - Benefits: 3%–5% O_2 and 10–15% CO_2 at 94%–96% RH
  - Tolerances: 3% O_2 and 15% CO_2 at 94%–96% RH
  - Injuries: $\leq$3% O_2 and $\geq$ 15% CO_2 at 98%–100% RH

Peeled beetroot is cut into ready-to-eat products and packaged in plastic films to maintain quality (Osornio and Chaves, 1998; Caleb et al., 2013). Rinsing and spin drying before packaging, use of packaging materials with high gas permeability rates, and storage at low temperature help to maintain quality and prevent off-odors and flavors, and microbial growth (Osornio and Chaves, 1998; Nicola and Fontana, 2014; Chandra et al., 2015; Edelenbos, 2016). Beneficial CA/MA reduces the respiration rate and prolongs shelf life (Osornio and Chaves, 1998; Watkins, 2000).

Problems under suboptimal conditions

Relative humidity

Fresh-cut beetroot loses very little weight (0.07%–0.12%) in MAP with and without perforations (Fig. 1). Packaging in plastic films without perforations maintains high humidity inside packages (98%–100% RH) and prevents development of white blush (Fig. 1A). In contrast, perforations reduce humidity (94%–96% RH) but promote development of white blush (Fig. 1B).

White blush originates from dehydration of damaged surfaces of pigmented, fresh, and fresh-cut root crops (Simõnes et al., 2010; Chandra et al., 2015; Edelenbos, 2016) such as fresh and fresh-cut carrots and fresh beetroots

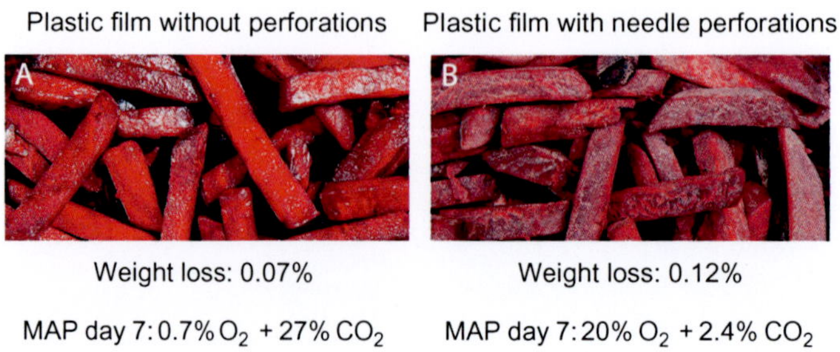

FIG. 1 Visual quality of fresh-cut beetroot stored for 7 days in modified atmosphere packaging (MAP) at 5°C without and with needle perforations. *(Photo: Jens Michael Madsen, Department of Food Science, Aarhus University, Denmark.)*

(Edelenbos et al., 2019a,b,c). White blush is undesirable and should be prevented, as consumers associate white blush with reduced freshness (Bolin and Huxsoll, 1991).

Gas composition

A gas composition of 3% O_2, 15% O_2 + 5% CO_2, or 6% O_2 + 14% CO_2 has positive effects on the respiration rates of fresh-cut beetroot as the rates are lower than in air. However, the effect of CA/MA is highly dependent on humidity. High RH in a condensing environment will facilitate microbial growth, which may lead to fermentation and exudate leakage, and thus increased risk of slimy surfaces, emission of fermentative volatiles such as acetaldehyde and ethanol, and development of sour off-odor notes (Osornio and Chaves, 1998). A CA down to 3% O_2 and up to 15% CO_2 at 94% RH had no effect on the sensory quality of fresh-cut beetroot except for the development of white blush (Fig. 2). In contrast, a MAP using films with no perforations giving O_2 down to 0.7% and CO_2 up to 27% increased product wetness and microbial growth, which resulted in slimy surfaces and formation of off-odors and off-flavors.

Comments

Fresh-cut beetroot spoils easily during storage at high RH (98%–100%). CA/MA in combination with high RH facilitate microbial growth and

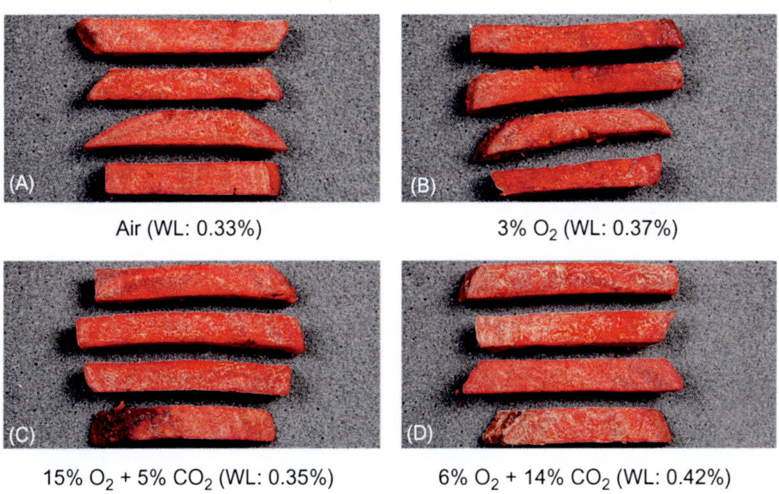

FIG. 2 Visual quality of fresh-cut beetroot stored for 7 days in air and CA at 5°C and ~ 94% RH. The weight loss (WL) during storage is given in parenthesis. *(Photo: Jens Michael Madsen, Department of Food Science, Aarhus University, Denmark.)*

formation of fermentative off-odors, and lead to development of slimy surfaces. The sensory quality of fresh-cut beetroot can be maintained by using perforated plastic films, which provide O_2 for aerobic respiration and give a slightly better humidity control. A lower humidity inside packages, however, may lead to dehydration and white blush on cut surfaces.

References

Bach, V., Mikkelsen, L., Kidmose, U., Edelenbos, M., 2015. Culinary preparation of beetroot (*Beta vulgaris* L.): The impact on sensory quality and appropriateness. J. Sci. Food Agr. 95, 1852–1859.

Bolin, H.R., Huxsoll, C.C., 1991. Control of minimally processed carrot (*Daucus carota*) surface discoloration caused by abrasion peeling. J. Food Sci. 56, 416–418.

Caleb, O.J., Mahajan, P.V., Fahad, A.A., Opara, U.L., 2013. Modified atmosphere packaging technology of fresh and fresh-cut produce and the microbial consequences—a review. Food Bioprocess. Technol. 6, 303–329.

Chandra, D., Choi, A.J., Kim, Y.P., Kim, J.G., 2015. Physicochemical, microbial and sensory quality of fresh-cut red beetroots in relation to sanitization method and storage duration. Ital. J. Food Sci. 27, 208–220.

Edelenbos, M., 2016. Preparation and handling of fresh-cut root vegetables. Acta Hortic. 1141, 77–89.

Edelenbos, M., Wold, A.-B., Wieczynska, J., Luca, A., 2019a. Beetroot: requirements and recommendations. In: Gil, M.I., Beaudry, R. (Eds.), Controlled and Modified Atmospheres for Fresh and Fresh-Cut Produce. first ed. Elsevier, Amsterdam, The Netherlands.

Edelenbos, M., Wold, A.-B., Wieczynska, J., Luca, A., 2019b. Carrot: requirements and recommendations. In: Gil, M., Beaudry, R. (Eds.), Controlled and Modified Atmospheres for Fresh and Fresh-Cut Produce. first ed. Elsevier, Amsterdam, The Netherlands.

Edelenbos, M., Wold, A.-B., Wieczynska, J., Luca, A., 2019c. Fresh-cut carrot: requirements and recommendations. In: Gil, M., Beaudry, R. (Eds.), Controlled and Modified Atmospheres for Fresh and Fresh-Cut Produce. first ed. Elsevier, Amsterdam, The Netherlands.

Nicola, S., Fontana, E., 2014. Fresh-cut produce quality: Implications for a systems approach. In: - Florkowski, W.J., Shewfelt, R.L., Brueckner, B., Prussia, S.E. (Eds.), Postharvest Handling. A systems Approach, third ed. Elsevier, London, UK, pp. 217–2273.

Osornio, M.M.L., Chaves, A.R., 1998. Quality changes in stored raw grated beetroots as affected by temperature and packaging film. J. Food Sci. 63, 327–330.

Simõnes, A.N., Ventrella, M.C., Moretti, C.L., Carnelossi, M.A.G., Puschmann, R., 2010. Anatomical and physiological evidence of white blush on baby carrot surfaces. Postharvest Biol. Technol. Postharvest Biol. Technol. 55, 45–52.

Watkins, C.B., 2000. Responses of horticultural commodities to high carbon dioxide as related to modified atmosphere packaging. HortTechnology 10, 501–506.

Chapter 24.3

Roots: Carrots

Merete Edelenbos*, Anne-Berit Wold†, Justyna Wieczynska* and Alexandru Luca*

*Department of Food Science, Aarhus University, Aarhus, Denmark, †Department of Plant Sciences, Faculty of Biosciences, Norwegian University of Life Sciences, As, Norway

Quality characteristics

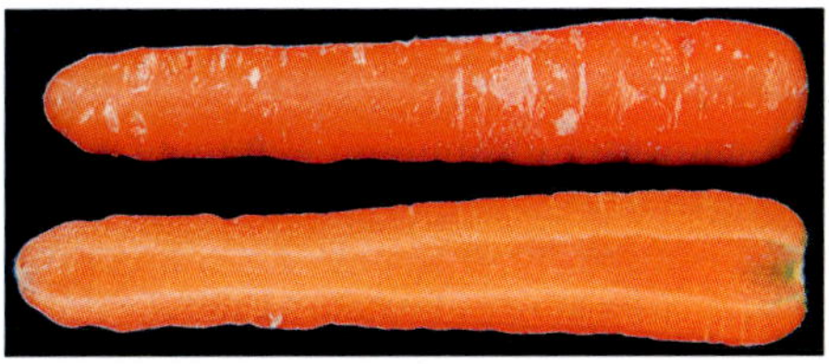

Photo: Jens Michael Madsen, Department of Food Science, Aarhus University, Denmark.

Carrots (*Daucus carota* L.) are classified based on root size, shape, uniformity, surface smoothness, surface and flesh color, firmness, sweet taste, the characteristic odor and flavor of terpenes, and lack of sprouts, cracks, decay, and bitterness (Varming et al., 2004; Kreutzmann et al., 2008a; Larsen and World, 2016).

Main causes of quality loss

(i) Moisture loss and desiccation
(ii) Spoilage caused by fungi and bacteria
(iii) Rooting and sprouting
(iv) Bitterness, harsh flavor, and lack of sweetness

Optimum storage conditions

- Optimum temperature and relative humidity: 0–1°C and 98%–100% RH.
- CA/MA considerations:
  - Benefits: 17%–18% O_2 and 3%–4% CO_2

Controlled and Modified Atmospheres for Fresh and Fresh-Cut Produce
https://doi.org/10.1016/B978-0-12-804599-2.00056-9

- Tolerances: $\geq$10% O_2 and $\leq$5% CO_2
- Injuries: <10% O_2 and >5% CO_2

Low temperature, high RH, and an atmosphere free from ethylene are required to maintain carrot quality (van den Berg and Lentz, 1974; Luo et al., 2016). Carrots can be stored for 6–8 months at 0–1°C and 98%–100% RH or for 5–6 months at 0–5°C and 90%–95% RH (Cantwell, 2002; Luo et al., 2016).

Storage at low temperature ($\leq$2°C) may increase sweetness due to enzymatic degradation of starch to free sugars (Bufler, 2013). Low temperature delays the onset of fungal growth (Korsten and Wehner, 2003), but it does not inhibit bacterial growth (Bartz and Wei, 2003). Microbial growth and ethylene ($\sim$0.2 $\mu L L^{-1}$) may induce bitterness through biosynthesis of polyacetylenes and 6-methoxymellein (Lafuente et al., 1996; Seljåsen et al., 2001; Kreutzmann et al., 2008b; Louarn et al., 2012), while pretreatment with 1-methylcyclopropene before storage in an ethylene-enriched air will prevent formation of bitter constituents (Kramer et al., 2012). Peeling and cooking will reduce bitterness as the peel contains higher concentrations of polyacetylenes and 6-methoxymellein than the cortex and core (Kreutzmann et al., 2008b), and polyacetylenes and 6-methoxymellein are readily leached out into boiling water during cooking (Mercier et al., 1994; Rawson et al., 2013).

CA is not recommended for long-term storage of carrots as the benefits are limited (Cantwell, 2002; Luo et al., 2016). CA with 1%–10% O_2 and 3%–4% CO_2 inhibits rooting and sprouting, but may promote fungal and bacterial growth (Apeland and Hoftun, 1969; Weichmann, 1969; Abdel-Rahman and Isenberg, 1974; Edelenbos, 2010). Additionally, low O_2 (<3%) and elevated CO_2 ($\geq$5%) concentrations enhance off-odor and off-flavor formation through fermentation (Weichmann, 1969; Watkins, 2000; Seljåsen et al., 2004; Edelenbos et al., 2010; Mahajan et al., 2016).

Washed and brushed carrots are packaged in plastic films with low water vapor permeability to prevent desiccation and development of white blush (Avena-Bustillos et al., 1993; Simõnes et al., 2010). White blush develops on the surface of damaged carrots by dehydration and structural changes of the superficial layers of the skin (Simõnes et al., 2010). Packaging may create inappropriate MA through inadequate O_2 and CO_2 transmission through the film (Seljåsen et al., 2004; Caleb et al., 2013; Larsen and World, 2016). Modified atmosphere packaging (MAP) should be used with caution as <10% O_2 and >5% CO_2 are injurious to carrots (Abdel-Rahman and Isenberg, 1974; Watkins, 2000).

Problems under suboptimal conditions

Variable temperature and relative humidity

Low RH (<95%) for prolonged times has negative effects on carrot quality (Fig. 1) due to moisture loss and desiccation, which is directly related to weight loss (Apeland and Baugerød, 1969; Shibairo et al., 1997; Herppich et al., 1999; Afek and Kays, 2004). Carrots can tolerate losing up to 8% of their original weight and still be marketable (Robinson et al., 1975). During prolonged storage in air at low RH, carrots may lose more than one-fourth of their original weight and roots will desiccate, wilt, shrivel, and become rubbery with brown and white patches on the skin (Fig. 1A and B). The brown color originates from enzymatic degradation of phenolic compounds in the presence of O_2 at the periderm (Toivonen and Brummell, 2008).

Storage in high RH (98%–100%) prevents desiccation and preserves root turgidity and firmness, but it also facilitates microbial growth and spoilage (Fig. 1D and F). As a result, carrots may be covered with brown spots from fungi growing on damaged surfaces (Bartz and Wei, 2003; Toivonen and Brummell, 2008) or slime produced by pectinolytic bacteria growing on the wet and wounded surfaces and tips of the root (Bartz and Wei, 2003; Jayani et al., 2005).

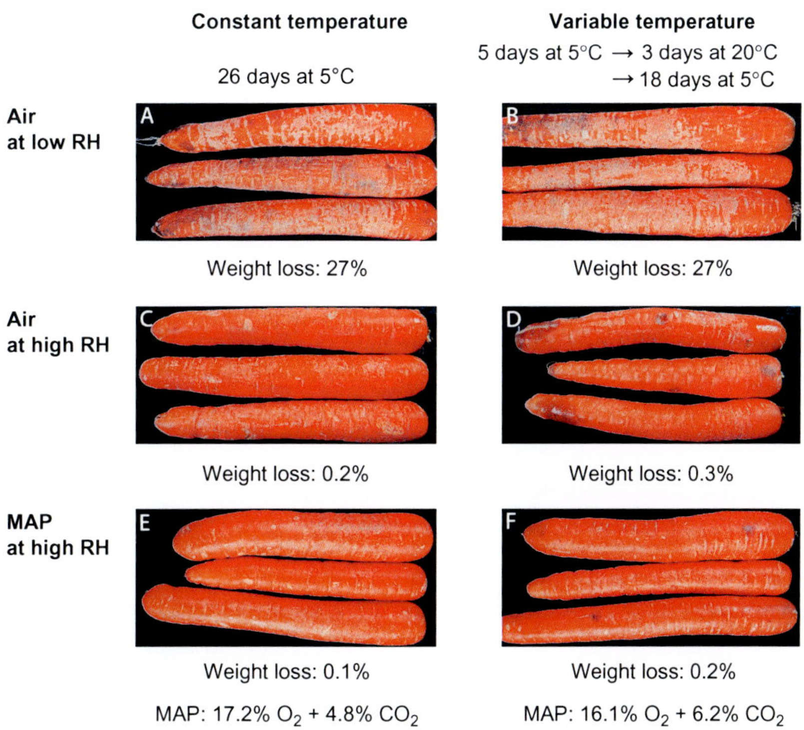

FIG. 1 Visual quality of carrots stored for 26 days at constant or variable temperatures in air or in modified atmosphere packaging (MAP). Low RH: <95%; high RH: >95%. *(Photo: Jens Michael Madsen, Department of Food Science, Aarhus University, Denmark.)*

Gas composition

Low O_2 (3%) increases and elevated CO_2 decreases (5%–14%) microbial decay (Figs. 1E and 2B–D) and affects the respiration rate. The respiration rate was 0.5 times lower in 3% O_2, and 2.2 and 5.6 times higher in 15% O_2 + 5% CO_2 and 6% O_2 + 14% CO_2, respectively, compared to the rate in air, indicating that carrot physiology was influenced by elevated CO_2 concentrations ($\geq$5%). Storage in MAP at variable temperatures (Fig. 1F) or in CA at 3% O_2 or 6% O_2 + 14% CO_2 (Fig. 2B and D) may increase acetaldehyde and ethanol emission and lead to development of off-odors (Edelenbos et al., 2010). Acetaldehyde and ethanol develop in response to injurious CA/MA concentrations (Watkins, 2000) and microbial growth (Rhee et al., 2011). Formation of acetaldehyde and ethanol in the carrots with 16% O_2 and 6% CO_2 (Fig. 1F) most probably originate from fermentation (Watkins, 2000) and/or growth of lactic acid bacteria (*Leuconostoc mesenteroides*) or other pectinolytic bacteria such as *Pseudomonas* spp. on the surfaces of carrots during storage (Bartz and Wei, 2003; Snowdon, 2010). Lactic acid bacteria are naturally present on the surfaces of fresh fruit and vegetables, and they readily grow on wet and wounded tissue forming CO_2, ethanol, esters, and acids (Rhee et al., 2011).

Comments

Excessive condensation during storage in plastic film should be avoided as it facilitates fungal and bacterial growth on the surfaces of washed, brushed and

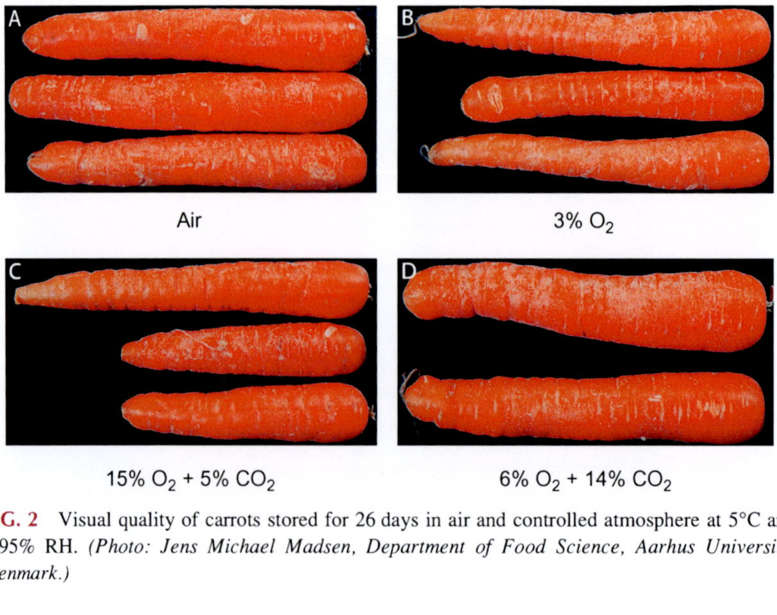

FIG. 2 Visual quality of carrots stored for 26 days in air and controlled atmosphere at 5°C and >95% RH. *(Photo: Jens Michael Madsen, Department of Food Science, Aarhus University, Denmark.)*

wounded carrots. Packaging prevents desiccation and surface dehydration and, to some extent, development of white blush. MAP up to 3%–4% CO_2 is beneficial as it reduces rooting and sprouting and has no negative effect on the sensory quality. However, higher CO_2 concentrations should be avoided to prevent fermentation and changes in the sensory characteristics of carrots.

References

Abdel-Rahman, M., Isenberg, F.M.R., 1974. Effects of growth regulators and controlled atmosphere on stored carrots. J. Agric. Sci. 82, 245–249.

Afek, U., Kays, S., 2004. Postharvest physiology and storage of widely used root and tuber crops. Hortic. Rev. 30, 253–316.

Apeland, J., Baugerød, H., 1969. Factors affecting weight loss in carrots. Acta Hortic. 20, 92–97.

Apeland, J., Hoftun, H., 1969. Physiological effects of oxygen on carrots in storage. Acta Hortic. 20, 108–114.

Avena-Bustillos, R.J., Cisneros-Zevallos, L.A., Salveit, M.E., 1993. Application of casein-lipid edible film emulsions to reduce white blush on minimally processed carrots. Postharvest Biol. Technol. 4, 319–329.

Bartz, J.A., Wei, C.-I., 2003. The influence of bacteria. In: Bartz, J.A., Brecht, J.K. (Eds.), Postharvest Physiology and Pathology of Vegetables. Marcel Dekker, AG, Basel, Switzerland, pp. 519–541.

Bufler, G., 2013. Accumulation and degradation of starch in carrot roots. Sci. Hortic. 150, 251–258.

Caleb, O.J., Mahajan, P.V., Fahad, A.A., Opara, U.L., 2013. Modified atmosphere packaging technology of fresh and fresh-cut produce and the microbial consequences—a review. Food Bioproc. Tech. 6, 303–329.

Cantwell, M., 2002. Appendix: summary table of optimal handling conditions for fresh produce. In: Kader, A.A. (Ed.), Postharvest Technology of Horticultural Crops. In: Vol. 3311, pp. 511–518. University of California, Agriculture and Natural Resources, Publication.

Edelenbos, M., 2010. How do alternative storage methods affect the storage potential of carrots? Acta Hortic. 858, 287–294.

Edelenbos, M., Balasubramaniam, M., Pedersen, H.T., 2010. Effects of minimal processing and packaging on volatile compounds and other sensory aspects in carrots. Acta Hortic. 876, 269–277.

Herppich, W.B., Mempel, H., Geyer, M., 1999. Effects of postharvest mechanical and climatic stress on carrot tissue water relations. Postharvest Biol. Technol. 16, 43–49.

Jayani, R.S., Saxena, S., Gupta, R., 2005. Microbial pectinolytic enzymes: a review. Process Biochem. 40, 2931–2944.

Korsten, L., Wehner, F.C., 2003. Fungi. In: Bartz, J.A., Brecht, J.K. (Eds.), Postharvest Physiology and Pathology of Vegetables. Marcel Dekker, AG, Basel, Switzerland, pp. 485–518.

Kramer, M., Bufler, G., Ulrich, D., Leitenberger, M., Conrad, J., Carle, R., Krammerer, D.R., 2012. Effect of ethylene and 1-methylcyclopropene on bitter compounds in carrots (*Daucus carota* L.). Postharvest Biol. Technol. 73, 28–36.

Kreutzmann, S., Thybo, A.K., Edelenbos, M., Christensen, L.P., 2008a. The role of volatile compounds on aroma and flavour perception in coloured raw carrot genotypes. Int. J. Food Sci. Technol. 43, 1619–1627.

Kreutzmann, S., Christensen, L.P., Edelenbos, M., 2008b. Investigation of bitterness in carrots (*Daucus carota* L.) based on quantitative chemical and sensory analyses. LWT- Food Sci. Technol. 41, 193–205.

Lafuente, T., López-Gálvez, G., Cantwell, M., Fa Yang, S., 1996. Factors influencing ethylene-induced isocoumarin formation and increased respiration in carrots. J. Amer. Soc. Hort. Sci. 121, 537–542.

Larsen, H., World, A.-B., 2016. Effect of modified atmosphere packaging on sensory quality, chemical parameters and shelf life of carrot roots (*Daucus carota* L.) stored at chilled and abusive temperatures. Postharvest Biol. Technol. 144, 76–85.

Louarn, S., Nawrocki, A., Edelenbos, M., Jensen, D.F., Jensen, O.N., Collinge, D.B., Jensen, B., 2012. The influence of the fungal pathogen *Mycocentrospora acerina* on the proteome and polyacetylenes and 6-methoxymellein in organic and conventionally cultivated carrots (*Daucus carota*) during post harvest storage. J. Proteomics 75, 962–977.

Luo, Y., Suslow, T., Cantwell, M., 2016. The Commercial Storage of Fruits, Vegetables, and Florist and Nursery Stocks. In: Gross, K.C., Wang, C.Y., Saltveit, M.E. (Eds.), Agriculture Handbook 66. U.S. Department of Agriculture, Beltsville, MD, pp. 264–266.

Mahajan, P.V., Luca, A., Edelenbos, M., 2016. Development of a small and flexible sensor-based respirometer for realtime determination of respiration rate, respiratory quotient and low O_2 limit of fresh produce. Comput. Electron. Agr. 121, 347–353.

Mercier, J., Arul, J., Julien, C., 1994. Effect of food preparation on the isocoumarin, 6-methoxymellein, content of UV-treated carrots. Food Res. Int. 27, 401–404.

Rawson, A., Brunton, N.P., Rai, D.K., McLoughlin, P., Tiwari, B.K., Tuohy, M.G., 2013. Stability of falcarinol type polyacetylenes during processing of Apiaceae vegetables. Trends Food Sci. Technol. 30, 133–141.

Rhee, S.J., Lee, J.-E., Lee, C.-H., 2011. Importance of lactic acid bacteria in Asian fermented foods. Microb. Cell Fact. 10, 1–13. Suppl 1, S5.

Robinson, J.E., Browne, K.M., Burton, W.G., 1975. Storage characteristics of some vegetables and soft fruits. Ann. Appl. Biol. 81, 399–408.

Seljåsen, R., Hoftun, H., Bengtsson, G.B., 2001. Sensory quality of ethylene-exposed carrots (*Daucus carota* L., cv 'Yukon') related to the contents of 6-methoxymellein, terpenes and sugars. J. Sci. Food Agr. 81, 54–61.

Seljåsen, R., Hoftun, H., Selliseth, J., Bengtsson, G.B., 2004. Effects of washing and packing on sensory and chemical parameters in carrots (*Daucus carota* L.). J. Sci. Food Agric. 84, 955–965.

Shibairo, S.I., Upadhyaya, M.K., Toivonen, P.M.A., 1997. Postharvest moisture loss characteristics of carrot (*Daucus carota* L.) cultivars during short-term storage. Sci. Hortic. 71, 1–12.

Simões, A.N., Ventrella, M.C., Moretti, C.L., Carnelossi, M.A.G., Puschmann, R., 2010. Anatomical and physiological evidence of white blush on baby carrot surfaces. Postharvest Biol. Technol. 55, 45–52.

Snowdon, A.L., 2010. Post-Harvest Diseases and Disorders of Fruits and Vegetables: Volume 2: Vegetables. CRC-Press, London.

Toivonen, P.M.A., Brummell, A., 2008. Biochemical bases of appearance and texture changes in fresh-cut fruit and vegetables. Postharvest Biol. Technol. 48, 1–14.

van den Berg, L., Lentz, C.P., 1974. High humidity storage of some vegetables. Can. Inst. Food Sci. Technol. J. 7, 260–262.

Varming, C., Jensen, K., Moller, S., Brockhoff, P.B., Christiansen, T., Edelenbos, M., Bjorn, G.K., Poll, L., 2004. Eating quality of raw carrots—correlations between flavour compounds, sensory profiling analysis and consumer liking test. Food Qual. Prefer. 15, 531–540.

Watkins, C.B., 2000. Responses of horticultural commodities to high carbon dioxide as related to modified atmosphere packaging. HortTechnology 10, 501–506.

Weichmann, J., 1969. Experiments concerning the CO_2 production and storage of carrots in small store boxes. Acta Hortic. 20, 98–107.

Chapter 24.4

Roots: Fresh-cut carrots

Merete Edelenbos*, Anne-Berit Wold[†], Justyna Wieczynska* and Alexandru Luca*

*Department of Food Science, Aarhus University, Aarhus, Denmark, [†]Department of Plant Sciences, Faculty of Biosciences, Norwegian University of Life Sciences, As, Norway

Quality characteristics

Photo: Jens Michael Madsen, Department of Food Science, Aarhus University, Denmark.

Fresh-cut carrots (*Daucus carota* L.) include peeled and cut products that are diced, shredded, or cut into disks or sticks. Important quality characteristics are surface color, juiciness, moistness, crunchiness, sweet taste, characteristic carrot odor and flavor, and lack of green tissue, harsh flavor and bitter taste, fermentative odors, and slimy surfaces.

Main causes of quality loss

(i) White blush
(ii) Slimy surfaces and fermentative odors
(iii) Loss of juiciness, sweetness, and characteristic carrot odor and bitter taste

Optimum storage conditions

- Optimum temperature and relative humidity:
  - 0–5°C and 98%–100% RH to minimize white blush
  - 0–5°C and 94%–96% RH to minimize slimy surfaces
- CA/MA considerations:
  - Benefits: 2%–5% O_2 and 15%–20% CO_2
  - Tolerances: $\geq$ 2% O_2 and $\leq$20% CO_2
  - Injuries: $<$ 2% O_2 and $>$20% CO_2

Controlled and Modified Atmospheres for Fresh and Fresh-Cut Produce
https://doi.org/10.1016/B978-0-12-804599-2.00057-0

Peeled carrot is cut into ready-to-eat products and packaged in plastic films to maintain quality (Caleb et al., 2013). The severity of processing and the geometry of pieces determine the respiration rate and hence the MAP gas composition, which influence shelf life (Izumi et al., 1996; Barry-Ryan et al., 2000; Cliff-Byrnes et al., 2007; Edelenbos et al., 2010; Edelenbos, 2016). Abrasive peeling and cutting of carrots into carrot discs result in higher respiration rate, microbial load, pH, and weight loss compared to hand peeling and cutting (Barry-Ryan and O'Beirne, 2000). Mechanical slicing with blunt or sharp blades also increases the respiration rate more than manual slicing with sharp blades (Barry-Ryan and O'Beirne, 1998). Rinsing and spin drying, use of packaging materials with high gas transmission rates, and storage at low temperatures maintain carrot quality (Klaiber et al., 2004; Edelenbos et al., 2010).

A CA of 0.5% O_2+10% CO_2, 2% O_2+15% CO_2, or 5% O_2+5% CO_2 reduces the respiration rate and prolongs shelf life (Izumi et al., 1996; Simõnes et al., 2011). A CA of <2% O_2 and>20% CO_2 accelerates microbial growth, and promotes lactic acid bacteria growth, slimy appearance, and formation of fermentative odors and flavors (Charlin et al., 1990; Kato-Noguchi and Watada, 1997; Kato-Noguchi, 1998). Fresh-cut carrot slices, sticks, and shreds produce ethylene (<0.1 μL/kg/h); however, the levels are too low to induce bitterness (Izumi et al., 1996).

Problems under suboptimal conditions

Relative humidity

Fresh-cut carrot loses very little weight (0.09%–0.12%) in MAP with or without perforations (Fig. 1). Packaging in plastic films without perforations maintains high humidity inside packages (98%–100% RH) and prevents white blush (Fig. 1A), but the high humidity may induce microbial decay and lead to fermentation (Charlin et al., 1990). Increasing the gas permeability by use of needle perforations will only reduce the RH inside packages slightly, but improve the gas composition, and thus help maintain the initial sensory quality except for white blush (Fig. 1B). White blush develops on damaged surfaces

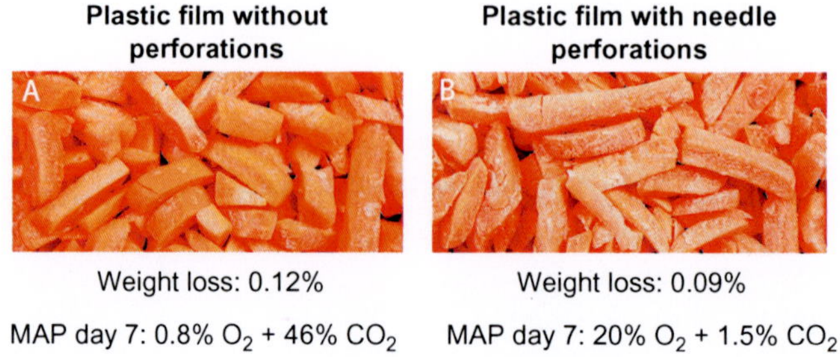

FIG. 1 Visual quality of fresh-cut carrot stored for 7 days in modified atmosphere packaging (MAP) at 5°C without and with needle perforations. *(Photo: Jens Michael Madsen, Department of Food Science, Aarhus University, Denmark.)*

after processing due to dehydration and is mainly observed on pigmented root crops (Cisneros-Zevallos et al., 1995; Izumi et al., 1996; Simões et al., 2010; Edelenbos, 2016) such as fresh beetroot and carrot, and fresh-cut beetroot (Edelenbos et al., 2019a,b,c). White blush develops more severely on fresh-cut carrot stored in needle-perforated than nonperforated film, as the needle perforations increase dehydration from the cut surfaces (Fig. 1B) regardless of CA concentration (Fig. 2). White blush is undesirable and should be prevented, as consumers associate white blush with reduced freshness (Bolin and Huxsoll, 1991).

Gas composition

A CA concentration of 3% O_2 reduces the respiration rate to 0.6 times the rate in air. CA concentrations with elevated CO_2 such as 15% O_2 + 5% CO_2 or 6% O_2 + 14% CO_2 have no effect on the respiration rate. However, storage in 6% O_2 + 14% CO_2 at 5°C and ~94% RH had a slight negative effect on the sensory quality, as sticks developed off-odor and emitted acetaldehyde and ethanol. Fresh-cut carrots reaching 0.8% O_2 and 46% CO_2 in MAP (Fig. 1A) were severely damaged after 7 days' storage at 5°C. The tissue was wet, sticks produced a strong sour off-odor, and elevated acetaldehyde and ethanol concentrations were observed.

Comments

Fresh-cut carrot easily spoils after processing. Low humidity and beneficial MA facilitate dehydration and development of white blush, but these conditions also help to maintain the characteristic odor and flavor of fresh-cut carrot. Severe processing, inadequate film gas permeability, and high humidity in a condensing environment facilitate microbial growth, fermentative odors, and slimy surfaces.

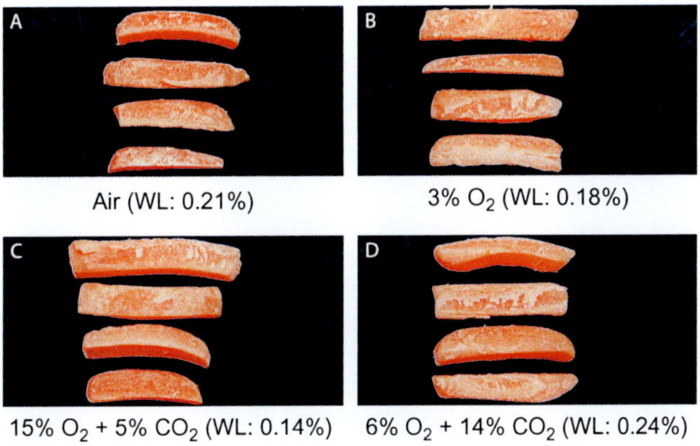

FIG. 2 Visual quality of fresh-cut carrot stored for 7 days in air and CA at 5°C and ~94% RH. The weight loss (WL) during storage is given in parenthesis. *(Photo: Jens Michael Madsen, Department of Food Science, Aarhus University, Denmark.)*

References

Barry-Ryan, C., O'Beirne, D., 1998. Quality and shelf-life of fresh cut carrot slices as affected by slicing method. J. Food Sci. 63, 1–6.

Barry-Ryan, C., O'Beirne, D., 2000. Effect of peeling method on the quality of ready-to-use carrot slices. Int. J. Food Sci. Technol. 35, 243–254.

Barry-Ryan, C., Pacussi, J.M., O'Beirne, D., 2000. Quality of shredded carrots as affected by packaging film and storage temperature. J. Food Sci. 65, 726–730.

Bolin, H.R., Huxsoll, C.C., 1991. Control of minimally processed carrot (*Daucus carota*) surface discoloration caused by abrasion peeling. J. Food Sci. 56, 416–418.

Caleb, O.J., Mahajan, P.V., Fahad, A.A., Opara, U.L., 2013. Modified atmosphere packaging technology of fresh and fresh-cut produce and the microbial consequences—a review. Food Bioprocess. Technol. 6, 303–329.

Charlin, F., Nguyen-The, C., Hilbert, G., Chambroy, Y., 1990. Modified atmosphere packaging of fresh, "ready-to-use" grated carrots in polymeric films. J. Food Sci. 55, 1033–1038.

Cisneros-Zevallos, L., Saltveit, M.E., Krochta, J.M., 1995. Mechanism of surface white discoloration of peeled (minimally processed) carrots during storage. J. Food Sci. 60, 320–323. 333.

Cliff-Byrnes, V., Brennan, L., O'Bernie, D., 2007. The effect of preparatory procedures and storage temperature on the quality of carrot discs packaged in modified atmospheres. Int. J. Food Sci. Technol. 42, 482–494.

Edelenbos, M., 2016. Preparation and handling of fresh-cut root vegetables. Acta Hortic. 1141, 77–89.

Edelenbos, M., Balasubramaniam, M., Pedersen, H.T., 2010. Effects of minimal processing and packaging on volatile compounds and other sensory aspects in carrots. Acta Hortic. 876, 269–277.

Edelenbos, M., Wold, A.-B., Wieczynska, J., Luca, A., 2019a. Beetroot: requirements and recommendations. In: Gil, M.I., Beaudry, R. (Eds.), Controlled and Modified Atmospheres for Fresh and Fresh-Cut Produce. first ed. Elsevier, Amsterdam, The Netherlands.

Edelenbos, M., Wold, A.-B., Wieczynska, J., Luca, A., 2019b. Carrot: requirements and recommendations. In: Gil, M., Beaudry, R. (Eds.), Controlled and Modified Atmospheres for Fresh and Fresh-Cut Produce. first ed. Elsevier, Amsterdam, The Netherlands.

Edelenbos, M., Wold, A.-B., Wieczynska, J., Luca, A., 2019c. Fresh-cut beetroot: requirements and recommendations. In: Gil, M., Beaudry, R. (Eds.), Controlled and Modified Atmospheres for Fresh and Fresh-Cut Produce. first ed. Elsevier, Amsterdam, The Netherlands.

Izumi, H., Watada, A.E., Ko, N.P., Douglas, W., 1996. Controlled atmosphere storage of carrot slices, sticks and shreds. Postharvest Biol. Technol. 9, 165–172.

Kato-Noguchi, H., 1998. Effects of a low oxygen atmosphere on lactic fermentation in shredded carrot root tissue. J. Plant Physiol. 152, 368–371.

Kato-Noguchi, H., Watada, A.E., 1997. Effects of low-oxygen atmosphere on ethanolic fermentation in fresh-cut carrots. J. Amer. Soc. Hort. Sci. 122, 107–111.

Klaiber, R.G., Baur, S., Magel, L., Hammes, W.P., Carle, R., 2004. Quality of shredded, packaged carrots as affected by different washing treatments. J. Food Sci. 69, SNQ161–SNQ166.

Simõnes, A.D.N., Ventrella, M.C., Moretti, C.L., Carnelossi, M.A.G., Puschmann, R., 2010. Anatomical and physiological evidence of white blush on baby carrot surfaces. Postharvest Biol. Technol. 55, 45–52.

Simõnes, A, D.N., Allende, A., Tudela, J.A., Puschmann, R., Gil, M.I. 2011. Optimum controlled atmospheres minimize respiration rate and quality losses while increase phenolic compounds in baby carrots LWT—Food Sci. Technol. 44, 277–283.

Chapter 24.5

Roots: Parsnips

Merete Edelenbos*, Anne-Berit Wold†, Justyna Wieczynska*
and Alexandru Luca*
*Department of Food Science, Aarhus University, Aarhus, Denmark, †Department of Plant Sciences, Faculty of Biosciences, Norwegian University of Life Sciences, As, Norway

Quality characteristics

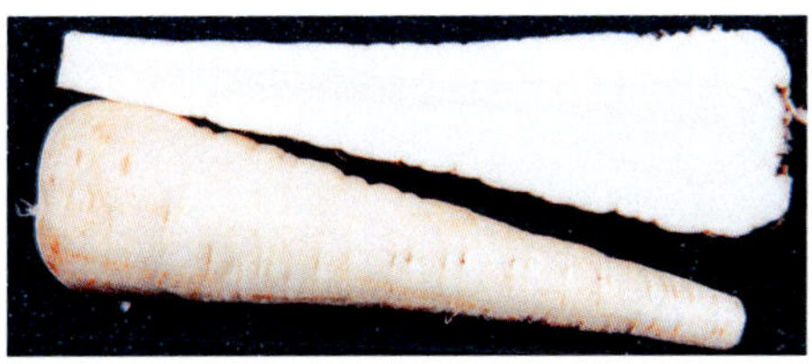

(Photo: Jens Michael Madsen, Department of Food Science, Aarhus University, Denmark)

Parsnips (***Pastinaca sativa*** **L.**) are classified based on root size, shape, uniformity, and surface smoothness, lack of sprouts, damages, wounds and decay, surface color, firmness, sweet taste, a characteristic parsnip odor and flavor, and lack of bitterness.

Main causes of quality loss

(i) Moisture loss and desiccation
(ii) Surface browning
(iii) Spoilage caused by fungi and bacteria
(iv) Bitterness

Optimum storage conditions

- Optimum temperature and relative humidity: 0–1°C and 98%–100% RH.
- CA/MA considerations:
  - Benefits: 20% O_2 and 1% CO_2
  - Tolerances: 19% O_2 and <2% CO_2
  - Injuries: <19% O_2 and >2% CO_2

Controlled and Modified Atmospheres for Fresh and Fresh-Cut Produce
https://doi.org/10.1016/B978-0-12-804599-2.00058-2

Parsnips can be stored for 4–6 months at 0–1°C and 98%–100% RH (van den Berg and Lentz, 1973). Storage at low temperature (<1°C) increases sweetness due to enzymatic degradation of starch to free sugars (Shattuck et al., 1988; Bufler and Horneburg, 2013) and delays the onset of fungal growth (Korsten and Wehner, 2003), but does not inhibit bacterial infection (Bartz and Wei, 2003).

Parsnips may become bitter during storage from accumulation of furocoumarins and polyacetylenes (Johnson and Brannon, 1973; Ostertag et al., 2002; Kreutzmann et al., 2008; Kramer et al., 2012; Shattuck et al., 1988). Bitterness develops in response to fungal and bacterial growth and from storage in an ethylene-enriched atmosphere ($5\,\mu L\,L^{-1}$) (Shattuck et al., 1988). Peeling and cooking reduce bitterness as the peel contains more furocoumarins and polyacetylenes than the cortex and the core (Shattuck et al., 1988; Kreutzmann et al., 2008), and polyacetylenes are readily degraded and leached out into boiling water during cooking (Rawson et al., 2013). However, furocoumarins are heat stable and will not be degraded during cooking (Ivie et al., 1981).

CA is not recommended for long-term storage of parsnips as the benefits are limited (Cantwell, 2002; Stoll and Weichmann, 1987). Packaging in plastic films helps to maintain firmness and turgidity of the root as the parsnip periderm is thin and has a low resistance to evapotraspiration in storage. Thus, parsnips easily lose moisture to the surrounding atmosphere (Afek and Kays, 2004; Espelie et al., 1980; Warning, 1934). Surface browning is a significant problem in the handling of parsnips (Kaldy and Markakis, 1972), as phenolic compounds of the damaged and wounded periderm are oxidized by polyphenol oxidase to brown-colored pigments in the presence of O_2 (Toivonen and Brummell, 2008). MAP can preserve the white color of washed and brushed parsnips by reducing the formation of brown pigments, but MAP should be used with caution. Packaging films may create unfavorable gas and humidity conditions due to inadequate water vapor and gas transmission rates, resulting in anaerobic respiration and enhanced bacterial growth (Caleb et al., 2013).

Problems under suboptimal conditions

Variable temperature and relative humidity

Low RH (<95%) for a prolonged time has a negative effect on parsnip quality (Fig. 1) as the cell turgor decreases, the tissue loses moisture, the root dehydrates, and the product loses weight (van den Berg and Lentz, 1973; Afek and Kays, 2004). Parsnips can tolerate losing up to 7% of their original weight and still be marketable (Robinson et al., 1975). During prolonged storage in air at low RH, parsnips may lose more than one-fourth of their original weight and roots will desiccate, wilt, shrivel, and become rubbery with brown and rotten patches (Fig. 1A and B). Storage in high RH (>95%) prevents desiccation and preserves the root turgidity and firmness (Fig. 1C–F).

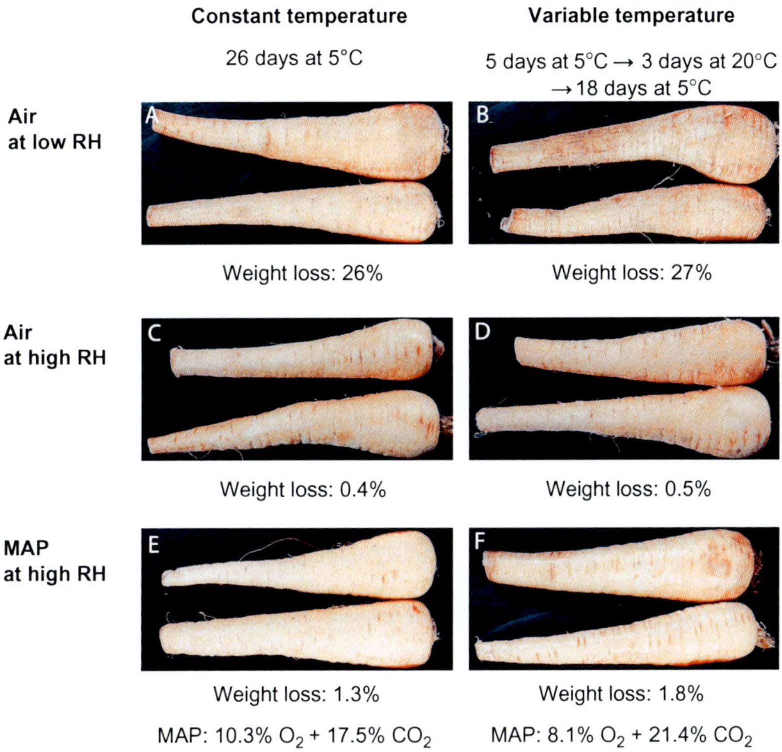

FIG. 1 Visual quality of parsnips stored for 26 days at constant or variable temperatures in air or in modified atmosphere packaging (MAP). Low RH: <95%; high RH: >95%. *(Photo: Jens Michael Madsen, Department of Food Science, Aarhus University, Denmark.)*

Gas composition

Storage in CA/MA has variable effects on parsnip quality (Figs. 1 and 2). CA/MA prevents sprouting (Fig. 1E and F) and preserves the white color (Figs. 1E, F, and 2D), especially at 8%–10% O_2 and 18%–21% CO_2 in MAP, and in 6% O_2 and 14% CO_2 in CA. However, CA may have negative effects on parsnip physiology, as the respiration rate may increase (Watkins, 2000). The respiration rate of parsnips shown in Fig. 2 was 1.5 times higher at 3% O_2 and 15% O_2+5% CO_2, and 3 times higher at 6% O_2+14% CO_2 than in air. Storage in MAP at constant or variable temperatures (Fig. 1E and F) or in CA at 3% O_2, 15% O_2+5% CO_2, or 6% O_2+14% CO_2 (Fig. 2B, C, D) resulted in increased emission of acetaldehyde and ethanol, and formation of off-odors. High relative humidity in storage, e.g., in MAP, may increase the wetness of surfaces due to condensation of water vapor and promote growth of lactic acid bacteria (*Leuconostoc mesenteroides*) and other pectinolytic bacteria such as *Pseudomonas* spp. (Bartz and Wei, 2003; Snowdon, 1990). As these bacteria

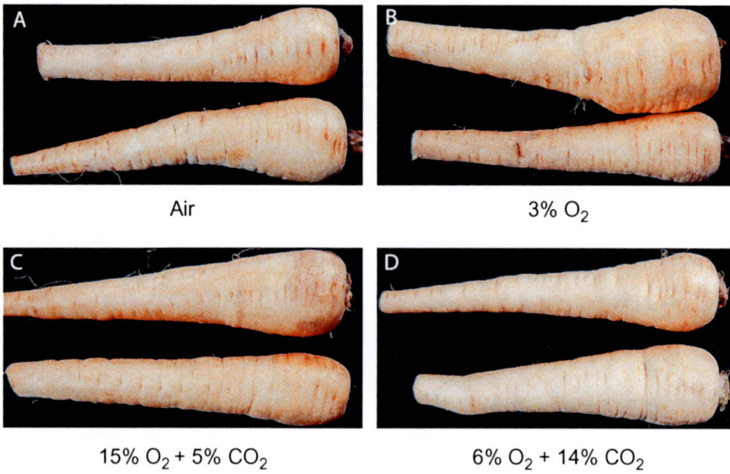

FIG. 2 Visual quality of parsnips stored for 26 days in air and controlled atmosphere at 5°C and >95% RH. *(Photo: Jens Michael Madsen, Department of Food Science, Aarhus University, Denmark.)*

hydrolyze pectin substances between the cell walls (Jayani et al., 2005), surfaces readily become slimy at high humidity (Fig. 1F). Moreover, it has been reported that the sensory quality is affected by increased volatile emission (Edelenbos et al., 2010) due to increased emission of ethanol, esters, and acids from growth of lactic acid bacteria (Rhee et al., 2011).

Comments

Fresh parsnips should be dry at the time of packaging as bacteria readily proliferate and grow on wet and damaged surfaces. CA/MA can prevent surface browning and sprouting, but inappropriate O_2 and CO_2 concentrations should be avoided as they promote fermentation and alter the sensory profile of parsnips.

References

Afek, U., Kays, S., 2004. Postharvest physiology and storage of widely used root and tuber crops. Hortic. Rev. 30, 253–316.

Bartz, J.A., Wei, C.-I., 2003. The influence of bacteria. In: Bartz, J.A., Brecht, J.K. (Eds.), Postharvest Physiology and Pathology of Vegetables. Marcel Dekker, AG, Basel, Switzerland, pp. 519–541.

Bufler, G., Horneburg, B., 2013. Changes in sugar and starch concentrations in parsnip (*Pastinaca sativa* L.) during root growth and development and in cold storage. J. Hortic. Sci. Biotechnol. 88, 756–761.

Caleb, O.J., Mahajan, P.V., Fahad, A.A., Opara, U.L., 2013. Modified atmosphere packaging technology of fresh and fresh-cut produce and the microbial consequences—a review. Food Bioprocess Technol. 6, 303–329.

Cantwell, M., 2002. Appendix: summary table of optimal handling conditions for fresh produce. In: Kader, A.A. (Ed.), Postharvest Technology of Horticultural Crops. In: Vol. 3311. University of California, Agriculture and Natural Resources, Publication, pp. 511–518.

Edelenbos, M., Balasubramaniam, M., Pedersen, H.T., 2010. Effects of minimal processing and packaging on volatile compounds and other sensory aspects in carrots. Acta Hortic. (876) 269–277.

Espelie, K.E., Sadek, N.Z., Kolattukudy, P.E., 1980. Composition of suberin-associated waxes from the subterranean storage organs of seven plants. Parsnip, carrot, rutabaga, turnip, red beet, sweet potato and potato. Planta 148, 468–476.

Ivie, G.W., Holt, D.L., Ivey, M.C., 1981. Natural toxicants in human foods: psoralens in raw and cooked parsnip root. Science 213, 909–910.

Jayani, R.S., Saxena, S., Gupta, R., 2005. Microbial pectinolytic enzymes: a review. Process Biochem. 40, 2931–2944.

Johnson, C., Brannon, D.R., 1973. Xanthotoxin: a phytoalexin of pastinaca sative root. Phytochemistry 12, 2961–2962.

Kaldy, M.S., Markakis, P., 1972. The browning of parsnips. Can. Inst. Food Sci. Technol. J. 5, 37–38.

Korsten, L., Wehner, F.C., 2003. Fungi. In: Bartz, J.A., Brecht, J.K. (Eds.), Postharvest Physiology and Pathology of Vegetables. Marcel Dekker, AG, Basel, Switzerland, pp. 485–518.

Kramer, M., Bufler, G., Nothnagel, T., Carle, R., Kammerer, D.R., 2012. Effects of cultivation conditions and cold storage on the polyacetylene contents of carrot (*Daucus carota* L.) and parsnip (*Pastinaca sativa* L.). J. Hortic. Sci. Biotechnol. 87, 101–106.

Kreutzmann, S., Christensen, L.P., Edelenbos, M., 2008. Investigation of bitterness in carrots (*Daucus carota* L.) based on quantitative chemical and sensory analyses. LWT—Food Sci. Technol. 41, 193–205.

Ostertag, E., Becker, T., Ammon, J., Bauer-Aymanns, H., Schrenk, D., 2002. Effect of storage conditions on furocoumarin levels in intact, chopped, or homogenized parsnips. J. Agric. Food Chem. 50, 2565–2570.

Rawson, A., Brunton, N.P., Rai, D.K., McLoughlin, P., Tiwari, B.K., Tuohy, M.G., 2013. Stability of falcarinol type polyacetylenes during processing of Apiaceae vegetables. Trends Food Sci. Techn. 30, 133–141.

Rhee, S.J., Lee, J.-E., Lee, C.-H., 2011. Importance of lactic acid bacteria in Asian fermented foods. Microb. Cell Fact. 10 (Suppl 1, S5), 1–13.

Robinson, J.E., Browne, K.M., Burton, W.G., 1975. Storage characteristics of some vegetables and soft fruits. Ann. Appl. Biol. 81, 399–408.

Shattuck, V.I., Yada, R., Lougheed, E.C., 1988. Ethylene-induced bitterness in stored parsnips. Hortscience 23, 912.

Snowdon, A., 1990. A Color Atlas of Post-Harvest Diseases of Fruits and Vegetables. Vol. 2. Vegetables. CRC Press, Boca Raton.

Stoll, K., Weichmann, J., 1987. Root vegetables. In: Weichmann, J. (Ed.), Postharvest Physiology of Vegetables. Marcel Dekker, New York, pp. 541–553.

Toivonen, P.M.A., Brummell, A., 2008. Biochemical bases of appearance and texture changes in fresh-cut fruit and vegetables. Postharvest Biol. Technol. 48, 1–14.

van den Berg, L., Lentz, C.P., 1973. High humidity storage of carrots, parsnips, rutabagas and cabbage. J. Amer. Soc. Hort. Sci. 98, 129–132.

Warning, W.C., 1934. Anatomy of the vegetative organs of the parsnip. Bot. Gaz. 96, 44–72.
Watkins, C.B., 2000. Responses of horticultural commodities to high carbon dioxide as related to modified atmosphere packaging. HortTechnology 10, 501–506.

Further reading

Shattuck, V.I., Kakuda, Y., Yada, R., 1989. Sweetening of parsnip roots during short-term cold storage. Can. Inst. Food Sci. Technol. J. 22, 378–382.

Chapter 24.6

Roots: Fresh-cut parsnips

Merete Edelenbos*, Anne-Berit Wold[†], Justyna Wieczynska*
and Alexandru Luca*
*Department of Food Science, Aarhus University, Aarhus, Denmark, [†]Department of Plant Sciences, Faculty of Biosciences, Norwegian University of Life Sciences, As, Norway

Quality characteristics

(Photo: Jens Michael Madsen, Department of Food Science, Aarhus University, Denmark.)

Fresh-cut parsnips (*Pastinaca sativa* L.) include peeled and cut products that are diced, shredded, or cut into disks or sticks. Important quality characteristics are white color, juiciness, crunchiness, sweet taste, a characteristic parsnip odor and flavor, and lack of bitterness, fermentative odors, and slimy surfaces.

Main causes of quality loss

(i) Browning of surfaces
(ii) Fermentative odors and slimy surfaces
(iii) Lack of juiciness, sweetness, and a characteristic parsnip odor and flavor

Optimum storage conditions

- Optimum temperature and relative humidity: 0–5°C and 94%–96% RH.
- CA/MA considerations:
  - Benefits: $\geq 3\%$ O_2 and $\leq 10\%$ CO_2
  - Tolerances: $\geq 3\%$ O_2 and $\leq 15\%$ CO_2
  - Injuries: $< 3\%$ O_2 and $> 15\%$ CO_2

Controlled and Modified Atmospheres for Fresh and Fresh-Cut Produce
https://doi.org/10.1016/B978-0-12-804599-2.00059-4

Peeled parsnip is cut into ready-to-eat products and packaged in plastic films to maintain quality (Caleb et al., 2013; Edelenbos, 2016). Rinsing and spin drying before packaging, use of packaging materials with high gas permeability rates, and storage at low temperatures maintain quality and reduce microbial growth (Edelenbos et al., 2010; Caleb et al., 2013; Nicola and Fontana, 2014; Edelenbos, 2016). Beneficial RH and CA/MA reduce the respiration rate, maintain sensory quality, and prolong shelf life (Watkins, 2000).

Problems under suboptimal conditions

Relative humidity

Fresh-cut parsnip loses very little weight (0.26%–0.35%) in MAP with and without perforations (Fig. 1). Packaging in plastic films without perforations maintains high humidity inside packages (98%–100% RH), alters the gas composition, and prevents enzymatic browning (Fig. 1A). In contrast, perforations increase gas transmission rates, maintain sensory odor and flavor, but result in oxidation and browning of the product (Fig. 1B) due to enzymatic browning of phenolic compounds on surfaces (Toivonen and Brummell, 2008).

Gas composition

CA/MA have some positive effects on the white color of parsnips (Figs. 1 and 2) but CA/MA should be used with caution as injurious gas compositions are devastating for sensory aroma and flavor (Watkins, 2000; Caleb et al., 2013). A gas concentration of 3% O_2 or 15% O_2+5% CO_2 reduces the respiration rates as compared to air. Storage in 6% O_2+14% CO_2 and MAP down to 0.8% O_2 and up to 63% CO_2 preserve the white color (Figs. 1B and 2D), but result in sour off-odors and high concentrations of acetaldehyde and ethanol.

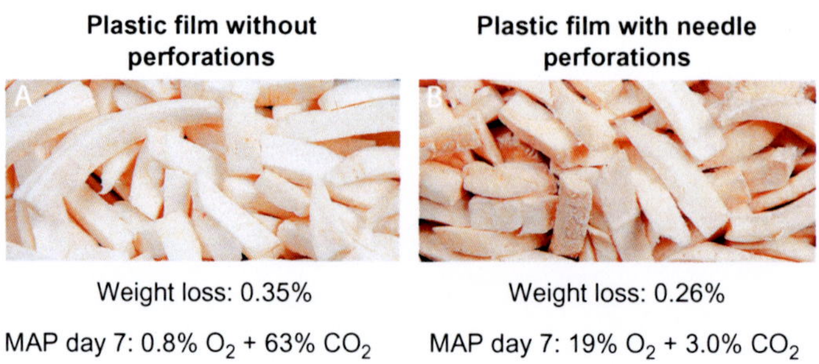

FIG. 1 Visual quality of fresh-cut parsnip stored for 7 days in modified atmosphere packaging (MAP) at 5°C without and with needle perforations. *(Photo: Jens Michael Madsen, Department of Food Science, Aarhus University, Denmark.)*

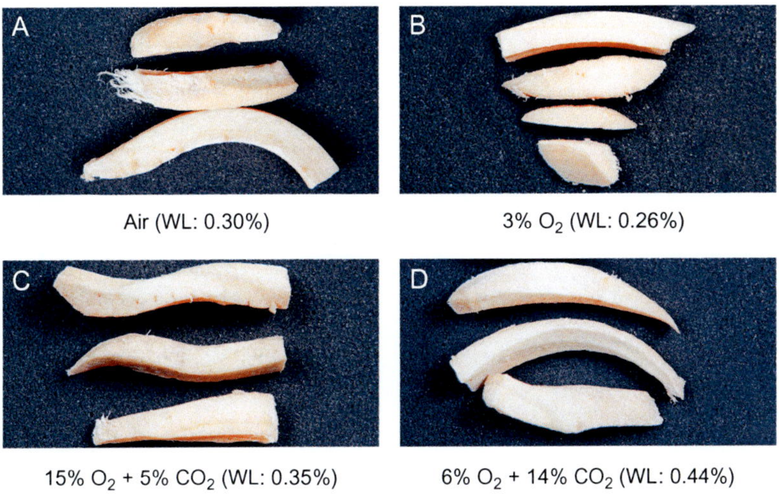

FIG. 2 Visual quality of fresh-cut parsnip stored for 7 days in air and CA at 5°C and ~ 94% RH. The weight loss (WL) during storage is given in parenthesis. *(Photo: Jens Michael Madsen, Department of Food Science, Aarhus University, Denmark.)*

Comments

Fresh-cut parsnip loses quality after processing due to enzymatic browning. CA/MA may preserve color, but caution should be taken to maintain beneficial gas compositions as injurious levels will create fermentative odors and flavors and influence the sensory quality.

References

Caleb, O.J., Mahajan, P.V., Fahad, A.A., Opara, U.L., 2013. Modified atmosphere packaging technology of fresh and fresh-cut produce and the microbial consequences —a review. Food Bioprocess Technol. 6, 303–329.

Edelenbos, M., 2016. Preparation and handling of fresh-cut root vegetables. Acta Hortic. 1141, 77–89.

Edelenbos, M., Balasubramaniam, M., Pedersen, H.T., 2010. Effects of minimal processing and packaging on volatile compounds and other sensory aspects in carrots. Acta Hortic. 876, 269–277.

Nicola, S., Fontana, E., 2014. Fresh-cut produce quality: Implications for a systems approach. In: Florkowski, W.J., Shewfelt, R.L., Brueckner, B., Prussia, S.E. (Eds.), Postharvest Handling. A Systems Approach, third ed. Elsevier, London, UK, pp. 217–2273.

Toivonen, P.M.A., Brummell, A., 2008. Biochemical bases of appearance and texture changes in fresh-cut fruit and vegetables. Postharvest Biol. Technol. 48, 1–14.

Watkins, C.B., 2000. Responses of horticultural commodities to high carbon dioxide as related to modified atmosphere packaging. HortTechnology 10, 501–506.

Chapter 24.7

Roots: Radishes

Merete Edelenbos*, Anne-Berit Wold†, Justyna Wieczynska*
and Alexandru Luca*
*Department of Food Science, Aarhus University, Aarhus, Denmark, †Department of Plant Sciences, Faculty of Biosciences, Norwegian University of Life Sciences, As, Norway

Quality characteristics

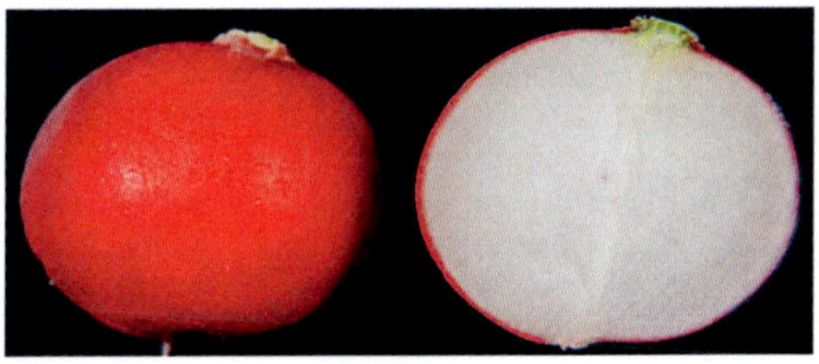

Photo: Jens Michael Madsen, Department of Food Science, Aarhus University, Denmark)

Radishes (*Raphanus sativus* L.) are classified based on size, shape, uniformity, surface and flesh color, firmness, a balanced odor and taste, pungency, and lack of sprouts, decay, and cracks (Coogan et al., 2001; Hassell, 2016).

Main causes of quality loss

(i) Moisture loss and desiccation
(ii) Sprouting and cracking
(iii) Spoilage caused by fungi and bacteria
(iv) Pithiness and browning of the flesh

Optimum storage conditions

- Optimum temperature and relative humidity: 0–1°C and 98%–100% RH.
- CA/MA considerations:
  - Benefits: 1%–3% O_2 and 2%–3% CO_2

Controlled and Modified Atmospheres for Fresh and Fresh-Cut Produce
https://doi.org/10.1016/B978-0-12-804599-2.00060-0

Tolerances: 3%–5% CO_2
Injuries: >5% CO_2

Storage at low temperature and high relative humidity (RH) preserves the texture and maintains the appearance of topped radishes (Hassell, 2016). Topped radishes can be stored for 3–4 weeks at 0°C (Hassell, 2016). Storage at elevated temperatures (5°C) shortens the shelf life as it accelerates top sprouting and the formation of adventitious roots from the base (Wang, 1998). Sprouting can be prevented by trimming off the growing point a few millimeters above the swollen hypocotyl (Figs. 1, 2), by dipping or fumigating with methyl jasmonate, or by use of CA (Hassell, 2016; Lipton, 1972; Wang, 1998).

Storage in an atmosphere with 1%–2% O_2 and 2%–3% CO_2 prevents top sprouting and maintains quality (Hassell, 2016; Cantwell, 2002). Topped radishes are packaged in plastic films with a low water vapor transmission rate to prevent desiccation and surface dehydration (Caleb et al., 2013), as the skin is thin and has little resistance to water loss (Afek and Kays, 2004; Zaki et al., 2012).

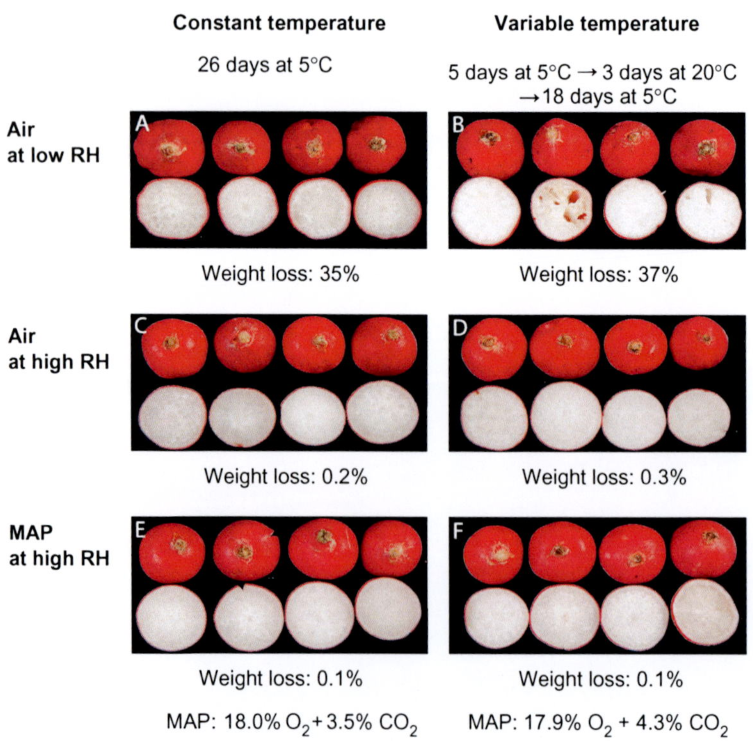

FIG. 1 Quality of radish stored for 26 days at constant or variable temperatures in air or in modified atmosphere packaging (MAP). Low RH: <95%; high RH: >95%. *(Photo: Jens Michael Madsen, Department of Food Science, Aarhus University, Denmark.)*

Problems under suboptimal conditions

Variable temperature and relative humidity

Low RH (<95%) has negative effects on radish quality (Fig. 1) due to loss of turgor pressure and loss of weight (Afek and Kays, 2004). During 26 days of storage at constant or variable temperatures, radishes may lose more than one-third of their original weight (Fig. 1A, 1B). Roots desiccate and shrivel, and the inner tissue becomes pithy due to collapsed cells and disintegrated membranes, which facilitate enzymatic browning and a grayish-brown apperance (Toivonen and Brummell, 2008; Wood et al., 2005). High RH (>95%) preserves the texture and maintains the original shape of the root (Fig. 1C, 1E) but roots sprout, crack, and rot more easily (Fig. 1). Cracking originates from absorption of moisture through the surface of the skin and an influx of water to the underlying tissue (Glenn et al., 2005). Sprouting and cracking should be avoided as consumers perceive top sprouts >3 mm and surface cracks >6 mm as negative quality traits (Hassell, 2016; Lipton, 1972). Cracking can be prevented by the use of packaging films with high water transmission rates (Caleb et al., 2013), which reduce the risk of condensation of water on surfaces during handling and storage.

Gas composition

Storage in CA/MA has variable effects on the quality of radishes (Figs. 1; 2). Low O_2 (3%) prevents top sprouting (Fig. 2B), and elevated CO_2 concentrations (≥5%) postpone bacterial rot, but high CO_2 concentrations also alter skin color

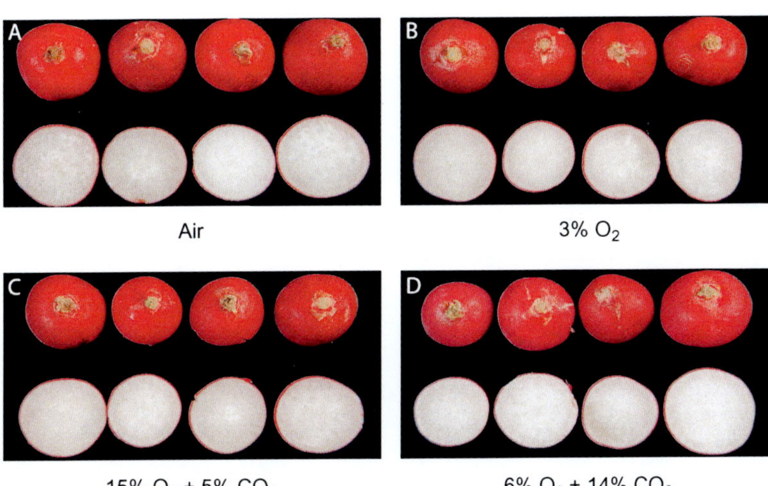

FIG. 2 Radish quality after 26 days storage in controlled atmosphere at 5°C and >95% RH. *(Photo: Jens Michael Madsen, Department of Food Science, Aarhus University, Denmark.)*

(Giusti and Wroldstad, 1996) (Fig. 2C and D). Bacterial rot caused by *Xanthomonas vesicatoria* causes slightly sunken black spots on the surface of the skin (Snowdon, 2010). Low O_2 and elevated CO_2 concentrations had different effects on the respiration rate, as rates were lower in 3% O_2 and higher in 15% O_2+5% CO_2 and 6% O_2+14% CO_2 than in air. A higher respiration rate with elevated CO_2 indicates that the radishes were responding negatively to elevated CO_2 concentrations (Edelenbos et al., 2010; Watkins, 2000); however, not to such an extent that acetaldehyde and ethanol were emitted. Similar findings were obtained for beetroots, which did not emit fermentative odors in response to CA/MA (Edelenbos et al., 2019). Instead, other notes having sulfur in their chemical structure developed in radishes in response to CA/MA.

Comments

The surfaces of radish should be dried before packaging in plastic films as the skin easily absorbs moisture and cracks during storage. Storage in low O_2 (3%) inhibits sprouting and maintains quality, while storage in elevated CO_2 ($\geq$5%) increases respiration and alters the skin color.

References

Afek, U., Kays, S., 2004. Postharvest physiology and storage of widely used root and tuber crops. Horticultural Rev. Vol. 30. 253–316.

Caleb, O.J., Mahajan, P.V., Fahad, A.A., Opara, U.L., 2013. Modified atmosphere packaging technology of fresh and fresh-cut produce and the microbial consequences—a review. Food Bioprocess Technol. 6, 303–329.

Cantwell, M., 2002. Appendix: summary table of optimal handling conditions for fresh produce. In: Kader, A.A. (Ed.), Postharvest Technology of Horticultural Crops. In: Vol. 3311. University of California, Agriculture and Natural Resources, Publication, pp. 511–518.

Coogan, R.C., Wills, R.B.H., Nguyen, V.Q., 2001. Pungency levels of white radish (*Raphanus sativus* L.) grown in different seasons in Australia. Food Chem. 72, 1–3.

Edelenbos, M., Balasubramaniam, M., Pedersen, H.T., 2010. Effects of minimal processing and packaging on volatile compounds and other sensory aspects in carrots. Acta Hortic 876, 269–277.

Edelenbos, M., Wold, A.-B., Luca, A., 2019. Beetroot: requirements and recommendations. In: Gil, M., Beaudry, R. (Eds.), Controlled and Modified Atmospheres for Fresh and Fresh-Cut Produce. first ed. Elsevier, Amsterdam, The Netherlands.

Giusti, M.M., Wroldstad, R.E., 1996. Characterization of red radish anthocyanins. J. Food Sci. 61, 322–326.

Glenn, G.M., Chiou, B.-S., Imam, S.H., Wood, D.F., Orts, W.J., 2005. Role of cuticles in produce quality and preservation. In: Lamikanra, O., Imam, S., Ukuku, D. (Eds.), Produce Degradation: Pathways and Prevention. CRC Press, Boca Raton, FL, pp. 19–53.

Hassell, R.L., 2016. Radish. In: Gross, K.C., Wang, C.Y., Saltveit, M.E. (Eds.), The Commercial Storage of Fruits, Vegetables, and Florist and Nursery Stocks. Agriculture Handbook. In: 66, U.S. Department of Agriculture, Beltsville, Maryland, pp. 524–526.

Lipton, W.J., 1972. Market quality of radishes stored in low O_2-atmospheres. Amer. Soc. Hort. Sci. J. 97, 164–167.

Snowdon, A.L., 2010. Post-Harvest Diseases and Disorders of Fruits and Vegetables: Volume 2: Vegetables. CRC-Press, London.

Toivonen, P.M.A., Brummell, A., 2008. Biochemical bases of appearance and texture changes in fresh-cut fruit and vegetables. Postharvest Biol. Technol. 48, 1–14.

Wang, C.Y., 1998. Methyl jasmonate inhibits postharvest sprouting and improves storage quality of radishes. Postharvest Biol. Technol. 14, 179–183.

Watkins, C.B., 2000. Responses of horticultural commodities to high carbon dioxide as related to modified atmosphere packaging. HortTechnology 10, 501–506.

Wood, D.F., Imam, S.H., Sabellano, G.P., Eyerly, P.R., Orts, W.J., Glenn, G.M., 2005. Microstructure of produce degradation. In: Lamikanra, O., Imam, S., Ukuku, D. (Eds.), Produce Degradation: Pathways and Prevention. CRC Press, Boca Raton, FL, pp. 529–561.

Zaki, H.E.M., Takahata, Y., Yokoi, S., 2012. Analysis of the morphological and anatomical characteristics of roots in three radish (*Raphanus sativus*) cultivars that differ in root shape. J. Hortic. Sci. Biotechnol. 87, 172–178.

Chapter 25.1

Tubers: Fresh-cut potatoes

Juan A. Tudela and Maria Isabel Gil
Group on Quality and Safety, Department of Food Science and Technology, CEBAS-CSIC, Murcia, Spain

Quality characteristics

Good quality of fresh potato slices, strips, or cubes is related to bright appearance, firm texture, slightly moist, with the flesh color typical of the variety and without any symptoms of darkening or dehydration (Fig. 1).

To maintain these quality characteristics, usually a combination of antibrowning agents such as sodium bisulfite or organic acids and a very low O_2 from modified atmosphere (MA) packaging or vacuum packaging is normally used.

Some of the main factors that influence quality are cultivar, harvesting and handling conditions, tuber maturity, and peeling system.

Controlled and Modified Atmospheres for Fresh and Fresh-Cut Produce
https://doi.org/10.1016/B978-0-12-804599-2.00062-4

FIG. 1 Examples of quality loss of fresh potato strips due to browning (left), dehydration (center), and exudation (right).

Main causes of quality loss

- Enzymatic browning
- Dehydration and whitening
- Off-odors
- Exudation

Optimum storage conditions

Optimum temperature

The optimum storage temperature is 0°C (Fig. 2). Even though this temperature represents a problem during long storage for whole potatoes, in the case of fresh-cut potatoes, it does not cause sweetening.

Optimum relative humidity

To prevent whitening, dehydration, and loss of firmness, fresh-cut potatoes must be stored at very high RH (98%–100%). When MA or vacuum packaging is used, permeability of film to water vapor should be very low.

CA/MA considerations

Modified atmosphere packaging with low levels of O_2 (1%–3%) at 0–5°C reduces cut surface browning, but it is not enough to prevent browning (Gorny, 2003).

A shelf life of 3 weeks at 2°C can be achieved by applying an active MA with an initial N_2 flush. Under these conditions, a gas composition of 2%–5% O_2 + 3%–5% CO_2 was obtained during storage within the packages (Gunes and Lee, 1997). These gas concentrations have no effect on microbial population compared to air (Gunes et al., 1997).

FIG. 2 Packages of fresh-cut potatoes during storage at 4°C.

A controlled atmosphere of 3% O_2 +12% CO_2 combined with a dip in 0.025% sodium bisulfite preserved potato slices for 8d at 5°C (Ma et al. 2010). Moreover, a shelf life of more than 7d at 5°C can be obtained when bisulfite is replaced by citric and ascorbic acids and packed under a MA of 20% CO_2 + 80 N_2 (Laurila et al., 1998).

A superatmospheric oxygen composition of 80% O_2 + 10%–20% CO_2 controls browning development and suppresses the respiration rate compared to a CA of 2.5% O_2 + 10% CO_2 during 14d of storage at 4°C (Angós et al., 2008).

An active MA with initial O_2 levels of 0.5%–1.0% using a 35 μm thick PP film can preserve appearance of fresh-cut potatoes for 10 d at 4°C, although it does not avoid development of slight off-odors (Fig. 3). These off-odors can be eliminated with 1 microperforation in the film, but this alteration reduces shelf life to 8 d (M.I. Gil, unpublished results).

FIG. 3 Appearance of fresh potato strips stored in passive MA (left) versus active MA (right).

Problems under suboptimal conditions

Very low levels of O_2 (<1%) can induce off-odors.

Furthermore, MA without O_2 (ex. 30% N_2 + 70% CO_2) can promote production of botulinum toxin in sliced fresh potatoes with temperature abuse (Solomon et al., 1998), so complete elimination of O_2 should be avoided.

References

Angós, I., Virseda, P., Fernández, T., 2008. Control of respiration and color modification on minimally processed potatoes by means of low and high O_2/CO_2 atmospheres. Postharvest Biol. Technol. 48, 422–430.

Gorny, J.R., 2003. A summary of CA and Ma requirements and recommendations for fresh-cut (minimally processed) fruits and vegetables. Acta Hortic. (600), 609–614.

Gunes, G., Lee, C.Y., 1997. Color of minimally processed potatoes as affected by modified atmosphere packaging and antibrowning agents. J. Food Sci. 62, 572–576.

Gunes, G., Splittstoesser, D.F., Lee, C.Y., 1997. Microbial quality of fresh potatoes: effect of minimal processing. J. Food Prot. 60, 863–866.

Laurila, E.K., Hurme, E.U., Ahvenainen, R.T., 1998. Shelf life of sliced raw potatoes of various cultivar varieties—substitution of bisulfites. J. Food Prot. 61, 1363–1371.

Ma, Y., Wang, Q., Hong, G., Cantwell, M., 2010. Reassessment of treatments to retard browning of fresh-cut Russet potato with emphasis on controlled atmospheres and low concentrations of bisulphite. Int. J. Food Sci. Technol. 45, 1486–1494.

Solomon, H.M., Rhodehamel, E.J., Kauttert, D.A., 1998. Growth and toxin production by *Clostridium botulinum* on sliced raw potatoes in a modified atmosphere with and without sulfite. J. Food Prot. 61, 126–128.

Further reading

Barth, M.M., Zhuang, H., Saltveit, M.E., 2016. Fresh-cut vegetables. In: Gross, K.C., Wang, C.Y., Saltveit, M.E. (Eds.), USDA Handbook 66—The Commercial Storage of Fruits, Vegetables, and Florist and Nursery Crops, pp. 624–641. http://www.ba.ars.usda.gov/hb66/contents.html.

Chapter 25.2

Tubers: Fresh and fresh-cut sweetpotatoes

Jeffrey K. Brecht
Horticultural Sciences Department, University of Florida, Gainesville, FL, United States

Quality characteristics

Sweetpotato quality is mainly determined by physical traits such as firm texture, smooth surface, uniform shape, size and color, and freedom from defects such as cracks, cuts, bruises, abrasions, decay and insect damage. Culinary quality is determined mainly by starch and sugar concentrations, and these are affected by cultivars and storage conditions. The flesh of most orange types of sweetpotato

Controlled and Modified Atmospheres for Fresh and Fresh-Cut Produce
https://doi.org/10.1016/B978-0-12-804599-2.00063-6

is moist when cooked, while tropical or boniato-type sweetpotatoes have white or cream colored flesh that is dry when cooked.

Main causes of quality loss

Maintenance of sweetpotato quality during storage is dependent on proper curing and avoidance of water loss and chilling injury (CI). Curing involves placing the roots immediately after harvest under temperature and humidity conditions that are conducive to wound healing, typically about 29°C or 30°C and >90% relative humidity (RH) for 4–8 days (Kays, 2016; Ryall and Lipton, 1979). Curing results in formation of a periderm layer under the peel, the cell walls of which are suffused with waxy suberin, greatly increasing the sweetpotatoes' resistance to attack by decay organisms. The periderm also decreases water loss during subsequent storage, but an RH of 85%–90% is still required to avoid shriveling, which limits storage life in the absence of decay or CI.

Chilling injury occurs at temperatures below the chilling threshold of about 13°C and can develop rapidly below 10°C although cured roots are less susceptible than uncured roots (Ryall and Lipton, 1979). Chilling injury symptoms include "hardcore" in cooked roots that otherwise exhibit no injury symptoms, surface pitting followed by growth of saprophytic microbes, internal breakdown, and flesh darkening and off-flavors when cooked (Picha, 1987). Temperatures above about 15°C may allow sprouting and flesh pithiness and development of decay organisms such as *Rhizopus* and *Fusarium* species and others (Kays, 2016). Fresh-cut sweetpotato is handled at <5°C to ensure food safety, but the exposure time required for marketing does not appear to be sufficiently long to result in development of CI symptoms.

Optimum storage conditions

Optimum temperature

The optimum temperature for sweetpotato storage is the same as the chilling threshold temperature, i.e., 13°C. Fresh-cut sweetpotato is ideally handled at <5°C from processing through distribution to retail display for food safety reasons.

CA/MA considerations

Benefits

Sweetpotatoes that have been properly cured can be stored for 6–10 months at 13°C and 85%–90% RH, with little additional benefit possible from CA/MA. Storage in 7% O_2 plus 2%–3% CO_2 reportedly minimized storage losses due to decay and weight loss of sweetpotato (Mattus and Hassan, 1968).

McConnell et al. (2005) reported that holding fresh-cut (shredded) sweetpotatoes in a modified atmosphere package (MAP) at 5% O_2 + 4% CO_2 for 14 days at 4°C maintained a smaller microbiological load and higher tissue firmness, dry matter, ascorbic acid, and starch compared with storage in air, but there were no changes in color, β-carotene, and sugars; however, headspace ethanol increased to 20 mL/L during the 14-day storage.

Tolerances

CA storage of sweetpotatoes at above 10% CO_2 or below 7% O_2 has generally been discouraged due to the production of alcoholic or other off-flavors (Mattus and Hassan, 1968). However, sweetpotatoes have been found to be somewhat tolerant to extreme CA conditions during short exposures at high temperatures. Chang and Kays (1981) found that 'Jewel' sweetpotato respiration at 25°C was reduced in 5%–15% O_2 after 7 days, but increased in 2.5% O_2, indicating induction of fermentative metabolism. In a study on the potential for use of insecticidal atmospheres for tropical sweetpotatoes, Delate and Brecht (1989) found that 2% or 4% O_2 inhibited curing at 30°C, causing sunken necrotic areas and decay, but the roots tolerated 8% O_2; however, exposure to 60% CO_2 in the presence of 8% or 21% O_2 during curing resulted in decay within 1 week. Cured sweetpotatoes were more tolerant of extreme atmospheres, with 2% or 4% 0_2 plus 40% CO_2 or 4% O_2 plus 60% CO_2 for 1 week at 25°C having little effect on postharvest quality, but exposure to 2% O_2 plus 60% CO_2 resulted in increased decay, less sweetpotato flavor, and more off-flavor (Delate and Brecht, 1989).

Erturk and Picha (2002) reported induction of fermentative metabolism in fresh-cut sweetpotato after 2–3 days at 2 or 8°C when the O_2 concentration was $<1\%$.

Injuries

Atmosphere injury of sweetpotatoes is mainly exhibited as elevated ethanol and proliferation of saprophytic decay organisms on necrotic areas on the root surface (Fig. 1).

Problems under suboptimal conditions

Reduced O_2 and elevated CO_2 can inhibit sweetpotato curing as shown by Delate and Brecht (1989). Due to sweetpotato tissue density and the necessity for storage at the nonchilling temperature of 13°C, sweetpotatoes may be somewhat prone to induction of fermentative metabolism if gas exchange is restricted during long-term storage.

FIG. 1 Superficial decay, identified as *Rhizopus* and *Diplodia* spp., following atmosphere injury of 'Picadita' boniato (tropical) sweetpotato roots stored for 1 week at 25°C and 75% relative humidity in a potential insect quarantine atmosphere of 1% O_2 plus 60% CO_2.

Comments

Innate extended storage potential and propensity for induction of fermentative metabolism at relatively high O_2 concentrations mitigates the use of CA storage for whole sweetpotatoes, but it appears that MAP can be beneficial for fresh-cut products.

References

Chang, L.A., Kays, S.J., 1981. Effect of low oxygen storage on sweet potato roots. J. Amer. Soc. Hort. Sci. 106, 481–483.

Delate, K.M., Brecht, J.K., 1989. Quality of tropical sweetpotatoes exposed to controlled-atmosphere treatment for postharvest insect control. J. Amer. Soc. Hort. Sci. 114, 963–968.

Erturk, E., Picha, D.H., 2002. Modified atmosphere packaging of fresh-cut sweet potatoes. Acta Horticult. 583, 223–230.

Kays, S.J., 2016. Sweetpotato. In: Gross, K.C., Wang, C.Y., Saltveit, M.E. (Eds.), The Commercial Storage of Fruits, Vegetables, and Florist and Nursery Stocks. U.S. Dept. Agric., Agric. Handbook No. 66, Washington, DC, pp. 566–570.

Mattus, G.E., Hassan, F.M., 1968. Controlled atmosphere storage studies with sweet potatoes. HortScience 3, 90 (Abstr.).

McConnell, R.Y., Truong, V.D., Walter Jr., W.D., McFeetersF, R.F., 2005. Physical, chemical and microbial changes in shredded sweet potatoes. J. Food Proc. Pres. 29, 246–267.

Picha, D.H., 1987. Chilling injury, respiration and sugar changes in sweet potatoes stored at low temperature. J. Amer. Soc. Hort. Sci. 112, 497–502.

Ryall, A.L., Lipton, W.J., 1979. Handling, transportation and storage of fruits and vegetables. In: Vegetables and Melons. second ed. Vol. 1. AVI Publ. Co., Westport, CT

Index

Note: Page numbers followed by *f* indicate figures and *t* indicate tables.

B

C

F

G

H

I

K

P

Q

R

S

T

CPI Antony Rowe
Eastbourne, UK
February 23, 2020

SHAH Book - DMS
HOUSE

(Nitin Sir)